Biological Control
in Agricultural IPM Systems

ACADEMIC PRESS RAPID MANUSCRIPT REPRODUCTION

Proceedings of the Symposium on Biological Control in Agricultural Integrated Pest Management Systems held at the Citrus Research and Education Center, University of Florida, at Lake Alfred, June 4–6, 1984.

Biological Control in Agricultural IPM Systems

Edited by

Marjorie A. Hoy
Department of Entomological Sciences
University of California
Berkeley, California

and

Donald C. Herzog
Department of Entomology and Nematology
University of Florida
Gainesville, Florida
and
Agricultural Research and Education Center
Quincy, Florida

1985

ACADEMIC PRESS, INC.
(Harcourt Brace Jovanovich, Publishers)
Orlando San Diego New York London
Toronto Montreal Sydney Tokyo

ACADEMIC PRESS, INC.
Orlando, Florida 32887

United Kingdom Edition published by
ACADEMIC PRESS INC. (LONDON) LTD.
24–28 Oval Road, London NW1 7DX

LIBRARY OF CONGRESS CATALOGING-IN-PUBLICATION DATA

Main entry under title:

Biological control in agricultural IPM systems.

Proceedings of a conference held at the Citrus
Research and Education Center of the University of Florida
at Lake Alfred on June 4-6, 1984.
Includes index.
1. Agricultural pests—Biological control—Congresses.
2. Agricultural pests—Biological control—United
States—Congresses. 3. Agricultural pests—Integrated
control—Congresses. 4. Agricultural pests—
Integrated control—United States—Congresses.
I. Hoy, Marjorie A. II. Herzog, D. C. (Donald C.)
SB933.3.B548 1985 632'.96 85-9157
ISBN 0-12-357030-1

PRINTED IN THE UNITED STATES OF AMERICA

85 86 87 88 9 8 7 6 5 4 3 2 1

Contents

Biological Control of Arthropod Pests

Contributors and Conference Participants*

Numbers in parentheses indicate the pages on which the authors' contributions begin.

George S. Abawi (433), *Department of Plant Pathology, New York State Agricultural Experiment Station, Cornell University, Geneva, New York 14456*

*Perry L. Adkisson (41), *Deputy Chancellor for Agriculture, Texas A&M University System, College Station, Texas 77843*

*William W. Allen (572), *Department of Entomological Sciences, University of California, Berkeley, California 94720*

*Lloyd A. Andres (388), *Biological Control of Weeds Laboratory, USDA/ARS, Albany, California 94607*

*Ralph T. Baker (25), *Department of Plant Pathology, Colorado State University, Ft. Collins, Colorado 80973*

*Carl S. Barfield (88, 101, 121), *Department of Entomology and Nematology, University of Florida, Gainesville, Florida 32611*

Steven K. Beckendorf (167), *Department of Molecular Biology, University of California, Berkeley, California 94720*

Elizabeth H. Beers (103), *Department of Entomology, Pennsylvania State University, University Park, Pennsylvania 16802*

*Elizabeth A. Bernays (373), *Division of Biological Control, University of California, Albany, California 94706*

*Drion G. Boucias (263, 280), *Department of Entomology and Nematology, University of Florida, Gainesville, Florida 32611*

*Leopoldo E. Caltagirone (189), *Division of Biological Control, University of California, Berkeley, California 94720*

*James R. Cate (537), *Department of Entomology, Texas A&M University, College Station, Texas 77843*

*Raghaven Charudattan (347), *Department of Plant Pathology, University of Florida, Gainesville, Florida 32611*

---

*Those names with asterisks were conference participants; others contributed to various presentations at the conference and chapters to this proceedings.

Brian A. Croft (123), *Department of Entomology, Oregon State University, Corvallis, Oregon 97331*
*Louis A. Falcon (229), *Department of Entomological Sciences, University of California, Berkeley, California 94720*
*Peter Faulkner (263), *Department of Microbiology, Queen's University, Kingston, Ontario K7L 3N6, Canada*
*Donald L. Flaherty (501), *Agricultural Building, County Civic Center, Visalia, California 93277*
*Raymond E. Frisbie (41), *Integrated Pest Management Coordinator, Texas A&M University, College Station, Texas 77843*
*Joseph E. Funderburk (67), *Department of Entomology and Nematology, University of Florida, North Florida Research and Education Center, Quincy, Florida 32351*
*Daniel Gonzalez (50), *Division of Biological Control, University of California, Riverside, California 92521*
Joseph G. Hancock (415), *Department of Plant Pathology, University of California, Berkeley, California 94720*
*Joseph C. Headley (53), *Department of Agricultural Economics, University of Missouri, Columbia, Missouri 65201*
*Donald C. Herzog (3, 67, 557), *Department of Entomology and Nematology, University of Florida, North Florida Research and Education Center, Quincy, Florida 32351*
Harvey C. Hoch (433), *Department of Plant Pathology, New York State Agricultural Experiment Station, Cornell University, Geneva, New York 14456*
Keith R. Hopper (201), *Southern Field Crops, Insect Management Research Laboratory, USDA/ARS, Stoneville, Missouri 38776*
*Marjorie A. Hoy (3, 151, 167), *Department of Entomological Sciences, University of California, Berkeley, California 94720*
*Carl B. Huffaker (13), *Division of Biological Control, University of California, Berkeley, California 94720*
*Larry A. Hull (103), *Department of Entomology, Pennsylvania State University, Fruit Research Laboratory, Biglerville, Pennsylvania 17307*
*Carlo M. Ignoffo (243), *Biological Control of Insects Research Laboratory, USDA/ARS, Columbia, Missouri 65201*
*Harry K. Kaya (283), *Division of Nematology, University of California, Davis, California 94616*
Hiroshi Kido (501), *Department of Entomology, University of California, Davis, California 95616*
*Edgar G. King (201), *Southern Field Crops Insect Management Research Laboratory, USDA/ARS, Stoneville, Missouri 38776*
*W. Joe Lewis (89), *Insect Biology and Population Management Research Laboratory, USDA/ARS, Tifton, Georgia 31793*
*Steven E. Lindow (395), *Department of Plant Pathology, University of California, Berkeley, California 94720*

*James J. Marois (461), *Department of Plant Pathology, University of California, Davis, California 95616*

*S. Bruce Martin (433), *Department of Plant Pathology, North Carolina State University, Raleigh, North Carolina 27650*

*Clayton W. McCoy (481), *Department of Entomology and Nematology, University of Florida, Citrus Research and Education Center, Lake Alfred, Florida 32850*

*Christopher A. Mullin (123), *Department of Entomology, Pennsylvania State University, University Park, Pennsylvania 16802*

*Carolyn Napoli (455), *Department of Plant Pathology, University of California, Berkeley, California 94720*

Donald A. Nordlund (89), *Insect Biology and Population Management Research Laboratory, USDA/ARS, Tifton, Georgia 31793*

*Robert J. O'Neil[1] (323), *Department of Entomology and Nematology, University of Florida, Gainesville, Florida 32611*

Janine E. Powell (201), *Southern Field Crops Insect Management Research Laboratory, USDA/ARS, Stoneville, Missouri 38776*

*Reece I. Sailer (166, 187, 200, 226), *Department of Entomology and Nematology, University of Florida, Gainesville, Florida 32611*

*Milton N. Schroth (415), *Department of Plant Pathology, University of California, Berkeley, California 94720*

Merle Shepard[2] (557), *Department of Entomology and Zoology, Clemson University, Clemson, South Carolina 29631*

*Grover C. Smart (302, 478), *Department of Entomology and Nematology, University of Florida, Gainesville, Florida 32611*

Brian J. Staskawicz (455), *Department of Plant Pathology, University of California, Berkeley, California 94720*

Vernon M. Stern (501), *Department of Entomology, University of California, Riverside, California 92521*

*Jerry L. Stimac (323), *Department of Entomology and Nematology, University of Florida, Gainesville, Florida 32611*

*Ronald E. Stinner (62, 150, 322, 344), *Department of Entomology, North Carolina State University, Raleigh, North Carolina 27650*

*Maurice J. Tauber (3), *Department of Entomology, Cornell University, Ithaca, New York 14853*

*Seymour D. Van Gundy (467), *Department of Nematology, University of California, Riverside, California 92507*

*L. Theodore Wilson (303, 501), *Department of Entomology, University of California, Davis, California 95616*

*Kenneth V. Yeargan (521), *Department of Entomology, University of Kentucky, Lexington, Kentucky 40546*

[1]Present address: Department of Entomology, Purdue University, West Lafayette, Indiana 47907.
[2]Present address: Department of Entomology, International Rice Research Institute, Manila, Philippines.

Preface

This volume on "Biological Control in Agricultural IPM Systems" evolved, at least in part, as an outgrowth of a trip to the Peoples' Republic of China in 1982. As members of a ten-person National Academy of Sciences Team on biological control of arthropod pests, we were able to contrast the use of biological control in Chinese agriculture and in the United States, and we subsequently asked the question, How can biological control be used more extensively and effectively in U.S. agriculture? Further impetus to organization of the conference from which this volume arose came during a National Interdisciplinary Biological Control Conference held in Las Vegas, Nevada, in February, 1983. The large number of participants attending that conference suggested that interest in biological control was expanding.

Originally, we outlined "Biological Control in Agricultural IPM Systems" as a one-half day symposium for presentation at the 1983 or 1984 annual meeting of the Entomological Society of America. However, our half-day symposium quickly grew into a three-day conference and we realized that a separate work conference was needed to cover the subject adequately, and that participation of and input was needed from entomologists, plant pathologists, weed scientists, nematologists, toxicologists, and economists. Thus the conference was conceived and held on June 4–6, 1984, at the Citrus Research and Education Center of the University of Florida at Lake Alfred.

Participants in the conference worked diligently to review critically the status of biological control in their areas of expertise and as an implementable component in agricultural IPM systems. We are not aware of any comparable review of the subject matter in print. While there are several important reference volumes on both biological control and on IPM, none have focused specifically on the role and status of biological control within the text of agricultural IPM systems. Because of the complexity of the subject matter, we deemed it desirable to limit our discussions to agriculture in the United States and to exclude animal, rangeland, urban, and glasshouse IPM systems.

Many students in recent years have been taught that biological control and plant resistance are cornerstones of IPM. Despite the supposed, and often-quoted status of biological control as a basic subdiscipline underlying IPM, research to incorporate biological control into agricultural cropping management systems is still relatively uncommon, and funding for such effort is scarce. It is our perception that IPM program managers have often tended to support projects that focused on research designed to time pesticide applications effectively, to model plant/pest interactions, to identify and/or refine economic injury levels, and to predict pest populations. These subjects are certainly important and deserve the attention that they have received. However, funds for biological control—whether by importation, augmentation, or conservation—probably constitute 20% or less of the funding allocated to IPM programs. Biological control research has therefore coped with extraordinarily limited resources, often not keeping up with inflation.

This perception raised in our minds a series of questions relative to biological control programs and resources (sources and adequacy of funding) for such programs. The following questions were articulated in the introduction to the conference and were subsequently discussed, both overtly and covertly, by participants during the three days:

- Is biological control currently an integral part of agricultural IPM programs in U.S. agriculture? Are the costs, risks, and benefits associated with and accruing to biological control sufficient to justify the statement that it is, along with plant resistance, a foundation upon which IPM can rest?
- If biological control is not fully implemented and/or utilized in U.S. agriculture, should more research and especially more funding be directed toward such implementation and utilization?
- What factors limit the expanded use of biological control in agriculture? Is it limited only because of a lack of necessary funding? Or, are new attitudes, approaches, or technologies needed to fully exploit the potential of biological control? For example, can genetic engineering technologies be useful tools to improve biological control of arthropod pests, weeds, plant pathogens, or nematodes?
- Integrated pest management is described as an interdisciplinary approach to the solution of agricultural problems. Is such cooperation only painfully, and too rarely, achieved? If so, are interdisciplinary meetings such as this conference fruitful? Or, are our individual disciplines so filled with incomprehensible jargon, highly detailed research needs, and divergent interests that our efforts to communicate are a waste of time? Can we exchange ideas, concepts, and perspectives in order to provide a new focus in our approach to biological control in agricultural IPM systems?

During our conference, we attempted to scrutinize very carefully the current status of biological control in our agricultural IPM systems in a broad sense. We also provided perspectives and perceptions of some of the more specific research needed to enhance the use of biological control of arthropod pests, weeds, plant pathogens, and nematodes in agricultural IPM systems. We hope that in doing so we have fostered an expanded horizon for biological control. We recognize that new beginnings do not develop overnight, but we do hope that our conference and this proceedings volume will be a constructive contribution to the development of a new horizon for biological control.

That there have been two large national meetings in the last two years on biological control suggests that we may be at the threshold of a new era relative to biological control. Integrated pest management has undergone evolutionary changes during the past 20 years. Many important advances have been made in managing pests. The important role of biological control organisms in regulating pests, especially insects and mites, of agricultural crops was not fully recognized until they were decimated by applications of pesticides harmful to the beneficial species, but not to the pest. Overt utilization of biological control agents has only just begun; much yet remains to be done to make biological control of arthropod pests, weeds, plant pathogens, and nematodes reliable and accepted components in agricultural IPM programs. Reliance on pesticides as the major, in some cases sole, suppression tactic in IPM is seen as having an increasingly limited future. Thus, while pesticides will continue to be, for the forseeable future, important in IPM programs, problems surrounding their continued use are expanding and intensifying. Pesticide resistance among pests is a major concern. Increased restrictions on pesticide use and uncertainty about the discovery and development of replacement chemicals gives great cause for concern. Biological control is highly compatible with quarantines, plant resistance, and cultural controls. The appropriate and full deployment of biological control in agricultural IPM systems remains to be achieved. We hope that this volume will assist in this endeavor.

Finally, we must thank many people, not least of whom are the conference participants, who unselfishly supported the aims and goals of the conference. Crucial financial support was provided by the University of Florida, Institute of Food and Agricultural Sciences; the University of California; and the National Science Foundation Office of Interdisciplinary Research, which provided funds to "foster a comprehensive, interdisciplinary view of research needed to advance biological control as a strategy in integrated pest management." The editors express special appreciation to the following employees of the North Florida Research and Education Center, University of Florida, who worked tirelessly, spending many long hours preparing the various versions of this volume: Ms. J.M. Walden, Ms. K.M. Lurding, and Ms. J.V. Smith. Without their assistance and cooperation, completion of this project would have been extremely difficult.

Marjorie A. Hoy
Donald C. Herzog

Executive Conference Summary

BIOLOGICAL CONTROL IN AGRICULTURAL IPM SYSTEMS: A BRIEF OVERVIEW OF THE CURRENT STATUS AND FUTURE PROSPECTS

M.J. Tauber, M.A. Hoy and D.C. Herzog

Our Conference was concerned with biological control in agricultural IPM systems, a subject of considerable current interest because of a perceived urgency to develop and adopt safe and efficient methods for managing agricultural pests. Problems associated with pest suppression (including environmental pollution, deleterious effects of pesticides on non-target organisms, pesticide resistance, resurgence of target pests, secondary pest outbreaks, escalating costs of developing, producing and applying pesticides) all affect the vitality and profitability of agriculture and the well-being of our society.

General challenges of the Conference were to: 1) provide up-to-date reviews of past work on biological control of arthropods, weeds, nematodes and plant pathogens in the context of agriculture, 2) identify the status of biological control in current IPM programs, and 3) identify research required to accomplish greater integration of biological control into agricultural IPM systems.

In many, if not most cases, biological control by itself does not provide economically acceptable pest suppression in agricultural cropping systems. Therefore, biological control must be developed and implemented as a component of IPM. However, if it is to be an integral part of IPM (along with plant resistance, cultural methods and pesticidal controls) biological control must be nurtured to become a strong, vital entity. This is especially true, as stressed in many of the following chapters, if biological control is to be a vital component of agricultural IPM systems.

NATIONAL ISSUES

The papers presented and subsequent discussions at the Conference addressed crucial biological control and integrated pest management issues and issued specific recommendations. These are presented in individual papers and discussion summaries; they will not be repeated here in detail. In this summary, emphasis will

ISBN 0-12-357030-1

be placed on a synthesis of general needs in biological control. We will consider a general framework within which specific problems and recommendations can be successfully and effectively addressed.

Biological control and IPM, together and separately, face serious obstacles. These originate from the lack of fundamental biological data and the lack of knowledge necessary to develop economically, environmentally and socially sound crop and animal production systems. The problems that we face have three dimensions: 1) developing efficient and effective biological control methods, 2) expanding and integrating biological control into IPM systems, and 3) providing the scientific, social, economic and educational framework for public acceptance of biologically-based IPM techniques in agricultural production. Thus, developing biological control as a reliable tactic is a major undertaking at both the scientific and technological levels. To develop it to the state where it can be integrated into IPM systems on a larger scale depends on resources for research and development that are greater that those that biological control has ever received.

How do we justify an expansion of resource allocation to biological control? An essential consideration is that our society must recognize the economic, social and health-related long-range costs and consequences of present agricultural crop and animal production systems and their practices. We can no longer avoid an assessment of the true cost of inadequate support for agricultural research on the one hand, or of either maintaining or increasing investment in agricultural research on the other. Proponents of biological control must emphasize that the actual costs associated with many current pest control practices, including environmental and social costs, is very high indeed, and that the cost of biological control, even considering its associated risks, are by comparison lower.

How does this approach differ from the approaches of the past? Past objectives in agricultural research have focused heavily on increasing yield without providing an assessment of the full cost of achieving this goal. Our revised objectives must concentrate on improving net benefits to society as well as to agricultural producers. To do this, we must consider values important to society (employment, quality and stability of food supply, effects on the environment), as well as the monetary costs of agricultural production. Research designed to achieve these objectives requires investment to expand both basic and applied research, and particularly interdisciplinary efforts.

Papers presented at the Conference demonstrated that our scientific needs form a continuum from the very basic to the applied. Given the state of present and probable future agricultural problems, basic sciences are not sufficiently well-developed to provide a broad base upon which to build effective IPM programs. Similarly, applied sciences are being weakened because of diminishing support and understaffing relative even to present research requirements. Finally, the processes involved in transfer and application of new scientific knowledge are in need of additional support.

In general, our institutions have not supported many aspects of long-range agricultural research to the extent that is necessary to meet present and future demands; instead, they have tended to emphasize and respond to short-term objectives and solutions. Emphasis has been on planning new programs and reorganizing state and national institutions. However, despite the fact that a comprehensive approach to pest management requires that all necessary disciplines (in the biological, social and systems sciences) be involved from the onset of program planning, incentives necessary for stimulating and maintaining interdisciplinary research have not been forthcoming. Furthermore, reorganizational efforts have not generally been accompanied by increased funding that is focused on crucial or critical areas; there has been reshuffling of organizational charts, but no broadly-based increased investment, particularly for biological control of exotic and native pests.

It is necessary to establish how biological control, plant resistance, cultural controls, semiochemicals and other tactics compare with each other and with chemical control, in terms of resources available. Scientists must work with imaginative administrators to establish new national and regional priorities. There should follow selective increases in support for agricultural research along a broad front. First, we need increased state and federal support for sustained disciplinary and interdisciplinary research and technological development programs. Second, we need realistic and expanded competitive grants programs to open up new areas of basic research, as well as to provide sufficient funding for applied, interdisciplinary projects.

The keys to progress in research in agricultural production and crop and animal protection lie in a balance among and increases in the following: 1) state and federal support, 2) long- and short-term research, 3) disciplinary and interdisciplinary research, 4) basic and applied research and technological development, 5) formula and competitive funding, and 6) greater staffing in selected areas within both applied and basic disciplines.

Only a balanced, well-formulated policy can attract the needed talent to agriculturally-related research, biological control and IPM. Such a policy can provide the incentives for scientists to maintain their primary disciplinary association while participating in the planning, development and delivery of interdisciplinary research for the integration of biological control into agricultural IPM.

RESEARCH NEEDS IN BIOLOGICAL CONTROL

General Perceptions

To become an integral part of IPM, which is an applied science with a quantitative basis, biological control must become more quantitative. Lack of precision and predictability are prevailing general problems in biological control. Nearly every paper that follows stresses the necessity for quantifying some aspect of the biology and ecology of agroecosystem components. The problem of quantification is not just a problem of methodology, but in the outlook of biological control scientists. Some biological control practitioners still retain the view that biological control as a discipline is, and will remain, largely an art. Presentations at the Conference and various chapters in this proceedings effectively countered this view and stressed that unless biological control is both more reliable and predictable, it will not be generally adopted and integrated into IPM systems. Not only should we be able to predict when biological control will work, we must also be able to determine under what conditions it will not work. Furthermore, we must assess the degree of risk involved in relying on biological control as a tactic within the overall production/management strategy.

Research

The presentations and discussions in our Conference analyzed research needs for expanding the role of biological control in agricultural IPM systems. There was general agreement that full utilization of biological control in this context requires four components: 1) research, 2) development, 3) delivery and education, and 4) evaluation. Although this Conference focused on research needs for agricultural cropping systems, implementation of biological control in IPM systems cannot be achieved without concomitant strong efforts in all four areas. All papers stressed that research in biological control must be quantitative and interdisciplinary in nature; it must concentrate on mechanisms; it must span the basic to the applied; and it must involve both field and laboratory approaches.

A major weakness in developing predictive capabilities relative to natural enemy effectiveness is that we do not have adequate information concerning the natural enemies/antagonists that are being introduced or manipulated. We do not adequately understand the biology, ecology, physiology, behavior, biochemistry or genetics of those organisms. We also lack knowledge of the inherent variability that exists within and among species groups, species and populations. Furthermore, standardization of program design, rearing/culturing, testing, release and evaluation is lacking so that comparisons are difficult, and reliability and repeatability of results are low. There are insufficient resources and coordination for identifying and selecting effective strains of natural enemies/antagonists or for developing standardized and efficient rearing/culturing, storage, handling and release procedures. Often there is no standardization of evaluation methods. Solutions to these problems require long-term investment in monetary and human resources, and no short cuts appear possible.

In the shorter term, two areas of research show promise of providing potential solutions: 1) selectivity of pesticides and 2) genetic engineering of natural enemies. A major conclusion reached is that pesticides need not be incompatible with the use of natural enemies. However, to achieve compatibility, pesticides must be used with precision, on the basis of extensive knowledge of the biology, ecology, and behavior of both the target pests and thier natural enemies. This must be done in a holistic context requiring coordination among entomologists, plant pathologists, weed scientists, nematologists and crop production specialists. It is probable that the most dramatic increase in the utilization of biological control in agricultural IPM systems could come through the judicious use of selective pesticides in conjunction with effective natural enemies in specific cropping systems, in specific geographic regions. While we have some knowledge of pesticide selectivity, it is woefully inadequate to generally allow such precise usage.

Genetic engineering has been perceived by many to offer promise in the development of improved strains of natural enemies or in the elucidation of mechanisms involved in biological control. Development and/or selection of pesticide-resistant natural enemies is perceived as highly desirable. Resistant strains could be developed through natural selection in the field and be identified by screening programs. Alternatively, artificial laboratory selection or genetic engineering techniques could produce resistant natural enemy strains in the laboratory. At the present time, genetic engineering techniques are inadequate for use with characters that are polygenically determined. Thus, single gene traits such as pesticide resistance may be the most important in genetic engineering studies. Furthermore, such strains can be easily evaluated, readily identified as improved, and implemented for use.

Development

Development, in our context, refers to the adaptation of a particular tactic (biological control within IPM) to specific agricultural conditions. Therefore, it involves multiple pest situations and the integration of biological control with multiple tactics (cultural, plant resistance, chemical control) into a unified strategy. This requires an interdisciplinary approach (vertical and horizontal integration of disciplines) and it involves a vast amount of data, the interpretation of which requires complex models.

To expand the role of biological control in agricultural IPM systems, it is essential to focus on building IPM around mortality, and other adverse effects on pests, caused by biological control agents. This can be done reliably and predictably only if biological control is integrated with other subdisciplines bearing on the IPM effort. Data must be gathered and presented in a form that contributes to models and component submodels (crop, pest, beneficial species, weather, economics) that are relevant to IPM, and thus must be sensitive to economic goals, including economic thresholds.

Delivery, Education and Evaluation

Even if an operational pest management system is developed, acceptance of biologically-based pest control methods by agricultural producers and agricultural industry may be difficult to attain. Four aspects are important: 1) supply (availability of a standardized, economically attractive product that is readily stored, shipped, and used), 2) education (through demonstration), 3) sale (economic motives), and 4) service (availability of advice, aid, and follow-up). These steps require efficient rearing/culturing of natural enemies (including artificial diets/ media, standardization, and reliable methodology).

Evaluation involves establishment of biological control effectiveness in terms of economics. Several factors come into play: 1) efficiency, 2) complexity/simplicity of use, and 3) attractiveness of the product. Biological control projects must include risk analyses that establish the trade-offs involved in reliance on biological control. Projects should be planned from the beginning with this in mind. Such evaluations should be long-term, conducted over many cropping seasons.

SUMMARY

To expand the role of biological control in agricultural IPM systems, the following must be considered:

National Policy

1. Establish a national effort to determine economic, environmental, and social costs of agricultural production and its components, including costs measured in monetary terms, as well as those which are not.
2. Assess and strengthen national investment in the biological, physical, and social sciences at both the basic and applied levels.
3. Include active scientists (from the biological, physical, and social disciplines) when initiating the planning of national science policy.

4. Strengthen and focus funding as follows:

   a. Provide increased institutional (state, federal, private) funding for long-term disciplinary research.
   b. Expand the competitive grants funding program framework in support of specific project funding.
   c. Increase staffing of scientific and support personnel in specific disciplines.

5. Establish incentives for scientists to initiate and maintain interdisciplinary activities through special funds, rewards, etc.

IPM Context

1. Include biological control specialists when initiating IPM projects so that biological control can have a central role.
2. Strengthen biological control as a full-fledged subdiscipline, by providing support for its component processes (systematics, exploration, rearing, conservation, evaluation).
3. Quantify the approach to biological control research and delivery, so that its methods and results will be predictable, reliable, and more easily evaluated.

General Introduction

BIOLOGICAL CONTROL IN INTEGRATED PEST MANAGEMENT: AN ENTOMOLOGICAL PERSPECTIVE

C.B. Huffaker

This conference is concerned with the practical use of biological control in integrated pest management (IPM). I have been asked to introduce the area of biological control of insects and mites in this context. To do this, I wish briefly to describe what I consider the proper scope of IPM, as well as that of biological control. I wish also to correct some misconceptions about the financial support biological control has experienced. Some workers have looked at each in somewhat different terms than the more conventional to which I adhere. Some European workers (Mathys 1980) have wanted to synonymize biological control with integrated control (or IPM). Various others have wanted to include as biological control various biological forms of pest control besides the use of natural enemies (Maxwell & Harris 1974; Sweetman 1936, 1958; USDA 1978).

I will treat the broader area, IPM, first and then biological control, which can be either a part of an IPM program or, at times, of great practical utility standing alone, as has been the case with many of the outstanding examples of classical biological control.

Among the forms some would include as biological control are the use of resistant varieties or strains, use of pheromones as a means of controlling a pest whether or not involving action of a natural enemy, use of insect hormones to disrupt the target pest's activities, and, finally, use of various techniques to interfere genetically with the reproduction of the target species, especially by release of sterile insects to disrupt successful mating (Maxwell & Harris 1974, USDA 1978).

Integrated insect control was used in the early part of this century before the terms "integrated control" or "integrated pest management" were thought of. Entomologists W.D. Hunter and B.R. Coad of the USDA advocated the basic principles in 1923 (Hunter & Coad 1923), for control of the cotton boll weevil in the U.S. Residue destruction and pest-tolerant varieties were the primary tactics. Insecticides were considered supplementary, to be used only when weevil damage reached a measured level. Plant pathologists, likewise, had developed disease management programs incor-

ISBN 0-12-357030-1

porating varietal selection and cultural procedures, again employing only minimal use of the early fungicides (such as Bordeaux mixture) (see further, Bottrell & Adkisson 1977).

Use of pesticides dramatically expanded in this century, at first slowly, but following the insect pest suppressions achieved with DDT, with propitious momentum. Synthetic organic pesticides production had increased from less than 0.5 million pounds in 1951 to ca. 1.4 billion pounds--or 3,000 times as much--in 1977 (Bottrell 1979). Ecologically oriented pest control techniques were being replaced by pesticides. Although weaknesses in a policy of unilateral reliance on pesticides had become obvious, benefits to human health and food and fiber production were persuasive arguments to use them.

In 1959, the ecological approach received a significant thrust through the formal definition, statement of principles, and a description of the techniques required for its research and implementation by a group of California workers--V.M. Stern, R.F. Smith, R. van den Bosch and K.S. Hagen (Stern et al. 1959). The undesirable consequences of a unilateral reliance on chemicals were assessed. These workers noted that pesticide interference with natural enemies of pest species has a two-fold effect for the future: 1) The suppressive role of these enemies is greatly reduced or eliminated, allowing rapid rebounds by target pests, and 2) potential or minor pests can be "unleashed" from control by their natural enemies and become major pests.

Rachel Carson's (1962) Silent Spring not only cited some of these crop-centered problems with use of chemicals, but also dared to consider the broader implications for humans and, in fact, for all organisms on earth. Events then moved rapidly because Silent Spring served to crystalize in the public mind and the Halls of Congress a consciousness of the inherent dangers. The list of chemicals now being challenged include, not only pesticides, but also various other air, water, and soil pollutants, even those chemical and radioactive wastes that had been stored away in concentrated "caches" and largely forgotten as we now know even from the media outlets.

An even more significant factor forcing consideration of integrated control was the widespread development of resistances to the pesticides, one after another, in the target insect species. With several hundred such examples, entomologists began to think they might soon be unable to control various major pests by use of chemicals (National Academy of Sciences 1975).

In the late 1960's and early 1970's, both the USDA and the National Science Foundation, later joined by the Environmental Protection Agency, launched major efforts to look into insect control in a more holistic manner--that is, to concentrate on a given cropping system in a specific farming area--and to seed an in-depth understanding of the ecosystems, in regard especially to the impact of pests on crop growth and the cost-benefit relationships. This, in essence, is putting into practice what some researchers had been urging for years (Stern et al. 1959)--in fact, as early as 1923 (Hunter & Coad 1923). These national efforts have been underway now for more than a decade and have been broadened to consider the whole complex of crop pests, not just the insects, and, furthermore, all agronomic considerations as an integrated system of crop and farm management. Frisbie & Adkisson (this volume) deal more specifically with these developments.

I will deal now more specifically with biological control. With no offense to those who think differently, the discipline for many years, as represented by such leaders as W.R. Thompson of the Commonwealth Institute of Biological Control, H.S. Smith of California, Phillipe Sylvestri of Italy, R.C.L. Perkins of Hawaii, R.J. Tillyard and A.P. Dodd of Australia, and G.C. Ullyett of South Africa, has meant the suppression of a host or prey species by its natural enemies. The premise on which this biological control rests is that in certain circumstances many populations are held at low densities by their natural enemies (i.e., by parasitoids, predators, pathogens, and other such forms known as "antagonists" that control plant pathogens) (Schroth & Hancock, this volume). For many pests or potential pests of crops, such enemies provide ample protection. This premise rests on an observed "balance of nature;" populations are necessarily restricted in abundance; with an increase in density they use up resources, defile places where they live, or generate increased intensity of various adverse factors, such as predators, parasites, pathogens, or antagonists.

Such enemies thus have a unique role, in that a host's regulation by its natural enemies arises as a host (or prey) and natural enemy <u>population</u> <u>interaction</u>, often in a complex environmental matrix. Action of hormones or pheromones, or introduction of sterile or genetically deranged insects of the <u>pest</u> <u>species</u> <u>itself</u>, is not at all the same. Neither is the development of resistant crop or animal varieties, valuable as these methods are. So the discipline, by its unique nature and by tradition, is best confined to action of parasites, predators, pathogens, and "antagonists," in my view. There must, of course, be supportive conditions and tactics to create favorability.

Classical or conventional biological control tries to establish, in a new area, a biological control link existing in the home area, or one related to it. Observed rarity of a pest species in its native home and devastating status in an invaded area which lacks effective natural enemies suggests the <u>possibility</u> that control might be achieved by introducing its associated enemies. This theory follows the general theory of the roles of many natural enemies in the balance of nature; it does not explain specifically how they achieve or maintain such a role, or the necessary <u>conditions</u>.

This conference seeks to bring together examples of the use of biological control agents in practical IPM programs. It has often been said that biological control and use of resistant varieties or strains are the cornerstones of good IPM. But relatively few examples have been developed and adequately documented. There are a number of examples wherein certain manipulations that have been shown experimentally to improve or achieve biological control have been improperly interpreted as practical and implemented programs, or simply misinterpreted. For example, the use of small blackberry patches near grapes in California's Central Valley for increasing effectiveness of the mymarid egg parasite, *Anagrus epos*, on the grape leafhopper has not proven practical and implemented commercially. Neither has the deliberate introduction of cyclamen mites into strawberry fields to establish and sustain a continuity of predator effectiveness against this pest species (see below). Moreover, Flaherty (1969) was misinterpreted by Hussey & Huffaker (1976) to suggest that johnsongrass could be grown as a "cover crop" in grapes to foster better biological control of spider mites in grapes. Flaherty (1969) did show a fostering effect from presence of johnsongrass as weeds, and he and colleagues experimented with related grasses for use as a cover crop, but there was no implementation of any such procedure. Also, the use of strip planting of alfalfa and cotton in California to foster biological control has never been practical or practiced on a substantial scale. By way of introduction, I will discuss some of the older employments of biological control as well as a few newer ones. Some of the early work is summarized in a number of chapters in the book "Biological Control" which I edited (Huffaker 1971).

Far more has been said about the possibilities than about proven, large-scale commercial utilization. Despite the influx of more than $20 million to develop IPM programs for some major U.S. crops, only a small amount has gone to develop the use of parasites, predators, pathogens and antagonists. In fact, the so-called "Huffaker" project, shortly after it was well under way, was stripped of funds for this specific area. For more than a decade now this "cornerstone", and I would add that of using pest-resistant cultivars, has suffered a general neglect, while overall IPM has had extensive support. This must be changed. Programs in biological control, even more so where considered in conjunction with necessary use of pesticides, demand extensive support for efforts in the basic, classical, augmentative, and manipulative areas, including various conservation measures concerning use of the pesticides themselves.

We need large amounts of funds, appreciation of the time element necessary to obtain results for complex systems, and better general understanding of the field itself. Regarding the latter, biological control of Klamath weed has been viewed by some as damaging to the "balance of nature" relative to one or more native *Hypericum* species of very limited occurrence. Yet, the National Park Service was so concerned over the massive disturbance in Yosemite wrought by the weed itself that it made exception to its own long-standing policy against deliberate introduction of alien species and introduced *Chrysolina*, after which the Yosemite meadows were returned more nearly to their pristine state, only through the action of this alien insect.

Such costs and benefits, of course, must always be carefully weighed.

In considering this whole question, one should distinguish between different types of cropping systems. Classical biological control, or even periodic augmentation of natural enemies, has historically been most effective standing alone, in systems where, for economic or other reasons, large-scale use of insecticides or other disturbing practices (e.g., fire for refuse destruction) are minimal. Range and forest systems and small farm plots where pesticide use is avoided to meet a special market (health food stores) or in situations where the economics of using pesticides is prohibitive, are examples. Moreover, the more promising conditions for classical biological control usually include crops that already have most species under natural control. On the other hand, large-scale commercial agriculture involving crops that have a major complex of pests are most likely to require the integration of chemical and cultural methods with the use of natural enemies; the latter does not stand alone but needs to be fostered by IPM. Most of our extensive rosaceous tree crops and many of our vegetable and row crops in the United States, for example, fall in this category. Alfalfa and other hay crops had for years an economic picture discouraging costly programs of insecticide use, and the central pests have been alfalfa weevils. Biological control has been emphasized (Huffaker et al. 1971, Smith & Huffaker 1972). In the following pages, I note exceptions or give some typical examples.

Effective biological control can sometimes be obtained and maintained on a grand scale despite the use of insecticides, as has been true of the biological control of cottony cushion scale by the Vedalia beetle in citrus in various countries where intensive use is made of insecticides for control of other citrus insects and mites. On the other hand, even here, the use of DDT in or near California citrus disrupted this control and occasioned catastrophic outbreaks of the scale in 1946; DDT could not be used in citrus (DeBach 1974).

In various other instances, the benefit of powerful biological control agents, introduced or natural ones, could be maintained only under no use or very cautious use of pesticides, including fungicides, insecticides and acaricides. There are instances where integration has been unnecessary, as in the programs of biological control of the weedy St. Johnswort and of prickly pears in various parts of the world, simply because in general pesticides are not used in these range environments. But the question of how we can best use these biological control agents normally brings to hand the question of how we can utilize them effectively when pesticides are needed to control other pest species for which adequate natural enemies are not at hand or available. We can, of course, try to protect the enemies by using less disturbing practices, or we may conserve them or augment their numbers at critical times; we may also develop strains of the natural enemies that are more tolerant of a spectrum of pesticides. We may even attempt to achieve continuity and a more reliable biological action by adding the pest species itself to the habitat at critical times.

APPLES IN NOVA SCOTIA

An early apple pest management program was initiated in Nova Scotia by A.D. Pickett, F.T. Lord, and colleagues during the Depression years (1930's) when the market picture for apples there was so poor that pesticide resistance, induced-pest problems, and the high costs of insecticides threatened the industry (Flint & van den Bosch 1981; MacPhee et al. 1976; A.D. Pickett, personal communication). These workers termed their program one of "harmonizing" chemical and biological controls. The program of harmonizing the effective use of natural enemies and chemicals began earlier and involved not only the European red mite (and other mites) but also the notorious codling moth, a lecanium scale, *Lecanium coryli*, the black-headed budworm and the oyster shell scale, *Lepidosaphes ulmi*. Natural enemy control of spider mites was recognized after DDT and other synthetic materials were incorporated into the spray programs because it was then that spider mites became a major problem, as they did throughout the world (Huffaker et al. 1969).

The oyster shell scale had been held under control, presumably by its natural enemies (mainly *Aphytis mytilaspidis* and *Hemisarcoptes malus*). But by 1930 it had become a pest. Lord (1947) reported that sulfur-based sprays for disease control rendered both enemy species ineffective; substitution of copper-based or ferbam fungicides has eliminated the problem since the 1940's (MacPhee et al. 1976, Pickett 1965).

In Nova Scotia apples, the European red mite is also normally controlled by a complex of predators, the complex varying considerably from orchard to orchard unless the predators/red mites balance is disrupted by use of pesticides. The solution has been to use selective materials (e.g., Ryania for codling moth) or to achieve adequate selectivity by timing and reduced dosage (MacPhee & MacLellan 1971).

The codling moth proved to be an unusual problem because of the very high degree of control demanded commercially. Also, in Nova Scotia it has a single generation per year, whereas in most other apple growing areas it has at least two. The natural enemy complex thus is often sufficient to maintain a high level of control if protected from adverse chemicals and where Ryania has been used when treatment has been necessary. The other enemies of other serious apple pests (listed above) are also thus protected. MacLellan, in a series of papers (1963, 1971) reported a large complex of enemies of codling moth in Nova Scotia, among which woodpeckers cause substantial winter mortality (MacPhee et al. 1976). Similar programs in other regions of the world where codling moth has two generations a year have commonly failed (MacPhee & MacLellan 1971, MacPhee et al. 1976).

Fig. 1 (from Flint & van den Bosch 1981, data from Patterson 1966) shows the consequences of a Nova Scotia IPM program in experimental plots from its initiation in 1950 to 1962. Flint &

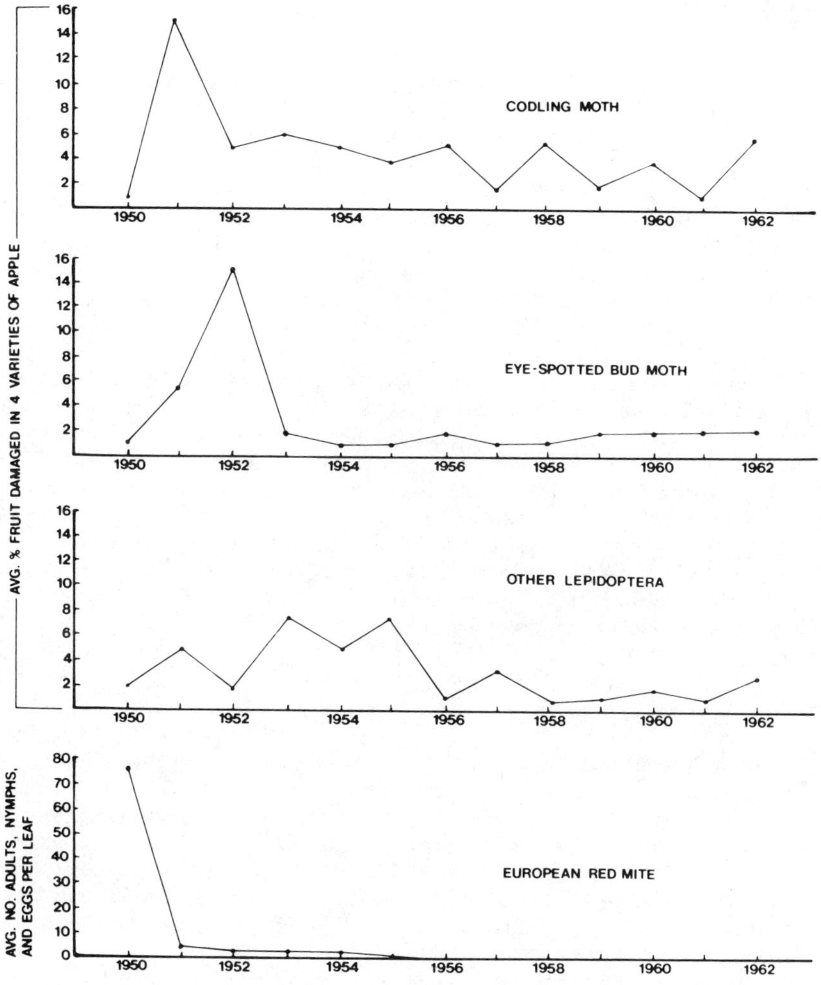

Fig. 1. Populations of pest and potential pest arthropods in Nova Scotia experimental apple orchard plots. Reduction of pest population lagged one to two years behind instigation of IPM program in 1950, representing time required for natural enemy populations to recover from effects of pre-1950 heavy pesticide use (after Flint & van den Bosch 1981; data from Patterson 1966).

van den Bosch (1981) comment: "The integrated pest management program has been so successful that 90% of Nova Scotia apple growers have been using the program continuously since the 1950's".

COTTON IN TEXAS

The IPM program for cotton in Texas has been a major success with which Cate and Frisbie & Adkisson deal (this volume). Nevertheless, I wish here to call your attention to this program because it shows how cultural, chemical, varietal breeding and biological controls can be combined to give good insect pest control, high yields and improved profits.

Natural enemies of bollworms and budworms (*Heliothis* spp.) are essential to this program, although the central focus of the research has been to control the boll weevil. The latter has involved development and employment of dwarf cottons grown under closer planting, such that yields are maintained, yet harvesting is advanced. Destruction of the crop residues, on which boll weevils can continue to breed, can thus be done before a large brood is ready to overwinter (diapause), and the residual diapausing populations can be decimated further by fall diapause treatments. This enables boll weevil treatments during the following crop year to be considerably delayed and thus the natural enemies of *Heliothis* can and do keep those pests under control (Cate, this volume; Croft et al. 1984). In the absence of such natural enemy controls, because of the phenomenal resistance of the budworm in particular, catastrophe could be much greater than from unchecked boll weevils.

MITES IN CALIFORNIA

Two areas of special interest are represented by work in California on pest mites using predatory mites. Work on predatory phytoseiids feeding on cyclamen mites in strawberries (Huffaker & Kennett 1956, and references cited therein) was so phenomenally demonstrative of the potential value of these predators as to stimulate worldwide work on phytoseiids, especially as predators of tetranychids (see Huffaker et al. 1969, McMurtry 1983, for reviews).

Huffaker & Kennett (1956) also showed experimentally that release of a pest itself may be used to achieve the continuity of the prey (or host) species necessary to a long-term and continuing effective biological control balance. This again was in work with cyclamen mites on strawberries, a pest so feared by growers that they would not spread them in their fields, despite advantages shown experimentally. Huffaker & Kennett (1953) had also shown that *Metaseiulus occidentalis* in strawberries had a relatively good tolerance to parathion. Recent genetic improvement of this same trait in this phytoseiid, and their mass production and successful colonization and use in IPM programs in almonds, is a major achievement (Hoy 1983). Dr. Hoy and colleagues developed strains of M. occidentalis resistant to a variety of the pesticides needed for control of mites and other pests on almonds. This is especially noteworthy in light of the pessimism that formerly existed relative to such an attempt. The work represents a practical genetic improvement of a natural enemy and also documents that mass releases and colonization of such phytoseiids can be highly beneficial in a complex IPM program (Hoy 1983).

MICROBIAL AGENTS

Microbial agents would seem to be, potentially, rather ready-made substitutes for disruptive broad-spectrum insecticides for use in IPM programs. The wide use of various forms of Bacillus thuringiensis, and their manufacture and sale by pesticide manufacturers, attests to their potential utility, but they have not been extensively used (Falcon, this volume). A Special Study Team of the USDA in 1978 prepared a report "Biological Agents for Pest Control" in which the status and prospects for their use was reviewed and assessed.

I will not attempt to summarize this area; it can be accessed, for example, in the above publication, in the encyclopedic work edited by Burges & Hussey (1971), and in Falcon (1976), National Academy of Sciences (1975), Weiser et al. (1976), World Health Organization (1973), and in Falcon (this volume).

REFERENCES

Bottrell, D.G. 1979. Integrated pest management. President's Council Environmental Quality. U.S. Government Printing Office, Washington, D.C. 120 pp.

Bottrell, D.G., and P.L. Adkisson. 1977. Cotton insect pest management. Annu. Rev. Entomol. 22:451-481.

Burges, H.D., and N.W. Hussey (Ed.). 1971. Microbial control of insects and mites. Academic Press, N.Y. 861 pp.

Carson, Rachel. 1962. Silent spring. Houghton Mifflin, Boston, MA. 368 pp.

Croft, B.A., P.L. Adkisson, R.W. Sutherst and G.A. Simmons. 1984. Application of ecology for better pest control, pp. 763-795. In: C.B. Huffaker and R.L. Rabb (Eds.), Ecological entomology. John Wiley & Sons, N.Y.

DeBach, P. 1974. Biological control by natural enemies. Cambridge Univ. Press, London. 323 pp.

Falcon, L.A. 1976. Problems associated with the use of arthropod viruses in pest control. Annu. Rev. Entomol. 21:305-324.

Flaherty, D.L. 1969. Ecosystem trophic complexity and densities of the Willamette mite, Eotetranychus willamettei Ewing (Acarina: Tetranychidae). Ecology 50:911-916.

Flint, M.L., and R. van den Bosch. 1981. Introduction to integrated pest management. Plenum Press, N.Y. 240 pp.

Hoy, M.A. 1983. Opportunities for genetic improvement of mites as biological control agents, pp. 141-146. In: M.A. Hoy, G.L. Cunningham and L. Knutson (Eds.), Biological control of pests by mites. Univ. Calif. Agric. Expt. Sta. Special Publ. 3304.

Huffaker, C.B. (Ed.). 1971. Biological control. Plenum Press, N.Y. 511 pp.

Huffaker, C.B., and C.E. Kennett. 1953. Differential tolerance to parathion of two *Typhlodromus* predatory on cyclamen mites. J. Econ. Entomol. 96:707-708.

Huffaker, C.B., and C.E. Kennett. 1956. Experimental studies on predation: Predation and cyclamen mite populations on strawberries in California. Hilgardia 26:191-222.

Huffaker, C.B., R.F. Smith, E.J. Armbrust and associates. 1971. The principles, strategies and tactics of pest population regulation and control in major crop ecosystems: 1. Alfalfa. University of California International Center for Integrated and Biological Control, Berkeley, California. (Mimeo).

Huffaker, C.B., M. van de Vrie and J.A. McMurtry. 1969. The ecology of tetranychid mites and their natural control. Annu. Rev. Entomol. 14:125-174.

Hunter, W.D., and B.R. Coad. 1923. The boll weevil problem. USDA Farm. Bull. 1329, 30 pp.

Hussey, N.W., and C.B. Huffaker. 1976. Spider mites, pp. 178-228. In: V.L. Delucchi (Ed.), Studies in biological control. International Biological Programme 9. Cambridge Univ. Press, London.

Lord, F.T. 1947. The influence of spray programs on the fauna of apple orchards in Nova Scotia. II. Oystershell scale, *Lepidosaphes ulmi* (L.). Can. Entomol. 79:196-209.

MacLellan, C.R. 1963. Predator populations and predation on the codling moth in an integrated control orchard--1961. Mem. Entomol. Soc. Can. 32:45-54.

MacLellan, C.R. 1971. Woodpecker ecology in the apple orchard environment. Proc. Tall Timbers Conf. Ecol. Anim. Control Habitat Mgt. 2:273-284.

MacPhee, A.W., and C.R. MacLellan. 1971. Cases of naturally-occuring biological control in Canada, pp. 312-328. In: C.B. Huffaker (Ed.), Biological control. Plenum Press, N.Y.

MacPhee, A.W., L.E. Caltagirone, M. van de Vrie and E. Collyer. 1976. Biological control of pests of temperate fruits and nuts, pp. 337-358. In: C.B. Huffaker and P.S. Messenger (Eds.), Theory and practice of biological control. Academic Press, N.Y.

Mathys, G. 1980. Comments as Secretary-General on 5 August before the General Assembly of the International Organization of Biological Control (IOBC) at Kyoto, Japan.

Maxwell, F.G., and F.A. Harris (Eds.). 1974. Proceedings of the summer institute on biological control of plant insects and diseases. Univ. Press Miss., Jackson. 647 pp.

McMurtry, J.A. 1983. Phytoseiid predators in orchard systems: A classical biological control success story, pp. 21-26. In: M.A. Hoy, G.L. Cunningham and L. Knutson (Eds.), Biological control of pests by mites. Univ. Calif. Agric. Expt. Sta. Spec. Publ. 3304.

National Academy of Sciences. 1975. Pest control: An assessment of present and alternative strategies. (5 volumes). Washington, D.C.

Patterson, N.A. 1966. The influence of spray programs on the fauna of apple orchards in Nova Scotia. XVI. The long-term effect of mild pesticides on pests and their predators. J. Econ. Entomol. 59:1430-1435.

Pickett, A.D. 1965. The influence of spray programs on the fauna of apple orchards in Nova Scotia. XIV. Supplement to II. Oystershell scale, Lepidosaphes ulmi (L.). Can. Entomol. 102:586-591.

Smith, R.F., and C.B. Huffaker. 1973. Integrated control strategy in the United States and its practical implementation. EPPO Bull. 3(3):31-49.

Stern, V.M., R.F. Smith, R. van den Bosch and K.S. Hagen. 1959. The integration of chemical and biological control of the spotted alfalfa aphid: The integrated control concept. Hilgardia 29:81-101.

Sweetman, H.L. 1936. The biological control of insects. Comstock Publ. Co., Ithaca, N.Y. 461 pp.

Sweetman, H.L. 1958. The principles of biological control. Wm. C. Brown, Co., Dubuque, IA. 560 pp.

USDA. 1978. Biological agents for pest control. Status and prospects. (Report of a special study team of the Office of Environ. Quality Activities) USDA. U.S. Government Printing Office, Washington, D.C. 138 pp.

Weiser, J., G.E. Bucher and G. O. Poinar. 1976. Host relationships and utility of pathogens, pp. 169-185. In: C.B. Huffaker and P.S. Messenger (Eds.), Theory and practice of biological control. Academic Press, N.Y.

World Health Organization. 1973. The use of viruses for control of insect pests and disease vectors. WHO Tech. Rept. Ser. 531, 48 pp.

BIOLOGICAL CONTROL OF PLANT PATHOGENS: DEFINITIONS

Ralph Baker

"If you wish to converse with me," Voltaire said, "define your terms." It is desirable to adhere to this dictum especially when establishing bases for discussion in the discipline of biological control of plant pathogens. First, interactions may be much more complex than in other disciplines, e.g. entomology; thus, the meanings of terms may be finely drawn and encompass concepts that are more superficial in nature. Again, scientists have adopted various definitions for the same term; therefore, in a workshop such as this, it is essential that we understand the meanings of our language to prevent confusion.

DEFINING BIOLOGICAL CONTROL

Johnson & Curl (1972) contributed a "broad concept of biological control of plant disease [that] includes any type of disease reduction or decrease in inoculum potential of a pathogen brought about directly or indirectly by other biological agencies." They were not clear as to what the "other" is, and therein lies the basis for the different definitions associated with biological control in plant pathology. Garrett (1965) was quite specific in the constitution of the "other"--"we are bound to mean a living microorganism or macroorganism other than the diseased or damaged plant acting as host and the pathogen or pest causing the disease or damage." As in all definitions currently employed, he exempted man himself as the biological agent and, when excluding the pathogen, it was assumed that he did not preclude control through use of avirulent strains.

Baker & Cook (1974) excluded no agent other than man in their definitions. A shortened version is found in their recently published book (Cook & Baker 1983): "Biological control is the reduction of the amount of inoculum or disease-producing activity of a pathogen accomplished by or through one or more organisms other than man." This includes genetic manipulation of the higher-plant host and cultural controls as, for example, close-spacing of crop plants to discourage growth of weeds.

ISBN 0-12-357030-1

To provide a working definition for use in delimiting discussion at the USDA Interdisciplinary Biological Control Conference held at Las Vegas in 1983, an ad hoc group composed of entomologists, plant pathologists, and weed scientists developed a definition (Baker 1983) based on "pest supression with biotic agents, excluding the process of breeding for resistance to pests, sterility techniques, and chemicals modifying pest behavior." This is close, if not identical, to Garrett's (1965) concept and will be used as a working definition in this paper to delimit the boundaries of the subject.

An attempt was made (Baker 1975) to determine the degree of acceptance of the various concepts of what constitutes biological control. Response to a survey by 138 members of the American Phytopathological Society working with soil-borne pathogens indicated 49% preferred Baker & Cook's (1974) definition of biological control, 34% preferred Garrett's (1965), whereas 17% felt a "good" definition had not been published. It is obvious that scientists engaged in teaching, research, and publishing in this subject area do not have uniform concepts.

SOME CLASSIC CASE HISTORIES

Although Potter demonstrated as early as 1908 that the activity of a plant pathogen might be inhibited by its own metabolic products, it was not until 1926 that Sanford suggested that the control of the potato scab by green manuring was due to biological control brought on by increase in population density of certain saprophytic bacteria. A review of the exciting developments that followed are provided by Garrett (1965) and more recently by Cook & Baker (1983).

Early in this century it was known that carbon disulphide fumigation of soil infested with *Armillaria mellea* in California orchards killed the pathogen in a short period of two to three days. Moderate doses killed it after 30-50 days. Bliss (1951) reported that *Trichoderma viride* attacked the pathogen in the latter case after it was "weakened" by the fumigant.

Several examples of control in which the biological control agent interacts with the pathogen in host tissue are available. *Heterobasidion* (*Fomes*) *annosum* attacks conifers. In newly-planted plantations, stumps provide substrates for an increase in inoculum potential of the pathogen. Stumps were inoculated with a prolific antagonistic fungus, *Peniophora gigantea*, with limited ability to cause rot (Rishbeth 1963). This treatment precluded colonization by *H. annosum* so that it could not establish a food base to attack living roots. The antagonist is supplied to lumberjacks as dehydrated pellets and can be applied in water or in the oil used to lubricate saws.

Pathogenic strains of *Agrobacterium tumefaciens* (causal agent of crown gall) may be controlled by application of an avirulent strain (K84) of A. radiobactor to wounds incurred in handling and transplanting of susceptible hosts (Kerr 1980).

Chestnut blight, induced by Endothia parasitica, is no longer a problem in cultivated chestnuts in Italy. Grente & Saute (1969) found that the pathogen has become hypovirulent. Such hypovirulent strains may be inoculated into chestnut trees and the trait transmitted to healthy isolates of the pathogen by hyphal anastomoses. Anagnostakis (1982) provided a recent review on the subject.

While it has been known for some time that soils may be suppressive to certain pathogens, renewed interest in the subject was recently initiated by Baker & Cook's (1974) first book on biological control, which clearly stated the potentials for use of such soils in biological control. For example, they witnessed attempts in Australia to artificially duplicate nutrition status, organic matter content, and understory composition comparable to those of the undisturbed rain forest for control of phytophthora root rot of avocado induced by Phytophthora cinnamomi. Soils, so treated, became suppressive.

Suppressive soils may be classified in two categories: those that are induced, and those that apparently are consistently suppressive in nature over a long period of time. When new lands were planted to potatoes after the second world war, devastating epidemics of common scab of potato (induced by Streptomyces scabes) occurred. Where long-term culture of potatoes had been practiced, the disease was not present. Menzies (1959) determined that potato monoculture induced the soils to become suppressive. A small amount of this soil transferred to previously conducive soil amended with alfalfa induced suppressiveness.

A classic example of induced suppressiveness is furnished by the take-all decline phenomenon. When the Dutch established the polders by draining the Zuider Zee, the first grain crops grown in these areas were severely infected by Gaeumannomyces graminis var. tritici, causal agent of take-all. Any people with less fortitude than the Dutch might have given up; however, by repeated cropping of the hosts, disease declined over a five- to seven-year period to relatively low values. Gerlagh (1968) and Vojinovic (1972) established that suppressiveness of soil associate with take-all decline is induced by microbial antagonists.

A more stable form of suppressiveness is found in the Salinas Valley of California. In that area, fusarium wilt (induced by f. sp. of Fusarium oxysporum) is not found, even though susceptible hosts are consistently grown (Smith & Snyder 1972). Fusarium-suppressive soils also are reported eslewhere, as for example, in the Chateaurenard region in France (Alabouvette et al. 1977).

THE PECULIAR MECHANISMS OF BIOLOGICAL CONTROL IN PLANT PATHOLOGY

The classical concept of biological control used by entomologists is comparatively simple in that only "the action of parasites, predators or pathogens in maintaining another organisms' population density at a lower average than would occur in their absence" (DeBach 1964) is usually considered. In plant pathology, the overwhelming hypertrophy of the disease triangle involving the interactions of pathogen, host, environment and vectors provides a more complicated system. To the interactions among plant host, pathogen, and environmental factors is added the influence of biological control agents. These machinations and the pathways involved in biological control (sensu Garrett 1965) are presented in diagramatic form in Fig. 1.

In this scheme, there are two pathways through which the biological agent(s) may operate. Externally to the host, the control agent may be antagonistic and, thereby, reduce the activity, efficiency, and/or inoculum density of the pathogen through antibiosis, competition, or predation/hyperparasitism. This leads to a reduction in inoculum potential (Baker 1977) of the pathogen, which is defined as "the energy available for the colonization of a host or substrate at the surface of the substrate to be colonized" (Garrett 1956).

The biocontrol agent may operate primarily in the host tissue, thereby initiating a resistance response in the host (cross protection), becoming antagonistic, or transmitting factors rendering the pathogens avirulent (hypovirulent). This is called "protective inoculation" by Cook & Baker (1983). All of these interactions are mediated by environment, and this factor may

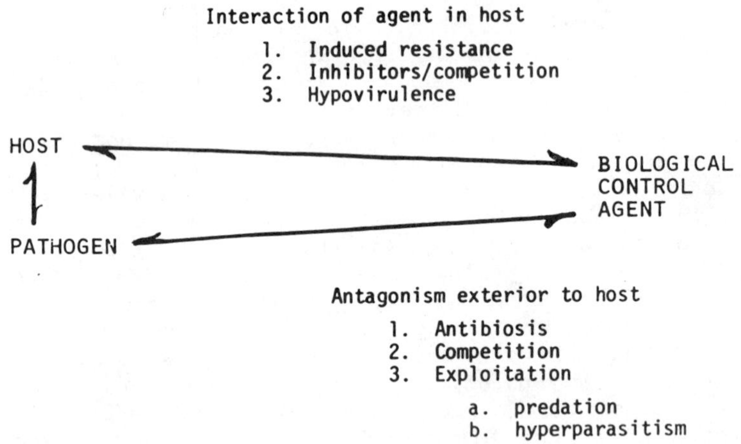

Fig. 1. Some pathways of mechanisms involved in biological control of plant pathogens.

have an overriding impact in determining whether biological control operates in a system.

Interaction of Biocontrol Agents in Host

Cross protection In this case, a biological control agent is inoculated into a plant host. The plant recognizes the presence of a foreign microorganism and initiates a resistance response, which can be chemical and/or morphological in character. Subsequent introduction of a virulent pathogen finds the host "forewarned" and its resistance increased.

Kuc (1982a) recently reviewed the mechanisms associated with resistance to viral, bacterial, and fungal diseases. Treatment with appropriate biocontrol agents initiates systemic immunization of the host, which is accompanied by enhanced peroxidase activity and lignification. The lignification process may be highly toxic to fungal growth, bacterial multiplication, and viral replication. The phenolic precursors are toxic, the free radicals highly reactive. Substrates may become unavailable, and infectious agents may be unable to bind to host cell walls (Kuc 1982b). The most interesting facet of this type of mechanism is that it is nonspecific. For example, resistance to anthracnose in cucumber (induced by *Colletotrichum lagenarium*, a fungal pathogen) was induced systemically by a localized infection with tobacco necrosis virus (Jenns & Kuc 1980). This suggests that the initiation of resistance is dependent upon a function common to infectious agents.

Production of inhibitors and/or competition for host receptor sites One of the few registered biological control agents currently available commercially is strain K84 of *A. radiobacter*. The agent occurs frequently in the rhizosphere of higher plants, but is avirulent. A related bacterium, *A. tumefaciens*, however, induces crown gall on a wide variety of hosts, but it requires a wound for penetration into host tissue. Tissues may be protected from the pathogen by pretreatment of wounds with strain K84. The agent may kill the pathogen by production of an antibiotic (agrocin 84) (Kerr 1980); however, strains not producing agrocin also may reduce gall size (Cooksey & Moore 1982). This suggests that strain K84 also may prevent attachment to the host receptor sites.

Hypovirulence Grente & Sauret (1969) coined the term "hypovirulent" to identify the avirulent strains of *E. parasitica* found in healing cankers of chestnut trees. Subsequent research (MacDonald et al. 1978) suggested that a double-stranded RNA (a mycovirus) was responsible for the avirulence. Upon introduction of an avirulent strain of appropriate compatibility into an infected chestnut tree, hyphal anatomoses occur with the (previously) virulent pathogen. Thus, hypovirulence is now used to indicate a transmissible trait that is distinguished from the non-transmissible loss of virulence induced by mutations.

Antagonism Exterior to the Host

Antibiosis The term is used here in the broad sense in which the metabolic product(s) of one organism directly inhibits or destroys another.

In the early history of the awakening to the possibility that microbial factors could influence activity of soilborne plant pathogens, the presence of antibiotics produced in culture gave researchers a convenient explanation for the observed antagonistic phenomena. This first hypothesis was not readily supported, however, and led Pramer (in Baker & Snyder 1965) to the conclusion that: "The amounts formed would be small...and their influence would be limited to the immediate area. There is, I believe, a lack of convincing evidence to support the thesis that antibiotics produced under natural conditions are capable of exerting a significant and generalized, rather than restricted and generally inconsequential, effect."

Basic research supports this statement as summarized by Baker (1968). For production of antibiotics, microorganisms require substantial amounts of substrate, particularly carbon. Once produced, most antibiotics have electrical charges on their molecules. Colloids with opposite charges abound in soil and inactivate these inhibitors. There is even evidence that antibiotics can be utilized as a sole carbon source by some of the more voracious saprophytes!

As mentioned by Pramer in his statement, the work of Wright (1956) demonstrated that the antibiotic, gliotoxin, produced by *Trichoderma viride* could only be detected in colonized seed coats or in bits of organic matter. This suggests, indeed, that antibiotics cannot exert a "generalized ... effect."

Cook & Baker (1983) recently have taken issue with Pramer's statement. They cite two lines of evidence in support of their contention that antibiotics play a substantial role in some biological control systems. The first, supplied by Bruehl et al. (1969), demonstrated that *Cephalosporium gramineum* produced a wide-spectrum antibiotic in wheat straw. This is a survival mechanism that enables the fungus to retain possession of the substrate for a long period of time, and confirms the conclusion by both Wright (1956) and Pramer that antibiotics function in localized habitats. The second line of evidence, supplied by Kloepper & Schroth (1981), indicates that ferric Pseudobactin-producing isolates of *Pseudomonas* spp. increased plant growth, whereas mutants not producing the compound did not. Pseudobactin is involved in competition for iron, is a siderophore (treated later), and has been mislabeled as an antibiotic (Neilands 1973). Again, biological control realized by the activity of these fluorescent pseudomonads operates in the rhizosphere, and this emphasizes the principle that inhibitory compounds operate in localized areas (Scher & Baker 1982).

In modern approaches to the controversy, a balanced view might only take issue with the idea that antibiotics exert a "generally inconsequential effect." In some systems, antibiosis could play a role in protection of a well-defined and treatable infection court. For example, an antibiotic-producing agent could be applied to a seed to protect against pre-emergence damping-off pathogens. The dead seed coat can supply the carbon necessary for antibiotic production, and this same substrate would not immobilize antibiotics once they are produced. Lif-shitz & Windham (personal communication) suggest that this mechanism operates in the biological control of pythium damping-off after seed treatment with _Trichoderma harzianum_. The biological control agent is a mycoparasite; however, the incubation period required to establish infection in hyphae of _Pythium_ spp. is too long by far to implicate this mechanism. Antibiosis in vitro between the agent and _Pythium_ spp. is not observed when conventional techniques are employed, but observations of interactions on a micro-level revealed a likely inhibitory mechanism. As _Pythium_ spp. approached the antagonist, the protoplasm in the hyphal tip retreated. Subsequently, various signs of stress occurred in the thallus of the _Pythium_ spp., such as hyphal distortion, vacuolation, and formation of septa. Such reactions occurred in a brief period of time and could explain the observed damping-off by seed treatment with _T. harzianum_. Since the protoplasm in the hyphal tip is induced to retreat, the causal agent of the phenomenon was termed "the routing factor."

Competition In a narrow sense, competition is "active demand in excess of immediate supply of material or condition on the part of two or more organisms" (Clark 1963). In soil, microorganisms compete almost exclusively for substrate (Baker 1981); however, competition also may be involved in occupation of potential infection sites by biocontrol agents (Cooksey & Moore 1982).

The basic principle involved in competition is that the pathogen must be deprived of a nutrient essential for pathogenesis. Therefore, research is usually directed toward determining factors essential to successful germination and penetration by the pathogen in the host infection court. Biological control agents, then, are used to create deficiencies of these essential elements. Factors found, so far, to be potentially essential are nitrogen, carbon, and recently, iron (Fe).

An example illustrating the principles involved was reviewed by Baker (1968). The causal agent of bean root rot is _Fusarium solani_ f. sp. _phaseoli_. The fungus survives in soil by producing chlamydospores. Carbon and nitrogen, essential for germination and penetration of the host by the pathogen, are supplied in host exudates, or are present in the soil. If cellulose or a crop residue such as barley straw are incorporated into soil, no symptoms of bean root rot are observed. The mechanism involved is competition for nitrogen. Substrates introduced into the soil support the activities of saprophytes. If carbon to nitrogen ratios are wide in the introduced substrate, however, the available nitrogen is immobilized and utilized by the biomass so that nitrogen becomes limiting. Any nitrogen exuded from the host also is utilized immediately by the rampant microbial community, supplied with abundant carbon, but deficient in essential nitrogen (Benson & Baker 1970). Thus, the pathogen is deprived of this essential element.

The system mentioned immediately above is an example of a mechanism involving the general soil microflora; however, it is possible to manage a specific component of the microflora to induce competition for a specific element. The addition of fluorescent pseudomonads to soil induces previously-conducive soils to become suppressive to fusarium wilt pathogens (Scher & Baker 1980). The mechanism in this case appears to be competition for iron.

The iron available in soil for plants and/or microorganisms is defined by the equilibrium reaction:

$$Fe(OH)_3 + H^+ \rightleftarrows Fe^{3+} + 3\ H_2O \qquad \text{(Lindsey 1974)}$$

The principle factor mediating the equilibrium of this equation is the pH of the soil. Thus, the influence of pH may be considerable in determining whether Fe^{3+} can be available to the pathogen, and, ultimately, whether a soil is suppressive or conducive since Fe^{3+} is necessary for germ tube elongation from spores of formae specialis of _Fusarium oxysporum_ (Scher & Baker 1982). Indeed, when the suppressive Metz sandy loam of the Salinas Valley in California (native pH ca. 8.0) was reduced to 7.0, about half of the suppressiveness was lost. At a pH of 6.0 the soil was conducive (Scher & Baker 1980). A theoretical basis for induction of suppressiveness through manipulation of Fe availability by use of chelates and/or siderophores (Kloepper et al. 1980, Scher & Baker 1982) was established.

When _Pseudomonas putida_, ethylenediaminedi-O-hydroxyphenylacetic acid (EDDHA), or the ferrated form (FeEDDHA), was added to conducive soil, the soil became suppressive to fusarium wilt of flax, cucumber, or radish. Suppressiveness was not induced by Fe-ethylenediamine-tetracetic acid (FeEDTA), or Fe-diethylenediamine-tetracetic acid (FeDTPA). The results may be explained by the mechanism of competition for Fe illustrated in Fig. 2.

The three ligands bind Fe^{3+} at various stability constants: EDDHA, $\log_{10} = 33.9 >$ DTPA, $\log_{10} K = 27.3 >$ EDTA, $\log_{10} K = 25$ (Lindsey 1974). Fe-binding siderophores have been detected for most microorganisms examined (Neilands 1974), and fusarium wilt causal agents produced siderophores of the hydroximate class under Fe-deficient conditions that have stability constants in the proximity of $\log_{10} K = 29$. This constant is many powers of ten less than most bacterial catechol siderophores, and _P. putida_

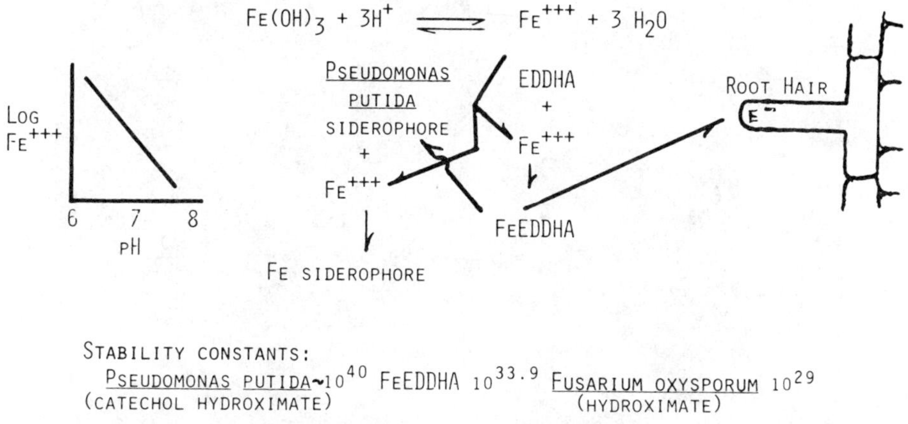

Fig. 2. Mechanisms and pathways involved in soil suppressive to fusarium wilt pathogens through competition for iron (Fe) on the rhizoplane (by permission, from Baker 1983).

produces a mixed hydroximate catechol siderophore (pseudobactin) as reported by Teintze et al. (1980).

Theory expansion suggests that adequate Fe^{3+} is available to fusarium wilt pathogens for germination and penetration through root tips of the host in conducive soil. Since many pseudomonads have high rhizosphere competence (Schmidt 1979), there should be intense competition for Fe^{3+} at the rhizoplane (infection court) when the P. putida produces siderophores, especially in alkaline soils. Therefore, Fe^{3+} is bound in such a way that it is unavailable to F. oxysporum.

The ligand EDDHA has a higher stability constant than the siderophore produced by the pathogen. Fe^{3+} complexed by this ligand in the rhizosphere would be unavailable to the pathogen; however, EDDHA-bound Fe can be utilized by the host. This frees EDDHA to bind more Fe^{3+} at the root surface (Lindsey 1974) and the soil is suppressive. This is a dynamic system in which the siderophores produced by pseudomonads, the EDDHA, and the root combine to render Fe^{3+} limiting for pathogenesis by the fusarium wilt pathogen.

Exploitation Mites and vampyrellid amoebae (Homma et al. 1979) prey on the thalli of fungi. The amoebae "bore" holes in the cell walls of fungi and digest the protoplasm within the cell (Fig. 3). Evidence, so far, suggests that these antagonists, however, do not substantially affect biological control. Among other limiting factors, the constant water content of soil must be well above that found in typical agricultural soils.

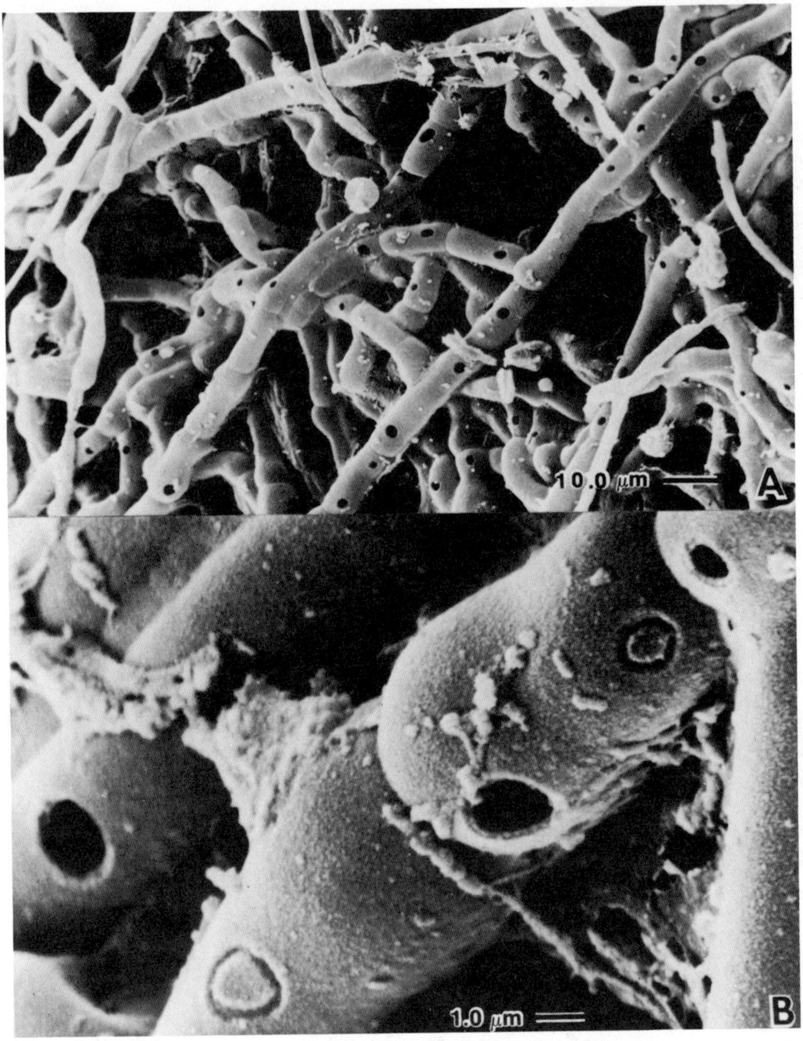

Fig. 3. Annular depressions and perforations of the pigmented hyphae of Gaeumannomyces graminis var. tritici caused by vampyrellid amoebae (by permission, from Homma et al. 1979).

Hyperparasitism has been implicated as a biological control mechanism in a number of systems. Perhaps the best documented example involves a model system in which radishes were repeatedly planted at weekly intervals in soil infested with *Rhizoctonia solani* (Henis et al. 1978, 1979). Typical interactions of host, pathogen, and biocontrol agent are illustrated in Fig. 4. In a typical experiment in acid soils, disease (damping-off) incidence was low initially, but by the third replanting, no plants emerged. In subsequent replantings, however, disease incidence decreased until, in some experiments, there was 100% emergence. The inoculum density of *R. solani* rose and fell in proportion to disease incidence. In the first planting, therefore, the soil was conducive. Such suppressiveness was correlated with an increase in population density of *Trichoderma* spp. Subsequently, a modification of Koch's postulates was employed to implicate *T. harzianum* as the causal agent of suppressiveness (Liu & Baker 1980). *Trichoderma* spp. are mycoparasites (Chet et al. 1981), and this provides an explanation for the observed decrease in inoculum density of *R. solani* during the generation of suppressiveness.

Other Mechanisms

The scheme in Fig. 1 does not illustrate all possible pathways for mechanisms of biocontrol. For example, Burke et al. (1972) provided an explanation for the observed decrease in disease severity resulting from barley rotations. Barley roots in a previous crop broke the plow sole, which was a barrier to deep rooting by the bean plants. In addition, decomposing barley residues reduced soil compaction. Such a situation allows a more extensive root volume for the bean in subsoil that contains less inoculum of the pathogen. In this case the biocontrol agent is barley.

PROBLEMS EXPLICIT TO BIOCONTROL OF PLANT PATHOGENS IN AGRICULTURE

There are few biocontrol agents of plant pathogens employed at present in agriculture in spite of long-term research efforts by laboratories, and the more recent dramatic increase in industrial interest. Obviously, there are problems.

Some of the problems are associated with economics. For example, chemical control is considered to be cost effective, is usually more efficient than biological control, and is relatively unaffected by environmental conditions. While biological control may be integrated with other measures to make the total effort more cost effective, this approach is in its infancy.

Cook & Baker (1983) list 41 agents with potential for immediate application in their antagonistic vitae. Thus, there is no lack of agents with potential for use in biological control. The chief problem in application is in managing host, environmental

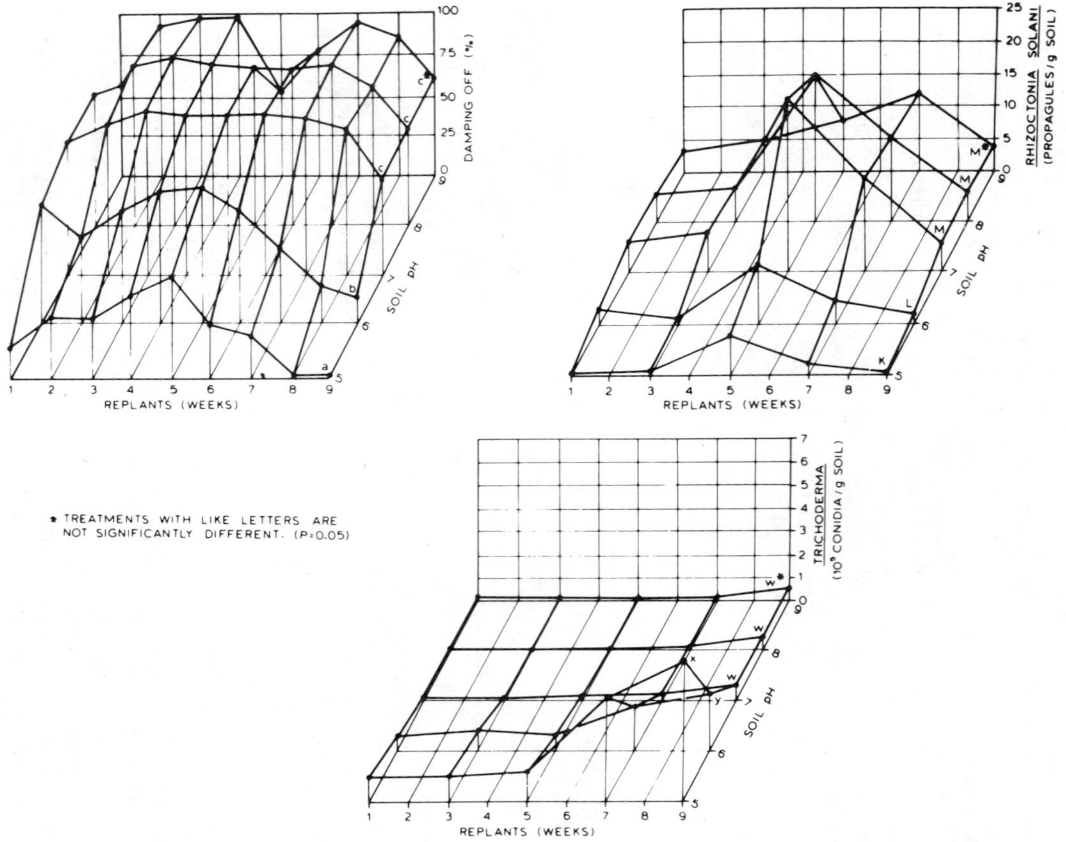

Fig. 4. Effect of soil pH on development of suppressiveness when it was infested with _Rhizoctonia solani_ and replanted at weekly intervals with radish (by permission, from Liu & Baker 1980).

conditions and the agent itself so that the activity of the biological agents are enhanced. For example, Trichoderma harzianum has been effective in a number of applications in the field where temperatures were high enough so that it could be active (Harmon et al. 1981). No effects were observed in cool climates or in early plantings.

At the 1971 annual meeting of the American Phytopathological Society, a discussion was held entitled "Biological Control--Mission Impossible?" This reflected a general pessimism that inherent problems were not likely to be solved. Attitudes over the past decade became more optimistic, but much still needs to be done to successfully complete the mission.

REFERENCES

Alabouvette, C., F. Rouxel and J. Louvet. 1977. Recherches sur la resistance des sols aux maladies III. Effects du rayonnement y sur la microflora d'un sol et sa resistance a la Jusariose vascularires du Melon. Annu. Phytopath. 9:467-471.

Anagnostakis, S.L. 1982. Biological control of chestnut blight. Science 215: 466-471.

Baker, R. 1968. Mechanisms of biological control of soil-borne pathogens. Annu. Rev. Phytopathol. 6:263-294.

Baker, K.F., and R.J. Cook. 1974. Biological control of plant pathogens. W.H. Freeman, San Francisco. 433 pp.

Baker, K.F., and W.C. Snyder (Eds.). 1965. Ecology of soil-borne plant pathogens. Univ. Calif. Press, Berkeley. 571 pp.

Baker. R. 1975. Biological control of plant pathogens by K.F. Baker and R.J. Cook. Mycologia 67:199-200.

Baker, R. 1977. Inoculum potential, pp. 137-157. In: J.G. Horsefall and E.B. Cowling (Eds.), Plant disease an advanced treatise, Vol. II. Academic Press, N.Y.

Baker, R. 1981. Ecology of the fungus, Fusarium: Competition, pp. 245-249. In: P.E. Nelson, T.A. Toussoun and R.J. Cook (Eds.), Fusarium: Diseases, biology, and taxonomy. Penn. State Univ. Press, University Park.

Baker, R. 1983. State of the art: Plant diseases, pp. 14-22. In: S.L. Battenfield (Ed.), Proceedings of the national interdisciplinary biological control conference. 15-17 Feb. 1983. Las Vegas, NV., CSRS/USDA, Washington, DC.

Benson, D.M., and R. Baker. 1970. Rhizosphere competition in model soil systems. Phytopathology 60:1058-1061.

Bliss, D.E. 1951. The destruction of Armillaria mellea in citrus soils. Phytopathology 41:665-683.

Bruehl, G.W., R.L. Millar and B. Cunfer. 1969. Significance of antibiotic production by Cephalosporium gramineum to its saprophytic survival. Can. J. Plant Sci. 49:235-246.

Burke, D.W., D.E. Miller, L.D. Holmes and A.W. Barker. 1972. Counteracting bean root rot by loosening the soil. Phytopathology 62:306-309.

Chet, I., G.E. Harman and R. Baker. 1981. Trichoderma hamatum: Its hyphal interactions with Rhizoctonia solani and Pythium spp. Micro. Ecol. 7:29-38.

Clark, F.E. 1963. The concept of competition in microbial ecology, pp. 339-345. In: K.F. Baker and W.C. Snyder (Eds.), Ecology of soil-borne plant pathogens. Univ. Calif. Press, Berkeley.

Cook, R.J., and K.F. Baker. 1983. The nature and practice of biological control of plant pathogens. Amer. Phytopathol. Soc., St. Paul, MN. 539 pp.

Cooksey, D.A., and L.W. Moore. 1982. High frequency spontaneous mutations to agrocin 84 resistance in Agrobacterium tumefaciens and A. rhizogenes. Physiol. Plant Pathol. 20:129-135.

DeBach, P. 1964. Biological control of insect pests and weeds. Reinhold, N.Y. 844 pp.

Garrett, S.D. 1956. Biology of root-infecting fungi. Cambridge Univ. Press, London. 294 pp.

Garrett, S.D. 1965. Toward biological control of soil-borne plant pathogens, pp. 4-17. In: K.F. Baker and W.C. Snyder (Eds.), Ecology of soil-borne plant pathogens. Univ. Calif. Press, Berkeley.

Gerlagh, M. 1968. Introduction of Ophiobolus graminis into new polders and its decline. Neth. J. Plant Pathol. 74(Suppl. 2):1-97.

Grente, J., and S. Sauret. 1969. L'hypovirulence exclusive phenomene original en pathologie vegetale. C.R. Acad. Sci. Paris 268:2347-2350.

Harmon, G.E., I. Chet and R. Baker. 1981. Factors affecting Trichoderma hamatum applied to seeds as a biocontrol agent. Phytopathology 71:569-572.

Henis, Y., A. Ghaffar and R. Baker. 1978. Integrated control of Rhizoctonia solani damping-off of radish: Effect of successive plantings, PCNB, and Trichoderma harzianum on pathogen and disease. Phytopathology 68:900-907.

Henis, Y., A. Ghaffar and R. Baker. 1979. Factors affecting suppressiveness to Rhizoctonia solani in soil. Phytopathology 69:1164-1169.

Homma, Y., W. Sitton, R.J. Cook, and K.M. Old. 1979. Perforation and destruction of pigmented hyphae of Gaeumannomyces graminis by vampyrellid amoebae from Pacific Northwest wheat field soils. Phytopathology 69:1118-1122.

Jenns, A.E., and J. Kuc. 1980. Characteristics of anthracnose resistance induced by localized infection of cucumber with tobacco necrosis virus. Physiol. Plant Pathol. 17:81-91.

Johnson, L.F., and E.A. Curl. 1972. Methods for research on the ecology of soil-borne plant pathogens. Burgess Publ. Co., Minneapolis, MN. 247 pp.

Kerr, A. 1980. Biological control of crown gall through production of agrocin 84. Plant Disease 64:25-30.

Kloepper, J.W., and M.N. Schroth. 1981. Relationship of in vitro antibiosis of plant growth-promoting rhizobacteria to plant growth and the displacement of root microflora. Phytopathology 71:1020-1024.

Kloepper, J. W., J. Leong, M. Teintze and M. N. Schroth. 1980. Pseudomonas siderophores: A mechanism explaining disease-suppressive soils. Curr. Microbiol. 4:317-320.

Kuc, J. 1982a. Induced immunity to plant disease. BioScience 32:854-860.

Kuc, J. 1982b. Plant immunization-mechanisms and practical implications, pp. 157-178. In: R.K.S. Wood (Ed.), Active defense mechanisms in plants. Plenum Press, N.Y.

Liu, S., and R. Baker. 1980. Mechanisms of biological control in soil suppressive to Rhizoctonia solani. Phytopathology 70:404-412.

Lindsey, W.L. 1974. Role of chelation in micronutrient availability, pp. 507-524. In: R.W. Carson (Ed.), The plant root and its environment. Univ. Press Va., Charlottesville.

MacDonald, W.L., F.C. Check, J. Luchuk and C. Smith (Eds.). 1978. Proceedings of the American chestnut symposium. West Virginia Univ. Books, Morgantown. 122 pp.

Mankau, R., and N. Prasad. 1977. Infectivity of Bacillus penetrans in plant parasitic nematodes. J. Nematol. 2:7-8.

Menzies, J.D. 1959. Occurrence and transfer of a biological factor in soil that suppresses potato scab. Phytopathology 49:648-652.

Neilands, J.B. 1973. Microbial ion transport compounds (siderophores), pp. 167-202. In: I.G.L. Eickhorn (Ed.), Inorganic biochemistry. Elsevier, Amsterdam.

Neilands, J.B. (Ed.). 1974. Microbial iron metabolism. Academic Press, N.Y. 597 pp.

Potter, M.C. 1908. On a method of checking parasitic diseases in plants. J. Agr. Sci. 3:102-107.

Rishbeth, J. 1963. Stump protection against Fomes annosus. III. Inoculation with Peniophora gigantia. Ann. Appl. Biol. 52:63-77.

Sanford, G.B. 1926. Some factors affecting the pathogenicity of Actinomyces scabes. Phytopathology 16:525-547.

Scher, F.M., and R. Baker. 1980. Mechanism of biological control in a Fusarium-suppressive soil. Phytopathology 70:412-417.

Scher, F.M., and R. Baker. 1982. Effect of Pseudomonas putida and a synthetic iron chelator on induction of soil suppressiveness to fusarium wilt pathogens. Phytopathology 72:1567-1573.

Schmidt, E.L. 1979. Initiation of plant root microbe interactions. Annu. Rev. Microbiol. 33:355-379.

Smith, S.N., and W.C. Snyder. 1971. Relationship of inoculum density and soil types to severity of fusarium wilt of sweet potato. Phytopathology 61:1049-1051.

Teintze, M., M.B. Hassain, C.L. Baines, J. Leong and D. van der Helm. 1980. Structure of ferric pseudobactin a siderophore from a plant growth promoting Pseudomonas. Biochemistry 20:6466-6457.

Vojinovic, A.D. 1973. The influence of microorganisms following Ophiobolus graminis Sacc. on its further pathogenicity. Org. Eur. Med. Prot. Plants Bull. 9: 91-101.

Wright, J.M. 1956. The production of antibiotics in soil. III. Production of gliotoxin in wheatstraw buried in soil. Ann. Appl. Biol. 44:461-466.

IPM: DEFINITIONS AND CURRENT STATUS IN U.S. AGRICULTURE

R.E. Frisbie and P.L. Adkisson

DEFINITIONS OF IPM

Integrated Pest Management (IPM) either by chance or circumstance was used as a descriptive term by entomologists as early as 1952. At the time the term commonly used for IPM was integrated control. Synonymy of the terms integrated control and IPM was not broadly applied until the late 1960's. Michelbacher & Bacon (1952) were the first to use the term integrated control when describing methodologies for the selection, timing and dosage of insecticide treatments for the control of walnut aphid and preservation of beneficial arthropods in California. Several scientists used applied ecological means in an attempt to control pests. These include the work by S.A. Forbes in Illinois, and C.W. Woodworth and Harry Smith in California (Metcalf 1930, Smith 1977). Several workers, well before the term integrated control or IPM was coined, developed ecologically based, multi-tactic approaches to the management of insect pests. For example, Professor Frederick Malley of Texas A&M University, and L.O. Howard, W.D. Hunter and W.E. Hines of the U.S. Department of Agriculture developed a multi-tactical strategy for the management of the boll weevil in the early 1900's (Smith 1977).

Although there have been many attempts to seek a common definition of IPM, it is not the intent here to review in detail all these wide ranging definitions. The definition of IPM that has been most used was coined by Smith & Reynolds (1966) in an FAO Symposium on Integrated Pest Control. They defined IPM as follows: "Integrated pest control is a pest population management system that utilizes all suitable techniques in a compatible manner to reduce pest populations and maintain them at levels below those causing economic injury. Integrated control achieves this ideal by harmonizing techniques in an organized way, by making control practices compatible, and by blending them into a multi-faceted, flexible, evolving system." It is generally agreed that IPM has its basis in applied ecology through detailed study of population dynamics to provide the biological basis for determining the appropriate management strategies and tactics for

ISBN 0-12-357030-1

the economic suppression or control of pest species. As stated earlier, IPM has been defined by many. It is probably one of the few scientific terms that ever has been redefined by a President of the United States. In the President's 1979 environmental message to Congress, President Jimmy Carter described IPM as "a systems approach to reduce pest damage to tolerable levels through a variety of techniques, including predators and parasites, genetically resistant hosts, natural environmental modifications, and when necessary and appropriate, chemical pesticides."

Although integrated pest control, and later IPM, had its roots in applied ecology in the 20th century, attention to IPM was intensified beginning in the early 1970's. As a matter of fact, the 1970's and the early 1980's may easily be termed the "Golden Era" of IPM. Although entomologists knew and began to focus on the adverse side effects created by the new waves of insecticides produced in the post World War II era, it was not necessarily the development of insecticide-resistant pest strains, the occurrence of secondary pests through the disruption of beneficial organisms, nor other pest related factors that drew attention to the IPM concept as the most rational approach to the management of pests. Instead, it was the public outcry generated by Rachel Carson in her famous book Silent Spring which drew attention to the adverse environmental side effects of pesticide use. Clearly, the pesticides that were producing the most adverse effects on non-target organisms were the chlorinated hydrocarbon insecticides. When the Environmental Protection Agency was formed in 1970, one of the first acts of this young regulatory agency was to ban DDT. Public interest and media coverage immediately focused on wildlife and human side effects caused by the inappropriate application of insecticides. Although the chlorinated hydrocarbons received most of the publicity, a great many other insecticides were also lumped into the environmentally hazardous group. It was in this emotion-charged atmosphere that IPM was brought forth as the best way to resolve the problem of environmental pollution and human risk caused by insecticides while maintaining crop yields. The Council on Environmental Quality, as a matter of policy, championed IPM as the best way to control pests without unduly harming the environment.

The mission of entomologists at the time, although they were interested in reducing adverse side effects of insecticides, was primarily to develop methods for economically controlling insect pests. At this time the environmental movement was well organized and there was much common ground between those wanting to minimize pesticide abuse and those needing to control insects; thus, an alliance was formed between the proponents of IPM and the environmentalists. Newsom (1974) captured the sentiment of this time perhaps better than anybody in the following statement: "Regrettably, a great many of us have oversold the proven and potential effectiveness of the alternatives to use of conventional pesticides. Substantial numbers of entomologists have jumped on the environmental band wagon and vicious attacks have been made on the use of DDT and related insecticides, seemingly convinced that once DDT is banned all the environmental ills related to pest control will be corrected. Others have stood supinely by without making any effort to help defend the sensible use of all pesticides, DDT included, apparently unaware that effective pest management systems for the vast majority of our major pests require the intelligent use of these chemicals." Newsom (1974) said entomologists involved in IPM have "painted themselves into a corner" and were being asked to develop programs that would not necessarily elminate pesticides but would significantly reduce the amounts used. These specialists were asked to develop pest control programs that would minimize pesticide pollution of the environment. Beginning in the early 1970's, the federal government provided substantial grant funds to initiate major programs for research and implementation of IPM.

SIGNIFICANT IPM PROGRAMS

Because insecticides were a major environmental pollutant, entomology dominated IPM in the early 1970's in terms of grant funds received. The IPM approach received significant support and publicity with the initiation of the "Huffaker Project" (1972-78) jointly funded by the National Science Foundation, EPA, and USDA. This program was highly successful in introducing systems science and modeling as a basis for studying insect populations and their management. This program made significant advances in developing methodologies for monitoring insect pests, showing the need for applying pesticides only when pests reach economically damaging levels, developing insect resistant crop varieties, gaining a better appreciation for biological control, and developing methodologies for economically and environmentally evaluating IPM programs. The emphasis was on arthropod pests and little attention was given to plant diseases, nematodes or weeds. During the same period funds were provided to the Cooperative Extension Service (CES) by USDA-Federal Extension to begin the development of pilot pest management programs. From 1971 to 1975, 39 federally-funded pilot pest management projects were initiated in 29 states by CES. The Cooperative Extension Service IPM programs were based on the use of field scouting to supply necessary data to determine the need for the application of insecticides. Field demonstrations were used as on-farm models for educational programs. Since their conception, many of CES pest management programs now include management of weeds, diseases, and nematodes, as well as insects. Today, virtually every state conducts an extension IPM program.

Following the Huffaker Project, an interdisciplinary project directed at studying the effects and minimizing damage of multiple pests (insects, weeds, diseases, and nematodes) to four major crops was developed by the Consortium for Integrated Pest Management (CIPM) in 1979. This 17-university project was funded initially by EPA under their mandate to protect the environment. The Consortium for Integrated Pest Management has made considerable progress in the past five years in designing and modeling systems that simulate mutiple pest effects on crop yields, upgrading methods of biological and environmental sampling, improving action thresholds, developing multiple-resistant crop varieties, improving methods of biological control, expanding economic analysis, and in expediting the implementation of research findings by the Cooperative Extension Service. Again, a major difference between the CIPM Project and the Huffaker Project was its focus on the multiple pest complexes attacking each of the crops, alfalfa, apple, cotton, and soybeans.

When the CIPM Project was in an early stage, a 1980 report to the President by the Council of Environmental Quality reviewed the progress of IPM (Steinhauer & Yarn 1980). Although this report was highly complimentary of the progress made by IPM in reducing pesticide use and increasing profits to farmers, it was critical of IPM as a system that dealt only with "the use of a single tactic against a single pest or a complex of pests." It was this statement, perhaps more than any other, that reflected the need to deal with all the pests of a crop in a systematic manner. This was the goal of the CIPM Project.

ECONOMIC AND ENVIRONMENTAL INCENTIVES

IPM during the 1970's and into the early 1980's has received considerable funding and publicity as a reasonable course of action for the management of pests, not only in crop production but for livestock, forestry, and urban areas. IPM is now established as the sensible approach to pest control. In the last three years, interest has shifted from environmental to economic concerns. Because of a significant economic recession in the United States and worldwide, questions have begun to be asked not in environmental, but in economic terms. A major question has been directed to IPM programs regarding whether or not they are a good financial investment. Do they increase farmer profitability? In an ever-increasing atmosphere of budget tightening at both federal and state levels, IPM has come under severe scrutiny for its productivity. Those states and agencies that were farsighted enough to develop economic evaluations of their IPM programs have found the evaluations to be of tremendous value in defining the profitability of practicing IPM on American farms. These studies have formed the data base for retaining funding for IPM research and extension projects (see Headley, this volume).

In nearly every instance where an economic evaluation has been conducted on an IPM program, it has shown increases in profits at the farm and regional levels. One only has to glance briefly at the literature to find out that IPM has significantly increased net profits to farmers. Lacewell & Taylor (1980) summarized several economic analyses that had been conducted on the implementation of cotton IPM programs in Texas. By using several techniques, including budget analysis, linear programming and statistical analysis, they clearly show the profitability of programs such as the Texas High Plains boll weevil (Anthonomus grandis) control program. This program showed an annual, regional impact (increase in economic activity) of $63 million and a state economic impact of $82 million, while farmers benefited by a $27 million increase in net profits. An evaluation of a short-season IPM cotton production system in south Texas showed an increase in direct farmer profits of $11 million/year and an increase in economic activity for the Coastal Bend region of Texas of $94 million/year which translated into an increase in economic activity for the state of $141 million/year. Shaunak et al. (1982) evaluated the economic impact and risk regarding the implementation of a short-season cotton IPM strategy in the Lower Rio Grande Valley. This study compared two sets of three-year periods, 1973-1975 and 1976-1978. The data base was derived from the Lower Rio Grande Valley cotton pest management program conducted by the Texas Agricultural Extension Service. By studying irrigated and dryland production in comparing short-season and conventional systems, it was determined that the dryland, short-season production system had the highest net return per acre, $93.42 in 1973-1975 and $117.31 in 1976-1978, when compared to conventional dryland systems that returned $35.73 in 1973-1975 and $68.06 in 1976-1978. Reduced insecticide costs and increased yield were the two primary factors that affected net returns. It was concluded that by shifting to a short-season production system there would be a substantial reduction in insecticide use, lint yield would be increased, and, perhaps as importantly, risk would be substantially reduced.

The economic implications of delayed uniform planting to control boll weevil in the Texas Rolling Plains was studied by Masud et al (1984). Net per acre returns were estimated to be increased by $21.36 for farmers using this cultural control technique as compared to conventional practices. The average annual economic impact of uniform delayed planting in the region of the Rolling Plains was estimated at $192 million. The annual economic impact to the state for this practice was estimated at $305.19 million. Risk estimated by coefficient of variation showed the uniform delayed planting date practice for production years 1972-1980 was less variable in comparison with conventional cotton production practices. There are several other economic evaluations that have been conducted in Texas that clearly demonstrate profitability of IPM systems. It would do well to restate that these increases are estimated on an annual basis and accrue over time.

Similar examples on profitability of the implementation of apple IPM systems are seen in Massachusetts (Whalon & Croft 1983). Increase in net profitability was observed in a cost/benefit analysis of arthropod pest control practices under IPM systems by comparing them with conventional practices. Major differences were observed in reduction of production costs related to dollars expended for the application of pesticides. There was an increase of net profit of $242.89/hectare in orchards under the IPM system when compared with conventional practices. In 1975, the difference was even greater at $314.26/hectare. White & Thompson (1982) evaluated the economic feasibility of tree fruit integrated pest management programs in the northeastern U.S. Using apples as an example, it was shown that profits were increased primarily through reduced pesticide use. An increase of $63.53/hectare was established for the northeastern apple producing region.

Economic evaluations of alfalfa weevil (*Hypera postica*) management through carefully timed harvests indicated that for a representative farm in the North Central region of the U.S. using this technology, returns to land management and risk increased per hectare returns by $61.75 (McGuckin 1983). Using this strategy, insecticide use was reduced by 45%. In Illinois, it was found that maximum profits were directly associated with a fall application of insecticide along the perimeter of fields to control the alfalfa weevil. Returns to land, management and risk increased profits by $69.16/hectare using this program in comparison with conventional control programs. As an added bonus, insecticide use for alfalfa was reduced by 75% (T. McGuckin, personal communication).

Integrated pest management programs, both research and extension, that were initiated in the early 1970's have had sufficient time to impact on production agriculture. The economic implications of increased profitability are clear. One of the original missions of IPM programs was to provide for the timely application of insecticides. It is time to see how well IPM has performed in this regard. Table 1 shows the planted hectares of corn, soybean, cotton, grain sorghum and peanuts for the years 1971, 1976, and 1982.

Table 2 provides an estimate of the total insecticide use on these crops for the same years. Table 2 also includes the pounds

Table 1. Planted hectares (millions).[a]

Crop	1971	1976	1982
Corn	30.0	34.2	33.1
Soybeans	17.6	20.3	28.9
Cotton	5.0	4.7	4.6
Grain Sorghum	8.4	7.4	6.5
Peanuts	0.6	0.6	0.5

[a] The crop production annual summary, SRS, Crop Reporting Board, USDA, 1971, 1976, 1982.

Table 2. Farm insecticide use: 1) active ingredients million pounds; 2) pounds/ hectare).[a]

	1971		1976		1982	
Crop	1	2	1	2	1	2
Corn	25.5	0.84	32.0	0.93	30.1	0.91
Soybeans	5.6	0.32	7.9	0.39	10.9	0.37
Cotton	73.4	14.62	64.1	13.53	16.9	3.70
Grain Sorghum	5.7	0.66	4.6	0.61	2.5	0.39
Peanuts	6.0	9.88	2.4	3.95	1.0	1.90

[a] Inputs-outlook of situation, USDA Economic Research Service, 1OS-2, October 1983.

of active ingredient applied per hectare. Although integrated pest management programs have been under way in all of these commodities, implementation was intensified beginning in 1972 and 1973 on cotton, grain sorghum, and peanuts. For cotton, there has been a significant reduction in the overall use of insecticide applied per hectare. The cancellation of DDT in 1972 may have had an impact on the amount of insecticide applied per hectare for cotton; however, this was not reflected in 1976 as the amount remained relatively constant. What is interesting to note is the dramatic decline from 1976 to 1982 when insecticides applied per/hectare were reduced from 13.5 to 3.7 pounds. A similar observation may be made for grain sorghum. Peanut IPM, which has been aggressively pursued throughout the South, has had highly impressive results; insecticide use has been reduced from 9.89 pounds per hectare in 1971 to 1.90 pounds by 1982. One may argue that this reduction in insecticide use results from the change in classes of insecticides used with a shift from chlorinated hydrocarbons to organophosphates. This is only partially true. The primary factors involved in the reduction of insecticide use on these commodities (cotton, grain sorghum, and peanuts) is due to integrated pest management programs either implemented through the Cooperative Extension Service, private consulting entomologists, or by individual farmers. Throughout the 1970's and into the 1980's the private consultant entomologist industry has expanded greatly. Many new professionals have entered the business and their impact has been felt. The data in Table 2 also show that cotton is no longer the crop of highest insecticide use. Corn has now taken its place as the crop that consumes the most insecticides. For example, in 1982 corn farmers applied 30 million pounds of insecticide while cotton farmers applied approximately 17 million pounds. Of all the insecticides applied to major field crops in 1982, corn and cotton accounted for 43% and 24% respectively. This is compared to 1976 when cotton accounted for 49% and corn for 25% of total insecticides applied.

Not only has there been a reduction in the amount of pesticide applied per hectare, there has been a substantial reduction in the proportion of hectares treated (Table 3). By observing the change in the proportion of hectares treated between 1971 and 1982, it can be seen that use on corn has remained relatively constant at ca. 36% of the hectares treated with a slight upward trend. Treatment of soybean, although remaining relatively constant in years 1971 and 1976, shows an upsurge to 12% of the hectares treated by 1982. A substantial trend reversal was observed in cotton, grain sorghum and peanuts. The percentage of hectares treated for cotton was reduced from ca. 60% in 1971 and 1976 to 36% by 1982. This represents a 46% reduction in the use of pesticides.

These results show the economic and environmental benefits produced by the investment made by federal and state governments for research and extension IPM programs. The original investments were made on an environmental basis to reduce the amount of pesticides entering the environment. This investment has clearly paid great dividends.

THE FUTURE OF IPM

The question now is: "What does the future hold for IPM for the remainder of the decade and into the 1990's?" The National IPM Coordinating Committee is to be congratulated for developing a national plan entitled "Research and Technology Transfer in

Table 3. Proportion of hectares treated with insecticides (percent).[a]

Crop	1971	1976	1982	% Change 1971-1982
Corn	35	38	37	+15
Soybeans	8	7	12	+60
Cotton	61	60	36	-46
Grain Sorghum	39	27	26	-48
Peanuts	87	55	48	-54

[a] Inputs-outlook of situation, USDA Economic Research Service, 10S-2, October 1983.

Agriculture: A National Plan for Integrated Pest Management". Cooperation between the State Agricultural Experiment Stations, Cooperative Extension Service, and the respective branches of USDA (ARS, FE, and ERS) have joined together with a national plan to conduct research and educational programs in IPM. The leadership for this plan came from the State Agricultural Experiment Stations based on input from administrative and technical committees in the four respective regions across the country. Many of these regional programs are now under way and the future success of IPM in the State Agricultural Experiment Stations largely depends upon how well organized, administered and scientifically sound these programs will be. It is of extreme importance that system scientists, agricultural engineers working closely with biological scientists, remain at the forefront of the emerging technology for IPM. We look at CIPM as an example where modeling and system science were the main integrative forces in bringing far-ranging disciplines (entomologists, plant pathologists, nematologists, weed scientists, agronomists, mathematicians and agricultural economists) together to concentrate on major production systems. Biological control, either by conserving and managing native beneficial species or by the introduction of exotic biological control agents, will remain an integral component of any IPM program. Emphasis should also be placed on the active role of agricultural economists in each of the regional IPM projects. Agricultural economists should not only be involved in the evaluation of specific projects, but should also be called upon to assist in the direction of research priorities that bear directly on variable cost factors in production agriculture. As recently as April 1984, U.S. Congressional hearings on agriculture research, extension, and higher education have related IPM to supply/demand projections. In the prevailing trend of supply side economics, IPM has an important role to play. It is pointed out that although IPM may not significantly increase yields in all cases, it does play a significant part in reducing the cost of production. If farmers can produce equal or slightly greater yields at a reduced cost, there will be an economic advantage. The bottom line for IPM is to increase farmer profit by reducing costs per unit of output. If producers lower their per unit costs through implementing IPM strategies they will be more competitive with producers that do not use this strategy within a region. If a region adopts this strategy, it will be more competitive with other regions and, furthermore, if these strategies are adopted nationally for a given commodity our farmers will be more competitive in the world market.

In closing, it is not only our responsibility as scientists to develop new research strategies and tactics, it is also our responsibility to clearly demonstrate and document the economic and environmental advantages of IPM, now and in the future.

REFERENCES

Lacewell, R.D., and C.R. Taylor. 1980. Benefit-cost analysis of integrated pest management programs. In: E.G.B. Gooding (Ed.), Proc. seminar and workshop--pest and pesticide management in the carribean. CICP/USAID 2:283-301.

Masud, S., R.D. Lacewell, E.P. Boring and T.W. Fuchs. 1984. Economic implications of a delayed uniform planting date for cotton production in the Texas Rolling Plains. Texas Agric. Exp. Sta. Bull. (In Press).

McGuckin, T. 1983. Alfalfa management strategies for Wisconsin dairy farms--application of stochastic dominance. North Central J. Agr. Econ. 5:43-49.

Metcalf, C.L. 1930. Obituary, Stephen Alfred Forbes, May 29, 1844-March 13, 1930. Entomol. News 41:175-178.

Michelbacher, A.E., and O.G. Bacon. 1952. Walnut insect and spider-mite control in northern california. J. Econ. Entomol. 45:1020-1027.

Newsom, L.D. 1974. Pest management: History, current status and future progress, pp. 1-19. In: F.G. Maxwell and F.A. Harris (Eds.), Proc. summer institute of biological control of plant insects and diseases. Univ. Press of Miss., Jackson.

Shaunak, R.K., R.D. Lacewell and J.W. Norman. 1982. Economic implications of alternative cotton production strategies in the Lower Rio Grande Valley of Texas, 1973-1978. Texas Agric. Exp. Sta. Bull. 1420, 25 pp.

Smith, R.F. 1977. History and complexity of integrated pest management. Presentation at: Pest control strategies--understanding and action symposium, Cornell Univ., N.Y., June 22-23.

Smith, R.F., and H.T. Reynolds. 1966. Principles, definitions and scope of integrated pest control. In: Proc. FAO symposium on integrated pest control, Rome, Italy 1:11-17.

Steinhauer, A., and S. Yarn. 1980. Report to the President--progress made by the Council on Environmental Quality federal agencies in the advancement of integrated pest management. Prepared by the Interagency IPM Coordinating Committee, pp. 7-9.

Whalon, M.E., and B.A. Croft. 1983. Implementation of apple IPM, pp. 411-447. In: B.A. Croft and S.C. Hoyt (Eds.), Integrated management of insect pests of pome and stone fruits. John Wiley & Sons, N.Y.

White, G.B., and P. Thompson. 1982. The economic feasibility of tree fruit integrated pest management programs in the Northeast. J. Northeast Agr. Econ. Council 11(2):39-46.

DISCUSSION: Biological control in integrated pest management: An entomological perspective.
Biological control of plant pathogens: Definitions.
IPM: Definitions and current status in U.S. agriculture.

DISCUSSION LEADER: D. Gonzalez

Several questions were discussed: 1) Is biological control a cornerstone of IPM? If yes, then why has it received so little funding in recent IPM programs? 2) If biological control is not a cornerstone in U.S. cropping systems, are IPM programs to be measured principally by yields and reduction in pounds of insecticide(s) applied? This may be acceptable on a short term basis, but what of the long term stability of IPM systems? There is merit to the concept that pest mortality as caused by natural enemies should be a basis for IPM strategies, with chemicals being supplementary rather than primary mortality factors. 3) How can biological control best be implemented into crop management systems? Is present knowledge sufficient? Do we need more education and extension efforts? If we need more information, will it require new research directions, such as biotype evaluation, genetic engineering, selective biorational pesticides? Can natural enemies be integrated with those pesticides that are truly necessary? 4) Is IPM truly an interdisciplinary crop management technique? Can methodologies be developed that will incorporate weed, arthropod, nematode and plant pathogen management?

Problems in implementing biological control in IPM systems were discussed and the following conclusions reached: 1) There is an urgent need to integrate predictive and reliable (quantitative) data involving natural enemies into IPM systems, particularly economic injury levels. IPM specialists should recognize that (particularly arthropod) natural enemies require food to insure their maintenance in the system. Studies on the impact of pesticides on natural enemies need to be expanded to include provision for these necessary food resources. 2) Enhanced educational and extension efforts are required; controversy exists with regard to whether more funds are needed to obtain more information and/or whether more funds, more information and new technologies are required (including genetic engineering) to provide new solutions to problems. 3) The reliance on pesticides as the cornerstone of IPM is only a short term solution to a long term problem. We need pesticides as a short term solution, but we also need long term, stable strategies to pest management. Such stable, long term management can be provided by expanding the use of biological control in agricultural IPM systems.

COST-BENEFIT ANALYSIS: DEFINING RESEARCH NEEDS

J.C. Headley

As the term implies, cost-benefit analysis is a measuring and comparing of the costs and benefits associated with some action. As one of the methods in economic planning, it has been most widely used and came to prominence in water resource planning and, in fact, was required under the Flood Control Act of 1936 (Eckstein 1958). Later it became a requirement for program evaluation by a number of federal agencies.

While the technique of cost-benefit analysis came to prominence as a tool in planning social investments, it derives simply from basic economic theory. The concept of profit maximization in neo-classical economic theory has as a necessary condition that the incremental costs must equal the incremental benefits. That is, the total costs of an economic activity minus the total benefits will be a maximum when the activity is enlarged to the point where the cost of the last unit of the activity added just equals the benefit from that unit of activity. Therefore, when a farmer decides to add another five kilos of nitrogen fertilizer, if he is rational, he must compare the expected added cost of that five kilos of fertilizer to the expected added benefit therefrom.

Cost-benefit analysis may be used on small investments or activities by an individual such as a farmer as well as for large corporate or public investments. There are numerous books about cost-benefit such as Anderson & Settle (1977), Sassone & Schaffer (1978), and Gittinger (1982).

A common use of cost-benefit analysis is to evaluate new technology to see if adoption will result in higher or lower profits. It is natural then that cost-benefit analysis is of interest to researchers who are developing new ways to control pests. Cost-benefit analysis can provide a systematic evaluation of the new technique compared to the old technique to indicate whether the new technique has any economic advantage and to estimate the magnitude of that advantage. The development of this kind of economic information can be useful in helping farmers to: 1) understand the new technique better, 2) to give farmers a basis for deciding whether to adopt the new technique

ISBN 0-12-357030-1

and 3) to justify research investment. In short, it provides estimates of the "bottom line" for a new technique.

SOME EXAMPLES

Since about 1960, agricultural economists have shown interest in the economics of pest control. Bibliographies such as those by McCarl (1981) and Osteen et al. (1981) provide a relatively complete list of work related to the economics of pest control. Much of the work on the economics of pest control has been directed at estimating the benefits of pesticides, especially those which have been subject to EPA registration cancellation and suspension. The costs and benefits of various cultural practices and resistant varieties used in IPM systems have also been analyzed. Examples, which are cited by Frisbie & Adkisson (this volume), include studies of diapause control programs for boll weevil and short season cotton in Texas, delayed uniform planting of cotton in Texas for boll weevil control, IPM programs for apples in the northeast and alfalfa weevil IPM programs in Wisconsin and Illinois.

One characteristic of the economics of pest control literature is that it represents work largely by agricultural economists using the results of biological research which are usually not suited for use in economic evaluation of control practices. That is, the work does not, on the whole, flow from interdisciplinary research. Much of it is oriented toward estimating the benefits of pesticides or alternatively, the costs of being without pesticides. There is very little research extant that deals with the economics of biological control.

One example of the evaluation of a biological control technology is a study by Reichelderfer (1979). This was a study of the economic feasibility of using the parasitic wasp *Pediobius foveolatus* to manage Mexican bean beetle on soybean. The study used a budgeting approach constructing cost and returns budgets for soybean in several different producing areas. Costs and returns were estimated for: 1) conventional insect control using either carbaryl or disulfoton, 2) biological control using the wasp, and 3) biological control using the wasp in conjunction with scouting.

Essentially what Reichelderfer did was to assume that the yield was the same for all of the control options and then to compute the change in cost of production for insecticides vs. biological control. The results showed that the cost of biological control of Mexican bean beetle was lower than the cost of insecticides giving a higher net return per hectare assuming equal yields. Reichelderfer found that the net economic advantage for biological control without scouting over conventional control ranged from $7.43 to $0.125 per hectare in the Delaware, Maryland, Virginia area and in North Carolina, respectively. The economic difference found suggests the yield for biological control could be reduced as much as 29.3 kilos per hectare below that for conventional economic control, before biological control would lose its economic advantage in Delaware, Maryland and Virginia.

This example is used not because it is an example of ideal cost-benefit analysis, nor because it is an example of a poor study. Rather it is representative of the kind of studies that one is forced to do when biological data must be collected from completed studies or even guessed at by the researcher. Scientific estimates of the relative effectiveness of the wasp and the insecticides accompanied by yield data for each would have made this study much more credible. Data indicating the degree of control to be expected with biological control compared to conventional control and the effect of the controlled pest population on yield would have freed the economist from the need to assume equal control and therefore equal yields. Under the circumstances, Reichelderfer had no choice but to construct what seemed to her to be the most likely and believable scenarios and do the best job possible.

Another study by Reis et al. (1983) analyzed alternative control methods for soybean cyst nematode in southern Illinois. Since the producer has to choose pest control strategies within the context of the total farm business rather than considering only the costs and benefits from a single enterprise, crop rotations, fertilization and labor and capital constraints were considered.

The study used experiments that had been conducted in Kentucky in an area similar to southern Illinois. The experiments included data on fertilization, nematicides, and resistant varieties. However, the data did not include information on crop rotation. Since the researchers wanted to include the rotation effect, they had to combine some data from another experiment in Illinois with the Kentucky data to estimate yields in crop rotations.

Reis et al. (1983) considered 24 different methods of controlling nematodes with combinations of varieties, fertilizer, nematicide and crop rotation and used linear programming to select the control method that maximized the income over variable costs from the farm. Linear programming is a technique for mathematically allocating fixed resources among competing production activities to, in this case, maximize net farm income. For details of linear programming see Osburn & Schneeberger (1978). The results showed that farms with differing amounts of operating funds available selected different crop rotations and different combinations of nematode controls as a means of maximizing their net returns. A sample of the solutions indicated that part of the farm would be in continuous soybeans with a resistant cultivar and the balance was to be in a corn-soybean rotation using a susceptible cultivar and nematicides.

The study by Reis et al. (1983) demonstrates some of the more complicated things one can do in a cost-benefit analysis. The economists did not have a single experimental data base which considered all of the control methods and the interactions among them. Consequently, it was necessary to synthesize a data set reflecting these interactions. It also demonstrates the need for high quality data to support the more complicated analysis.

A third example of cost-benefit in pest management is one by Headley (1983) which examined the profitability of orchard sanitation and early harvest to control navel orangeworm, *Amyelois transitella*, in almonds. Cost and return budgets were considered for a model orchard representing scenarios ranging from no control to a scenario utilizing orchard sanitation (destruction of mummy nuts in winter) and early harvest. The net returns of pest management costs were compared for the different scenarios. In general, the results showed winter cleanup to be a profitable substitute for a May spray of azinphosmethyl.

While this study is useful for almond growers, it suffers from the problems mentioned earlier in connection with the other two examples. That is, the economic evaluation was performed separately from biological research designed to measure the effectiveness of sanitation and early harvest. Therefore, it was necessary to make some assumptions about the efficacy of the alternate control methods and about the interaction between navel orangeworm controls and mite and scale populations. In addition, in the absence of data from experiments over a number of seasons, the analysis is static and gives no insight into yields and control costs over time. Had such data been available, costs and returns for each of several years might have been estimated rather than for just a single or average year. This sort of information over time could be very useful especially for a perennial crop such as almonds.

In summary then, these examples show: 1) how cost-benefit analysis can be used in pest management, 2) some of the methods used, such as budgeting and linear programming, and 3) the shortcomings of such analyses in the absence of interdisciplinary research. The next section will address data needs and methods used in cost-benefit analysis.

DATA NEEDS

Clearly, data needed to do an ideal cost-benefit analysis on a biological control strategy consist of two parts. First, data are needed which 1)indicate the crop damage in terms of quantity and quality resulting from varying levels of pest infestation; 2) the efficacy of biological control for different levels of infestation, if applicable; 3) the resource requirements of the biological control such as labor and equipment for application, the quantity of the biological control agent, and labor needed to gather information needed to use the agent (e.g., scouting); 4) effects in subsequent time periods on infestation, yield and other pest controls needed; and 5) interactive relationships among biological control and factors such as soil fertility, varieties and weather. Second, data are needed on the prices of product and inputs used in the production of the crop. Where the crop may be in competition with other enterprises on the farm, costs and returns associated with the competitive crops are also needed.

The data which the economist would like to have under ideal conditions require a more complicated and extensive experimental design than the biologist might develop in isolation. This may explain why this author is unable to cite a published study that meets the ideal conditions. The biologist tends to be interested in only one or two parameters of a system and will design an experiment to provide a high degree of statistical confidence in the estimates. To estimate more parameters such as for interactions requires a larger experiment and more resources and brings one face to face with the economics of resource allocation in research. This requires the cooperation of biologists and economists in defining the objectives of the research and the information needed by both to satisfy the objective. In the process of prioritization of research tasks some sacrifices will be necessary to achieve a modicum of success and not violate the budget.

Some biologists have argued that by the time they are sure that the biology will work, there are no resources left for economic evaluation. This might be dealt with by thinking of two levels of research. At one level is fundamental biological research which asks questions concerning biological feasibility. At the second level is interdisciplinary research where the objective is to establish the relation between various important biological variables and their interactions with other variables such as fertility, irrigation, and other farm operational variables that are important to determine technical and economic feasibility in the field.

In any event, when biologists believe that they have discovered a biological reality that shows promise for the development of a field practice, they should begin consulting with agronomists, engineers and farm management economists. Discussion of the research plan for development and evaluation of the new practice should reveal the important dimensions of the problem and serve to guide the research resource allocation. This should help to prevent the biologist coming to the economist with experimental results that do not lend themselves to economic evaluation.

Time Dimensions

An important aspect of all pest management research concerns the matter of the time dimension in pest management. Taylor (1972) discussed the need for a dynamic approach to selection of pest management strategies. Certainly, many, if not most farm pest control decisions are made on the basis of a myopic or short term static analysis. However, all biologists know that what happened last year predisposes to some degree the outcome of the next year.

For some insects, eggs in 1984 are a function of the population overwintering which in turn was influenced by the control measures taken in 1983. All who take the concept of pest management seriously must recognize that the time dimension has to be given more consideration in future research. Economists have methods of doing dynamic analysis such as dynamic programming (Shoemaker 1977), control theory (Hueth & Regev 1974) and simulation (Reichelderfer & Bender 1979). What economists who are interested in pest management lack is data from well designed experiments that describe and explain the population dynamics of both pests and beneficial species. The modeling work that is growing, especially in entomology, is beginning to fill that need.

Some excellent models exist in the work of Curry et al. (1978), Ruesink (1976), Waddill et al. (1976), Wang et al. (1977), and Watt (1961), to mention only a few. Most of the existing crop-pest models either do not include predation effects or model them poorly (see Stimac & O'Neil, this volume). However, these models can still be used for simulation experiments, freeing resources for research on the predation and parasitism effects.

Risk Dimensions

Decision-makers always face uncertain outcomes. If decision-makers in pest management have some knowledge of the distribution of possible outcomes and the factors that condition that distribution they should be in a better position to make plans that include their attitude toward risk.

In the examples given earlier, (Headley 1983, Reichelderfer 1978, Reis et al. 1983), the studies used parameters relating to yield loss, degree of infestation and control efficacy which were single values. If the variance in these parameters due to uncontrollable events, such as weather, is relatively small, then the use of single values such as the mean may serve reasonably well. However, if the variance is large (the standard deviation is 50% of the mean), this could cause large variation in the net benefits associated with the practice. Experienced farmers will be aware of this variability and will be suspicious of recommendations based on mean outcomes.

Biological controls are living controls. Because they are living, they have life cycles that are influenced by events both within and beyond the control of the decision-maker. A parasite, for example, requires an environment that includes a food supply and conditions under which it can grow and reproduce. Some of these factors such as food supply are at least partially under the control of the manager. Other factors such as temperature and natural enemies of the parasite are not completely controllable under field conditions. Therefore, research to ascertain the effectiveness of such a biological control must consider both the controllable and the uncontrollable conditions.

Biological experiments are usually controlled experiments. However, in every experiment there are still many variables which are not controlled such as temperature, rainfall, migration of competitive species, etc. Often the uncontrolled variables are not measured as a part of the experiment and the effects are included in experimental error. Some of the uncontrolled effects may be correlated with time, and if the experiment were to provide observations over time, the time path of these events could be modeled and related to other variables of interest in the experiment such as pest populations, biological control populations, yield and product quality. Given information such as this, it would be possible to simulate the effects of biological control over a period of time and to compute the costs and returns to the investment in the control over a planning horizon of more than one season or pest life cycle.

Usually the results of an experiment will exhibit variance replicated through time. The damage from a given pest infestation will not be the same even for the same crop in the same location for different time periods. The crop could be in a different stage of growth or health due to factors beyond the control of the experimenter. This variance may be random and, depending upon its magnitude, will change the cost-benefit ratio from period to period. However, all is not lost providing the measurements exist which allow one to construct a probability distribution of the outcomes.

The availability of the probability distribution of the outcomes of alternate pest management strategies makes economic evaluations more realistic. In comparing a biological control to a chemical, a risk-averse farmer would like to know something about the variability of the results. One alternative may provide the opportunity for a higher expected value of the outcome, but also have a higher variability of outcomes. Whether a farmer wants to adopt such a practice will depend on how much variability he is willing and able to withstand to acquire the higher average or expected net income. For example, a farmer who is in a very tight cash flow situation might decide to adopt the control practice with a lower average return in exchange for its relatively low season variability of income. There are a number of good references on the topic of decision analysis which are pertinent to pest management (Anderson et al. 1977).

The point of this discussion is that there is a definite need for research in pest management in general and biological control in particular that considers the decision framework that exists where the adoption decisions are made. Especially for perennial crops, farmers need evaluations that consider inter-year effects; better decisions can be made if they are known. For farmers who are highly risk-averse, either because of financial situations or because of personality, evidence of the probability distribution of outcomes from alternate strategies will allow them to select from among alternatives to achieve the appropriate trade-off between expected net income and net income variability.

SUMMARY AND CONCLUSIONS

This paper has attempted to do three things. First, to explain cost-benefit analysis and to show the need for economic evaluation as a means of encouraging farmers to adopt pest management practices that are feasible and profitable. Second, to demonstrate the need for data which allow the economic evaluation of practices that consider the influence of time and variability of outcomes. Third, to indicate some of the adjustments necessary in the research process if research is to be accomplished in an interdisciplinary setting.

It was shown that profitability will be the principal concern of farmers in deciding to adopt biological control practices. This profitability will likely be most convincing in a total farm business framework. Any effects that new pest control practices have on the economic relationship between farm enterprises must be evaluated. The evaluation of crop relationships will require more data and more complex experimental designs.

Because biological phenomena are related through time, experiments, either from plots or from computer simulations with models, are needed to bring these dynamic dimensions into consideration. For many pests and crops, myopic, one-season-at-a-time selection of control strategies will probably not achieve optimal economic results.

Since the outcomes of natural processes are usually not deterministic, there is a need for data describing the probability distributions of outcomes. Risk aversion on the part of farmers makes income variability very important. If farmers perceive that alternative control strategies display a range of variability in results, but are unable to quantify that risk, this introduces more uncertainty into the choice of practices.

Finally, all members of the research team must be involved from the beginning if the research objectives are to be assigned priorities. Trade-offs may have to be made on the part of all disciplines to provide new practices that are feasible and profitable.

Considerable progress has been made. Disciplines are communicating with each other, and the presence of this paper in this volume is evidence that many biologists agree that economic evaluation is necessary. Now, the task must be attempted.

RECOMMENDATIONS

1. Cost-benefit analysis of biological pest management practices require specific kinds of biological information. If information is not provided by a thorough research plan, the

economic evaluation becomes the victim of ad hoc procedures to estimate the values of the missing parameters.
2. Scientific economic evaluation requires experiments designed with economic evaluation as an objective. Research of this kind will be more complicated, take more time, be more expensive and must be interdisciplinary. All of the disciplines should be involved in the research from the beginning. Every member of the research team must understand the needs of the other disciplines.
3. For many pests and crops, one-season-at-a-time selection of control strategies will probably not achieve optimal results. For biological phenomena, plot experiments or computer simulation experiments through time are needed to bring dynamic dimensions into consideration.
4. Risk aversion on the part of farmers makes income variability an important consideration. There is a need for data describing the probability distributions of outcomes from biological pest control methods.
5. To be most useful to farmers, economic evaluation of biological control practices should show the effects that new control practices have in a total farm business framework, not just the enterprise directly impacted by biological control.

ACKNOWLEDGMENTS

The author acknowledges comments on an earlier draft by Drs. Marjorie A. Hoy and Kenneth V. Yeargan.

REFERENCES

Anderson, L.G., and R.F. Settle. 1977. Benefit-cost analysis. Lexington Books, Lexington, KY. 140 pp.

Anderson, J.R., J. Dillon and B. Hardaker. 1977. Agricultural decision analysis. Iowa State Univ. Press, Ames. 344 pp.

Curry, G.L., R.M. Feldman and K.C. Smith. 1978. A stochastic model of a temperature dependent population. Theor. Pop. Biol. 13:197-213.

Eckstein, O. 1958. Water resource development. Harvard Univ. Press, Cambridge. 300 pp.

Gittinger, J.P. 1982. Economic analysis of agricultural projects. Johns Hopkins Univ. Press, Baltimore, MD. 505 pp.

Headley, J.C. 1983. Economic analysis of navel orangeworm control in almonds. Calif. Agric. 37(5/6):27-29.

Hueth, D., and U. Regev. 1974. Optimal pest management with increasing pest resistance. Am. J. Agr. Econ. 56:543-552.

McCarl, B.A. 1981. Economics of integrated pest management. Oregon State Univ. Agr. Exp. Sta. Spec. Rept. 636, 142 pp.

Osburn, D.D., and K.C. Schneeberger. 1983. Modern agricultural management. Reston Publ. Co., Reston, VA. 415 pp.

Osteen, C.D., E.B. Bradley and L.F. Moffitt. 1981. The economics of agricultural pest control. USDA Ser. 14, 53 pp.

Reichelderfer, K.H. 1979. Economic feasibility of a biological control technology. Nat. Res. Econ. Div.; Econ., Stat., and Coop. Serv.; USDA Econ. Rept. 430, 20 pp.

Reichelderfer, K.H., and F.E. Bender. 1979. Application of a simulative approach to evaluating alternative methods for control of agricultural pests. Amer. J. Agr. Econ. 61:258-267.

Reis, R.P., G.R. Noel and E.R. Swanson. 1983. Economic analysis of alternative control methods for soybean cyst nematode in southern Illinois. Plant Disease 67:480-483.

Ruesink, W.G. 1976. Modeling pest populations in alfalfa ecosystems with special reference to alfalfa weevil, pp.80-89. In: R.L. Tumala, D.L. Haynes and B.A. Croft (Eds.), Modeling for pest management. Michigan State Univ., East Lansing.

Sassone, P.G., and W.A. Schaffer. 1978. Cost-benefit analysis. Academic Press, N.Y. 173 pp.

Shoemaker, C. 1977. Pest management models of crop ecosystems, pp. 545-574. In: C. Hall and J. Day (Eds.), Ecosystem modeling in theory and practice. Wiley-Interscience, N.Y.

Taylor, C.R. 1972. Dynamic economic evaluation of pest control strategies. Ph.D. thesis, Univ. Missouri, Columbia.

Waddill, V.H., B.M. Shepard, J.R. Lambert, G.R. Carner and D.N. Baker. 1976. A computer simulation model for populations of Mexican bean beetle on soybeans. South Carolina Agr. Exp. Sta. SB 590, 18 pp.

Wang, Y., A.P. Gutierrez, G. Oster, and R. Daxl. 1977. A general model for plant growth and development coupling plant-herbivore interactions. Can. Entomol. 109:1359-1374.

Watt, K.E.F. 1961. Mathematical models for use in pest control. Can. Entomol. 93(Suppl. 19), 62 pp.

DISCUSSION: Cost Benefit Analysis: Defining research needs.

DISCUSSION LEADER: R.E. Stinner

Since the results of relying on biological control are often more variable than relying on pesticides, what is the best strategy? The answer is one that provides outcomes that are consistent with a farmer's attitudes toward risk. Unfortunately, uncontrolled variables are rarely measured in experimental situations. Given our present weak quantitative understanding of natural enemy/pest interaction, we need to measure these uncontrolled variables so that at least correlations can be obtained to reduce the uncertainty often associated with the use of biological control or biorational pesticides.

When should economics be included in biological control research? Inclusion at the point when implementation is imminent is too late for several reasons: 1) agriculturalists are often poorly informed as to what are the actual current production practices and are often ignorant with respect to the economic and social impact of potential extensions of their research. 2) In many cases, the economic theory is in place, but the data to allow its implementation may be lacking. With relatively little effort these key data may be obtained if an economist is involved in the original planning. 3) Economists are often approached after the fact and asked to do basic bookkeeping, an unrewarding task, to say the least. Early collaboration could avoid research "white elephants" with little or no chance of implementation because of economic/cultural/risk constraints. Such collaboration also provides the economist the chance to develop challenging research on economic theory which may generate new technology for the agriculturalist to examine.

Biological Control of Arthropod Pests

PLANT RESISTANCE AND CULTURAL PRACTICE INTERACTIONS WITH BIOLOGICAL CONTROL

D.C. Herzog and J.E. Funderburk

In the context of production and management, the crop is the integrative entity (Huffaker 1982), the target of herbivory, the focus of protection and management. The two production activities that first impact, either directly or indirectly, on crop pests and their natural enemies are the species and variety of crop planted and the aggregate of cultural practices utilized in production or management of the surrounding crop and non-crop vegetation.

The focus of this review will be on the compatibility of plant resistance with biological control of arthropods by their arthropod and microbial natural enemies and on cultural practices that may be utilized to enhance biological control under field conditions. Certain references will be made to the deleterious effects of cultural practices on natural enemies. We will not address chemical insect control as a cultural practice as this is the subject of two other chapters in this volume (see Hull & Beers; Mullin & Croft, this volume).

The following statement, extracted from Smith et al. (1976), will set the tone for the intent of this review and much of the information contained herein. Scientific pest control requires a basic knowledge of ecological principles, biological intricacies of each pest, and natural factors that tend to regulate their numbers. "...we must understand Nature's methods of regulating populations and maximize their application."

PLANT RESISTANCE

The possible ecological interactions of herbivorous insects and their natural enemies with the plant substrate upon which they feed, reproduce, and/or reside are limitless. A substantial body of literature indicates that predator and parasite efficiency may be altered by the host plant of the target prey or host. Effects of crop species on the success, magnitude or timing of biological control processes will not be addressed in this review. The interested reader is referred to Bergman & Tingey (1979) for an

ISBN 0-12-357030-1

introduction to this literature. A generalized, hypothetical treatment of the relationship of variable plant defensive chemistry to the susceptibility of insects to natural enemies is provided by Schultz (1983). Though restricted to predators and parasites, the Schultz review, amply provided with citations to key literature, provides a good general introduction to the relationship of plants to mechanisms regulating herbivorous species. Despite the fact that different crop species produce the greatest range of response among natural enemies, cultivars of the same species of host (here used interchangably with variety) may also produce substantial differential effects on natural enemy success.

The following statement of Wilson & Huffaker (1976) reflects much of the paradigm upon which integrated pest management is built: "...biological control, together with plant resistance... are the core around which pest control in crops and forests should be built." Biological control may be successful only within the context of integrated control (integrated pest management). They then come full circle by asserting that integrated control usually depends upon a core of biological control and plant resistance. The proponents of plant resistance and integrated pest management have neglected an important responsibility. We often assert that plant resistance to pests is completely compatible with biological control in integrated pest management systems, but present little or no data to support this contention. Unfortunately, the following statement of Bergman & Tingey (1979) remains true: "Although the combined effectiveness of resistant cultivars and biological control has been studied in a few instances, the interactions between plant resistance and arthropod predators and parasites remain poorly known." Even this statement neglects the importance of microbial pathogens as biological control agents. A rigorous search of the literature uncovers few titles that deal with the compatibility of plant resistance and biological control.

The primary objective of programs on insect resistance in crop plants is to develop crop varieties resistant to one or more insect pests while maintaining or improving basic agronomic characteristics, usually yield. Insect resistance will be most optimally employed as an adjunct to other control measures (Ortman & Peters 1980), especially biological control. The use of these two control tactics within a unified management strategy brings together unrelated mortality effects, thereby reducing the rate of pest population genetic response to selection pressure of either resistance or natural biological control. Acting in concert, these two regulatory mechanisms provide density-independent mortality in times of low pest density and dynamic density-dependent mortality in times of pest increase (Bergman & Tingey 1979). Pest population development trajectory can be diminished even with low levels of plant resistance, thereby providing a relative advantage to biological control agents (DeBach 1974, Starks & Berry 1976, van Emden 1966).

A brief definition of the mechanisms of resistance seems in order to set the stage for the discussion that follows. Painter (1951) proposed the following categorization of the mechanisms of plant resistance to pests. Antixenosis is a nonpreference response of insects to plants that lack certain characteristics that allow them to serve as hosts which results from a "choice" on the part of individuals while in search of shelter, food or oviposition substrate. Kogan & Ortman (1978) proposed the term antixenosis to replace nonpreference as used by Painter (1951). Antibiosis describes adverse effects of a plant on insect survival, development and/or reproduction. Tolerance refers to the ability of a plant to withstand infestation and to support populations that would severely damage susceptible plants. Painter (1951) further proposed pseudoresistance to include apparent resistance which results from the expression of transitory characteristics of potentially susceptible hosts, and included host evasion, induced resistance, and escape.

Mechanisms by which plant resistance can influence the performance of natural enemies have been discussed by Painter (1951). First, reduction in prey populations may affect the success of some predators and parasites if prey density falls below the "optimum" searching capacity of the natural enemy. Secondly, host plant-induced changes in prey physiology and behavior may modify the success of natural enemies. Bergman & Tingey (1979) assert that allomonal or toxic resistance factors and morphological defense mechanisms may also limit populations of natural enemies that come into physical contact with the host plant or that use the plant for incidental feeding.

To our knowledge there has been no systematic effort to measure either long- or short-term effects of plant resistance on the magnitude or phenology of indigenous natural enemy populations, on potential success or failure of introduced exotic natural enemies, or on the level of mortality inflicted by biological control agents on the pest(s) targeted by plant resistance. Little information yet exists on elements defining compatibility of plant resistance with biological controls. The attempt of Bergman & Tingey (1979) at such a classification follows: indirect reduction in predator and parasite densities as a result of reduced prey populations (predators and parasites as a general rule are not adversely affected on resistant cultivars); undesirable side-effects in companion crops through reductions in a natural enemy reservoir; additivity of population suppression due to plant resistance and biological control.

Biological Control/Plant Resistance Interactions

Tingey (1981) proposed the following "functional categories" of plant resistance on the target pest: 1) impaired behavior; 2) impaired metabolism; 3) structural barriers; and 4) tolerance. This classification will be followed in a brief description of plant resistance/biological control interactions.

Impaired behavior This involves the behavioral component of host selection and utilization and presumes discriminatory activity of individuals, and includes much of Painter's "nonpreference." It is well known that cotton nectar serves as either a primary or secondary carbohydrate food source for a number of pest and beneficial insects (Butler et al. 1972). The reduction or elimination of extrafloral nectaries in cotton through genetic selection provides a mechanism of resistance that essentially deprives a spectrum of pests of an adequate food source (Maxwell et al. 1976). Studies by Laster & Meredith (1974) have shown that the nectariless trait produces no great deleterious effect on natural enemy populations. However, Schuster (1980) cautions that the utilization of this mechanism of resistance in cotton resistance breeding programs will produce potentially deleterious effects through reducing supplementary food for certain parasites and predators.

Impaired Metabolism This represents the nutritional elements of resistance and includes much of Painter's "antibiosis". The nutritional substrate offered by a host plant can indirectly influence predator and parasite fitness in several ways (Bergman & Tingey 1979). First, prey confined to feeding on resistant hosts commonly experience retarded growth rates, extended developmental rate, increased mortality, and reduced fecundity. With such alteration of fundamental physiological processes, the nutritional quality of prey as food for predators may be affected.

Secondly, natural enemies may be exposed to toxic resistance factors through their prey. Alkaloidal or other chemical substances that may be either antibiotic or innocuous to herbivorous prey when ingested may prove to be highly toxic or repellant to predators and/or parasites. α-tomatine, an alkaloid in tomato with antibiotic properties against *Heliothis zea*, has proven to be highly toxic to larval *Hyposoter exiguae* within its host (Campbell & Duffey 1979).

The effect of antibiotic resistance factors in soybean on parasites and microbial pathogens has perhaps been studied in greater depth than for many other crops. McCutcheon & Turnipseed (1981) were unable to demonstrate under field conditions any difference in parasitism of green cloverworm, *Plathypena scabra*, populations on resistant and susceptible soybean genotypes. Yanes & Boethel (1983) demonstrated adverse effects of antibiotic resistance against the soybean looper, *Pseudoplusia includens*. *Microplitis demolitor*, an exotic parasite, displayed decreased emergence from surviving hosts. Similarly, *Copidosoma truncatellum* suffered reductions in adult emergence from *P. includens* when that species was reared on resistant genotypes (R.M. Beach & J.W. Todd, personal communication). Powell & Lambert (1984) demonstrated incompatibility of certain resistant soybean genotypes with development and survival of *Microplitis croceipes*, an endemic parasite of *H. zea*. Dover (1984) demonstrated minimal effects of antibiotic factors on *Pediobius foveolatus* in the

Mexican bean beetle, *Epilachna varivestis*, reared on resistant genotypes.

Although field populations were low and not significantly different among genotypes, Gilreath (1977) examined field-collected larval cohorts of *P. scabra*, *H. zea*, *P. includens* and *Anticarsia gemmatalis* from resistant and susceptible soybean and was unable to demonstrate differential levels of infection by *Nomuraea rileyi*, *Entomophthora gammae* or the nuclear polyhedrosis viruses of *H. zea* or *P. includens*. Bell (1978) was unable to demonstrate differential mortality due to *N. rileyi* of *H. zea* larvae reared on resistant and susceptible leaves in the laboratory. In a similar experiment using a commercial formulation of *Bacillus thuringiensis*, Bell (1978) was able to demonstrate a 50% increase in mortality of larvae reared on resistant leaves over those reared on susceptible leaves. It is encouraging to note that Oliveira (1981) was able to demonstrate under field plot conditions that the infection of the velvetbean caterpillar, *Anticarsia gemmatalis*, by *N. rileyi* was greater in resistant than in susceptible plantings in the face of reduced larval populations. That would suggest that larvae stressed through antibiotic resistance may be more susceptible to infection by the fungus.

Structural Barriers Resistance due partially or solely to morphological characteristics of plants is often difficult to place in traditional definitions of antibiosis or antixenosis. Kogan (1975) collectively referred to impairment of oviposition, feeding and/or mobility by morphological characteristics as *phenetic resistance*. Such morphological features are of use as resistance factors principally among those insects which bore into plant tissue. The resistance mechanism involves the impairment or prohibition of penetration into the plant. Examples are husk tightness and length in corn conferring resistance to *H. zea* (Cameron & Anderson 1966, Douglas 1947, Luckmann et al. 1964); increased hardness of stalk tissues in rice and sugarcane conferring resistance to the striped rice borer, *Chilo suppressalis*, and the sugarcane borer, *Diatraea saccharalis*, respectively (Agarwal 1969, Djamin & Pathak 1967); tightness of leaf sheaths in sugarcane conferring resistance to the sugarcane borer (Kyle & Hensley 1970); density of stem pith and node tissues in wheat and various cucurbits conferring resistance to the wheat stem sawfly, *Cephus cinctus*, and to the squash vine borer, *Melittia cucurbitae*, respectively (Howe 1949, O'Keefe et al. 1960). The overall effect of this type of resistance is a prolongation of the time during which the pest remains in an exposed situation and susceptible to natural control agents.

In fact, morphological characters may be directly detrimental to arthropod natural enemies. Certain genotypes of tobacco with hooked and/or glandular trichomes have been shown to severely limit egg parasitism of the tobacco hornworm, *Manduca sexta*, by *Telenomus sphingus* and *Trichogramma minutum* (Rabb & Bradley 1968) and of *Heliothis virescens* (Elsey & Chaplin 1978). Similarly, Milliron (1940) found that the glandular trichomes of *Nicotiana glutinosa* entrapped *Encarsia formosa*, a parasite of whiteflies. However, studies by Obrycki & Tauber (1984) and Obrycki et al. (1983) indicate that glandular trichomes of resistant potatoes do not substantially reduce either predation on or parasitism of *Myzus persicae* or *Macrosiphum euphorbiae*. Other studies indicate an inverse relationship between cotton trichome density and the level of successful attacks on *H. zea* eggs by the parasite, *Trichogramma pretiosum* and the predator *Chrysopa rufilabris* (Treacy et al. 1983). Egg parasitism by *T. pretiosum* was greatest on glabrous, lowest on pilose, and intermediate on normally hirsute genotypes. However, the effect of trichome density on predation by *C. rufilabris* varied with stage of prey. For early instar larvae, predation was higher on heavily pubescent genotypes than on glabrous, with normal pilosity intermediate. For eggs, the situation was reversed. Predation was much reduced on the genotypes with heavier pubescence.

May (1951) proposed that the searching ability of aphid parasites and predators might be hampered on pilose cotton cultivars. Glabrous cottons are known to reduce the number of *Heliothis* spp. eggs, larvae and damaged fruit (Lukefahr et al. 1971). Hallman (1979) found that parasitism of cotton leafworm, *Alabama argillacea*, eggs by *Trichogramma* spp. was greater than that of *Heliothis* spp. eggs on cotton. He observed that *Trichogramma* walked more frequently over the relatively smooth leaf surfaces, where the majority of the *A. argillacea* eggs were found, than over the more pubescent shoot terminals, where most of the *Heliothis* spp. eggs were located.

Tolerance Although we could locate no literature to support the hypothesis, possibly the greatest potential for compatibility of plant resistance and biological control is with the joint utilization of pest tolerant varieties and microbial pathogens. Since most insect pathogens are heavily density-dependent, the ability of a tolerant crop to support substantial populations of a pest should allow a pathogen sufficient host substrate on which to develop devastating epizootics.

Other Interactions The utilization of insect resistant varieties has been criticized because of potentially unacceptable reductions in predator and parasite population densities resulting indirectly through reductions in prey populations (Casagrande & Haynes 1976, Kennedy et al. 1975, Pimentel & Wheeler 1973, Schuster et al. 1976). In general, these studies have indicated that, though natural enemy populations may be reduced, predator and parasite efficiency normally is not adversely effected. Bergman & Tingey (1979), however, cautioned that the expression of varietal resistance may vary with pest density, a parameter modified, at least in part, by the action of natural enemies. Proponents of plant resistance, on the other hand, argue that in the case of specific, targeted resistance, generalist predator and parasite species and certain microbial pathogens will be only minimally reduced due to limited impact of resistance on alternate prey species.

In many cases, mechanisms of resistance are not known or are poorly understood. Observations have been made which indicate both compatibility and incompatibility of plant resistance and biological control. Levels of parasitism by Campoletis chlorideae and Carcelia illota are reduced in populations of Heliothis armigera on resistant chickpea genotypes (Sithanantham et al. 1982). On the other hand, levels of parasitism of the sorghum midge, Contarinia sorghicola, by Aprostocetus diplosidis and Tetrastichus venustus was greater on resistant sorghum genotypes that supported lower levels of infestation (Wiseman & McMillian 1970). On cotton, Burleigh (1975) was unable to demonstrate differences in parasitism of Heliothis spp. larvae by M. croceipes on okra-leaf vs. full-canopied varieties. However, infection of larvae by N. rileyi was significantly higher on the closed canopy varieties. Aphidius matricariae is reported to be capable of exterminating populations of Myzus persicae on resistant varieties of chrysanthemums in glasshouses in Great Britain, while aphid population suppression by the parasite on susceptible chrysanthemums has not proven to be economically satisfactory (Wyatt 1970). This parasite has the added advantage of being resistant or tolerant to pesticides currently used against aphids on chrysanthemums.

Certain characteristics of otherwise insect-susceptible crop varieties may either aid or hinder biological control processes in the field. For example, plant volatiles of different commercial corn hybrids have been shown to differentially attract Lydella thompsoni (= L. grisescens), a parasite of the European corn borer, Ostrinia nubialis (Franklin & Holdaway 1966). Brussels sprouts characterized by glossy leaves have been shown to be more attractive to predators and parasites than those with waxy foliage (Chandler 1968, Way & Murdie 1965). Shelter provided by large leaf veins and pubescence has been shown to result in greater predatory mite abundance on certain apple cultivars (Downing & Moilliet 1967). Finally, natural enemies may be exposed to toxic substances through their prey. For example, Apanteles congregatus has been found to seldom parasitize larvae of the tobacco hornworm reared on dark tobacco cultivars, compared to burley tobacco. It has been suggested that the greater nicotine content of the dark tobaccos is passed to larval parasites through their prey (Gilmore 1938a, b).

Success Stories

Probably the best documented success story of compatibility of biological control and plant resistance is that of rhodesgrass scale, Antonina graminis, on Texas rangelands. Rhodesgrass scale is a cosmopolitan insect, feeding on over 100 hosts (Chada & Wood

1960), including 38 species of range grasses in Texas (Schuster 1967). Fortunately, two of the major climax dominant grasses of the region do not serve as hosts. Neodusmetia sangwani was introduced from India against rhodesgrass scale in 1959, was established (Dean et al. 1961), and its effectiveness under field conditions documented (Schuster et al. 1971). Two companion programs were established at the same time: a study of resistance of the native Texas range grasses and promising plant introductions (Schuster 1967); and the development of a rhodesgrass scale-tolerant variety of rhodesgrass which raised the damage threshold and increased the utility of biological control (Schuster & Dean 1973). The crucial period for control occurs during the June population peak at which time the parasite is just recovering from suppressed populations due to general host unsuitability during May. The resistant rhodesgrass variety "Bell" was found to tolerate twice the population of scales as other available varieties without appreciable damage. Where used, the resistant variety provides relief until the parasites effectively reduce scale populations (Schuster & Dean 1973).

In a second well-documented example, the resistant crop in fact enhances the operation of a naturally occurring parasite. Damage by the greenbug, Schizaphis graminum, to barley and sorghum is greatly reduced by the combined effects of resistant varieties and the parasite, Lysiphlebus testaceipes, compared to the unilateral implementation of either tactic (Schuster & Starks 1975, Starks et al. 1972).

CULTURAL PRACTICES (CULTURAL CONTROL)

Many diverse cultural practices are employed to plant, maintain, and harvest crops in all agroecosystems. These measures influence the ecology of the crop production system. Even slight modifications in crop production practices can substantially affect the dynamics of arthropod pest and natural enemy populations, as alterations of such cultural activities alter the microenvironment of the crop (Arkin & Taylor 1981, Barfield & Gerber 1979, Hatfield & Thomason 1982). Consequently, some measures may lead either directly or indirectly to increases or decreases in crop stress from pests. Huffaker (1982) stressed the need for ecological studies of the interrelationships of pests and other elements in crop ecosystems, and called for detailed studies to closely quantify these relationships.

Cultural control is the deliberate alteration of the agroecosystem, either the cropping system or specific crop production practices, to reduce pest numbers and crop injury. Cultural control activities historically have been an important aid to pest control (see also Schroth & Hancock; Van Gundy, this volume). Some successful cultural control methods operate by enhancing the benefits derived from natural enemies. Conversely, successful biological control methods often require alteration of common cultural practices. Cultural and biological control can be economically and ecologically desirable and compatible means of reducing crop stress from pests.

The agroecosystem can be modified in many ways to improve the benefits obtained from biological control organisms. Natural enemy populations can be enhanced by increasing the amount of necessary habitat and/or by altering habitats to be more suitable for their colonization or subsequent population development.

Increasing Natural Enemy Habitat Through Cultural Practices

Agroecosystems containing wild-vegetation habitats are usually characterized by a greater diversity of resident arthropod species. Many natural enemy species require food sources in the form of flower pollen, flower nectar, or innocuous arthropods that are not present in particular crop habitats. Such food requirements can sometimes be provided to support natural enemy populations by including specific wild-vegetation habitats within or near fields. Information about alternative food requirements of predators and parasites have been reviewed by Altieri & Whitcomb (1979) and Altieri et al. (1977).

Examples of the deliberate establishment or manipulation of non-crop vegetative habitats by cultural measures which result in increased natural enemy populations are summarized in Table 1. In some cases, benefits of biological control were derived from areas not producing crops by wild plants located in hedge rows or woodlands. One such example described by Doutt et al. (1966) involves blackberry plantings located outside vineyards increasing biological control of grape leafhoppers, *Erythroneura elegantula*, by an egg parasite, *Anagrus epos* (see Flaherty et al., this volume).

The native perennial stinging nettle, *Urtica dioica*, has not been reported to support economically important pests in England, but serves as a source of aphid natural enemies (Perrin 1975). Primarily, these natural enemies consist of anthocorid, mirid, and coccinellid predators; hymenopterous parasites; and pathogenic fungi. Populations of these organisms increase in the spring on aphid and psyllid species inhabiting perennial stinging nettle, and some species disperse to crops with the onset of summer. Cutting of perennial stinging nettle stands in mid-June is described as a potential method of culturally enhancing biological control in nearby crops.

Another potential method of enhancing natural enemy populations is the establishment of weeds directly into crop habitats. Several successful examples of this approach are presented in Table 1. Altieri (1981) noted that predator populations are greater in Georgia soybean fields infested with sicklepod, *Cassia obtusifolia*, than in weed-free fields. Consequently, soybean injury from insect pests was reported to be reduced, primarily because of lower populations of velvetbean caterpillar, *A. gemmatalis*, and southern green stinkbug, *Nezara viridula*. Weedy soybean fields in Indiana also harbor a greater diversity and abundance of natural enemies (Shelton & Edwards 1983). The predator *Coleomegilla maculata* has been shown to be more abundant in soybean with grassy and broadleaf weeds than in fields with no weeds. Other predators (*Orius* spp., *Nabis* spp., and *Harpalus* spp.) apparently prefer fields with grassy weeds or mixed grassy and broadleaf weeds. Weedy soybean fields have smaller populations of the Mexican bean beetle, an important phytophagous pest in Midwestern production areas, than weed-free fields.

In Colombia, populations of two major pest species, *Empoasca kraemeri* and *Diabrotica balteata*, are reduced in weedy plantings compared to those that are weed-free, partially due to the action of natural enemies (Altieri et al. 1977). In these fields, predator populations consist primarily of species of Dolichopodidae, Reduviidae, and Nabidae. Dolichopodids have not been shown to be affected by weed density; however, reduviid and nabid populations were greater in weedy habitats than in weed-free habitats. Parasitism of *E. kraemeri* (primarily by *Anagrus* sp.) is similar in weedy and weed-free plantings.

Table 1. Examples of increased populations of biological control organisms with establishment of wild plants or weeds in field margins (FM), adjacent woodlands (AW), or crop habitats (CH).

Crop Host	Location of Established Habitat	Biological Control Organism(s)	Reference(s)
Bean	CH, FM	Predators	Altieri et al. (1977)
Bean, Cabbage Cucumber, Watermelon,	CH, FM	Predators	Cochereau (1976)
Grape	CH	Predators	Flaherty (1969) Flaherty et al. (1971)
	AW	Parasitoids	Doutt & Nakata (1973) Doutt et al. (1966)
Hops	CH	Parasitoids	Copland (1979)
Soybean	CH	Predators	Shelton & Edwards (1983) Altieri (1981)
Wheat	CH	Predators	Speight & Lawton (1976)
Not reported	FM	Predators	van Emden (1965)
Not reported	CH	Predators	Bombosch (1966)
Not reported	FM	Parasitoids Pathogens Predators	Perrin (1975)

These and other studies provide evidence that manipulation of wild vegetation may be practical and even necessary to adequately manage pests by biological control in small farm culture under conditions of subsistence agriculture. But, the roles of wild vegetation in the biologies of natural enemies must be defined in additional cases involving production-level agriculture before generalizations can be made. Investigations involving wild vegetation and entomopathogens are particularly lacking. Other problems must also be addressed before such practices can be widely accepted. Methods must be developed to establish and maintain wild vegetation habitats in agroecosystems and to manipulate them so that the non-crop vegetation (weeds) does not compete unacceptably with crops. Altieri & Whitcomb (1979) discuss and review such problems.

Cropping systems may also be modified to include crop hosts that provide excellent habitat for beneficial insect populations. Benefits from these natural enemies can be obtained in some cases merely by introducing the crop host into the cropping system, but frequently interplanting, strip harvesting, trap cropping, and/or other practices are desirable or necessary. Examples where interplanting, trap cropping, or strip harvesting increased natural enemy habitat and thereby improved biological control of major insect pests are presented in Table 2. Some of these measures have been adopted by producers; others, for various reasons, have not. Several of these examples involve California agroecosystems in which alfalfa constitutes a major component crop. Alfalfa is apparently an ideal habitat for numerous natural enemy species that are important regulatory agents of pests in alfalfa and other crops, including *Lygus* spp., *H. zea*, *Trichoplusia ni*, *Spodoptera exigua*, spider mites, and others.

Harvesting of alfalfa by solid cutting of fields produces very unfavorable conditions for continued survival of natural enemy species (Stern et al. 1964, van den Bosch & Stern 1969). Nabid, chrysopid, anthocorid, and coccinellid populations decline sharply after solid cutting, while geocorid populations suffer relatively low mortality. Like most predators, parasites of aphid and lepidopterous pests also decline. Alfalfa habitats, however, remain favorable for all natural enemies when harvested by strip cutting. With the strip-cutting technique, alfalfa is harvested in alternate strips so that two different-aged hay growths occur in the field simultaneously. Strip cutting, therefore, enhances biological control directly within alfalfa fields. Additionally, adults of the natural enemies that rapidly proliferate in alfalfa readily disperse to other nearby fields, and injury is reduced in nearby crops, such as cotton, bean, tree fruits, etc. The practice of interplanting alfalfa in cotton further increases benefits of biological control (Stern 1969). Eggs and small larvae of numerous pests are attacked by both predators and parasites, and plant injury from arthropods is substantially reduced. Though technically sound, producers have

Table 2. Examples of improved biological control of major insect pests through the use of trap cropping, interplanting, and strip harvesting.

Cultural Practice	Biological Control Agent(s)	Reference(s)
Alfalfa strip-harvested	Predators Parasites	Stern et al. (1964) van den Bosch & Stern (1969)
Alfalfa interplanted in cotton	Predators	Stern (1969)
Clover or brome grass intercropped in grape	Predators	Flaherty et al. (1971)
Clover, spinach, or beans interplanted in Brassica crops	Predators	Coaker (1980)
Corn interplanted in cotton	Parasites	Chacon (1973)
Sesame interplanted in cotton	Parasites	Pair et al. (1982) Laster & Furr (1972)
Snap bean trap crops near soybean	Parasistoids	Rust (1977) Stevens et al. (1975)
Sorghum interplanted in cotton	Predators	Robinson et al. (1972a) Robinson et al. (1972b)
Early-maturing soybean trap crops in soybean	Parasitoids Predators Pathogens	Newsom & Herzog (1977)

refused to adopt this management tactic (C.B. Huffaker & L.A. Falcon, personal communication).

In other geographical areas, a variety of crops sometimes are interplanted in cotton to reduce pest injury in this intensively managed commodity (see Table 2). Such crops include sorghum, sesame, and corn, but others can provide benefits. Interplantings in these cases operate by acting as a trap crop (being more attractive than cotton to pests) and by serving as a reservoir for natural enemies. Other types of trap cropping have been used to augment or conserve benefits of natural enemies. For example, early maturing soybean varieties planted near main soybean plantings attract bean leaf beetle, *Cerotoma trifurcata*, and southern green stink bug, *N. viridula* (Newsom & Herzog 1977). Appropriately timed applications of insecticide to the trap crop controls the target pest in the trap reservoir, prevents spread of the pest to the main crop, and conserves natural enemies throughout the remainder of the ecosystem. Similarly, border rows of snap beans planted prior to soybean attract and function as a trap crop for the Mexican bean beetle (Rust 1977, Stevens et al. 1975). Releases of a parasite, *P. foveolatus*, into such a trap crop creates a reservoir from which parasites disperse and affect effective suppression of Mexican bean beetle in the soybean crop.

Cultural measures are easily modified to include or to manipulate crops so that the benefits of biological control are enhanced. Intercropping, strip cropping, and trap cropping are important ways to enhance natural enemy populations in integrated pest management systems.

Enhancing Natural Enemy Survival and Development Through Cultural Practices

Cultural practices employed within crop habitats can affect biological control directly by destroying arthropod populations or indirectly by altering microclimatic conditions within the crop habitat (see also Schroth & Hancock, this volume). Activities that produce such microenvironmental modifications include irrigation, fertilizer application, row spacing, seeding rate, planting date and tillage. These practices may lead to substantial improvement in benefits of biological control, especially in cases where changes are more favorable to natural enemy populations than to pest populations.

Seedbed preparations and cultivation practices modify soil habitats where many pest and natural enemy species reside during at least part of their life cycles. Post-harvest plowing of European small-grain fields has dramatic effects on biological control of cereal leaf beetle, *Oulema melanopus* (Carl 1979). Adults of this injurous species disperse from fields and are unaffected by plowing, but populations of its larval and pupal parasites remain in the soil and are destroyed. Disking or harrowing has fewer negative impacts on populations of these biological control agents than plowing.

Repeated cultivations to control weeds increase dusty conditions on California grape vines, and thereby foster pest mite, *Tetranychus pacificus* and *Eotetranychus williamettei*, buildups (Flaherty et al. 1971). The efficiency of natural enemies of these mites is reduced, mechanism(s) unknown, and outbreaks occur. An alternate weed-control program of only two spring cultivations followed by three or four summer mowings of between-row cover crops or natural weed species has alleviated this problem.

Ridging soil against potato vines in South Africa increases mortality of the potato tuber moth, *Phthorimaea operculella*, due to indigenous and imported parasitoid species (Watmough et al. 1973). This cultural practice, combined with one or two carefully-timed insecticide sprays, greatly reduces crop injury by the pest. Likewise, shallow tillage of wheat stubble infested by wheat stem sawfly increases the frequency of total parasitism (Holmes 1982).

Irrigation substantially alters relative humidity and temperature in and under crop canopies, and changes in these parameters can greatly alter benefits of biological control. Feeding, egg laying, and development of many phytophagous mites are reduced under conditions of high relative humidity. Consequently, irrigation of corn on the Texas High Plains reduces populations of Banks grass mites, Oligonychus pratensis (Chandler et al. 1977). Predator populations remain unaffected by the measure. Likewise, ample water applied to Brussels sprouts in England reduces fecundity and reproductive rate of the cabbage aphid, Brevicoryne brassicae, and green peach aphid, Myzus persicae, thereby enhancing the action of natural enemies (van Emden & Wearing 1965). Irrigation also is necessary for establishment of aphid parasitoid species in crops in arid geographical regions, such as the Middle East (Stary & Erdelen 1982).

High relative humidities are important for the dissemination and infectivity of some entomopathogenic microorganisms (especially fungal pathogens of arthropods), and irrigation can enhance such populations. For example, mortality of the pea aphid, Acyrthosiphon pisum, due to infection by Entomopthora spp. is greater in irrigated than unirrigated Argentine crops (Frezzi 1972). Additionally, many entomopathogens can be successfully enhanced by direct application of inoculum through irrigation systems (Hamm & Hare 1982).

Seeding rate, planting date, and row spacing influence the impacts of biological control within crop habitats, primarily by altering relative humidity, temperature, and/or light penetration. Natural enemies of phytophagous lepidopteran species in North Carolina soybean fields, are affected by planting date, seeding rate, and row spacing (Sprenkel et al. 1979). The predators, Geocoris punctipes, Nabis spp., and spiders and the entomopathogen, N. rileyi are affected by each cultural practices with seeding rate having the least effect and planting date the greatest effect. In North Carolina, benefits from these natural enemies are substantial when soybean is planted early, while biological control is maximized in double-cropped soybean (planted late) following wheat by planting in narrow rows at a high seeding rate. In Mississippi, planting soybean in narrow rows enhances spider and G. punctipes populations, but not Nabis spp. or lady beetle populations (Bushman et al. 1984). Planting cotton in narrow rows apparently enhances mortality of Heliothis spp. larvae by N. rileyi (Burleigh 1975).

Fertilizer applications also can affect pests and their natural enemies. Nitrogen, potassium, and phosphorus applications enhance plant growth of cotton, and population levels of some H. zea predators are increased, while others are not (Adkisson 1958). Selective fertilizer applications in Brussels sprouts influence plant nutritional levels, thereby altering developmental rate and fecundity of aphid pests and enhancing benefits of biological control (van Emden & Wearing 1965). Changes in pH and other parameters resulting from fertilization can directly impact on natural enemy populations. For example, very high concentrations of nitrogen, potassium, and phosphorus reduce survival of Steinernema carpocapsae (= Neoaplectana dutkyi), a nematode natural enemy of several rice pests (Rao et al. 1975). Concentrations of these nutrients typically applied in rice fields, however, have no detrimental effects on the organism. A unique example of a fortuitous effect of a cultural practice is provided by Chiang (1970), who found that the predatory efficiency of a mesostigmatid mite on corn rootworm larvae was dramatically increased by applications of manure to the field.

These examples illustrate the diverse ways that crop-production practices affect the ecology of agroecosystems, in particular, the development and survival of pest and natural enemy populations. Yet, the impacts of cultural measures on the ecology of arthropod species are rarely considered. This situation must be reversed if full benefits of biological control are to be realized in crop production systems.

RECOMMENDATIONS

The compatibility of biological control programs must ultimately be measured in the context of biological and economic evaluations and assessment of environmental impact. Where possible, crop production and management strategies should be tailored to be compatible with biological control. Systems models will become increasingly important as tools for analyzing cropping systems, evaluating compatibilities among alternative production and management tactics and strategies, and establishing guidelines for the integration of tactics into ecologically and economically sound systems for crop production and management (Herzog et al. 1985).

Host Plant Resistance

1. Since breeding for resistance is a long-term endeavor, perhaps it would be to our benefit to ascertain the compatibility of resistance with the endemic natural enemy complex using the genotypic (parental) source of resistance. While this will not be a panacea, it could potentially prevent the development of crop varieties which could be detrimental to biological control.
2. It is an unfortunate fact that the resistance/susceptibility of most crop varieties released for commercial production is unknown. Equally unfortunate is the fact that the potential interrelationships between commercial varieties, whether resistant or susceptible, and natural enemies of pests is unknown. Studies should be undertaken immediately to address these issues.
3. Most information on the relationship of biological control and host plant resistance is conjectural, speculative, or

based only on limited observation. Most of the available information deals solely with parasites, with very little attention being given to predators, microbial pathogens or other classes of natural enemies. Field-scale experimentation should be undertaken that will document compatibility of host plant resistance with biological control under "real world" conditions.

4. Plant-mediated elements of searching and host-finding behavior of parasites and predators should be investigated. Studies to elucidate the nutritive quality to predators and parasites of herbivorous species feeding on antibiotically resistant plants should be initiated.

Cultural Practices

1. Basic studies are urgently needed to determine ecological requirements of native and introduced biological control agents. Research is needed to determine microhabitat/microenvironmental effects of crop production practices. Then, a more precise definition of potential beneficial/detrimental impacts of cultural practices on biological control agents can be achieved.
2. Cultural practices designed to enhance the benefits of biological control should be economically, ecologically, and agronomically feasible. Therefore, such research efforts must be undertaken by an interdisciplinary team.
3. When explorations for exotic natural enemies are undertaken, the characteristics of natural enemy species needed for specific situations should receive early consideration. It is at this point that the cropping system and cultural practices employed at the sites of origin and the eventual destination are of great importance.
4. It is a practical impossibility to study all combinations of crops, pests, natural enemies, environments, and cultural practices to document their interrelationships. Therefore, continued development and refinement of system level crop/pest models should receive high priority to optimize coordination and implementation of research efforts.

REFERENCES

Adkisson, P.L. 1958. The influence of fertilizer applications on populations of Heliothis zea (Boddie), and certain insect predators. J. Econ. Entomol. 51: 757-759.

Agarwal, R.A. 1969. Morphological characteristics of sugarcane and insect resistance. Entomol. Exp. Appl. 12:767-776.

Altieri, M.A. 1981. Weeds may augment biological control of insects. Calif. Agric. 35(6):22-24.

Altieri, M.A., A. van Schoonhoven and J. Doll. 1977. The ecological role of weeds in insect pest management systems: A review illustrated with bean (Phaseolus vulgaris L.) cropping systems. PANS 23:185-206.

Altieri, M.A., and W.H. Whitcomb. 1979. The potential use of weeds in the manipulation of beneficial insects. HortScience 14:12-18.

Arkin, G.F., and H.M. Taylor (Eds.). 1981. Modifying the root environment to reduce crop stress. Amer. Soc. Agr. Engr., St. Joseph, MI. 407 pp.

Barfield, B.J., and J.F. Gerber (Eds.). 1979. Modification of the aerial environment of plants. Amer. Soc. Agr. Engr., St. Joseph, MI. 538 pp.

Beml, J.V. 1978. Development and mortality in bollworms fed resistant and susceptible soybean cultivars treated with Nomuraea rileyi or Bacillus thuringiensis. J. Georgia Entomol. Soc. 13:50-55.

Bergman, J.M., and W.M. Tingey. 1979. Aspects of interaction between plant genotypes and biological control. Bull. Entomol. Soc. Amer. 25:275-279.

Bombosch, S. 1966. Occurrence of enemies on different weeds with aphids, pp. 177- 179. In: I. Hodek (Ed.), Ecology of aphidophagous insects. Academia Publishing House, Prague.

Burleigh, J.G. 1975. Comparison of Heliothis spp. larval parasitism and Spicaria infection in closed and open canopy cotton varieties. Environ. Entomol. 4: 574-576.

Buschman, L.L., H.N. Pitre and H.F. Hodges. 1984. Soybean cultural practices: Effects on populations of geocorids, nabids, and other soybean arthropods. Environ. Entomol. 13:305-317.

Butler, G.D., G.M. Loper, S.E. McGregor, J.L. Webster and H. Margolis. 1972. Amounts and kinds of sugars in the nectars of cotton (Gossypium spp.) and the time of their secretion. Agron. J. 64:364-368.

Cameron, J.W., and L.D. Anderson. 1966. Husk tightness, corn earworm egg numbers, and starchiness of kernels in relation to resistance of corn to the corn earworm. J. Econ. Entomol. 59:556-558.

Campbell, B.C., and S.S. Duffey. 1979. Tomatine and parasitic wasps: Potential incompatibility of plant antibiosis with biological control. Science 205:700-702.

Carl, K.P. 1979. The importance of cultural measures for the biological control of the cereal leaf beetle Oulema melanopus (Col. Chrysomelidae). Mitt. Schweiz. Entomol. Ges. 52:443.

Casagrande, R.A., and D.L. Haynes. 1976. The impact of pubescent wheat on the population dynamics of the cereal leaf beetle. Environ. Entomol. 5: 153-159.

Chacon, R.C. 1973. Beneficios reportados por el control biologica de las plagas del algodonero a los agricultores de la Comarco Lagunera. Fitofilo 23:40-47.

Chada, H.L., and E.A. Wood. 1960. Biology and control of the rhodesgrass scale. USDA Tech. Bull. 1221.

Chandler, A.E.F. 1968. Some host plant factors affecting oviposition by aphidophagous Syrphidae (Diptera). Ann. Appl. Biol. 61:415-423.

Chandler, L.D., C.R. Ward, W.M. Lyle and E.D. Bynum. 1977. Effects of various irrigation schedules on population densities of Banks grass mite (Oligonychus pratensis) and its predatory arthropods in corn, the Texas High Plains. Texas Agr. Exp. Sta. Prog. Rep. 3443C, 17 pp.

Chiang, H.C. 1970. Effects of manure applications and mite predation on corn rootworm populations in Minnesota. J. Econ. Entomol. 63:934-936.

Coaker, T.H. 1980. Insect pest management in Brassica crops by inter-cropping. Bull. Intern. Organiz. Biol. Control 3:117-125.

Cochereau, P. 1976. Controle biologique, en Nouvelle Caledonie de Tetranychus urticae (Acarien: Tetranychidae) au moyen de Phytoseiulus persimilis (Acarien: Phytoseiidae), en cultures maraicheres. Entomophaga 21:151-156.

Copland, M.J.W. 1979. Hymenoptera in hop gardens, with particular reference to parasitoids associated with damsonhop aphid Phorodon humuli. Entomol. Exp. Appl. 25:146-152.

Dean, H.A., M.F. Schuster and J.C. Bailey. 1961. The introduction and establishment of Dusmetia sangawani on Antonina graminis in South Texas. J. Econ. Entomol. 54:952-954.

DeBach, P. 1974. Biological control by natural enemies. Cambridge Univ. Press, London. 333 pp.

Djamin, A., and M. D. Pathak. 1967. Role of silica in resistance to Asiatic rice borer, Chilo suppressalis (Walker), in rice varieties. J. Econ. Entomol. 60: 347-351.

Douglas, W.A. 1947. The effect of husk extension and tightness on earworm damage to corn. J. Econ. Entomol. 40:661-664.

Doutt, R.L., and J. Nakata. 1973. The Rubus leafhopper and its egg parasitoid: An endemic biotic system useful in grape pest management. Environ. Entomol. 2: 381-386.

Doutt, R.L., J. Nakata and F.E. Skinner. 1966. Dispersal of grape leafhopper parasites from a blackberry refuge. Calif. Agric. 20(10):14-15.

Dover, B.A. 1984. Development of the parasitoid, Pediobius foveolatus in Mexican bean beetles reared on susceptible and resistant soybean lines. M.S. Thesis. Clemson Univ., S.C.

Downing, R.S., and T.K. Moilliet. 1967. Relative densities of predacious and phytophagous mites on three varieties of apple trees. Can. Entomol. 99:738-741.

Elsey, K.D., and J.F. Chaplin. 1978. Resistance of tobacco introduction 1112 to the tobacco budworm and green peach aphid. J. Econ. Entomol. 71:732-725.

Flaherty, D.L. 1969. Ecosystem trophic complexity and densities of the Willamette mite, Eotetranychus willamettei Ewing (Acarina: Tetranychidae). Ecology 50:911-916.

Flaherty, D., C. Lynn, F. Jensen and M. Hoy. 1971. Influence of environment and cultural practices on spider mite abundance in southern San Joaquin Thompson seedless vineyards. Calif. Agric. 25(11):6-8.

Franklin, R.T., and F.G. Holdaway. 1966. A relationship of the plant to parasitism of European corn borer by the tachinid parasite Lydella grisescens. J. Econ. Entomol. 59:440-441.

Frezzi, M.J. 1972. Two fungi pathogenic to insects and three entomophagous insects, useful aids in Argentina in the biological control of the lucerne aphid (Acyrthosiphon pisum, Harris). Idia 291:21-30.

Gilmore, J.U. 1938a. Notes on Apanteles congregatus (Say) as a parasite of tobacco hornworms. J. Econ. Entomol. 31:712-715.

Gilmore, J.U. 1938b. Observation on the hornworms attacking tobacco in Tennessee and Kentucky. J. Econ. Entomol. 31:706-712.

Gilreath, M.E. 1977. Influence of soybean genotype on the susceptibility of lepidopterous larvae to pathogens. M.S. Thesis, Clemson Univ., S.C.

Hallman, G. 1979. Algunas ovservaciones sobre Trichogramma. Rev. Colombiana Entomol. 5:13-18.

Hamm, J.J., and W.W. Harf. 1982. Application of entomopathogens in irrigation water for control of fall armyworms and corn earworms. (Lepidoptera: Noctuidae) in corn. J. Econ. Entomol. 75:1074-1079.

Hatfield, J.L., and I.J. Thomason (Eds.). 1982. Biometeorology in integrated pest management. Academic Press, N.Y. 491 pp.

Herzog, D.C., J.L. Stimac, D.G. Boucias and V.H. Waddill. 1985. Compatibility of biological control in soybean insect management. In: P.L. Adkisson and Shijun Ma (Eds.). Proceedings of a Joint U.S. National Academy of Sciences/Chinese Academy of Sciences Symposium on Biological Control of Insect Pests. Sci. Sinica (In Press).

Holmes, N. D. 1982. Population dynamics of the wheat stem sawfly, Cephus cinctus (Hymenoptera: Cephidae), in wheat. Can. Entomol. 114:775-788.

Howe, W. L. 1949. Factors affecting the resistance of certain cucurbits to the squash borer. J. Econ. Entomol. 42:321-326.

Huffaker, C.B. 1982. Overall approach to insect problems in agriculture, pp. 1-71-192. In: J.L. Hatfield and I.J. Thomason (Eds.), Biometeorology in integrated pest management. Academic Press, N.Y.

Kennedy, G.G., A.N. Kishaba and G.W. Bohn. 1975. Response of several pest species to Cucumis melo L. lines resistant to Aphis gossypii Glover. Environ. Entomol. 4:653-657.

Kogan, M. 1975. Plant resistance in pest management, pp. 103-146. In: R.L. Metcalf and W.H. Luckmann (Eds.), Introduction to insect pest management. John Wiley & Sons, N.Y.

Kogan, M., and E.E. Ortman. 1978. Antixenosis--a new term proposed to replace Painter's "Nonpreference" modality of resistance. Bull. Entomol. Soc. Amer. 24:175-176.

Kyle, M.L., and S.D. Hensley. 1970. Sugarcane borer host plant resistance studies. Proc. Louisiana Acad. Sci. 33:55-67.

Laster, M.L., and R.E. Furr. 1972. Heliothis populations in cotton-sesame interplantings. J. Econ. Entomol. 65:1524-1525.

Laster, M.L., and W.R. Meredith. 1974. Nectariless cotton studies. Mississippi Agric. For. Exp. Sta. Res. Highlights 37(7):2-5.

Luckmann, W.H., A.M. Rhodes and E.V. Wann. 1964. Silk balling and other factors associated with resistance of corn to corn earworm. J. Econ. Entomol. 57:778-779.

Lukefahr, M.J., J.E. Houghtaling and H.M. Graham. 1971. Suppression of Heliothis populations with glabrous cotton strains. J. Econ. Entomol. 64: 486-488.

Maxwell, F.G., M.F. Schuster, W.R. Meredith and M.L. Laster. 1976. Influence of the nectariless character in cotton on harmful and beneficial insects, pp. 157- 161. In: T. Jermy (Ed.), The host plant in relation to insect behavior and reproduction. Plenum Press, N.Y.

May, A.W.S. 1951. Jassid resistance of the cotton plant. Queensland J. Agr. Sci. 8:43-68.

McCutcheon, G.S., and S.G. Turnipseed. 1981. Parasites of lepidopterous larvae in insect resistant and susceptible soybeans in South Carolina. Environ. Entomol. 10:69-74.

Milliron, H.E. 1940. A study of some factors affecting the efficiency of Encarsia formosa Gahan, an aphelinid parasite of the greenhouse whitefly, Trialeurodes vaporariorum (Westw.). Michigan Agr. Exp. Sta. Tech. Bull. 173, 23 pp.

Newsom, L.D., and D.C. Herzog. 1977. Trap crops for control of soybean pests. Louisiana Agric. 20(3):14-15.

Obrycki, J.J., and M.J. Tauber. 1984. Natural enemy activity on glandular pubescent potato plants in the greenhouse: An unreliable predictor of effects in the field. Environ. Entomol. 13:(In Press).

Obrycki, J.J., M.J. Tauber and W.M. Tingey. 1983. Predator and parasitoid interaction with aphid-resistant potatoes to reduce aphid densities: A two-year field study. J. Econ. Entomol. 76:456-462.

O'Keefe, L.E., J.A. Callenbach and K.L. Lebsock. 1960. Effect of culm solidness on the survival of the wheat stem sawfly. J. Econ. Entomol. 53:244-246.

Oliveira, E.B. 1981. Effect of resistant and susceptible soybean genotypes at different phenological stages on development, leaf consumption and oviposition of Anticarsia gemmatalis Hubner. M.S. Thesis, Univ. Florida, Gainesville.

Ortman, E.E., and D.C. Peters. 1980. Introduction, pp. 3-13. In: F.G. Maxwell and P.R. Jennings (Eds.), Breeding plants resistant to insects. John Wiley & Sons, N.Y.

Painter , R.L. 1951. Insect resistance in crop plants. Macmillan, N.Y. 520 pp.

Pair, S.D., M.L. Laster and E.F. Martin. 1982. Parasitoids of Heliothis spp. (Lepidoptera: Noctuidae) larvae in Mississippi associated with sesame interplantings in cotton, 1971-1974: Implications of host-habitat interaction. Environ. Entomol. 11:509-512.

Perrin, R.M. 1975. The role of perennial stinging nettle, Urtica dioica, as a reservoir of beneficial natural enemies. Ann. Appl. Biol. 81:289-297.

Pimentel, D., and A.G. Wheeler. 1973. Influence of alfalfa resistance on a pea aphid population and its associated parasites, predators and competitors. Environ. Entomol. 2:1-11.

Powell, J.E., and L. Lambert. 1984. Effects of three resistant soybean genotypes on development of Microplitis croceipes and leaf consumption by its Heliothis spp. hosts. J. Agr. Entomol. 1:169-176.

Rabb, R.L., and J.R. Bradley. 1968. The influence of host plants on parasitism of eggs of the tobacco hornworm. J. Econ. Entomol. 61:1249-1252.

Rao, P.S.P., P.K. Das and G. Padhi. 1975. Note on compatibility of DD-136 (Neoaplectanoma dutkyi), an insect parasitic nematode with some insecticides and fertilizers. Indian J. Agric. Sci. 45:275-277.

Robinson, R.R., J.H. Young and R.D. Morrison. 1972a. Strip-cropping effects on abundance of Heliothis-damaged cotton squares, boll placement, total bolls and yields in Oklahoma. Environ. Entomol. 1:140-145.

Robinson, R.R., J.H. Young and R.D. Morrison. 1972b. Strip-cropping effects on abundance of predatory and harmful cotton insects in Oklahoma. Environ. Entomol. 1:145-149.

Rust, R.W. 1977. Evaluation of trap crop procedures for control of Mexican bean beetle in soybeans and lima beans. J. Econ. Entomol. 70:630-632.

Schultz, J.C. 1983. Impact of variable plant defensive chemistry on susceptibility of insects to natural enemies, pp. 37-55. In: P.A. Hedin (Ed.), Plant resistance to insects. American Chemical Society, Washington, D.C.

Schuster, D.J., and K.J. Starks. 1975. Preference of Lysiphlebus testaceipes for greenbug resistant and susceptible small grain species. Environ. Entomol. 4: 887-888.

Schuster, M.F. 1967. Response of forage grasses to rhodesgrass scale. J. Range Manage. 20:307-309.

Schuster, M.F. 1980. Insect resistance in cotton, pp. 101-112. In: M.K. Harris (Ed.), Biology and breeding for resistance to arthropods and pathogens in agricultural crops. Texas Agr. Exp. Sta. Misc. Publ. 1451.

Schuster, M.F., J.C. Boling, and J.J. Morony. 1971. Biological control of rhodesgrass scale by airplane releases of an introduced parasite of limited dispersing ability, pp. 227-250. In: C.B. Huffaker (Ed.), Biological control. Plenum Press, N.Y.

Schuster, M.F., and H.A. Dean. 1973. Rhodesgrass scale resistance studies in rhodesgrass. J. Econ. Entomol. 66:467-469.

Schuster, M.F., M.J. Lukefahr and F.G. Maxwell. 1976. Impact of nectariless cotton on plant bugs and natural enemies. J. Econ. Entomol. 69:400-402.

Shelton, M.D., and C.R. Edwards. 1983. Effects of weeds on the diversity and abundance of insects in soybeans. Environ. Entomol. 12:296-298.

Sithanantham, S., V.R. Rao and W. Reed. 1982. The influence of host-plant resistance in chickpea on parasitism of Heliothis armigera Hb. larvae. Intern. Chickpea Newsl. 6(June):21-22.

Smith, R.F., J.L. Apple and D.G. Bottrell. 1976. The origins of integrated pest management concepts for agricultural crops, pp. 1-16. In: J.L. Apple and R.F. Smith (Eds.), Integrated pest management. Plenum Press, N.Y.

Speight, M.R., and J.H. Lawton. 1976. The influence of weed-cover on the mortality imposed on artificial prey by predatory ground beetles in cereal fields. Oecologia 23:211-213.

Sprenkel, R.K., W.M. Brooks, J.W. Van Duyn and L.L. Deitz. 1979. The effects of three cultural variables on the incidence of Nomuraea rileyi phytophagous lepidoptera, and their predators on soybeans. Environ. Entomol. 8:334-339.

Starks, K.J., and I.L. Berry. 1976. Evaluation of sorghum and small grain resistance to green bugs by population stimulations. Environ. Entomol. 5:205-209.

Starks, K.J., R. Muniappan and R.D. Eikenbary. 1972. Interaction between plant resistance and parasitism against the greenbug on barley and sorghum. Ann. Entomol. Soc. Amer. 65:650-655.

Stary, P., and C. Erdelen. 1982. Aphid parasitoids (Hym.: Aphidiidae, Aphelinidae) from the Yemen Arab Republic. Entomophaga 27:105-108.

Stern, V. 1969. Interplanting alfalfa in cotton to control lygus bugs and other insect pests. Proc. Tall Timbers Conf. Ecol. Anim. Control Habitat Mgt. 1:55-69.

Stern, V.M., R. Van den Bosch and T.F. Leigh. 1964. Strip cutting alfalfa for lygus bug control. Calif. Agric. 18(4):4-6.

Stevens, L.M., A.L. Steinhauer and T.C. Elden. 1975. Laboratory rearing of the Mexican bean beetle and the parasite, Pediobius foveolatus, with emphasis on parasite longevity and host-parasite ratios. Environ. Entomol. 4:953-957.

Tingey, W.M. 1981. The environmental control of insects using plant resistance, pp. 175-197. In: D. Pimentel (Ed.), Handbook of pest management in agriculture. Vol. 1. CRC Press, Boca Raton, FL.

Treacy, M.F., J.H. Benedict and J.C. Segers. 1983. Effect of smooth, hirsute and pilose cottons on the functional responses of Trichogramma pretiosum and Chrysopa rufilabris. Proc. Beltwide Cotton Conf. 1983, 3 pp.

van den Bosch, R., and V.M. Stern. 1969. The effects of harvesting practices on insect populations in alfalfa. Proc. Tall Timbers Conf. Ecol. Anim. Control Habitat Mgt. 1:47-54.

van Emden, H.F. 1965. The role of uncultivated land in the biology of crop pests and beneficial insects. Sci. Hortic. 17:121-136.

van Emden, H.F. 1966. Plant insect relationships and pest control. World Rev. Pest Control 5:115-123.

van Emden, H.F., and C.H. Wearing. 1965. The role of the aphid host plant in delaying economic damage levels in crops. Ann. Appl. Biol. 56:323-324.

Watmough, R.H., S.W. Broodryk and D.P. Annecke. 1973. The establishment of two imported parasitoids of potato tuber moth (Phthorimaea operculella) in South Africa. Entomophaga 18:237-249.

Way, M.J., and G. Murdie. 1965. An example of varietal variations in resistance in Brussels sprouts. Ann. Appl. Biol. 56:326-328.

Wilson, F., and C.B. Huffaker. 1976. The philosophy, scope, and importance of biological control, pp. 3-15. In: C.B. Huffaker and P.S. Messenger (Eds.), Theory and practice of biological control. Academic Press, N.Y.

Wiseman, B.R., and W.W. McMillian. 1970. Preference of sorghum midge among selected sorghum lines, with notes on overwintering midges and parasite emergence. USDA Prod. Res. Rep. 122, 8 pp.

Wyatt, I.J. 1970. The distribution of Myzus persicae (Sulz.) on year-round chrysanthemums. II. Winter season: The effect of parasitism by Aphidius matricariae Hal. Ann. Appl. Biol. 65:31-41.

Yanes, J., and D.J. Boethel. 1983. Effect of a resistant soybean genotype on the development of the soybean looper (Lepidoptera: Noctuidae) and an introduced parasitoid, Microplitis demolitor Wilkinson (Hymenoptera: Braconidae). Environ. Entomol. 12:1270-1274.

DISCUSSION: Plant resistance and cultural practice interactions with biological control.

DISCUSSION LEADER: C.S. Barfield

Discussion focused on what is known about the environmental requirements conducive to "effective" biological control. The concensus was that such requirements are known for particular species in a few specific locales. However, very little quantitative information on the biotic/abiotic requisites for effective biological control generally exists. Individuals from different disciplines noted that certain problems, such as diseases with a complex cycle, need scientists from several disciplines to attack the problem effectively, which led to the endorsement of interdisciplinary cooperation in biological control.

Another concensus reached was that biological control could make dramatic progress if a few target problems were selected and scientists from several disciplines worked together to solve them.

Concern was expressed over research which "worked," but which was never implemented. This emphasized again the need for effective communication, education, interdisciplinary research, and effective interaction between scientists and users.

Finally, the point was made that the plant is, ecologically, the "integrator" of plant stresses and that emphasis should be placed on understanding plant growth and response to stresses if we are to evaluate the role of biological control agents quantitatively.

BEHAVIOR-MODIFYING CHEMICALS TO ENHANCE NATURAL ENEMY EFFECTIVENESS

W.J. Lewis and Donald A. Nordlund

In the four trophic level system commonly found in terrestrial communities, there is a complex web of chemically mediated conversation (Price 1981). Plants produce chemicals that: inhibit the growth and development of other plants; attract, repel, or stimulate feeding and oviposition by herbivorous insects; and either attract or stimulate the activities of parasitic and predaceous insects. Herbivorous insects produce semiochemicals that influence other members of the same species, other herbivores, parasitic and predaceous insects, and even hyperparasitoids. Parasitic and predaceous insects produce semiochemicals that influence other parasitoids and predators as well as hyperparasitoids. Thus, even when considering an agricultural system based on only a single crop and only the most obvious and important chemically mediated interactions, the system is exceedingly complex. For example, Heliothis spp. are attacked by a number of parasitoids, many of which respond to chemical stimuli at various stages in their host selection behavior (Table 1).

The chemicals that mediate interactions between organisms are known as semiochemicals (Law & Regnier 1971). For a detailed review of the terminology used in discussions of semiochemicals, see Nordlund (1981) or Nordlund & Lewis (1976). For our purpose it should suffice to say that semiochemicals are divided into pheromones and allelochemics depending on whether the interaction is between members of the same species or members of different species. Pheromones are subdivided on the basis of the type of intraspecific interaction into sex pheromones, alarm pheromones, etc. Allelochemics are further divided into allomones, kairomones, or synomones depending on whether the releaser, receiver, or both releaser and receiver benefit in the interaction. We must stress that these semiochemical categories are not mutually exclusive but dependent on the specific interaction under discussion.

Semiochemicals have powerful biological activity and are often highly specific in nature. Thus, they offer great potential as effective and environmentally sound pest management tools. Most efforts with semiochemicals have centered around the

ISBN 0-12-357030-1

Table 1. Parasitoids of *Heliothis* spp. that have been shown to respond to synomones or kairomones.[a]

Species	Plant Produced Synomones	Kairomones		
		Long Range	Short Range	Oviposition
Cardiochiles nigriceps	X		X	X
Chelonus insularis		X	X	
Cotesia marginiventris	X	X	X	
Microplitis croceipes			X	
Campoletis sonorensis	X		X	
Telenomus heliothidis			X	
Trichogramma pretiosum	X	X	X	X
Trichogramma minutum				X
Trichogramma evanescens			X	
Trichogramma achaeae			X	
Archytas marmoratus				X
Eucelatoria sp.	X		X	

[a] After Nordlund et al. (1985a).

use of sex pheromones in mating disruption, trapping, and population monitoring (Mitchell 1981, Shorey & McKelvey 1977). However, chemically mediated behaviors associated with mating represent only a small fraction of the chemically mediated interactions that occur, even in a simple agroecosystem (Nordlund et al. 1981). There are numerous avenues for the use of semiochemicals in integrated pest management (IPM) programs either directly against the pest or through the enhancement of natural enemies (Greenblatt & Lewis 1983, Lewis & Nordlund 1984, Nordlund et al. 1985a).

In this paper we will review chemically mediated host or prey selection behaviors of entomophagous insects attacking herbivorous insect pests, document the vital role of semiochemicals in this behavior, and discuss how the effectiveness of entomophagous insects can be enhanced in biological control programs through behavioral manipulation.

SEMIOCHEMICAL INVOLVEMENT IN HOST OR PREY SELECTION

The ability of entomophagous insects to locate and recognize hosts or prey is dependent on stimuli from several sources. Though visual and tactical stimuli are often involved, recent studies have clearly demonstrated that semiochemicals generally play a major role in host or prey selection, with the possible exception of the prey selection behaviors of predaceous insects that rely on the ambush technique (Arthur 1981, Greany & Hagen 1981, Vinson 1981, Weseloh 1981). These semiochemicals may originate from plants, the host or prey, and from other entomophagous insects.

Plant-Produced Semiochemicals

Plants play an important role in the host or prey selection behavior of entomophagous insects attacking herbivorous insects. Habitat selection is an important step in the location of food or oviposition sites of most insects (Vinson 1981). Despite our awareness of the importance of plant produced semiochemicals, little research effort was expended in this area until recently, particularly regarding predators.

Herbivorous hosts may be attacked by a particular entomophagous insect on some plant species but not on others (Table 2). This variation in choice of plants is often due to differences in the semiochemicals produced by the various plants (Arthur 1962, Nettles 1980). For example, Nordlund et al. (1985b) found that extracts of tomato could be applied to corn to increase parasitization of *Heliothis zea* eggs by *Trichogramma pretiosum* (Table 3). Elzen et al. (1984a) found that the parasitoid *Campoletis sonorensis* is stimulated to orient to and search the growing terminals, flowers and buds of sorghum, cotton, and tobacco. They also identified several sesquiterpenes from cotton that influence this behavior (Elzen et al. 1984b). The presence or absence of weeds or other associated plants and the vegetation designs of

Table 2. Parasitization of *Heliothis zea* eggs by *Trichogramma* spp. in plots with different plant cultivars.[a]

Plant Cultivars	Mean % parasitization[b]
Tomato	42.9a
Corn-Tomato	33.8ab
Bean-Tomato	28.5bc
Corn-Bean-Tomato	28.5bc
Bean	19.6cd
Corn-Bean	13.4d
Corn	1.5e

[a] After Nordlund et al. (1984)
[b] Means followed by different letters are significantly different ($P < 0.05$) as determined by Duncan's Multiple Range Test.

Table 3. Percent parasitization of *Heliothis zea* eggs, by *Trichogramma pretiosum*: tomato plants treated with corn extract and corn plants treated with tomato extract.[a]

Treatment		Percent Parasitization
	Tomato Treated with Corn Extract	
Treated		53.4a
Control		52.6a
	Corn Treated with Tomato Extract	
Treated		37.7a
Control		28.5b

[a] From Nordlund et al. (1985b). Means, for each test, followed by different letters are significantly different ($P < 0.05$) as determined by ANOVA.

the habitat may influence the abundance of entomophages and the efficacy of their search behavior (Altieri et al. 1981).

Semiochemicals From Host or Prey

The term kairomone was coined by Brown et al. (1970) to describe allelochemics involved in interactions in which the receiver derives benefit. Studies of the host and prey selection behavior of entomophagous insects have revealed that many entomophages rely heavily on kairomones, both for location and acceptance of their hosts or prey (Arthur 1981, Greany & Hagen 1981, Weseloh 1981).

Long range chemoreception As early as the 1930's, researchers were reporting that the behavior of parasitoids was different when the parasitoids were exposed to odors of their host (Jacobi 1939, Murr 1930). In 1953, Ullyett reported that if one breaks open the pupae of a browntailed moth, *Euproctis terminalia*, in a forest, a swarm of *Pimpla bicolor* will appear within minutes. Green lacewing, *Chrysopa carnea*, adults select suitable oviposition sites on the basis of the presence of honey dew from their aphid prey, and the active chemical in the honey dew was found to be an oxidation product of the amino acid tryptophan (Hagen et al. 1976). Researchers have found that various parasitoids and predators are attracted to chemicals that also function as pheromones in their host species (Kennedy 1979; Lanier et al. 1972; Mitchell & Mau 1971; Nordlund et al. 1983; Rice 1968, 1969; Sternlicht 1973). Lewis et al. (1982) found, for example, that

Table 4. Mean percentage parasitization of *Heliothis zea* eggs by naturally occurring *Trichogramma* in field plots treated with Conrel fibers loaded with synthetic *Heliothis zea* sex pheromone.[a]

	Treated	Control
Reading 1 (day 1)	24.4	18.6
Reading 2 (day 2)	45.5	20.2
Means of both readings	35.6	22.6

[a] After Lewis et al. (1981). Means for Reading 1 are not significantly different ($P < 0.05$). Means for Reading 2 and means for both readings significantly different at $P < 0.01$ and $P < 0.001$, respectively.

parasitization of *H. zea* by *Trichogramma* increased when Conrel fibers (Pest-Select International, P.O. Box 278, 26405 West Highway 85, Buckeye, Arizona 85326), loaded with the sex pheromone of *Heliothis zea*, were applied in the field (Table 4). Generally, little is known about the chemical identity of long range semiochemicals, and even less is known about the behavioral mechanisms involved in the long range phase of host or prey selection by entomophages.

Close Range Chemoreception We know a good deal more about close range chemoreception than we know about long range chemoreception by entomophagous insects, primarily because the bioassays are easier to perform. Materials such as larval frass, moth scales, and secretions from accessory, mandibular, and labial glands have been shown to contain kairomones that stimulate host or prey finding or acceptance (Arthur 1981, Greany & Hagen 1981, Weseloh 1981). For example, Lewis et al. (1972) reported that a hexane extract of *H. zea* moth scales stimulated a host seeking response in *Trichogramma evanescens*. In a series of subsequent studies (Beevers et al. 1981; Gross et al. 1975; Lewis et al. 1975a, b, 1979; Nordlund et al. 1976, 1977a), several aspects of this interaction were examined. First, we demonstrated that applications of this extract in the field increased parasitization by *Trichogramma* where host density was high. The reason for the increase in parasitization was the stimulation and continuous reinforcement of an intensified searching behavior, not the direct attraction or guiding of the parasitoid directly to the host. Exposing the parasitoids to the extract for a short time immediately before their release stimulated a host seeking behavior resulting in increased rates of parasitization. In the laboratory, continuous exposure to this extract, in the presence of hosts, results in increased rates of parasitization, number of progeny produced, and longevity of *Trichogramma* females.

H. zea moth scales were found to be at least in part responsible for the higher rate of parasitization of naturally deposited *H. zea* eggs than occurred in naturally deposited *Trichoplusia ni* eggs. We later found that at low host population densities, constant exposure of *Trichogramma* to the moth scale extract was detrimental because movement from one oviposition site to another was not as efficient (since the search behavior elicited was intense). Thus, treatment pattern is very important (Lewis et al. 1979). These findings illustrate the complexity of behavioral responses of entomophages to semiochemicals and the importance of understanding the behavioral mechanisms governing the responses when designing techniques for utilization of semiochemicals in pest management programs.

Lewis et al. (1977) and Nordlund et al. (1977b) found that materials in the scales and apparently in the accessory gland secretion of *H. zea* moths elicited increased predation by *C. carnea* nymphs in petri dish arenas. Thus, it appears that several entomophagous insects respond to chemicals from the same source, if not the same chemicals.

Semiochemicals From Other Entomophagous Insects

As in other organisms (Prokopy 1981), entomophages optimize their distribution and use of resources by employing epideictic pheromones to detect the presence or previous activity of competing individuals of the same or different species. The ability of parasitoids to use epideictic pheromones to discriminate between parasitized and unparasitized hosts has been studied in depth (van Lenteren 1981). Price (1970) demonstrated that several ichneumonid cocoon parasitoids not only avoided their own epideictic pheromones, but recognized those of some congeneric and intergeneric individuals. Salt (1937) showed that *Trichogramma* uses epideictic pheromones on the substrate and directly on the egg to discriminate between healthy and parasitized hosts. He further demonstrated this parasitoid's use of chemoreceptors on its ovipositor to detect previous parasitization.

SEMIOCHEMICALS TO ENHANCE ENTOMOPHAGE ACTIVITY

Approaches for the use of semiochemicals to manipulate the behavior of entomophagous insects in pest control programs have recently been reviewed (Lewis & Nordlund 1985, Nordlund et al. 1985a). They discussed three major areas of potential use: release programs, manipulation of the environment, and rearing programs for beneficial insects.

Use of Semiochemicals in Entomophage Releases

Efficient releases of entomophagous insects (inoculative or inundative) require that a high proportion of the released insects actively search for hosts or prey in the target area. The observed tendency of containerized insects, however, is to disperse upon release. This response often results in an ineffective release. Several techniques to deal with this problem have been suggested including the use of semiochemicals to elicit an intensive search behavior just before or at the time of release (Gross 1981).

Use of Semiochemicals to Manipulate the Environment

Information on semiochemicals from plants (synomones) and from the host or prey (kairomones) may be useful for environmental manipulation to improve the effectiveness of released or indigenous natural enemies.

Synomones Information on the role and mechanisms by which synomones and other plant-produced semiochemicals influence the behavior of entomophagous insects may be used in several ways. It is obvious from demonstrations of the importance of these chemicals on host or prey selection behavior that different varieties of a particular crop may produce varying degrees of activity by the natural enemies. An understanding and consideration of the influences of these varietal changes on entomophages could be incorporated into plant breeding and choice of cultivars to maximize the effectiveness of natural enemy resources.

Since the semiochemical characteristics of plant assemblages are important to the abundance and performance of entomophages, this information could be used in the development of polycultural systems that improve the performance of entomophagous arthropods. Further, weed control and various other cultural practices that provide a habitat more conducive to the utilization of these biological agents could be developed (Altieri 1983).

Another avenue for improving the performance of entomophages is artificial provisioning of these important semiochemicals. Altieri et al. (1981) and Nordlund et al. (1985b) demonstrated that synomones from a preferred host plant could be applied to a less preferred plant to increase the activity of parasitoids. To date, however, there has been little research along these lines.

Kairomones The maintenance of an effective abundance and search activity of entomophagous insects within a habitat is dependent on their encountering kairomones in a sufficient quantity and frequency to indicate a suitably high population of hosts or prey. In many cases, the host or prey density that is suitable for the entomophage exceeds the economically threatening threshold of the producer. In experimental plots we have demonstrated that periodic contact with kairomones extracted from moth scales and impregnated on diatomaceous earth particles could be used to simulate the situation Trichogramma spp. would encounter in a high population of hosts (Lewis et al. 1979). However, proper treatment patterns are crucial when using these close range stimuli if an efficient trade off between retention and search efficiency is to be had (Beevers et al. 1981).

More recently, we discovered that T. pretiosum females respond to the sex pheromone of H. zea, and that field applications of a controlled-release formulation of a synthetic blend of this pheromone significantly increased parasitization by T. pretiosum (Lewis et al. 1982). We expect that the use of these more volatile kairomones will be more effective than the use of the kairomones in the moth scales. Further, this formulation can be used in a multitactic strategy to disrupt mating and enhance parasitization.

CONCLUSION

Chemical communication can be regarded as coded messages with some organisms possessing the key to the code and responding in an appropriate manner, while other organisms remain oblivious. Codes are always in danger of being broken, and as discussed by Price (1981), in evolutionary time, messages may be decoded by enemies. This apparently is the case with parasitic and predaceous insects and their responses to semiochemicals produced by their host or prey.

It is now time for man to decode the messages so that the behavior of beneficial insects can be manipulated to our advantage. The various findings discussed above indicate that manipulation of entomophagous insects is a reasonable goal with a great deal of potential for improving the effectiveness and safety of insect pest control.

Maximizing the effectiveness of natural enemies is the cornerstone of IPM programs. Host and prey selection behaviors have long been recognized as an important factor in the efficacy of entomophagous insects (Salt 1935). The preceeding discussion documents that semiochemicals play major roles in these behaviors. Only through understanding the roles of semiochemicals and the behavioral mechanisms by which entomophages use them will we be able to manipulate entomophagous insects and harness this important biological resource as a full and consistent component of our integrated pest management programs.

RECOMMENDATIONS

The development of technology for integrating the enhancement of natural enemy effectiveness with semiochemicals can best be accomplished with an intense systematic study of the overall foraging behavior of perhaps three or four selected entomophagous insects that are of economic significance and which lend themselves to experimentation.

The following research approaches and objectives are recommended, in order of priority, as research guidelines for the candidate organisms:

1. Develop bioassay techniques for elucidating behavioral responses to volatile semiochemicals and develop monitoring procedures for measuring their behavioral responses to plant and host insect volatiles during foraging among and within habitats.
2. Elucidate the roles and behavioral mechanisms governing insect responses to semiochemicals, determine their relative importance, and isolate and identify the major semiochemicals necessary for retention and efficient foraging of entomophages within target areas.
3. Determine the primary genetic and non-genetic (learning) characteristics influencing entomophage responses to the major semiochemicals.
4. Develop techniques and procedures for integrating knowledge of semiochemicals and their influences on entomophage foraging behavior into agronomic practices, such as plant breeding, cropping patterns, weed control, and irrigation, that will maximize the foraging efficiency of natural and released (imported and native) natural enemies.
5. Develop techniques and procedures for using behavioral modifying chemicals for monitoring and forecasting the density and performance of entomophage populations.

REFERENCES

Altieri, M.A. 1983. Vegetational designs for for insect-habitat management. Environ. Mgt. 7:3-7.

Altieri, M.A., W.J. Lewis, D.A. Nordlund, R.C. Gueldner and J.W. Todd. 1981. Chemical interactions between plants and Trichogramma sp. wasps in Georgia soybean fields. Prot. Ecol. 3:259-263.

Arthur, A.P. 1962. Influence of host tree on abundance of Itoplectis conquisitor (Say), a polyphagous parasite of the European pine shoot moth Rhyacionia buoliana (Schiff). Can. Entomol. 94:337-347.

Arthur, A.P. 1981. Host acceptance by parasitoids, pp. 97-120. In: D.A. Nordlund, R.L. Jones and W.J. Lewis (Eds.), Semiochemicals: Their role in pest control. John Wiley & Sons, N.Y.

Beevers, M., W.J. Lewis, H.R. Gross and D.A. Nordlund. 1981. Kairomones and their use for management of entomophagous insects: X. Laboratory studies on manipulation of host-finding behavior of Trichogramma pretiosum Riley with a kairomone extracted from Heliothis zea (Boddie) moth scales. J. Chem. Ecol. 7:635-648.

Brown, W.L., T. Eisner and R.H. Whittaker. 1970. Allomones and kairomones: Transspecific chemical messengers. BioScience 20:21-22.

Elzen, G.W., H.J. Williams and S.B. Vinson. 1984a. Response by the parasitoid Campoletis sonorensis (Hymenoptera: Ichneumonidae) to chemicals (synomones) in plants: Implications for host habitat location. Environ. Entomol. 13:(In Press).

Elzen, G.W., H.J. Williams and S.B. Vinson. 1984b. Isolation and identification of cotton synomones mediating searching behavior by the parasitoid Campoletis sonorensis. J. Chem. Ecol. 10:(In Press).

Greany, P.D., and K.S. Hagen. 1981. Prey selection, pp. 121-135. In: D.A. Nordlund, R.L. Jones and W.J. Lewis (Eds.), Semiochemicals: Their role in pest control. John Wiley & Sons, N.Y.

Greenblatt, J.A., and W.J. Lewis. 1983. Chemical environmental manipulation for pest insect control. Environ. Mgt. 7:35-41.

Gross, H.R. 1981. Employment of kairomones in the management of parasitoids, pp. 137-150. In: D.A. Nordlund, R.L. Jones and W.J. Lewis (Eds.), Semiochemicals: Their role in pest control. John Wiley & Sons, N.Y.

Gross, H.R., W.J. Lewis, R.L. Jones and D.A. Nordlund. 1975. Kairomones and their use for management of entomophagous insects: III. Stimulation of Trichogramma achaeae, T. pretiosum, and Microplitis croceipes with host-seeking stimuli at time of release to improve their efficiency. J. Chem. Ecol. 1:431-438.

Hagen, K.S., P.D. Greany, E.F. Sawall and R.L. Tassan. 1976. Tryptophane in artificial honeydews as a source of an attractant for adult Chrysopa carnea. Environ. Entomol. 5:458-468.

Jacobi, E.F. 1939. Uber Lebensweise, Auffinden des Wirtes and Regulierung der Individuenzahl von Mormoniella vitripennis Walker. Arch. Neerl. Zool. Leiden 3:197-282.

Kennedy, B.H. 1979. The effect of multilure on parasites of the European elm bark beetle, Scolytus multistriatus. Bull. Entomol. Soc. Amer. 25:116-118.

Lanier, C.N., M.C. Birch, R.F. Schmitz and M.M. Furniss. 1972. Pheromones of Ips pini (Coleoptera: Scolytidae). Variation in response among three populations. Can. Entomol. 104:1923-1973.

Law, J.H., and F.E. Regnier. 1971. Pheromones. Annu. Rev. Biochem. 40:533-548.

Lewis, W.J., M. Beevers, D.A. Nordlund, H.R. Gross, and K.S. Hagen. 1979. Kairomones and their use for management of entomophagous insects: IX. Investigations of various kairomone-treatment patterns for Trichogramma spp. J. Chem. Ecol. 5:673-680.

Lewis, W.J., R.L. Jones, D.A. Nordlund and H.R. Gross. 1975b. Kairomones and their use for management of entomophagous insects: II. Mechanisms causing increase in rate of parasitization by Trichogramma spp. J. Chem. Ecol. 1:349-360.

Lewis, W.J., R.L. Jones, D.A. Nordlund and A.N. Sparks. 1975a. Kairomones and their use for management of entomophagous insects: I. Evaluation for increasing rates of parasitization by Trichogramma spp. in the field. J. Chem. Ecol. 1:343-347.

Lewis, W.J., R.L. Jones and A.N. Sparks. 1972. A host-seeking stimulant for the egg parasite, Trichogramma evanescens: Its source an a demonstration of its laboratory and field activity. Ann. Entomol. Soc. Amer. 65:1087-1089.

Lewis, W.J., and D.A. Nordlund. 1985. Semiochemicals influencing fall armyworm parasitoids behavior: Implications for behavioral manipulation. Fla. Entomol. 68:(In Press).

Lewis, W.J., D.A. Nordlund, H.R. Gross, R.L. Jones and S.L. Jones. 1977. Kairomones and their use for management of entomophagous insects: V. Moth scales as a stimulus for predation of Heliothis zea (Boddie) eggs by Chrysopa carnea Stephens larvae. J. Chem. Ecol. 3:483-487.

Lewis, W.J., D.A. Nordlund and R.C. Gueldner. 1981. Semiochemicals influencing behaviors of entomophages: Roles and strategies for their employment in pest control, pp. 225-242. In: Les Mediateurs Chemiques. Les Colloques de l'INRA, 7. Inst. Nat. Rech. Agron. Paris.

Lewis, W.J., D.A. Nordlund, R.C. Gueldner, P.E.A. Teal and J.H. Tumlinson. 1982. Kairomones and their use for management of entomophagous insects. XIII. Kairomonal activity for Trichogramma spp. of abdominal tips, excretion, and a synthetic sex pheromone blend of Heliothis zea (Boddie) moths. J. Chem. Ecol. 8:1323-1331.

Mitchell, E.R. (Ed.). 1981. Management of insect pests with semiochemicals: Concepts and practice. Plenum Press, N.Y. 514 pp.

Mitchell, W.C., and R.F.L. Mau. 1971. Response of the female southern green stink bug and its parasite, Trichopoda pennipes, to male stink bug pheromones. J. Econ. Entomol. 64:856-859.

Murr, L. 1930. Uber den Geruchsinn der Mehmotienschluphwespe, Habrobracon juglandis zugleich Beitrag zum Orientierungsproblem. Z. Vgl. Physiol. 2:210-270.

Nettles, W. 1980. Adult Eucelatoria sp.: Response to volatiles from cotton and okra plants and from larvae of Heliothis virescens, Spodoptera eridania, and Estigmene acrea. Environ. Entomol. 9:759-763.

Nordlund, D.A. 1981. Semiochemicals: A review of the terminology, pp. 13-28. In: D.A. Nordlund, R.L. Jones and W.J. Lewis (Eds.), Semiochemicals: Their role in pest control. John Wiley & Sons, N.Y.

Nordlund, D.A., R.B. Chalfant and W.J. Lewis. 1984. Arthropod populations, yields, and damage in monoculture and polycultures of corn, beans, and tomatoes. Prot. Ecol. (In Press).

Nordlund, D.A., R.B. Chalfant and W.J. Lewis. 1985b. Response of Trichogramma pretiosum females to extracts of two plants attacked by Heliothis zea. Prot. Ecol. (In Press).

Nordlund, D.A., R.L. Jones and W.J. Lewis (Eds.). 1981. Semiochemicals: Their role in pest control. John Wiley & Sons, N.Y.

Nordlund, D.A., and W.J. Lewis. 1976. Terminology of chemical releasing stimuli in intraspecific and interspecific interactions. J. Chem. Ecol. 2:211-220.

Nordlund, D.A., W.J. Lewis and R.C. Gueldner. 1983. Kairomones and their use for management of entomophagous insects. XIV. Response of Telenomus remus to abdominal tips of Spodoptera frugiperda, (Z)-9-tetradecene-1-ol acetate and (Z)-9-dodecene-1-ol acetate. J. Chem. Ecol. 9:695-701.

Nordlund, D.A., W.J. Lewis, R.L. Jones and H.R. Gross. 1976. Kairomones and their use for management of entomophagous insects. IV. Effect of kairomones on productivity and longevity of Trichogramma pretiosum Riley (Hymenoptera: Trichogrammatidae). J. Chem. Ecol. 2:67-72.

Nordlund, D.A., W.J. Lewis, R.L. Jones, H.R. Gross and K.S. Hagen. 1977b. Kairomones and their use for management of entomophagous insects. VI. An examination of the kairomones for the predator Chrysopa carnea Stephens at the oviposition sites of Heliothis zea (Boddie). J. Chem. Ecol. 3:507-511.

Nordlund, D.A., W.J. Lewis, J.W. Todd and R.B. Chalfant. 1977a. Kairomones and their use for management of entomophagous insects: VII. The involvement of various stimuli in the differential response of Trichogramma pretiosum Riley to two suitable hosts. J. Chem. Ecol. 3:513-518.

Nordlund, D.A., W.J. Lewis, S.B. Vinson and H.R. Gross. 1985a. Behavioral manipulation of entomophagous insects for control of Heliothis spp. In: S. Johnson, E.G. King and J.R. Bradley (Eds.), Theory and tactics of Heliothis population management: I. Cultural and biological control. South. Coop. Ser. Bull. (In Press).

Price, P.W. 1970. Trail odors: Recognition by insects parasitic on cocoons. Science 170:546-547.

Price, P.W. 1981. Semiochemicals in evolutionary time, pp. 251-279. In: D.A. Nordlund, R.L. Jones and W.J. Lewis (Eds.), Semiochemicals: Their role in pest control. John Wiley & Sons, N.Y.

Prokopy, R.J. 1981. Epideictic pheromones that influence the spacing patterns of phytophagous insects, pp. 181-213. In: D.A. Nordlund, R.L. Jones and W.J. Lewis (Eds.), Semiochemicals: Their role in pest control. John Wiley & Sons, N.Y.

Rice, R.E. 1968. Observations on host selection by Tomicobia tibialis Ashmead (Hymenoptera: Pteromalidae). Contrib. Boyce Thompsom Inst. 24:53-56.

Rice, R.E. 1969. Response of some predators and parasites of Ips confusus (Lec.) (Coleoptera: Scolytidae) to olfactory attractants. Contrib. Boyce Thompson Inst. 24:189-194.

Salt, G. 1935. Experimental studies in insect parasitism. III. Host selection. Proc. Roy. Soc. London, Ser. B. 117:413-435.

Salt, G. 1937. Experimental studies in insect parasitism. V. The senses used by Trichogramma to distinguish between parasitized and unparasitized hosts. Proc. Roy. Soc. London, Ser. B. 122:57-75.

Shorey, H.H., and J.J. McKelvey (Eds.). 1977. Chemical control of insect behavior. John Wiley & Sons, N.Y. 414 pp.

Sternlicht, M. 1973. Parasitic wasps attracted by the sex pheromone of their coccid host. Entomophaga 18:339-342.

Ullyett, G.C. 1953. Biomathematics and insect population problems. Entomol. Soc. S. Afr. Mem. 2:1-89.

van Lenteren, J.C. 1981. Host discrimination by parasitoids, pp. 153-179. In: D.A. Nordlund, R.L. Jones and W.J. Lewis (Eds.), Semiochemicals: Their role in pest control. John Wiley & Sons, N.Y.

Vinson, S.B. 1981. Habitat location, pp. 51-77. In: D.A. Nordlund, R.L. Jones and W.J. Lewis (Eds.), Semiochemicals: Their role in pest control. John Wiley & Sons, N.Y.

Weseloh, R.M. 1981. Host location by parasitoids, pp. 79-95. In: D.A. Nordlund, R.L. Jones and W.J. Lewis (Eds.), Semiochemicals: Their role in pest control. John Wiley & Sons, N.Y.

DISCUSSION: Behavior-modifying chemicals to enhance natural enemy effectiveness.

DISCUSSION LEADER: C.S. Barfield

Communication in predators was discussed, as most of the presentation dealt with parasitoids and because it is clear that research on chemical communication in predators has received insufficient attention. It was assumed that the principles governing chemical communication of predators and parasitoids is the same.

Discussion turned to the possibility that other chemicals such as pesticides in the environment might interfere with chemical communication systems of parasitoids/predators. The use of pheromones to monitor for "presence or absence" of pest and/or natural enemy species shows promise but does not currently indicate in-field densities of either pest or natural enemy species. The caution was delivered that we need to develop monitoring methods on the basis of a thorough understanding of how these organisms behave. A certain density in a trap tells nothing about how the population from which these individuals were captured will perform as mortality agents against the pests. The group noted that crop variety (see Herzog & Funderburk, this volume) may influence chemical communication among biological control agents, but not much is currently known of this.

ECOLOGICAL SELECTIVITY: MODIFYING CHEMICAL CONTROL PRACTICES TO PRESERVE NATURAL ENEMIES

Larry A. Hull and E.H. Beers

The problems associated with the indiscriminate use of synthetic organic pesticides was first definitively outlined in the classic paper by Stern et al. (1959). Resistance, secondary pest outbreaks, and especially resurgence of pests, are now the standard teachings of integrated pest management (IPM), and may be found in any introductory text (Anonymous 1969, Metcalf & Luckmann 1982). Too often integrated control, specifically the integration of natural enemies and chemicals, has not been distinguished from classical biological control (Caltagirone 1981). The latter is exemplified by importation and manipulation of (primarily) exotic natural enemies bringing an (usually) introduced pest under complete control, such that no further action is necessary. Few agroecosystems, however, contend with only a single pest, and the pesticides used against a second or third pest destabilize the biological control system. Integrated control emphasizes the complementary action of insecticides and natural enemies in reducing pest numbers. Often the researchers working on classical biological control are not the same as those who evaluate pesticides to be used in the same crop. Integrated control requires that we combine pesticide efficacy tests with biological control research, so that the effects on the pest and its natural enemies are considered as a unified whole. Further, integration of management techniques requires that all pests, natural enemies, pesticides, and cultural practices be considered simultaneously in order for IPM to operate at the grower level.

SELECTIVITY OF PESTICIDES

Selectivity of pesticides has been grouped into two categories by several authors (Bartlett 1964; Metcalf 1972, 1982; Newsom et al. 1976; Watson 1975), viz., physiological and ecological selectivity. Since usage of these terms has varied (see Watson 1975 for review), a discussion of them is presented in this paper. Physiological selectivity is differentiated from ecological selec-

ISBN 0-12-357030-1

tivity in that the former is a characteristic of the compound itself, and the latter is a characteristic, i.e., conscious and knowledgable use, of the person applying it, and refers primarily to materials with a broad spectrum of activity. Physiological selectivity is the property of a compound which discriminates in terms of mortality between two taxa when applied at comparable rates of active ingredient under similar test conditions. It is based on differential toxicity at the site where the chemical interacts with the organism's biochemical processes. Physiological selectivity refers primarily to inherent differences between taxa, but does not exclude resistance acquired from repeated exposure to synthetic chemicals (see Mullin & Croft, this volume). Ecological selectivity is the judicious use of pesticides, based on critical selection, timing, dosage, placement and formulation of broad-spectrum pesticides. Its goal is to maximize pest mortality while minimizing beneficial arthropod mortality, and to alter the predator/prey ratio in favor of the former.

Despite the great promise for truly physiologically selective compounds, most of the materials in use today are broad spectrum in nature. Because of the difficulties in developing physiologically selective pesticides, it seems likely that broad-spectrum pesticides will remain the foundation of chemical tactics for years to come (Metcalf 1980, Newsom et al. 1976). Reason dictates that more attention and effort should be devoted to researching the optimal use of these valuable tools. The widespread vilification of these materials, beginning with the notable Silent Spring, caused a loss of interest in pesticide use research, and delayed the development of rational usage patterns. Negative public opinion ensuing from the anti-pesticide era precipitated a rush to explore non-chemical alternatives, and raised premature, if not false hopes for their potential on a commercial scale. As Metcalf (1982) succinctly points out, pesticides remain our cheapest, fastest, and most reliable curative tool to bring arthropod populations below economic injury levels (EIL). It is our opinion that work on broad-spectrum pesticide use in integrated control programs should not be relegated to the backwaters of scientific research. Such information is extremely vital to the successful development and implementation of IPM programs. This view is in no way meant to denigrate the excellent work done on alternative pest control strategies, nor does it imply endorsement of a return to previous patterns of massive pesticide overuse. Although progress in our understanding of biological, cultural, and genetic control in recent years has been substantial and long overdue, the optimism for pesticide independence should be replaced with realism concerning their fundamental and pivotal role in most extant IPM programs for the foreseeable future.

ROLE OF ECOLOGICAL SELECTIVITY IN AN INTEGRATED PEST MANAGEMENT PROGRAM

Although some success has been achieved in developing and using physiologically selective pesticides in pest management programs (Gruys 1976), the greatest potential for achieving selectivity with today's broad-spectrum pesticides appears to be in the ecological mode (Davis & Hoyt 1979, Watson 1975). The selective use of pesticides (ecological selectivity) can be obtained through such tactics as proper timing, selective placement on the plant or surrounding environment, dosage reduction, and application techniques. Developing information on these tactics can involve much less effort and expense than would be required for the development of physiologically selective chemicals (Glass 1975). However, modifying the use of broad-spectrum pesticides to achieve selectivity will be no easy task. The pest management specialist will be required to possess a greater understanding and knowledge of not only the key pest(s), but also secondary pests and associated natural enemies. Also required will be better assessment of the plant/ pest interaction (see Herzog & Funderburk, this volume), damage/yield relationships, dynamic economic thresholds and injury levels, and appropriate sampling techniques for both pests and natural enemies (see Wilson, this volume). Quantitative information on the interaction of pest/ beneficial organisms and agricultural chemicals and practices would allow precise, timely, and ecologically rational use of pesticides specific to the crop production system at hand.

SELECTION OF A PESTICIDE

Regardless of their spectrum of activity, pesticides are not equal in their toxicity to natural enemies, whether the inequality resulted from serendipity or the development of resistant strains. Pickett (1949) and Pickett & Patterson (1953) were the first to understand this phenomenon on deciduous tree fruits, i.e., that certain insecticides were more compatible with survival of natural enemies than others and should be used preferentially. Lead arsenate, for example, was quite efficacious against the foliage- and fruit-feeding Lepidoptera, but had very little effect on secondary pests (mites, aphids, and leafhoppers) and their natural enemies. Croft & Bode (1983) consider such insecticidal characteristics as an "ideal model" of insecticide selectivity for tree fruit IPM programs.

A priority objective for developing an IPM program is to assess the selectivity of pesticides on the pest/natural enemy complex using detailed laboratory and field experiments (Croft & Brown 1975, Metcalf 1980). Croft & Brown (1975) present an exhaustive review of the toxicities to the predatory mite, *Amblyseius fallacis*, of chemicals commonly applied to fruit crops. In areas across the United States where viable IPM programs exist that utilize natural enemies, researchers are currently spending as much time evaluating new experimental chemicals for their toxicity to natural enemies as for their effectiveness against pests.

The major fruit producing states now publish guidelines which contain information on the toxicity of currently-used pesticides on the pests and natural enemies associated with their respective programs (Table 1). With this information, additional selectivity can be achieved by applying these chemicals at optimal dosage and timing.

Table 1. Insecticide and acaricide efficiency guide for apple pests and natural enemies in Pennsylvania, 1984. (Taken in part from Anonymous 1984b).

	Pests[a]						Natural enemies[b]			
Pesticide	CM	PB	TM	PC	GA	WH	EM	SA	SL	AF
Azinphosmethyl 50WP	2	3	2	2	3	3	-	+	+	+
B. thuringiensis	3	4	2	4	4	4	-	0	0	0
Chlorpyrifos 50WP	1	2	1	2	2	4	-	+	+	+
Cyhexatin 50WP	1	-	-	-	-	-	1	+	+	+++
Dicofol 4F	-	-	-	-	-	-	2	+	+	++
Encapsulated Methyl Parathion 2FM	2	2	1	2	4	4	-	+	+	+
Fenvalerate 2.4EC	1	1	1	2	2	1	-	+++	+++	+++
Methomyl 1.8L	3	2	1	3	1	1	-	++	++	++
Oxamyl 2L	-	2	-	-	3	3	2	+	+	++
Parathion 8EC	2	2	2	3	2	3	-	+	+	-
Permethrin 3.2EC	1	1	1	2	2	1	-	++	+	+++
Phosalone 3E	2	2	2	2	2	2	3	+	+	++
Phosmet 50WP	2	3	2	2	3	4	-	+	+	+

[a] CM = Codling Moth, PB = Plant Bug, TM = Tufted Apple Budmoth, PC = Plum Curculio, GA = Green Apple Aphid, WH = White Apple Leafhopper, EM = European Red Mite. Pest control rating system: 1 = Excellent, 2 = Good, 3 = Fair, 4 = Poor, - = pesticide not rated for this insect or mite.

[b] SA = *Stethorus punctum* adults, SL = *S. punctum* larvae, AF = *Amblyseius fallacis*. Natural enemy rating system: 0 = nontoxic, + = slightly toxic, ++ = moderately toxic, +++ = highly toxic, - = no data available.

TIMING OF TREATMENTS

Newsom et al. (1976) stated that proper timing is often the most effective and economical method of achieving differential insecticide selectivity on the pest/natural enemy complex. Detailed, quantitative information on the biology and behavior of the pests and their natural enemies is required, as well as the use of dynamic economic thresholds on which to base pesticide applications. This information is rapidly being collected, but sound data on which to base management decisions are still lacking (Hoyt & Tanigoshi 1983).

The importance of proper timing has been shown for a number of pest/natural enemy complexes on a variety of crops (Croft 1975, Newsom et al. 1976, Watson 1975). Carbofuran is used to control the alfalfa weevil, *Hypera postica*, but it also kills most other insects in alfalfa at the time of application, including the weevil's parasite, *Bathyplectes curculionis*. If applied early in the spring as alfalfa growth begins, carbofuran causes minimal harm to *B. curculionis* since it is still in the pupal stage (Davis & Hoyt 1979). During the period of intensive pesticide use on deciduous tree fruit (April to mid-June), the predatory mite *A. fallacis* is protected in the ground cover, and materials toxic to this natural enemy may be used. The predator may be preserved by avoiding toxic materials when it is active and in the tree canopy during mid-summer, recommending toxic sprays only after it moves back to the ground cover in late summer (Croft 1975). Weires et al. (1982) recommended a prebloom application of oxamyl on apple against the spotted tentiform leafminer, *Phyllonorycter blandcardella*. This timing allowed greater survival of its parasite, *Apanteles ornigis*, which does not begin to emerge until after this spray. Hull & Starner (1983b) demonstrated that a single, early pesticide application provided control of the rosy apple aphid, *Dysaphis plantaginea*, while not disrupting predation by important natural enemies of mites and aphids, which become active later in the season.

Eliminating treatments of toxic materials during periods of high natural enemy activity presents "windows" for selectively timing sprays (Bartlett 1964). With the recent registration of the synthetic pyrethroid insecticides for tree fruits, much effort has centered on introducing this new class of compounds without harming successful IPM programs (Croft & Hoyt 1978). Selective timing of these materials is one method highly advocated for their introduction (Croft & Hoyt 1978, Hull & Starner 1983a). The leafroller, *Platynota idaeusalis*, is currently the most serious direct pest of apple in Pennsylvania (Hull et al. 1983), and the synthetic pyrethroids are presently the most effective insecticides available for control of this pest (Asquith et al. 1976; L.A. Hull, unpublished data). Season-long applications of these materials are devastating to the predator *Stethorus punctum*, and may cause outbreaks of the European red mite, *Panonychus ulmi*. The critical period to begin control of

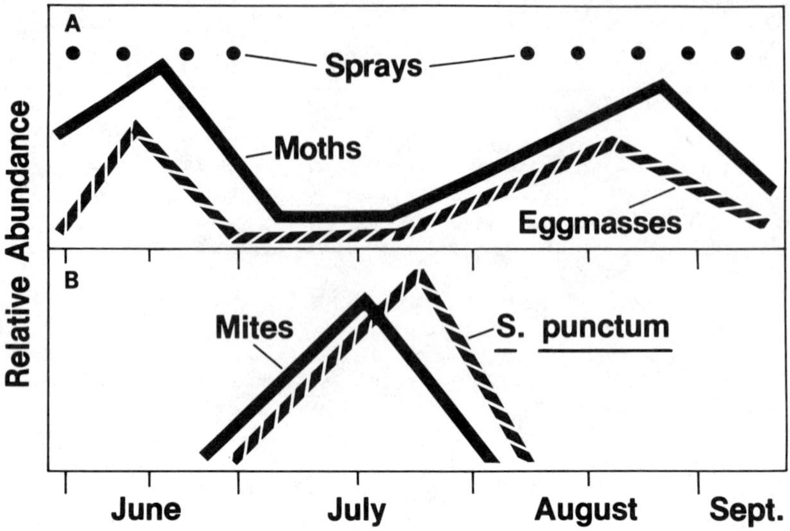

Fig. 1. Synchronizing insecticide applications for controlling the susceptible stages of the leafroller, P. idaeusalis, (A) while minimizing disruption of the typical seasonal interaction between S. punctum and the European red mite (B) in Pennsylvania apple orchards.

the second brood of P. idaeusalis is during early August. Organophosphorus insecticides, which are relatively nontoxic to S. punctum, should be used during July when the predator is rapidly colonizing the tree (Fig. 1B). By waiting until August to apply permethrin, a "window" occurs when the synthetic pyrethroids can be used successfully.

Different life stages of natural enemies may vary in their susceptibility to broad-spectrum pesticides, and these stages offer an opportunity for selective timing. In most holometabolous natural enemies the prepupal and pupal stages are often the most resistant stages. S. punctum pupae were the most tolerant to various organophosphate and carbamate insecticides, followed in order by eggs, adults, and larvae (Colburn 1971).

DOSAGE

The toxicity of any chemical substance is directly related to its dose. An initial step in the development of a successful IPM program is to determine the dosage necessary to prevent an economic loss (Metcalf 1982). At the same time data should be collected on the potential for natural enemies of the pest(s) in question to survive similar dosages. Reduced dosages double the opportunity for reducing pest control costs; directly, through reduced amounts of chemicals, and indirectly, if control exerted by natural enemies obviate the necessity for repeat applications.

Substantial progress has been made in developing IPM programs for soybeans by using the minimal dosage tactic (Newsom 1978). By substituting minimal rates of less persistent organophosphate and carbamate insecticides for the organochlorines, a high degree of selectivity was achieved. The amount of insecticide necessary for the control of soybean pests in 1976 was reduced by one-third to one-half of that recommended in 1973, and valuable populations of natural enemies were conserved (Newsom et al. 1980).

Hoyt (1969) found that azinphosmethyl at reduced rates, down to 50-60% of standard rates, still provided control of the codling moth, Cydia pomonella, on apple while allowing survival of moderate to high numbers of the predatory mite, Typhlodromus occidentalis. Collyer (1976) in New Zealand demonstrated similar selectivity, with reduced dosages of azinphosmethyl allowing sub-stantial survival of Typhlodromus pyri, which rendered treatment for P. ulmi unnecessary. In Pennsylvania, Travis (1977) demonstrated using laboratory assays that the adult, egg, and first two larval instars of P. idaeusalis were quite susceptible to very low doses of the five insecticides most commonly used in commercial orchards. Hull (unpublished data) found that insecticide dosages can be reduced to 75% below label rates and still control P. idaeusalis, as long as sprays are timed to hit these susceptible stages. Fig. 1A, B illustrate the concept of combining minimum rates with careful timing to apply alternate-row middle sprays of insecticides against the susceptible stages of P. idaeusalis, while not interfering with the interaction between S. punctum and the European red mite.

Criteria for inclusion of a new pesticide in an IPM program include: 1) effectiveness against pests; 2) the minimum dosages for keeping the pest(s) below economic injury levels; 3) phytotoxic effects, if any; 4) survival of natural enemies at pest-effective dosages (Hull & Baldwin 1982); and 5) impact on the arthropods that constitute the principal food source(s) for natural enemies (Gonzalez & Wilson 1982). Such investigations must be repeated as new compounds are developed. Evaluations cannot be done mechanically, but require insight, careful planning and observation, and meticulous collection and analysis of data (Asquith & Hull 1979). Precision in research techniques and reproducibility are essential in determining the accuracy of insecticide trials.

After passing the initial replicated, small scale tests where the materials are sprayed to run-off, the pesticide is then evaluated in larger orchard-size trials, where concentrations and application methods are similar to commercial operations (Asquith & Hull 1979). In Pennsylvania, the majority of pesticides are applied in concentrate form in volumes of 468 liters per hectare (Hull et al. 1983). Large-scale trials, where pest pressure better approximates commercial conditions, assist the researcher in better defining the proper dose. For example, fenvalerate and permethrin must be reduced to 5 and 6% of their respective label rates in order to allow substantial survival of S. punctum under commercial conditions when compared to organophosphate insecticides (L.A. Hull, unpublished data). Even at these low dosages, control of the codling moth and P. idaeusalis was only slightly inferior to the control provided by standard organophosphate insecticides (Hull & Starner 1983a).

One possible approach toward improving the level of control provided by synthetic pyrethroids at reduced dosages while still providing selectivity for natural enemies, is to combine them with a reduced dosage of a relatively selective organophosphate insecticide. The addition of azinphosmethyl or encapsulated methyl parathion to fenvalerate and permethrin did not increase mortality of S. punctum, but did significantly improve the level of pest control (Hull & Starner 1983a; L.A. Hull, unpublished data).

The use of insecticide mixtures to control many of the key pests in Pennsylvania (e.g., P. idaeusalis) is a practice commonly recommended to growers (Anonymous 1984b). Georghiou (1980) and Ripper (1956) advocated the employment of mixtures in an integrated control program to reduce the likelihood of pest organisms developing resistance to a single chemical. Others (Brown 1977, Keiding 1977) have refuted this idea, stating that simultaneous resistance will occur to both compounds. However, Tabashnik & Croft (1982) recently demonstrated with computer simulation that a low pesticide use strategy is likely to reduce the potential for pest resistance while maximizing the potential for biological control. Thus, the low dosage/combination approach has the potential for providing effective pest control, preserving natural enemies, and allowing the inclusion in an IPM program of an otherwise ecologically disruptive chemical.

APPLICATION TECHNIQUES AND PLACEMENT

The only purpose for applying pesticides is to kill pests. However, during this process many adjacent areas of ecological significance receive inadvertant spray drift, killing nontarget organisms and contaminating the environment. The method of application, the topography, and the microclimate all play important roles in the dispersal of pesticides outside their target area. For example, Ware et al. (1969) demonstrated that on cotton, aerial applications resulted in four to five times more drift downwind than applications made with conventional high-clearance ground sprayers. Currier et al. (1982) found detectable levels of carbaryl as far as 3.5 km away from apple orchards sprayed by aircraft during a moderate wind (8 to 12 km/h). Metcalf & Luckmann (1982) estimated that only 25-50% of the insecticides applied ever reach the plant surface of most crops, and that less than 1% reach the target pest.

Definition of the Biological Target

If the pesticide could be solely directed at its target (pest), drift contamination could be avoided and dose could be reduced. However, according to Joyce et al. (1977) the definition of the biological target for many agricultural pests is usually ambiguous. They state that "...it is too often assumed that pesticide application to crops is efficient only when the pesticide is evenly distributed over all plant surfaces..." Himel (1974) described this concept of "efficiency" as the most inefficient industrial process ever practiced.

The definition of the biological target is the first important parameter requiring description if pesticides are to be applied in the most efficient and least disruptive manner. Prerequisites for defining this target are knowledge of the pests' behavior and damage, and the pesticides' route of entry (Joyce et al. 1977). For example, Travis (1977) found the eggs of the leafroller, P. idaeusalis, are easily killed when hit by direct sprays of azinphosmethyl. In the orchard the eggs are deposited on the upper leaf surface. Thus, the eggs become the target and not the entire surface of the plant. However, the larvae hatching from the eggs cut the leaf petiole and use the leaf for protection and food. Since the larvae are well protected, the leaf thus constitutes the biological target. The impact of the applied pesticide on the natural enemy complex must also be considered when defining the biological target. After the biological target is described, detailed evaluation work will then allow the target dosage to be selected.

In tree fruit production areas across the United States, significant changes are occuring which will redefine the biological target depending on the pest involved. The fruit industry is rapidly changing from large trees and low numbers per hectare to much smaller trees and increased densities. In light of these changes, including the recent advances in low volume spraying, Hall & Reichard (1978) have advocated the use of the tree-row volume concept, a method of estimating the total foliar target area, for recommending pesticide dosages. Most pesticide recommendations, however, presently provide only one rate regardless of tree size and density. Only recently have some states (Michigan) incorporated the tree-row volume concept into their extension bulletins (Anonymous 1984a). Adjusting the rate for a well-defined target area will allow a reduction in total pesticides used, especially in high-density plantings.

Application Methods

With the target site defined, variables such as concentration, pressure, method of application and spraying system design can be manipulated to achieve a more efficient, less disruptive application process. For example, Croft (1975) recommended a concentrate application technique (2-10X) in apple orchards during the early season. Concentrate spraying reduces the amount of pesticides falling on the ground, in comparison to dilute (applied to runoff) sprays. In this instance, reducing pesticide contact with the ground cover allowed a greater number of A. fallacis to survive.

The method of applying concentrate sprays can also have a profound effect on the survival and potential utility of natural enemies. Lewis & Hickey (1964) demonstrated the benefits of using the alternate-row middle technique vs. both sides spraying for deciduous tree fruits (Fig. 2).

Reduced rates of pesticides may be used because the deposit is renewed approximately twice as often, depending upon pest pressure. It is especially suited for use in an IPM program because the spray coverage on the offside of the trees is comparatively effective at low rates of pesticides and it tends to leave refugia in which natural enemies can develop. S. punctum populations have been shown to increase to a greater degree on trees sprayed with the alternate-row middle system than on those sprayed from both sides. An acaricide application can then selectively change the predator:prey ratio in favor of the predators, thus reducing the need for additional sprays (L.A. Hull, unpublished data). In Pennsylvania, this method of application is currently used in over 95% of all sprays applied to apples (Hull et al. 1983). Ripper (1956) demonstrated similar improvement in the biological control of scale insects attacking California citrus by use of this technique at six month intervals.

Application Technology

Changes now occurring in application technology offer additional opportunities for increasing the effectiveness and selectivity of a pesticide. Barrat et al. (1981) found the ultra-low volume technique deposited at least as much pesticide on leaves as the dilute or low volume method, with less drift, less water, and less chemical than the conventional methods. Electrostatic sprayers generate positively-charged droplets which are attracted to negatively-charged plant surfaces. More droplets impinge on the crop, and the increased coverage allows for reduced pesticide dosages. Herzog et al. (1983) found that comparable levels of insect control and cotton yields could be obtained using one-half the rate of active ingredient. In addition, the reduced drift minimizes the impact of the pesticide on nontarget organisms.

Selective Placement

Restricting a pesticide to a specific plant part or location within a field minimizes its impact on nontarget areas. Watson (1975) applied azinphosmethyl to only the lower two-thirds of cotton plants, allowing predatory Orius species to increase in the tops. Hoyt (1969) demonstrated effective placement by applying the insecticide/fruit thinner carbaryl, which is highly toxic

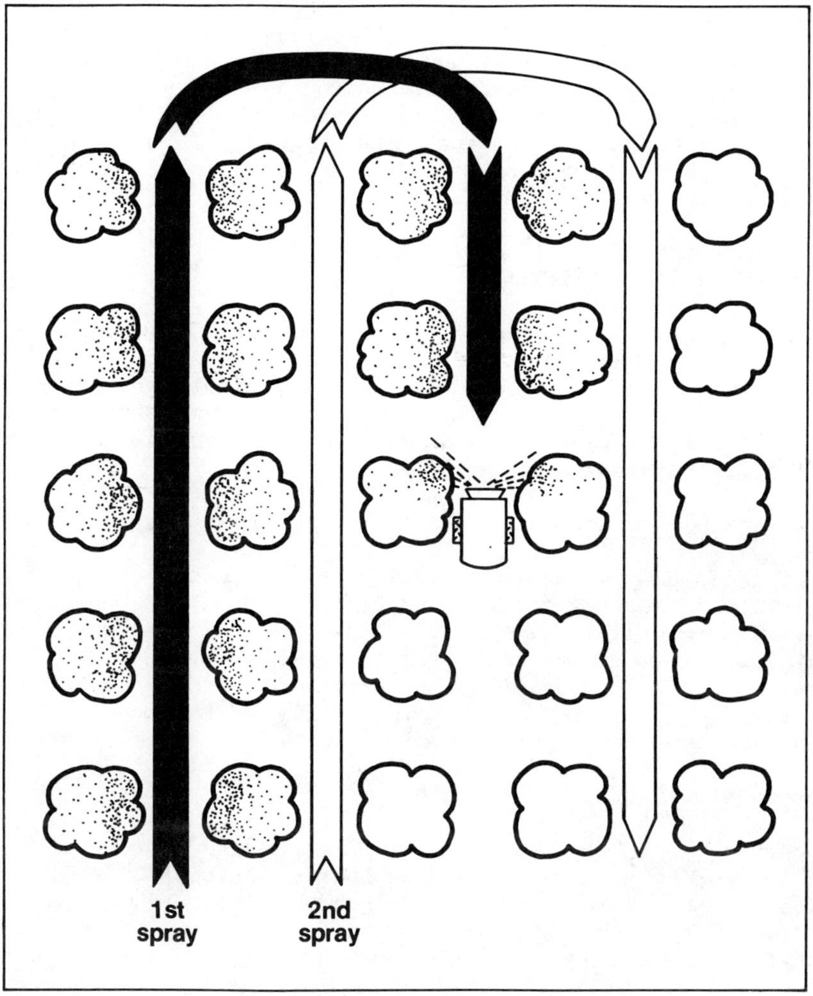

Fig. 2. Diagram of the alternate-row middle technique of applying pesticides and spray distributional pattern throughout the tree in Pennsylvania apple orchards.

to T. occidentalis, to only the upper and peripheral areas of apple trees. This tactic allowed conservation of the predatory mite which is primarily restricted to the inner regions of the tree during the fruit thinning period. Lawson et al. (1961) found that treatment of the top portion of tobacco for tobacco budworm, Heliothis virescens, and hornworms, Protoparce sp., allowed a useful degree of predation by Polistes wasps of lighter hornworm populations in the lower plant parts.

FORMULATION AND PERSISTENCE

Pesticide formulation can have a pronounced influence on its effectiveness and degree of selectivity (Newsom et al. 1976). The pesticide industry has directed much effort toward ensuring compatibility with other chemicals, reduced toxicity to nontarget organisms, and longer residual activity, via formulation improvements (Glass 1975).

Baits and Seed Treatments

Baits and seed treatments typically require a low dosage and are minimally disruptive to the environment. Mirex was used to control imported fire ants, Solenopsis species, in the southern U.S., using a very low rate (4.2 grams AI/ha), and a slow release formulation (Anonymous 1972). The material is unattractive to a majority of nontarget organisms, and therefore much more selective than broadcast applications of heptachlor or dieldrin.

Encapsulation

Ripper et al. (1948) were among the first to successfully coat the particles of a nonselective pesticide with a material in order to limit its contact with non-phytophagous species. For example, the encapsulation of the nonselective insecticide, methyl parathion, has made this material much safer for natural enemies. Asquith et al. (1976) and Hull (1979) found the chemical virtually nontoxic to the mite predators S. punctum and A. fallacis, whereas the standard emulsifiable concentrate formulation at an equivalent rate was highly toxic.

While the microcapsules help to make methyl parathion less dangerous for humans and natural enemies, the potential hazards to honey bees are increased. Burgett & Fischer (1977) demonstrated that foraging honey bees readily become contaminated with the microcapsules which are about the same size as pollen grains (30-50 u). The microcapsules are stored in the pollen pellets and ultimately in the hive where severe brood losses occur. Honey bees are also highly susceptible to other formulations of methyl parathion. However, due to the short residual activity, applications made during the early evening hours prevent unacceptably high mortality levels (Newsom et al. 1976).

Systemics

The exciting discovery of the systemic nature of the insecticide schradan in 1949 uncovered another technique of using insecticides in a selective manner (Ripper 1956). Schradan was rapidly absorbed by the foliage, thus becoming essentially unavailable to most organisms except phytophagous pests. It was highly toxic to a number of plant-sucking pests such as aphids, whiteflies, scale insects, and spider mites, but did not harm their predators, coccinellids and syrphids, nor the parasite *Diaeretus rapae* (Ripper et al. 1951). Lecrone & Smilowitz (1980) found the systemic carbamate insecticide pirimicarb highly toxic to the green peach aphid, *Myzus persicae*, but not harmful to its natural enemies, *Coleomegilla maculata lengi* and *Chrysopa oculata*.

Persistence

Natural enemies can often survive pesticide applications in resistant stages, protected sites or untreated reservoirs, if broad-spectrum short residual chemicals are timed and placed selectively, and no residues are present to prevent colonization (van den Bosch & Stern 1962). The persistence of the chlorinated hydrocarbon insecticides made them especially undesirable in this respect. For example, repeated mist applications of DDT to citrus trees remained toxic to certain parasitic Hymenoptera a full year after sprays ceased (Bartlett 1964). If the limited range of dispersal of the natural enemy does not allow escape from pesticide contact, then reducing residual activity is of no value. For example, Hull et al. (1978) demonstrated how even alternate-row middle applications of methomyl, a highly toxic short-residual carbamate insecticide, eliminated populations of *A. fallacis*, which is limited in its dispersal capabilities. In contrast, the highly mobile *S. punctum* adult population continued to increase, and controlled the European red mites, even though the methomyl sprays reduced the numbers of sedentary larvae. The long-residual synthetic pyrethroid fenvalerate, however, restricted the movement of adult *S. punctum* to sprayed apple trees for at least one month after application due to repellency or toxic effects (L.A. Hull, unpublished data).

EFFECTS OF ECOLOGICALLY SELECTIVE USE OF BROAD-SPECTRUM PESTICIDES ON NATURAL ENEMY RESISTANCE

The use of broad-spectrum pesticides in an ecologically selective manner theoretically can prevent the development of resistance in pest arthropods, but the impact on natural enemies has been little explored. Croft (1982) introduced the concept of "resistance management", which he describes as minimizing resistance in key pests while maximizing possibilities for exploiting resistance in secondary pests and natural enemies.

Tabashnik & Croft (1982), using computer simulation, developed a flow chart decision scheme for devising strategies to retard resistance in pests and promote resistance in natural enemies. They identified two main pesticide use strategies for managing resistance: low pesticide use and high pesticide use. The principle behind the first strategy is to reduce the rate at which susceptible genes are removed from the pest population, thereby reducing the rate of resistance development. The important operational factors are pesticide dose and spray frequency, the number of life stages killed, and pesticide half-life. This strategy is likely to maximize the role of natural enemies.

The principle behind the high pesticide use strategy is to kill pests that are heterozygous for pesticide resistant genes in order to maximize the impact of susceptible immigrants, a prerequisite for delaying resistance (Tabashnik & Croft 1982, Taylor & Georghiou 1979). However, the high dose strategy may accelerate the rate of pest resistance development because it is more likely to suppress resistance in natural enemies (Tabashnik & Croft 1982).

Considering the present economic barriers to developing physiologically selective pesticides and the difficulty in finding broad-spectrum pesticides showing activity for key pests only, resistance management reinforces the goals of ecological selectivity by prolonging the useful life of pesticides as well as promoting stability in agroecosystems.

RECOMMENDATIONS FOR FUTURE RESEARCH

It is evident from the discussion presented here that substantial gaps still exist in our knowledge of the role of natural enemies in the agroecosystem, and how best to modify chemical control practices to preserve them. Our recommendations are to expand and intensify research on the following:

1. **Degree and Relevance of Control Exerted by Natural Enemies** Evaluations of percent parasitism or percent mortality due to parasites/predators are starting points, but are insufficient in themselves, especially if a complex of natural enemies is involved. Life tables for natural enemies and pests alike would be valuable tools for identifying points in time where survivorship of natural enemies is low. Additional research is needed for natural enemies on the effects of food supplies, chemicals, stage susceptibility, and superparasitism. Life tables are also useful for determining life stage vulnerability of pests. If the economic threshold is very low, biological control may not be relevant, i.e., "biologically successful but economically unsuccessful" (Caltagirone 1981).
2. **Improvement of Economic Injury Levels** The chances for successful use of natural enemies are greatly improved by quantifying the EIL, particularly in the case of indirect pests

where the threshold may be higher than was formerly assumed. In order for growers to accept the use of natural enemies, there must be convincing information that their use will not result in economic losses, and that rapid, large-scale sampling will be reliable and effective.

3. **Development of Ecologically Selective Uses of Pesticides** Currently-registered and experimental insecticides should continue to be evaluated for improvement of timing, reduction in dosage, and lethal/sublethal effects on pests and natural enemies. Fungicides, herbicides and plant growth regulators should not be neglected in these evaluations. Scientists should be strongly encouraged to publish their studies in refereed journals, and not bury them in extension bulletins, newsletters and personal communiques. Although this type of research receives little, if any, consideration from large granting agencies, it will probably continue to provide the information upon which present and future IPM programs are built.
4. **Development of New Application Technology** Better descriptions of each biological target in an agroecosystem is the first important requirement. New approaches are needed to restrict pesticides to only certain plant parts. Also, more research, such as that on electrostatic spraying, is needed to minimize pesticide drift onto nontarget areas.
5. **Persistence and Formulation of Pesticides** Additional information is needed on the effects of pesticide degradation on natural enemies, especially in the area of minute quantities affecting susceptible individuals. With an increase in highly concentrated and aerial spraying (ultra-low volume) more research is required to determine the effects of essentially undiluted pesticides on natural enemy populations.
6. **Education and Training** We must continue to encourage, train, and educate individuals in the philosophy and practice of IPM if natural enemies are to be successfully utilized in most crops. We must also continue to educate the grower on the pest management approach to crop protection, especially in external costs and long-term implications of pesticide use.
7. **Interdisciplinary Research** Natural enemy populations are influenced by pesticides and cultural practices that are traditionally the domain of academic disciplines other than entomology. Fungicides, herbicides, ground cover management and plant nutrition are among some of the many factors which affect natural enemies either directly or through pest population dynamics. Scientists, like growers, must begin to approach problem-solving in a holistic manner.

ACKNOWLEDGMENTS

Some of the research reported herein has been financed in part with federal funds from the Environmental Protection Agency under grant CR-806277-020 and the U.S. Department of Agriculture under agreement 82-CSRS-2-1000. Published as article 6934 in the Journal Series of the Pennsylvania Agricultural Experiment Station. Thanks are expressed to M.A. Hoy, J.C. Schultz and R.A.J. Taylor for their critical reviews of the manuscript.

REFERENCES

Anonymous. 1969. Insect-pest management and control. In: Principles of plant and animal pest control, Vol. 3. National Academy of Sciences Publ. 1965, 508 pp. Washington, D.C.

Anonymous. 1972. Report of the Mirex Advisory Committee to William D. Ruckelshaus, Administrator. p. 70. Environmental Protection Agency, Washington, D.C.

Anonymous. 1984a. Fruit pesticide handbook. Michigan State Univ. Ext. Serv. Ext. Bull. E-154, 103 pp.

Anonymous. 1984b. Tree fruit production guide. Pennsylvania State Univ. Coop. Ext. Serv. 80 pp.

Asquith, D., and L.A. Hull. 1979. Integrated pest management systems in Pennsylvania apple orchards, pp. 203-220. In: D.J. Boethel and R.D. Eikenbary (Eds.), Pest management programs for deciduous tree fruits and nuts. Plenum Press, N.Y.

Asquith, D., L.A. Hull, J.W. Travis and P.D. Mowery. 1976. Apple, test of insecticides. 1975. Insectic. Acaric. Tests 1:17-19.

Barrat, R.E., J.L. Mass, H.J. Retzer and R.E. Adams. 1981. Comparison of spray droplet size, pesticide deposition, and drift with ultralow-volume, low-volume, and dilute pesticide application on apple. Plant Disease 65:872-875.

Bartlett, B.R. 1964. Integration of chemical and biological control, pp. 489-511. In: P. DeBach (Ed.), Biological control of insect pests and weeds. Reinhold Publ. Corp., N.Y.

Brown, A.W.A. 1977. Epilogue: Resistance as a factor in pesticide management, pp. 816-824. In: Proc. 15th intern. congr. entomol. Entomol. Soc. Amer., College Park, MD.

Burgett, M., and G. Fischer. 1977. The contamination of foraging honey bees and pollen with Penncap-M. Amer. Bee J. 117:626-627.

Caltagirone, L.E. 1981. Landmark examples in classical biological control. Annu. Rev. Entomol. 26:213-232.

Colburn, R.B. 1971. The predator *Stethorus punctum* and its relationship to *Panonychus ulmi*. Ph.D. thesis. Pennsylvania State Univ., University Park, PA.

Collyer, E. 1976. Integrated control of apple pests in New Zealand. 6. Incidence of European red mite, *Panonychus ulmi* (Koch), and its predators. N. Zeal. J. Zool. 3:39-50.

Croft, B.A. 1975. Integrated control of apple mites. Michigan State Univ. Ext. Serv. Ext. Bull. E-825, 12 pp.

Croft, B.A. 1982. Arthropod resistance to insecticides: A key to pest control failures and successes in North American apple orchards. Entomol. Exp. Appl. 31:88-109.

Croft, B.A., and A.W.A. Brown. 1975. Responses of arthropod natural enemies to insecticides. Annu. Rev. Entomol. 20:285-335.

Croft, B.A., and W.M. Bode. 1983. Tactics for deciduous fruit IPM, pp. 219-270. In: B.A Croft and S.C. Hoyt (Eds.), Integrated management of insect pests of pome and stone fruit. John Wiley & Sons, N.Y.

Croft, B.A., and S.C. Hoyt. 1978. Considerations in the use of pyrethroid insecticides for deciduous fruit pest control in the U.S.A. Environ. Entomol. 7: 627-630.

Currier, W.W., G.B. MacCollom and G.L. Baumann. 1982. Drift residues of air-applied carbaryl in an orchard environment. J. Econ. Entomol. 75: 1062-1068.

Davis, D.W., and S.C. Hoyt. 1979. Selective pesticides, pp. 72-79. In: D.W. Davis, S.C. Hoyt, J.A. McMurtry and M.T. AliNiazee (Eds.), Biological control and insect pest management. Univ. California, Div. Agric. Sci. Publ. 4096.

Gonzalez, D., and L. T. Wilson. 1982. A food-web approach to economic thresholds: A sequence of pests/predaceous arthropods on California cotton. Entomophaga 27:31-43.

Georghiou, G.P. 1980. Insecticide resistance and prospects for its management. Residue Rev. 76:131-145.

Glass, E.H. 1975. Integrated pest management: Rationale, potential, needs and implementation. Entomol. Soc. Amer. Special Publ. 75-2, 126 pp.

Gruys, P. 1976. Development and implementation of an integrated control program for apple orchards in the Netherlands, pp. 823-835. In: Proc. 8th British insectic. fungic. conf., 1975.

Hall, F., and D. Reichard. 1978. Keeping pesticide application on target. Amer. Fruit Grower 98(6):21, 30-31.

Herzog, G.A., W.R. Lambert, S.E. Law, W.E. Seigle, and D.K. Giles. 1983. Evaluation of an electrostatic spray application system for control of insect pests in cotton. J. Econ. Entomol. 76:637-640.

Himel, C.M. 1974. Analytical methodology in ULV. Pesticide application by ULV methods. BCPC Mono. No. 11.

Hoyt, S.C. 1969. Integrated chemical control of insects and biological control of mites on apples in Washington. J. Econ. Entomol. 62:74-86.

Hoyt, S.C., and L.K. Tanigoshi. 1983. Economic injury levels for apple insect and mite pests, pp. 203-218. In: B.A. Croft and S.C. Hoyt (Eds.), Integrated management of insect pests of pome and stone fruit. John Wiley & Sons, N.Y.

Hull, L.A. 1979. Apple, test of insecticides, 1978. Insectic. Acaric. Tests 4: 20-22.

Hull, L.A., D. Asquith and P.D. Mowery. 1978. Integrated control of the European red mite with and without the mite suppressant Dinocap. J. Econ. Entomol. 71: 880-885.

Hull, L.A., and D.M. Baldwin. 1982. Evaluation of insecticides for use in an integrated pest management program for apple. Down to Earth 38:9-11.

Hull, L.A., K.D. Hickey and W.W. Kanour. 1983. Pesticide usage patterns and associated pest damage in commercial apple orchards of Pennsylvania. J. Econ. Entomol. 76:577-583.

Hull, L.A., and V.R. Starner. 1983a. Impact of four synthetic pyrethroids on major natural enemies and pests of apple in Pennsylvania. J. Econ. Entomol. 76:122-130.

Hull, L.A., and V.R. Starner. 1983b. Effectiveness of insecticide applications timed to correspond with the development of rosy apple aphid on apple. J. Econ. Entomol. 76:594-598.

Joyce, R.J.V., Uk Solang, and C.S. Parkin. 1977. Efficiency in pesticide application, pp. 199-216. In: D.L. Watson and A.W.A. Brown (Eds.), Pesticide management and insecticide resistance. Academic Press, N.Y.

Keiding, J. 1977. Resistance in the housefly in Denmark and elsewhere, pp. 261-302. In: D.L. Watson and A.W.A. Brown (Eds.), Pesticide resistance and insecticide management. Academic Press, N.Y.

Lawson, F.R., R.L. Rabb, F.E. Guthrie and T.G. Bowery. 1961. Studies of an integrated control system for hornworms in tobacco. J. Econ. Entomol. 54:93-97.

Lecrone, S., and Z. Smilowitz. 1980. Selective toxicity of pirimicarb, carbaryl and methamidophos to green peach aphid, (*Myzus persicae*) (Sulzer), *Coleomegilla maculata lengi* (Timberlake) and *Chrysopa oculata* Say. Environ. Entomol. 9:752- 755.

Lewis, F.H., and K.D. Hickey. 1964. Pesticide application from one side on deciduous fruit trees. Penn. Fruit News 43:13-24.

Metcalf, R.L. 1972. Selective use of insecticides in pest management, pp. 72-91. In: Implementing practical pest management strategies. Proc. Natl. Ext. Insect-Pest Mgmt. Workshop, March 14-16, Purdue Univ., Lafayette, IN.

Metcalf, R.L. 1980. Changing role of insecticides in crop protection. Annu. Rev. Entomol. 25:219-256.

Metcalf, R.L. 1982. Insecticides in pest management, pp. 217-277. In: R.L. Metcalf and W.H. Luckmann (Eds.), Introduction to insect pest management. John Wiley & Sons, N.Y.

Metcalf, R.L., and W.H. Luckmann. 1982. Introduction to insect pest management. 2nd ed. John Wiley & Sons, N.Y. 577 pp.

Newsom, L.D. 1978. Progress in integrated pest management of soybean pests, pp. 157-180. In: E.H. Smith and D. Pimentel (Eds.), Pest control strategies. Academic Press, N.Y.

Newsom, L.D., R.F. Smith and W. H. Whitcomb. 1976. Selective pesticides and selective use of pesticides, pp. 565-591. In: C.B. Huffaker and P.S. Messenger (Eds.), Theory and practice of biological control. Academic Press, N.Y.

Newsom, L.D., M. Kogan, F.D. Miner, R.L. Rabb, S.G. Turnipseed and W.H. Whitcomb. 1980. General accomplishments toward better pest control in soybeans, pp. 51-98. In: C.B. Huffaker (Ed.), New technology of pest control. John Wiley & Sons, N.Y.

Pickett, A.D. 1949. A critique on insect chemical control methods. Can. Entomol. 81:67-76.

Pickett, A.D., and N.A. Patterson. 1953. The influence of spray programs on the fauna of apple orchards in Nova Scotia IV. A review. Can. Entomol. 85:472-498.

Ripper, W.E. 1956. Effect of pesticides on balance of arthropod populations. Annu. Rev. Entomol. 1:403-438.

Ripper, W.E., R.M. Greenslade and G.S. Hartley. 1951. Selective insecticides and biological control. J. Econ. Entomol. 44:448-459.

Ripper, W.E., R.M. Greenslade, J. Heath, and C.H. Barber. 1948. New formulations of DDT with selective properties. Nature 161:484-485.

Stern, V.M., R.F. Smith, R. van den Bosch and K.S. Hagen. 1959. The integrated control concept. Hilgardia 29:81-101.

Tabashnik, B.E., and B.A. Croft. 1982. Managing pesticide resistance in crop-arthropod complexes: Interactions between biological and operational factors. Environ. Entomol. 11:1137-1144.

Taylor, C.E., and G.P. Georghiou. 1979. Suppression of insecticide resistance by alteration of dominance and migration. J. Econ. Entomol. 72:105-109.

Travis, J.W. 1977. A bioassay study of commercial formulations of several insecticides on Platynota idaeusalis. M.S. thesis, Pennsylvania State Univ., University Park.

Ware, G.W., B.J. Estesen, W.P. Cahill, P.D. Gehardt and K.R. Frost. 1969. Pesticide drift. I. High-clearance vs. aerial application of sprays. J. Econ. Entomol. 62:840-843.

Watson, T.F. 1975. Practical consideration in use of selective insecticides against major crop pests, pp. 47-65. In: J.C. Street (Ed.), Pesticide selectivity. Marcel Dekker, N.Y.

Weires, R.W., J.R. Leeper, W.H. Reissig and S.E. Lienk. 1982. Toxicity of several insecticides to the spotted tentiform leafminer (Lepidoptera: Gracillariidae) and its parasite, Apanteles ornigis. J. Econ. Entomol. 75:680-684.

van den Bosch, R., and V.M. Stern. 1962. The integration of chemical and biological control of arthropod pests. Annu. Rev. Entomol. 7:367-386.

DISCUSSION: Ecological selectivity: Modifying chemical control practices to preserve natural enemies.

DISCUSSION LEADER: C.S. Barfield

Evaluating the effects of pesticides on biological control agents in the field is difficult. One difficulty relates to the size of the field plot; if plots are too small, then movement of natural enemies from untreated refuges can occur, particularly with pesticides with short residual activity, and the impact can appear to be minimal. Detrimental impacts may then become evident only after area-wide usage ensues after pesticide registration.

The group discussed the possibility of legislation requiring information on natural enemies when new pesticides are registered. Advantages and disadvantages of such a requirement were noted; most felt it would be worthwhile and highly desirable for such information to be included, but felt that such information should not be required. The costs of obtaining such information, particularly for small crops, could be high. Some pesticide companies currently provide some of this information, which should be evaluated for inclusion in IPM programs wherever possible.

Information on pest and natural enemy movements is critical in determining how best to utilize pesticides without disruption. Virtually nothing is known quantitatively about such movements. Nor can we provide quantitative data on the level of natural control expected in specific crop systems.

Selective use of pesticides can be achieved, although much remains to be studied. The key point made was that pesticides and natural enemies need not be incompatible. Proper and intelligent usage could allow the incorporation of much more biological control into IPM systems, thereby reducing the need for pesticides to control "secondary" pests. In the short term, learning how to use pesticides selectively would provide a rapid return on research investment. Selectivity through timing, application method, application site, and choice of pesticide offers substantial opportunities for enhancing the use of biological control in IPM systems.

AN UPDATE ON DEVELOPMENT OF SELECTIVE PESTICIDES FAVORING ARTHROPOD NATURAL ENEMIES

Christopher A. Mullin and Brian A. Croft

One tool of integrated pest management (IPM) is effective pesticides that do not impair biological control of pests. Unfortunately, most chemical pesticides used in agriculture are more injurious to natural enemies than the targeted pests. Occasionally chemicals toxic to pest species, but innocuous to beneficial organisms, have been serendipitously discovered by field studies. Industry, with few exceptions (Sechser 1981), does not directly screen promising pesticides on natural enemies, or design selective chemicals with biological control in mind. Thus, identification of selective pesticides is limited to test chemicals chosen solely as toxicants for economic pests and for mammalian safety. Economic restraints greatly limit directed synthesis of selective toxicants. A biorational approach to chemical development has recently received more attention as environmental and human impacts due to pesticide misuse forces cancellation or restricted use of many pesticides. Indeed, selectivity can be economical to industry provided that the chemical is targeted for pests of broad scope, such as prime herbivores on major crops, and yet designed to be safe to a range of biological agents of equal importance such as the major entomophagous arthropods.

Hence a clearer understanding of the physiological and biochemical basis for responses of pests and natural enemies to endogenous and exogenous toxicants is necessary. Aspects of their distinct biologies should provide insight into unique biochemical/ecological mechanisms that will accelerate the design and development of more broad spectrum, selective pesticides for IPM.

This review compares two major arthropod components of agricultural IPM, that of herbivorous pests and their entomophagous natural enemies relative to differences in nutrition, endocrinology, and other physiological attributes; their selective responses to chemicals that affect behavior and fitness, and the ensuing biochemical differences that are the basis for pesticidal selectivities. Extensive reviews on selective toxicology of mammals and arthropods are available elsewhere (Brooks 1978; Hollingworth 1975, 1976; Metcalf 1964; O'Brien 1961, 1967; Winteringham 1969).

ISBN 0-12-357030-1

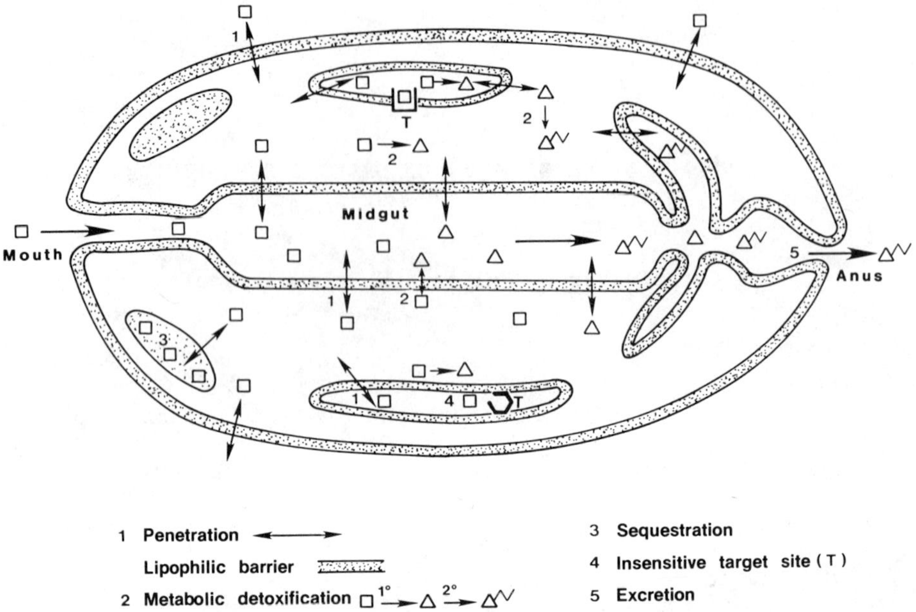

Fig. 1. Dynamics of toxicant selectivity in an animal.

A secure understanding of the major distinctions between arthropodicides and mammalicides requires knowledge of the defensive strategies animals use to survive an impending toxicant. Included among these are avoidance of toxicosis by, for example, sequestering the chemical in insensitive storage tissues such as fat or integument; presenting a penetration barrier such as an integument or internal lipid barriers; rapid excretion of the toxicant; metabolic detoxification of the toxicant; or retaining an insensitive target site for the toxicant. We emphasize that these strategies depicted in Fig. 1 provide the basis of physiological selectivity as coined by Ripper et al. (1951) to describe selectivitiy when the compound is administered directly to the organism. They are contrasted by ecological selectivity where behavior and habitat differences between pest and beneficial arthropods provide the basis of selectivity (Hull & Beers, this volume). The compatibility of chemical pesticides with other natural enemies in agricultural IPM including entomopathogens

(Ignoffo, this volume), entomogenous nematodes (Kaya, this volume), and pathogens of weeds (Charudattan, this volume) is discussed elsewhere.

NUTRITIONAL RESTRAINTS FOR THE HERBIVORE AND CARNIVORE CONDITIONS

Herbivores and carnivores consume foods that are nutritionally different. For example, foods of animal origin are higher in protein and lower in carbohydrate than plant diets. Also, animal-derived foods usually provide well-balanced amino acid levels, and sufficient essential fatty acids and vitamins for a carnivore. It is thus not surprising that carnivorous arthropods have about twice the assimilation efficiency (approximate digestibility) of a comparable herbivore (House 1977, Slansky & Scriber 1982). If "you are what you eat," then the contrasting nutrition of herbivorous pests versus their entomophagous enemies may have ramifications of importance to selective pesticide development, however this has seldom been studied. Food quality depends on digestibility, the kinds and amounts of nutrients available upon digestion, and the ability of these nutrients to meet the nutritional needs of the consumer. Although quantitative aspects of nutrient content differ between animal and plant foods, nutrient requirements of insects appear to be qualitatively similar. The most comprehensive work has been on the dietary requirements of the dipteran parasitoid *Agria housei*, hymenopteran parasitoids *Exeristes roborator* and *Itoplectis conquistor* and the neuropteran predator *Chrysopa carnea* (House 1977, Thompson 1981). In general, immature parasitoids and predators required the same ten essential amino acids, water-soluble vitamins, carbohydrates, nucleic acids, minerals, and a sterol such as cholesterol that immatures of herbivorous insects would require. However, the polyunsaturated fatty acid requirements typical of most phytophagous insects were not apparent in these carnivores. A discussion of the lipid composition of entomophagous arthropods may explain the apparent divergence of fatty acid content between herbivores and carnivores.

More long-chain fatty acids (beyond C_{18}) and higher degrees of unsaturation are typical of the structural fats of mammalian carnivores relative to herbivores (Crawford 1970). In insects, higher degree of polyunsaturated fatty acids is not correlated with increasing carnivory (Fast 1964, 1970; Thompson 1973). Polyunsaturated fatty acids of greater than C_{18} have only recently been studied in insects because of their rarity compared to vertebrates. Nevertheless, $C_{20:4}$ and greater polyunsaturated fatty acids have been documented in several herbivores and in an owlfly, a predaceous neuropteran (Stanley-Samuelson & Dadd 1983). Present data do not indicate any strong differences between these two ecological groups. Perhaps of significance is the generally higher fat content of phytophagous versus entomophagous insects

(Fast 1964). It is well known that higher lipid content can confer resistance to lipophilic insecticides such as DDT because of their sequestering in nonessential fat depots.

Fatty acid composition of dipteran and hymenopteran parasitoids is a duplicate of their host insects (Bracken & Barlow 1967, Thompson & Barlow 1974). For example, braconids, ichneumonids, pteromalids and eulophids contain fatty acids that are remarkably similar qualitatively and quantitatively to their hosts. Exeristes comstockii, a generalist icheumonid ectoparasite, duplicates the diverging fatty acid profiles regardless whether its larval host is dipteran, lepidopteran or hymenopteran. Similarly, the generalist endoparasite Itoplectis conquisitor adopts the fatty acid composition of its lepidopteran host. Initially, it was thought that these two ichneumonids might lack the capability for de novo fatty acid synthesis, but later studies (Jones et al. 1982, Thompson 1981) revealed enzyme pathways for both fatty acid synthesis and turnover in these parasitoids. As with host insects, parasitoids retain only monodesaturases, and cannot synthesize polyunsaturated fatty acids. By comparing the lipid metabolism within parasitic Hymenoptera, and specifically those species not greatly influenced by the fatty acid composition of their host, it appears that regulation of fatty acid content lies in esterification of glycerol phosphate and diglyceride to give phospholipids and triglycerides, respectively. Seemingly many parasitoids are deficient in these fatty acyl transferases or substrate specifity for these complex lipid synthetases has been altered relative to their host insect (Jones et al. 1982). Man and other animals use these same lipid synthetic pathways. It is difficult to envision how these metabolic differences may be exploited to design pesticides that will conserve natural enemies, yet not impact on mammals.

Arthropods require dietary or exogenous sources of sterol for normal growth and development because of their inability to biosynthesize them. Sterols function in insects as membrane constituents and as precursors to molting hormones (ecdysteroids) and many defensive secretions (Kircher 1982). Cholesterol usually meets the sterol needs of insects, but the predominant sterols (sitosterol, stigmasterol and campesterol) within the food of phytophagous insects, must be metabolically converted to cholesterol to satisfy this need. Most phytophagous insects remove the methyl and ethyl groups at C_{24} in plant sterols and convert them to cholesterol; however carnivores lack this capability and require dietary cholesterol (Svoboda et al. 1978). This differential capability to dealkylate sterols is observed within the Coccinellidae which includes phytophagous and the more common carnivorous members. Hence, the Mexican bean beetle, Epilachna varivestis, will dealkylate phytosterols while the entomophage Coccinella septempunctata is deficient in this regard (Svoboda & Robbins 1979, Svoboda et al. 1978). This biochemical selectivity has recently been used to design insecticides for herbivorous pests (Prestwich et al. 1983). Thus, 29-fluorophytosterols undergo lethal dealkylation in phytophagous insects to the highly poisonous fluoroacetate, whereas other entomophages deficient in these enzymes should not be affected. Moreover, vertebrates undergo de novo synthesis of steroids and do not require the ability to dealkylate sterols; therefore these proinsecticides should have low toxicity for mammals and other nontargets.

Other nutritional and physiological differences between natural enemies and their phytophagous hosts should prove useful in design of selective pesticides. For example, dietary triglycerides have little value as an energy source for larvae of some hymenopteran parasitoids (Thompson 1981). Interestingly, Diptera and Hymenoptera, which comprise 90% of the entomophagous parasitoids, rely on carbohydrate reserves such as trehalose rather than lipid as fuel for flight. Many phytophagous pests use lipids for flight energy (Fast 1970). It is not surprising that entomophagous insects, relative to herbivores, retain 3 to 4-fold higher levels of key regulatory enzymes for carbohydrate catabolism, including hexokinase, glycogen phosphorylase and phosphofructokinase (Crabtree & Newsholme 1975). Conceivably, chemicals that interfere with lipid mobilization could have a greater effect on courtship and oviposition fitness of phytophagous pests than on adult parasitoids and predators.

Predaceous insects respire faster than phytophagous prey presumably due to the increased search for prey. Oxygen consumption in a carnivorous mite (Thurling 1980) and a coccinellid (Tanaka & Ito 1982) were about double that of herbivores from the same taxonomic group, yet the carnivore can better survive starvation by markedly reducing its respiratory rate. Entomophages appear to have distinct advantages over phytophages relative to energy allocation. Development of selective toxicants based on these differences have not been exploited perhaps because of belief that mammals and insects are too similar in nutritional physiology and biochemistry to warrant research investigation. Interestingly, one of the most selective miticides available for IPM is cyhexatin. Its high toxicity for all phytophagous mites and lower toxicity for most predaceous mites and insect predators (Croft 1981) may be explained by its suggested mode of action. Cyhexatin inhibits mitochondrial $Mg2^{+}$-dependent ATPase, and at higher concentrations, Na^{+}-K^{+} ATPase (Ahmad & Knowles 1972, Desaiah et al. 1973), and thus functions as a respiratory inhibitor. Based on the discussion above, predaceous mites would be more refractory to respiratory inhibitors than herbivorous mites. Certainly these physiological selectivities can be made commercially tenable.

ENDOCRINE EVENTS CONNOTING SELECTIVE BIOREGULATION OF ARTHROPODS

Arthropods are endocrinologically distinct from vertebrates. The acyclic sesquiterpenoidal juvenile hormones and the steroidal molting hormones (ecdysteroids) which regulate metamorphosis, molting, reproduction, and diapause in insects have no comparable structures in the vertebrates (Coudron et al. 1981, Gilbert et al. 1980, Kircher 1982). Therefore much effort has focused on development of control agents that affect these hormone systems (Bowers 1982, Menn & Henrick 1981, Svoboda et al. 1978). Little is known about divergent hormone physiology and metabolism in herbivores and their natural enemies.

Developmental effects of parasitoids on their lepidopteran hosts include marked alterations in the electrophoretic profiles for hemolymph protein of the host during parasitization (Smilowitz 1973, Vinson & Barras 1970). Selected protein bands absent in parasitized larvae correlate to the arrested larval to pupal ecdysis of the host (Smilowitz & Smith 1977). Data suggest a biochemical basis for endocrine control of the host by the parasitoid. Parasitoids' involvement with the endocrine system of their host has long been documented (Thompson 1983, Vinson & Iwantsch 1980b). A key degradative enzyme regulating larval to pupal molt in Lepidoptera is juvenile hormone esterase, which exhibits an age-activity profile remarkably similar to ecdysis "proteins" noted above (Hammock & Quistad 1981). The possible coincidence of these biochemical events is further indicated by the abnormally depressed juvenile hormone esterase at molt in parasitized insects (Beckage & Riddiford 1982, Thompson 1983). In addition, parasitoids depend on ecdysteroids of their host as initiation cues for development (Baronio & Sehnal 1980).

Insect bioregulators having juvenile hormone action have received extensive study recently because of their safety to nontargets. Juvenile hormone is present in insects and, probably, the Acari, and absent in other arthropods and animals. Hence a large selectivity is possible as exemplified by the 1,700,000-fold higher toxicity of methoprene to cockroaches than to the rat (see Hollingworth 1976). Screening of over 5000 analogs for insect morphogenecity or toxicity has led to commercial development of the juvenile hormone mimics methoprene and kinoprene (Menn & Henrick 1981). Juvenoids of distinct embryogenetic and metamorphogenetic activities for Lepidoptera, Diptera and aphids have been found (Henrick 1982). Selectivities to beneficial and injurious arthropods have mixed effects. Certain diaryl juvenoids are inactive on parasitoids of homopterans (Henrick 1982), yet hydroprene, although of low activity on some hymenopteran parasitoids is deleterious to other dipteran and hymenopteran parasitoids (Smilowitz et al. 1976, Vinson & Iwantsch 1980a) at doses required for pest control. While endocrinological impediment of parasitoids by juvenile hormone mimics could be expected, it is gratifying that relative to their prey these mimics have diminished developmental effects on both insect (Bull et al. 1973) and mite predators (El-Banhawy 1980). Since juvenoids prolong the pestiferous larval or nymphal stages during which feeding is deemed injurious, their primary application has been to control pests in the adult stage, such as mosquitoes and other dipteran pests of mammals.

Because of these limitations, considerable effort has gone into discovering compounds that impede or disrupt juvenile hormone action in insects. Juvenile hormone antagonists or anti-juvenile hormones cause precocious metamorphosis, curtail larval damage, or stop reproduction of most insects that require juvenile hormone for vitellogenesis. The best studied juvenile hormone antagonists include the botanical precocenes from Ageratum sp. (Bowers 1982) and fluoromevalonolactone (Menn & Henrick 1981), which interfere with biosynthesis of juvenile hormone. Their narrow activity spectrum among insect species limits commercial development. Inadequate information is available to predict if these compounds will be selective for herbivores relative to entomophages. An alternative to anti-juvenile hormone effects would be to stimulate juvenile hormone degradation. Two major pathways for juvenile hormone catabolism are ester hydrolysis and epoxide hydration. Preliminary information does not indicate major differences for juvenile hormone degradation between herbivorous and carnivorous arthropods (Ajami & Riddiford 1973, Hammock & Quistad 1981). Also, endocrinological control of insects may arise from the inhibitors of ecdysone 20-monooxygenase, a key enzyme for ecdysteroid degradation, which disrupts larval molting (Koolman 1982).

OTHER PHYSIOLOGICAL AND BEHAVIORAL STRATEGIES FOR SELECTIVE INSECT CONTROL

Endoparasitoids often evade host immune mechanisms by preventing encapsulation. Melanization involving phenoloxidases appears to evoke encapsulation, and it has been suggested that disruption of phenol oxidase activity in the host confers this immunity (Soderhall 1982). Conflicting data, however, indicate that phenol oxidase may not suppress encapsulation in all endoparasitoids (Sroka & Vinson 1978). It is interesting that herbivores consume much larger amounts of phenolics in their food than parasitoids and predators, and that these natural defensive chemicals could be used to circumvent their consumption by natural enemies.

Defensive chemicals are semiochemicals that communicate information between organisms. They may be produced by exocrine glands, or simply sequestered internally so as to not elicit autotoxicity. Many herbivores use protective chemicals to fend off predation; also many predators and parasitoids use similar mechanisms to escape invertebrates and higher animals (Blum 1981, Rodriguez & Levin 1976). Major distinctions between these two groups are lacking, although entomophages seem to biosynthesize a higher percentage of defensive chemicals from acetate or nitrogenous precursors. This may be explained by the high protein and lipid intake of carnivores. Examples include the norsesquiterpenoids and steroids of predaceous aquatic beetles, and the alkaloidal coccinellines in Coccinellidae which are polyacetate derived (Blum 1981, Miller et al. 1975). Alternatively, many defensive toxins are sequestered intact from the food source (see Duffey 1980), and may become autotoxic during stress. For example, a lipid depot of toxin would be mobilized in a time of starvation. This may in general, but not always (see below), have greater consequences for the herbivore than for the natural enemy because most toxins are plant derived.

Other behavior modifying chemicals are pheromones, which primarily mediate communication among individuals of the same species. Hundreds of sex attractant pheromones have been identified for phytophagous species (Baker & Bradshaw 1979, Roelofs 1981, Silverstein 1981); only rudimentary information exists for entomophagous arthropods (Nordlund et al. 1981). Nevertheless, entomophages readily respond to host or prey as well as habitat-derived chemistry (Lewis & Nordlund, this volume). Even inorganic cations such as potassium and magnesium, which occur at high levels in herbivore hemolymph, can be oviposition stimulants for parasitoids (Nettles et al. 1982). A conclusion emerging from these studies is that a given chemical may elicit quite different behavioral responses from a wide variety of species. This might be counterproductive to attempts to develop ecologically selective semiochemicals for control of arthropods.

Research has recently focused on development of broad-spectrum feeding suppressants for herbivores. Most antifeedants come from natural plant sources (Kubo & Nakanishi 1979, Schoonhoven 1982). Azadirachtin, the potent, widely studied antifeedant of the neem tree, prevents feeding in at least 35 economic arthropods (Jacobson 1982). One drawback of many antifeedants is that in order to block consumption of a crop, a full coverage of the host plant including the meristem is required. However more potent compounds are being developed that impair the herbivores' ability to perceive and accept otherwise suitable hosts (Hollingworth 1975, Schoonhoven 1982). Many synthetic pesticides at sublethal dosages deter arthropod feeding, including pyrethroid insecticides (Ross & Brown 1982), formamidine acaricides (Hollingworth 1975), and organotin acaricides and fungicides (Hare et al. 1983). Chlordimeform has anorectic effects at about 0.1% of the acute dosage, and apparently interferes with biogenic amines (Beeman 1982, Beeman & Matsumura 1973). Although feeding deterrents should be selective for herbivorous pests, this may not always be the case. For example, chlordimeform is a more potent acaricide for a predatory mite of the housefly (*Macrocheles muscadomesticae*) than to the spider mite *Tetranychus kanzawai* or cheese mite *Tyrophagus putrescentiae* (Kuwahara 1978).

DIFFERENTIAL RESPONSES TO FOOD AND NUTRIENTS

Host and prey suitability for parasitoids and predators depends on both nutrient and anutrient components. These factors may be phagostimulants or deterrents, and it is the balance of both that determines final acceptability of the food (Bernays & Simpson 1982). Physical attributes of the diet are important (Vinson & Iwantsch 1980a), but perception of key chemical factors is often overriding. Most studies on diet breadth in entomophages are by manipulation of host or prey through controlled alterations in their respective plant food. Interaction studies among three trophic levels have revealed that nutrient deficiencies in the herbivore's diet may seriously effect entomophage development (Zohdy 1976). However, it is clear that plant toxins, as with herbivores (Rosenthal & Janzen 1979), are major disruptors of entomophage fitness (Barbosa et al. 1982, Campbell & Duffey 1981, Price et al. 1980).

Plant secondary chemicals are usually introduced to predators and parasites via passive accumulation in nonessential tissues of the herbivore. This sequestration benefits the herbivore because the phytochemicals are often more toxic to the potential entomophage than to the protected host (Duffey 1980). For example, the alkaloid α-tomatine, acquired by Hyposoter exiguae from Heliothis zea, seriously disrupts the endoparasitoid's development presumably by interfering with dietary sterol utilization (Campbell & Duffey 1981). Similarly, nicotine reduces survival of parasitoids on tobacco-feeding Lepidoptera, including the tobacco horn-worm, Manduca sexta, and fall armyworm, Spodoptera frugiperda (Barbosa et al. 1982, unpublished data). Deleterious effects of plant toxins on natural enemies may present a dilemma to control of herbivorous pests through chemical antibiosis in plants (Herzog & Funderburk, this volume). Alternatively, a variable distribution of defensive chemistry throughout an individual host plant rather than a homogeneous loading may increase searching of a herbivorous pest, and possibly increase random encounters with natural enemies (Schultz 1983). More often investigators have found plant resistance based on antibiosis to be compatible with biological control. Resistant sorghum and small grains to greenbug, Schizaphis graminum, were compatible with parasitism by the braconid Lysiphlebus testaceipes (Starks et al. 1972). These resistances, in part, are due to phenolic and flavonoid factors (Dreyer & Jones 1981, Juneja et al. 1975), and suggests that phenolic defenses against phloem-sucking insects may complement biological controls.

Predaceous insects (Kuo 1977, Pasteels 1978) are also affected by toxic components in their prey. While usually it is thought that the toxins are obtained from the host plant of the prey, this is not necessarily the case. To illustrate, only Megoura viciae, out of five species of aphids raised on broad bean, was found to be toxic to the coccinellid Adalia bipunctata (Blackman 1967), suggesting the toxin was biosynthesized or not detoxified in the aphid.

Clearly, differences based on nutritional, anutritional, hormonal and behavioral chemicals occur between herbivorous pests and their natural enemies. It is not surprising that pesticidal selectivities based on these features also exist. However, multicomponent and tri-trophic interactions may complicate the implementation of selective pesticides. For example, pesticides, and in particular herbicides (Komives & Casida 1983, Williams & James 1983), can increase the levels of toxic chemicals in crops, and may limit use of natural enemies.

PESTICIDE SELECTIVITY THAT FAVORS BIOLOGICAL CONTROL

The compatibility of commercial pesticides with entomophagous arthropods has been reviewed elsewhere (Croft & Brown 1975, Croft & Whalen 1982, Metcalf 1980, Newsom et al. 1976). Susceptibility to many insecticides increases in the order of phytophagous pest to predator to parasite. Among the organophosphates, methyl esters such as trichlorfon and thions (P = S) such as azinphosmethyl are more favorably selective than the respective ethyl esters and oxons (P = O). These better selectivities seem associated with metabolic events involving glutathione transferases and oxidative activation. Pyrethroids such as flucythrinate (Rajakulendran & Plapp 1982) owe their selectivity in part to the carboxylester group. It is fortunate that these chemical features also allow mammalian safety (Casida et al. 1983, Hollingworth 1976).

Metcalf (1982) has devised a pesticide rating system incorporating mammalian acute toxicity, environmental persistence, and nontarget toxicities. Unfortunately, because of data limitation, this assessment only utilizes one beneficial arthropod, the honeybee, which generally is more tolerant to pesticides than parasitic Hymenoptera (Franz et al. 1980). Nevertheless, insecticides with the best pest management ratings often correlate with selectivity for entomophagous arthropods, and include diflubenzuron, the newer pyrethroids, and trichlorfon.

ROLE OF DETOXIFICATION ENZYMES IN CHEMICAL SELECTIVITIES OF PESTS AND THEIR NATURAL ENEMIES

Metabolic transformation of lipophilic toxicants to excretable products usually proceeds by enzymatic events that can be grouped into phase I and II reactions. Phase I involves largely oxidative and hydrolytic reactions in which a polar group is introduced into the toxicant. Phase II enzymes generally conjugate an endogenous substrate such as glutathione or monosaccharides onto the polar group resulting from phase I metabolism (Kulkarni & Hodgson 1980). While these metabolic pathways function to ultimately detoxify toxicants including pesticides and plant allelochemicals to polar and thus excretable products, many of the initial reactions generate products more toxic than the parent xenobiotic. Included among these are selected oxidative reactions catalyzed by the cytochrome P-450 monooxygenases (mixed-function

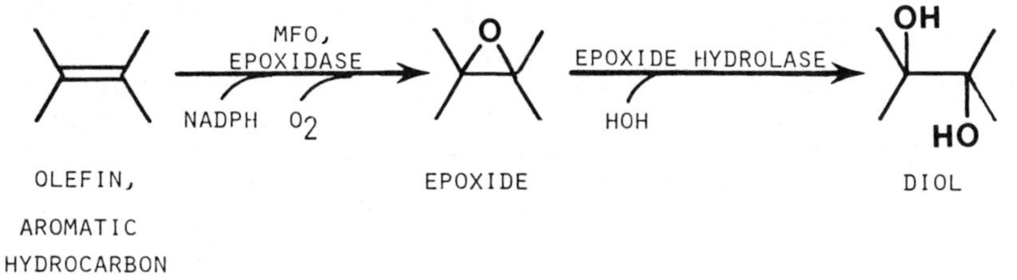

Fig. 2. A toxification-detoxification pathway for olefin and aromatic hydrocarbon metabolism.

oxidases) exemplified by desulfuration at the phosphorous bond (P = S to P = O), thioether oxidations, and epoxidation (Kulkarni & Hodgson 1980). A detoxification reaction working in concert can then dispose of the intermediate toxicant. An example of such a pathway is shown in Fig. 2. Thus the balance of activating or toxification reactions that concurs with degradatory or detoxification reactions determines the net consequence of a xenobiotic to an organism.

Feeding antecedents alter pesticide susceptibilities in herbivorous arthropods, and it was originally suggested that elevated detoxification enzymes due to prolonged exposures to plant toxicants were responsible (Gordon 1961). Biochemical evidence was provided by Krieger et al. (1971) who found increasing aldrin epoxidase (MFO) levels in lepidopteran larvae capable of consuming more plant hosts. The confirmation of MFO as one explanation for host-related tolerances of noctuids to pesticides comes from recent studies where up to 45-fold enhancement of midgut MFO occurs in larvae fed various herbaceous plants relative to a phytochemically benign plant (Brattsten 1979a, Yu 1983, Yu et al. 1979). A concomitant larval increase in insecticide tolerance of up to five-fold was also noted. The association of enzyme induction with plant defensive chemistry was shown through incorporation of purified allelochemicals into artificial diets of moths. These host plant influences on arthropod detoxification capabilities have been extended to hydrolytic enzymes including esterases and epoxide hydrolases (Dowd et al. 1983, Mullin & Croft 1983), and the conjugating enzyme glutathione S-transferase (Yu 1983). Clearly dietary plant allelochemicals may preadapt herbivores to more efficiently detoxify pesticides.

Herbivores may retain high detoxification fitness against chemicals because they must detoxify plant allelochemicals, whereas carnivores probably are less exposed to dietary toxicants and thus may have lower detoxification capabilities. This hypothesis was elaborated to explain the higher susceptibilities of natural enemies to pesticides relative to herbivorous pests. It was suggested that a "natural enemy may be able to express its capacity to adapt to a natural plant toxicant only after the pest had exploited its genetic plasticity in resisting the toxicant" (Croft & Brown 1975). Expression of enhanced detoxification may occur within a generation (induction) or over a longer term (tolerance or resistance development). Support for this hypothesis is provided by Brattsten & Metcalf (1973) where the susceptibilities of a wide variety of arthropod species to carbaryl were compared in the presence and absence of an MFO inhibitor, piperonyl butoxide. This inhibitor synergizes pesticide potency by eliminating a major defensive system, since carbaryl is known to be detoxified by oxidative metabolism. Thus a synergistic ratio (SR), which is the ratio of the LD_{50} for carbaryl in the absence of piperonyl butoxide to that in the presence of the synergist, provides an estimate of the MFO available for detoxification in the poisoned animal. These data were reanalyzed by Croft & Morse (1979) and further adapted here where insects were grouped into phytophagous or entomophagous species. Although a high variation of response within each ecological group is apparent, herbivores (32 species) had a ten times higher mean tolerance to carbaryl ($p < 0.001$) and a two-fold higher availability of MFO than carnivores (17 species). Some intragroup variation may be due to opposing reactions on carbaryl that are catalyzed by various MFO's since both toxification and detoxification pathways are known (Oonnithan & Casida 1968).

Although predators and parasitoids may lack well-developed MFO systems, they may be just as adapted as herbivores to use of nonoxidative detoxification mechanisms. Plapp and coworkers observed that insecticides detoxified largely by oxidative routes were more toxic to entomophagous arthropods than to herbivores, whereas insecticides predominately detoxified hydrolytically (esterases) are safer for carnivores (Plapp 1981, Plapp & Bull 1978).

These preliminary observations led us to examine detoxification enzymes in a herbivorous pest and its natural enemy (Mullin et al. 1982). The broadly polyphagous spider mite _Tetranychus urticae_ and one of its major acarine predators _Amblyseius fallacis_ were chosen as the herbivore-predator model because of similar biologies and morphologies. Oxidative (aldrin epoxidase, MFO), hydrolytic (esterases and epoxide hydrolases), and conjugating (glutatione transferase) enzymes that typify the metabolism of lipophilic toxicants to excretable products were measured. Major differences were found (Table 1). The herbivore had a five-fold higher MFO and a six-fold higher trans-hydrolase than the carnivore. Since both elevated MFO (Brattsten 1979b) and trans-epoxide hydrolase (Mullin & Croft 1984) have been ascribed to encounters with defensive plant chemistry, these results support the allelochemical preadaptation hypothesis. Also, the herbivore and carnivore had similar general esterase activities for α-naphthyl acetate (Table 1) supporting the hypothesis that arthropods should have similarly developed hydrolytic detoxification pathways. This is reasonable since esteratic pathways are

Table 1. Detoxification enzyme levels in a predatory mite *Amblyseius fallacis* and its prey mite *Tetranychus urticae* relative to a braconid ectoparasite *Oncophanes americanus* and its tortricid host *Argyrotaenia citrana*.

Species	Enzyme Activity (pmol/min-mg protein)[a]					
		Epoxide Hydrolase				
	Aldrin Epoxidase	trans	cis	trans/cis	Glutathione Transferase	Esterase (X 10^{-3})
A. fallacis	0.27	310	431	0.72	1095	318
T. urticae	1.44[b]	1713[b]	117[b]	14.6[b]	102[b]	389
O. americanus	0.85	407	727	0.56	135	307
A. citrana	0.29	501	239	2.10[b]	127	411
with midgut	26.8	1340[b]	536	2.50[b]	288	593

[a] Composite values for whole body preparations of both adult and immature stages presented. Data for mites is from Mullin et al. (1982); for *A. citrana*, a value including midgut enzyme levels is also given (Croft & Mullin 1984).

[b] Interspecific difference, herbivore-entomophage -- $p < 0.001$.

important in both detoxification and basal metabolism, and carboxylesterase, lipase, amidase, proteinase, cholinesterase and thioesterase activities may all be incorporated into this general esterase measurement (Heymann 1980).

For certain enzymes a carnivore can have an advantage over a herbivore in potential detoxification capability. Thus *A. fallacis* had 11-fold higher glutathione transferase and four-fold higher cis-hydrolase activities than its prey (Table 1). These enhanced levels may be due to the higher respiration rate of the predator which can lead to autooxidation metabolites that require detoxification by these enzyme systems (Mullin et al. 1982).

These findings on a predator-pest complex led us to examine a representative parasitoid-herbivore complex, that of the polyphagous tortricid *Argyrotaenia citrana* and its braconid ectoparasite *Oncophanes americanus* (Croft & Mullin 1984). Again, MFO and trans-epoxide hydrolase were elevated in the herbivore relative to the entomophage, whereas hydrolytic esterase levels were similar (Table 1), supporting the hypotheses relating detoxification potentials with feeding ecology. In contrast to the predator-prey situation, the glutathione transferase activity was similar between the parasite and its host. This is perhaps related to the different modes of life of the highly mobile predator relative to the more sessile parasitoid, where the parasitoid might require less of a protective enzyme associated with toxicants formed during high oxygen consumption.

A useful comparative index emerges from these studies, that of the ratio of trans- to cis-epoxide hydrolase (EH) activities. The herbivorous prey consistently had a higher trans/cis ratio than the respective entomophage (Table 1). One explanation for this selectivity may be the distribution of epoxides within plant and animal tissue. Plants biosynthesize trans- and higher substituted olefins including fatty acids, phenolics, alkaloids and terpenoids that are rare or absent in animals. Many are allelochemical defenses against herbivory. By contrast, cis-olefins usually have constitutive and homeostatic functions in both plants and animals (Mullin & Croft 1984). Epoxidation of olefins either within the producer or the consumer would produce dietary epoxides requiring detoxification by the EH of appropriate selectivity.

To explore these associations in greater detail, EH in 32 species of macro- and microarthropods were measured (Mullin & Croft 1984). The 18 herbivorous pests of economic importance surveyed here (Table 2) had on the average a 21-fold higher trans-EH and a nine-fold higher trans/cis EH ratio than the seven entomophagous arthropods (Table 3). In addition, a phytophagous coccinellid *Epilachna varivestis* (2.22 ± 0.17) had a much higher EH ratio than the predaceous coccinellids (0.73 ± 0.20) in the survey. These highly significant biochemical differences suggest that EH may be an appropriate biochemical lesion for development of a broad-spectrum bioregulator of herbivorous pests that will have little impact on entomophages. Presumably the crop pest should be more susceptible than its natural enemies to inhibitors of trans-EH.

Other detoxification enzyme selectivities between herbivores and carnivores have been observed. Lacewing (*Chrysopa carnea*) larvae are tolerant of the highly insecticidal cis-isomers of pyrethroids such as deltamethrin and permethrin presumably because they retain esterases that detoxify the cis-isomers faster than the less insecticidal trans-isomers (Ishaaya & Casida 1981). Herbivorous insects, by contrast, only poorly hydrolyze the cis-isomers and preferably detoxify the less insecticidal trans-pyrethroids (Casida et al. 1983, Soderlund et al. 1983). Replacement of the ester functionality by the analogous oxime either produces a highly insecticidal pyrethroid for both *C. carnea* and the coccinellid *Cryptolaemus montrouzeri*, which is also tolerant of pyrethroid esters, and again implicates a key role for esterases within entomophages for pyrethroid detoxification (Brown & Casida 1983). Tolerance of *C. carnea* larvae to the organophosphate trichlorfon relative to a prey, *H. virescens*, appears to be influenced primarily by poor penetration in the former rather than overt differences in metabolism (Bull & Ridg-

Table 2. Epoxide hydrolase in herbivorous pests.

ORDER Species	Epoxide Hydrolase trans	cis	trans/cis
ACARI			
Tetranychus urticae	1.71±0.07	0.12±0.01	14.6 ±0.7
COLEOPTERA			
Diabrotica barberi	38.5 ±7.3	2.59±0.35	14.9 ±3.7
D. virgifera	25.7 ±5.7	1.79±0.33	14.4 ±1.2
D. undecimpunctata howardi	28.0 ±6.6	2.46±0.58	11.4 ±0.6
Acalymma vittata	38.0 ±6.1	4.57±0.52	8.32±0.67
Epicauta pennsylvanica	16.1 ±4.1	4.4 ±1.25	3.67±0.34
Leptinotarsa decemlineata	18.2 ±0.9	8.04±0.11	2.26±0.12
Epilachna varivestis	0.78±0.06	0.35±0.04	2.22±0.17
Crioceris asparagi	2.84±0.84	1.46±0.23	1.94±0.36
Altica woodsi	3.23±0.16	1.89±0.13	1.71±0.20
Plagiodera versicolora	4.00±0.96	7.42±0.96	0.55±0.06
LEPIDOPTERA			
Hypantria cunea	1.80±0.20	0.10±0.20	18.0 ±3.2
Spodoptera frugiperda	5.58±1.11	0.71±0.05	7.85±1.13
Argyrotaenia citrana	3.85±0.08	1.61±0.03	2.52±0.07
Choristoneura rosaceana	1.00±0.22	0.49±0.05	2.06±0.26
Pieris rapae	2.73±0.15	2.18±0.07	1.25±0.09
Malacosoma americanum	6.09±0.30	5.24±0.52	1.16±0.05
DIPTERA			
Delia antiqua	13.0 ±2.1	8.41±1.20	1.55±0.07

[a] Activity in nmol diol formed/min-mg protein for adult (Coleoptera only) or larval midgut preparations except where otherwise indicated; $\bar{x} \pm$ S.E.M. for 3-12 determinations.
[b] Whole body preparations of adult and late instars.

way 1969). Juvenile hormone esterases, oxidases and epoxide hydrolases were much more active in the omnivorous housefly than the carnivorous dipterans *Sarcophaga bullata* and *Phormia regina* (Yu & Terriere 1978). Isopropyl parathion, which is much more toxic to the oriental fruit fly *Dacus dorsalis* than to either hymenopteran parasitoids *Opius longicaudatus* and *O. persulcatus* or the honeybee, *Apis mellifera*, presumably owes its selectivity at least for the honeybee to the overall combination of lowered

Table 3. Epoxide hydrolase in beneficial arthropods.

ORDER Species	Epoxide Hydrolase[a] trans	cis	trans/cis
ACARI			
Amblyseius fallacis[b]	0.31±0.04	0.43±0.04	0.72±0.07
NEUROPTERA			
Chrysopa carnea	0.85±0.27	0.81±0.24	1.05±0.99
COLEOPTERA			
Coccinella novemnotata	0.43±0.04	1.85±0.10	0.23±0.02
Hippodamia convergens	0.77±0.06	1.39±0.11	0.55±0.07
Coleomegilla maculata	0.92±0.33	0.68±0.05	1.35±0.50
HYMENOPTERA			
Pediobius foveolatus[b]	0.20±0.06	0.42±0.10	0.48±0.08
Onocophanes americanus[b]	0.41±0.03	0.73±0.05	0.56±0.07

[a] Activity in nmol diol formed/min-mg protein for adult midgut preparations except where otherwise indicated; $\bar{x}$ ± S.E.M. for six determinations.

[b] Whole body preparations used; for *A. fallacis* and *O. americanus* all life stages analyzed.

oxidative activation and an insensitive cholinesterase in the bee (Metcalf & Frederickson 1965). Since the MFO system of honeybees is unexpectedly efficient relative to other insects (Gilbert & Wilkinson 1974), it may be interesting to determine the relative contribution of activation and detoxification reactions within this pollinator.

TARGET SITE AND PESTICIDE RESISTANCE AS FACTORS FOR SELECTIVITY BETWEEN PESTS AND NATURAL ENEMIES

Comparative studies of target site differences between herbivorous pests and their entomophagous enemies have been limited. Substrate preferences for cholinesterases, the target for organophosphate and carbamate pesticides, showed no distinctive substrate-activity profiles between herbivorous, omnivorous and carnivorous insects, the latter group including a syrphid, blowfly and tiger beetle (Metcalf et al. 1955). Selective toxicity of nea was found to correlate with differences in receptor binding (Chang & Plapp 1983). Most study has concentrated on the mechanisms conferring resistance to pesticides, therefore the long term (over many generations) populational response to pesticide exposure.

There are two major mechanisms by which arthropods become intrinsically resistant to pesticides; enhanced rates of detoxification or decreased sensitivity of the target site (Georghiou & Saito 1983, Plapp 1976). The most intensively studied pest-beneficial complex of arthropods has been the acarines T. urticae and A. fallacis.

Organophosphate resistance in T. urticae is apparently imparted by a combination of the decreased sensitivity of acetylcholinesterase to organophosphate inhibition (Smissaert et al. 1970, Voss & Matsumura 1965), and to enhanced detoxification by either a carboxylesterase (Matsumura & Voss 1965) or by glutathione transferase (Mullin et al. 1982). By contrast, the predator A. fallacis develops organophosphate resistance predominately by increased degradation through the action of glutathione transferases (Motoyama et al. 1971, 1977) and nonspecific esterase (Mullin et al. 1982, Scott et al. 1983). Although esteratic degradation predominates over oxidative pathways in conferring pyrethroid resistance to A. fallacis (Mullin et al. 1982; Scott et al. 1983; Whalon et al. 1982; Mowry et al., unpublished data), MFO is apparently available to the predaceous mite, Metaseiulus occidentalis for developing resistance to carbamates (Roush & Plapp 1982). Another carnivore, the blowfly Crysomya putoria, has reportedly a carboxylesterase involvement in organophosphate resistance (Motoyama & Dauterman 1974).

Permethrin resistance in A. fallacis appears to be due in part to a knockdown factor which in other insects is due to a nerve insensitivity (Scott et al. 1983). Clearly more species of herbivores and carnivores must be compared under the same laboratory conditions wherein a broader scope of plausible mechanisms are analyzed.

The above results agree with the hypothesis of Plapp (1981) who argues that insecticides detoxified by nonoxidative pathways are safer for beneficial arthropods and thus more useful for IPM. Entomophagous arthropods appear well endowed to respond over a long term to pesticide exposure by increasing hydrolytic (esterase) and conjugating (glutathione transferases) enzyme activities. Herbivores and omnivores, on the other hand, rely heavily on oxidative pathways (MFO) for developing resistance to pesti-cides (Georghiou & Saito 1983) including even the newer insect growth regulators represented by the juvenoids (Sparks & Hammock 1983). Thus, the design of insecticides that are primarily detoxified nonoxidatively or activated oxidatively should yield chemicals that allow for selectivity to natural enemies in an agroecosystem. This is possibly why methyl esters of phosphorothionates, the newer synthetic pyrethroids, and diflubenzuron have enjoyed some success in IPM programs.

Little can be said about the role for target site insensitivity because of insufficient study. Nevertheless, enhanced detoxification capability should require less genetic alteration than changes in a pesticide target site. In the field, resistance in entomophages tends to develop only after it occurs in the prey or host because of the carnivore's limited food supply (Croft & Strickler 1983). Since enhanced detoxification enzymes may develop faster than target site alterations, natural enemies may rely more on the former strategy than their prey in order to circumvent chemical isolation due to insufficient food. Release of genetically improved entomophages that are more chemically fit should aid the concurrent implementation of non-selective pesticides and natural enemies (Croft 1982; Hoy 1979, this volume) provided that other fitness traits such as responsiveness to behavioral chemicals (Dawson et al. 1983) are not impaired.

SELECTIVE PEST BIOREGULATORS FOR THE FUTURE

Most pesticides inhibit the action of a biomolecule and thereby interfere with a critical event in an animal's life. With organophosphates and carbamates, detoxification of the neurotransmitter acetylcholine is inhibited, leading to an uncontrolled train of nerve impulses. Pyrethroids and some organochlorines antagonize ion gating across nerve membranes. Benzoylphenyl ureas such as diflubenzuron, and juvenoids interfere respectively with cuticle deposition and JH action. Unfortunately, the most highly successful and preferred pesticides used within the agrichemical community have been the nerve biocides which elicit immediate, tangible results in the field. Discriminating aspects of the infrastructure and neurochemistry of the nervous system within insects, and between insects and mammals is limited. Neurotoxicant selectivity largely results from metabolic differences between organisms, or, more fundamentally, from the obvious differences in morphology which dictate exposure. For example, insects have a large surface to volume ratio compared to mammals, and consequently are more accessible to toxicants applied to the cuticle or ingested with the food.

It seems that more selective bioregulators of pests, that is agents that act to control pest populations, would arise from chemicals that interfere with the organism's ability to deal with dietary toxicants. Useful targets such as the degradatory enzymes MFO and trans-EH, which are generally well developed in herbivorous pests, should be exploited. Inhibitors of these enzymes should disrupt the ability of the herbivore to detoxify plant toxicants, and thereby synergize the efficacy of plant defenses. Presumably entomophages would be less effected since their diet is less laden with natural xenobiotics. Hence synergists that impact on the detoxification capabilities of herbivorous pests could predictably augment host plant resistance and natural enemies, as well as other chemical components of pest control. However, these inhibitors might be slow-acting (delayed) toxicants evoking most likely decreased developmental fitness in populations. Nevertheless, inhibitors so designed will be biorationally directed at the feature that defines the pest (i.e., its feeding on an economic crop).

Insecticide synergists that interfere with MFO detoxification have been intensively studied because of their utility in increasing the efficacy and extending the field life of many nerve toxicants (Wilkinson 1976). However, their high cost has hitherto limited use to the more expensive pesticides such as the natural pyrethrins. While the MFO synergists such as piperonyl butoxide have been long registered for commercial use, their study has been largely devoted to their potentiation of other synthetic toxicants. It is time to investigate the intrinsic activity of these MFO inhibitors on the long term fitness of herbivore populations. Indeed, potent in vivo inhibitors that are selective for one class of cytochrome P-450 monooxygenases are available (Wilkinson et al. 1983), and it is likely that these chemicals will impact on the pest's fitness for dietary toxicants. Alternatively, in vivo inhibitors of trans-epoxide hydrolase should be sought. Perhaps the appropriate analog of available in vitro inhibitors of this enzyme such as chalcone oxides and other flavonoids (Mullin & Hammock 1982) would provide desirable in vivo synergists of plant toxicants.

Alternative slow-acting toxicants can be envisioned including inhibitors of the biosynthesis or degradation of important insect endocrine hormones such as the JH and ecdysteroids. Potent inhibitors of JH detoxification by a specific esterase are known (Hammock et al. 1984), and these thiotrifluoropropanones, if sufficiently active in vivo, should disrupt the feeding of the usually more pestiferous larval stages of crop pests. Anti-JH compounds have been discussed above. One drawback to the more biorational slow-acting toxicants of low mammalian toxicity is that, relative to the fast-acting neurotoxicants, these pest control agents are more likely candidates for carcinogenic activity in mammals. The reason for this is the low acute toxicity of these chemicals allows their chronic testing at higher dosages in mammals for subacute effects including carcinogenesis, mutagenesis and teratogenicity. Subacute effects to mammals even at biologically irrelevant dosages could cancel registration of a potential pest control agent under the existing toxic substances legislation. This factor together with the limited patent life available for innovative bioregulators of pests due to the time required for development and registration tends to retard investment into biorational chemicals (Geissbuhler et al. 1983, Hedin 1982, Menn & Henrick 1981).

Obtaining chemicals that will augment natural enemies can be furthered by incorporating the appropriate chemical features into the candidate molecules. Decreased nonoxidative detoxification and enhanced oxidative activation should be encouraged in the targeted pests. Hydrolytic pathways will not only favor natural enemies, but reduce the environmental persistence of the toxicant since hydrolytic processes are well-developed in the microbial world (Matsumura & Benezet 1978).

RECOMMENDATIONS

1. Prospective pest bioregulators should be routinely screened by industry on a representative group of natural enemy species. A few major agrichemical companies are already promoting this after acceptable mammalian toxicity is attained (Sechser 1981).
2. Research on ascertaining the biological, physiological and biochemical differences between pests and natural enemies should be vigorously pursued. There has been an excessive reliance on experimental arthropod species that are easily maintained in the laboratory, but are irrelevant models for agriculture.
3. Development of slow-acting toxicants that will promote other elements of IPM including plant antibiosis, natural enemies, and concurrently used pesticides, should be encouraged. Potential targets include the enzymes critical for active feeding by the pests on anutrient-laden diets, as well as behavioral targets that determine the ability of a pest to perceive and accept a food of economic importance. The goal will be broadspectrum, selective bioregulators of primary pests that have little impact on beneficial arthropods.
4. The development of resistance by the pest to these delayed toxicants should be impeded by their intrinsically slow action on a fundamental process of life (feeding), the incorporation of molecular features that favor detoxification by nontarget organisms, and by the multifaceted approach to control of the pest.
5. A re-education of the public to adopt chemical bioregulators of pest populations instead of immediate biocides will be necessary. Also, some modification of legislation for toxic substance control will be desirable so that registration of delayed toxicants of low mammalian toxicity would be favored over that of fast-acting neurotoxicants.

ACKNOWLEDGMENTS

This is Paper No. 7017 of the Pennsylvania Agricultural Experiment Station and Paper No. 7249 of the Oregon Agricultural Experiment Station.

REFERENCES

Ahmad, S., and C.O. Knowles. 1972. Biochemical mode of action of tricyclohexylhydroxytin. Comp. Gen. Pharmac. 3:125-133.

Ajami, A.M., and L.M. Riddiford. 1973. Comparative metabolism of the Cecropia juvenile hormone. J. Insect. Physiol. 19:635-645.

Baker, R., and J.W.S. Bradshaw. 1979. Biological chemistry. Part (iii). Insect Chemistry. Ann. Repts. Prog. Chem., London 76B:404-432.

Barbosa, P., J.A. Saunders and M. Waldvogel. 1982. Plant-mediated variation in herbivore suitability and parasitoid fitness, pp. 63-71. In: J.H. Visser and A.K. Minks (Eds.), Proceedings of the 5th international symposium on insect-plant relationships. Pudoc, Wageningen.

Baronio, P., and F. Sehnal. 1980. Dependence of the parasitoid Gonia cinerascens on the hormones of its lepidopterous hosts. J. Insect Physiol. 26:619-626.

Beckage, N.E., and L.M. Riddiford. 1982. Effects of parasitism by Apanteles congregatus on the endocrine physiology of the tobacco hornworm Manduca sexta. Gen. Comp. Endocrinol. 47:308-322.

Beeman, R.W. 1982. Recent advances in mode of action of insecticides. Annu. Rev. Entomol. 27:253-81.

Beeman, R.W., and F. Matsumura. 1973. Chlordimeform: A pesticide acting upon amine regulatory mechanisms. Nature 242:273-274.

Bernays, E.A., and S.J. Simpson. 1982. Control of food intake. Adv. Insect Physiol. 16:59-118.

Blackman, R.L. 1967. The effects of different aphid foods on Adalia bipunctata L. and Coccinella 7-punctata L. Ann. Appl. Biol. 59:207-219.

Blum, M.S. 1981. Chemical defenses of arthropods. Academic Press, N.Y. 562 pp.

Bowers, W.S. 1982. Endocrine strategies for insect control. Entomol. Exp. Appl. 31:3-14.

Bracken, G.K., and J.S. Barlow. 1967. Fatty acid composition of Exeristes comstockii (Cress) reared on different hosts. Can. J. Zool. 45:57-61.

Brattsten, L.B. 1979a. Ecological significance of mixed-function oxidations. Drug. Metab. Rev. 10:35-58.

Brattsten, L.B. 1979b. Biochemical defense mechanisms in herbivores against plant allelochemicals, pp. 199-270. In: G.A. Rosenthal and D.H. Janzen (Eds.), Herbivores, their interaction with secondary plant metabolites. Academic Press, N.Y.

Brattsten, L.B., and R.L. Metcalf. 1973. Synergism of carbaryl toxicity in natural insect populations. J. Econ. Entomol. 66:1347-1348.

Brooks, G.T. 1978. The metabolism of xenobiotics in insects, pp. 151-214. In: J.W. Bridges and L.F. Chasseaud (Eds.), Progress in drug metabolism, Vol. 3. John Wiley & Sons, N.Y.

Brown, M.A., and J.E. Casida. 1983. Oxime ether analogs of pyrethroids help determine the contribution of pyrethroid esterases toward selective toxicity. 185th National Meeting of the American Chemical Society, Seattle, WA.

Bull, D.L., and R.L. Ridgway. 1969. Metabolism of trichlorfon in animals and plants. J. Agric. Food Chem. 17:837-841.

Bull, D.L., R.L. Ridgway, W.E. Buxkemper, M. Schwarz, T.P. McGovern and R. Sarmiento. 1973. Effects of synthetic juvenile hormone analogues on certain injurious and beneficial arthropods associated with cotton. J. Econ. Entomol. 66:623-626.

Campbell, B.C., and S.S. Duffey. 1981. Alleviation of -tomatine-induced toxicity to the parasitoid, Hyposoter exiguae, by phytosterols in the diet of the host, Heliothis zea. J. Chem. Ecol. 7:927-946.

Casida, J.E., D.W. Gammon, A.H. Glickman and L.J. Lawrence. 1983. Mechanisms of selective action of pyrethroid insecticides. Annu. Rev. Pharmacol. Toxicol. 23:413-438.

Chang, C.P., and F.W. Plapp. 1983. DDT and synthetic pyrethroids: Mode of action, selectivity, and mechanism of synergism in the tobacco budworm (Lepidoptera: Noctuidae) and a predator, Chrysopa carnea Stephens (Neuroptera: Chrysopidae). J. Econ. Entomol. 76:1206-1210.

Coudron, T.A., J.H. Law and J.K. Koeppe. 1981. Insect hormones. Trends Biochem. Sci. 6:248-251.

Crabtree, B., and E.A. Newsholme. 1975. Comparative aspects of fuel utilization and metabolism by muscle, pp. 405-500. In: P.N.R. Usherwood (Ed.), Insect muscle. Academic Press, N.Y.

Crawford, M.A. 1970. The progression of long-chain fatty acids from herbivore to carnivore and the evolution of the nervous system. Biochem. J. 119:1-47.

Croft, B.A. 1981. Use of crop protection chemicals for integrated pest control. Phil. Trans. R. Soc. London 295B:125-141.

Croft, B.A. 1982. Arthropod resistance to insecticides: A key to pest control failures and successes in North American apple orchards. Entomol. Exp. Appl. 31:88-110.

Croft, B.A., and A.W.A. Brown. 1975. Responses of arthropod natural enemies to insecticides. Annu. Rev. Entomol. 20:285-335.

Croft, B.A., and J.G. Morse. 1979. Research advances on pesticide resistance in natural enemies. Entomophaga 24:3-11.

Croft, B.A., and C.A. Mullin. 1984. Comparison of detoxification enzyme systems in Argyrotaenia citrana (Lepidoptera: Tortricidae) and the ectoparasite, Oncophanes americanus (Hymenoptera: Braconidae). Environ. Entomol. 13:1330-1335.

Croft, B.A., and K. Strickler. 1983. Natural enemy resistance to pesticides: Documentation, characterization, theory and application, pp. 669-702. In: G.P. Georghiou and T. Saito (Eds.), Pest resistance to pesticides. Plenum Press, N.Y.

Croft, B.A., and M.E. Whalon. 1982. Selective toxicity of pyrethroid insecticides to arthropod natural enemies and pests of agricultural crops. Entomophaga 27: 3-21.

Dawson, G.W., D.C. Griffiths, J.A. Pickett and C.M. Woodcock. 1983. Decreased response to alarm pheromone by insecticide-resistant aphids. Naturwissenschaften 70:254-255.

Desaiah, D., L.K. Cutkomp and R. B. Koch. 1973. Inhibition of spider mite ATPases by plictran and three organochlorine acaricides. Life Sci. 13:1693-1703.

Dowd, P.F., C.M. Smith and T.C. Sparks. 1983. Influence of soybean leaf extracts on ester cleavage in cabbage and soybean loopers (Lepidoptera: Noctuidae). J. Econ. Entomol. 76:700-703.

Dreyer, D.L., and K.C. Jones. 1981. Feeding deterrency of flavonoids and related phenolics towards Schizaphis graminum and Myzus persicae: Aphid feeding deterrents in wheat. Phytochemistry 20:2489-2493.

Duffey, S.S. 1980. Sequestration of plant natural products by insects. Annu. Rev. Entomol. 25:447-477.

El-Banhawy, E.M. 1980. Comparison between the response of the predacious mite Amblyseius brazilli and its prey Tetranychus desertorum to the different IGRs methoprene and dimilin (Acari: Phytoseiidae, Tetranychidae). Acarologia 2-1:221- 227.

Fast, P.G. 1964. Insect lipids: A review. Mem. Entomol. Soc. Can. 37, 50 pp.

Fast, P.G. 1970. Insect lipids. Prog. Chem. Fats Other Lipids 11:181-242.

Franz, J.M., H. Bogenschutz, S.A. Hassan, P. Huang, E. Naton, H. Suter and G. Viggiani. 1980. Results of a joint pesticide test programme by the working group: Pesticides and beneficial arthropods. Entomophaga 25:231-236.

Geissbuhler, H., U. Muller, J.P. Pachlatko and H.R. Waespe. 1983. Biorational design of chemicals, pp. 643-656. In: L.W. Shemilt (Ed.), Chemistry and world food supplies: The new frontiers. Chemrawn II. Pergamon Press, N.Y.

Georghiou, G.P., and T. Saito. 1983. Pest resistance to pesticides. Plenum Press, N.Y. 825 pp.

Gilbert, L.I., W.E. Bollenbacher, W. Goodman, S.L. Smith, N. Agui, N. Granger and B.J. Sedlak. 1980. Hormones controlling insect metamorphosis. Rec. Prog. Horm. Res. 36:401-449.

Gilbert, M.D., and C.F. Wilkinson. 1974. Microsomal oxidases in the honey bee, Apis mellifera (L.). Pestic. Biochem. Physiol. 4:56-66.

Gordon, H.T. 1961. Nutritional factors in insect resistance to chemicals. Annu. Rev. Entomol. 6:27-54.

Hammock, B.D., and G.B. Quistad. 1981. Metabolism and mode of action of juvenile hormone, juvenoids, and other insect growth regulators, pp 1-83. In: D.H. Hutson and T.R. Roberts (Eds.), Progress in pesticide biochemistry, Vol. 1. John Wiley & Sons, N.Y.

Hammock, B.D., Y.A.I. Abdel-Aal, C.A. Mullin, T.N. Hanzlik and R.M. Roe. 1984. Substituted thiotrifluoropropanones as potent selective inhibitors of juvenile hormone esterase. Pestic. Biochem. Physiol. 22:209-223.

Hare, J.D., P.A. Logan and R.J. Wright. 1983. Suppression of Colorado potato beetle, Leptinotarsa decemlineata (Say), (Coleoptera: Chrysomelidae) populations with antifeedant fungicides. Environ. Entomol. 12:1470-1477.

Hedin, P.A. 1982. New concepts and trends in pesticide chemistry. J. Agric. Food Chem. 30:201-215.

Henrick, C.A. 1982. Juvenile hormone analogs: Structure-activity relationships, pp. 315-402. In: J.R. Coats (Ed.), Insecticide mode of action. Academic Press, N.Y.

Heymann, E. 1980. Carboxylesterases and amidases, pp. 291-323. In: W.B. Jakoby (Ed.), Enzymatic basis of detoxication, Vol. 2. Academic Press, N.Y.

Hollingworth, R.M. 1975. Strategies in the design of selective insect toxicants, pp. 67-111. In: J.C. Street (Ed.), Pesticide selectivity. Marcel Dekker, N.Y.

Hollingworth, R.M. 1976. The biochemical and physiological basis of selective toxicity, pp. 431-506. In: C.F. Wilkinson (Ed.), Insecticide biochemistry and physiology. Plenum Press, N.Y.

House, H.L. 1977. Nutrition of natural enemies, pp. 151-182. In: R.L. Ridgway and S.B. Vinson (Eds.), Biological control by augmentation of natural enemies. Plenum Press, N.Y.

Hoy, M.A. 1979. The potential for genetic improvement of predators for pest management programs, pp. 106-115. In: M.A. Hoy and J.J. McKelvey (Eds.), Genetics in relation to insect management. Rockefeller Foundation, N.Y.

Ishaaya, I., and J.E. Casida. 1981. Pyrethroid esterase(s) may contribute to natural pyrethroid tolerance of larvae of the common green lacewing. Environ. Entomol. 10:681-684.

Jacobson, M. 1982. Plant, insects, and man--their interrelationships. Econ. Bot. 36:346-354.

Jones, D., J.S. Barlow and S.N. Thompson. 1982. Exeristes, Itoplectis, Aphaereta, Brachymeria, and Hyposoter species: In vitro glyceride synthesis and regulation of fatty acid composition. Expl. Parasitol. 54:340-351.

Juneja, P.S., S.C. Pearcy, R.K. Gholson, R.L. Burton and K.J. Starks. 1975. Chemical basis for greenbug resistance in small grains. II. Identification of the major neutral metabolite of benzyl alcohol in barley. Plant Physiol. 56: 385-389.

Kircher, H.W. 1982. Sterols and insects, pp. 1-50. In: J. Dupont (Ed.), Cholesterol systems in insects and animals. CRC Press, Boca Raton, FL.

Komives, T., and J.E. Casida. 1983. Acifluorfen increases the leaf content of phytoalexins and stress metabolites in several crops. J. Agric. Food Chem. 3-1: 751-755.

Koolman, J. 1982. Ecdysone metabolism. Insect Biochem. 12:225-250.

Krieger, R.I., P.P. Feeny and C.F. Wilkinson. 1971. Detoxication enzymes in the guts of caterpillars: An evolutionary answer to plant defenses? Science 172: 579-581.

Kubo, I., and K. Nakanishi. 1979. Some terpenoid insect antifeedants from tropical plants, pp. 284-294. In: H. Geissbuhler (Ed.), Advances in pesticides science, Part 2. Pergamon Press, N.Y.

Kulkarni, A.P., and E. Hodgson. 1980. Metabolism of insecticides by mixed-function oxidase systems. Pharmac. Ther. 8:379-475.

Kuo, H.L. 1977. Auswirkungen zweier wirtspflanzen von Myzus persicae (Sulz.) auf den rauberischen blattlausfeind Aphidoletes aphidimyza (Rond.) (Diptera: Cecidomyiidae). Z. Angew. Entomol. 82:229-233.

Kuwahara, M. 1978. Toxicity of chlordimeform and its analogues and MFO-inhibitors to three species of mites, and improvement of toxicity by certain synergists. Appl. Entomol. Zool. 13:296-303.

Matsumura, F., and H.J. Benezet. 1978. Microbial degradation of insecticides, pp. 623-667. In: I.R. Hill and S.J.L. Wright (Eds.), Pesticide microbiology. Academic Press, N.Y.

Matsumura, F., and G. Voss. 1965. Properties of partially purified malathion carboxyesterase of the two-spotted spider mite. J. Insect Physiol. 11:147-160.

Menn, J.J., and C.A. Henrick. 1981. Rational and biorational design of pesticides. Phil. Trans. R. Soc. London 295B:57-71.

Metcalf, R.L. 1964. Selective toxicity of insecticides. World Rev. Pest Control 3:28-43.

Metcalf, R.L. 1980. Changing role of insecticides in crop protection. Annu. Rev. Entomol. 25:219-256.

Metcalf, R.L. 1982. Insecticides in pest management, pp. 217-277. In: R.L. Metcalf and W.H. Luckmann (Eds.), Introduction to insect pest management. John Wiley & Sons, N.Y.

Metcalf, R.L., and M. Frederickson. 1965. Selective insecticidal action of isopropyl parathion and analogues. J. Econ. Entomol. 58:143-147.

Metcalf, R.L., R.B. March and M.G. Maxon. 1955. Substrate preferences of insect cholinesterases. Ann. Entomol. Soc. Amer. 48:222-228.

Miller, J.R., L.B. Hendry and R.O. Mumma. 1975. Norsesquiterpenes as defensive toxins of whirligig beetles (Coleoptera: Gyrinidae). J. Chem. Ecol. 1:59-82.

Motoyama, N., and W.C. Dauterman. 1974. The role of nonoxidative metabolism in organophosphorus resistance. J. Agric. Food Chem. 22:350-356.

Motoyama, N., W.C. Dauterman and G.C. Rock. 1977. Toxicity of O-alkyl analogues of azinphosmethyl and other insecticides to resistant and susceptible predaceous mites, Amblyseius fallacis. J. Econ. Entomol. 70:475-476.

Motoyama, N., G.C. Rock, and W.C. Dauterman. 1971. Studies on the mechanism of azinphosmethyl resistance in the predaceous mite, Neoseiulus (T.) fallacis (Family: Phytoseiidae). Pestic. Biochem. Physiol. 1:205-215.

Mullin, C.A., and B.A. Croft. 1983. Host-related alterations of detoxification enzymes in Tetranychus urticae (Acari: Tetranychidae). Environ. Entomol. 12: 1278-1282.

Mullin, C.A., and B.A. Croft. 1984. Trans-epoxide hydrolase: A key indicator enzyme for herbivory in arthropods. Experientia 40:176-178.

Mullin, C.A., and B.D. Hammock. 1982. Chalcone oxides-potent selective inhibitors of cytosolic epoxide hydrolase. Arch. Biochem. Biophys. 216: 423-439.

Mullin, C.A., B.A. Croft, K. Strickler, F. Matsumura and J.R. Miller. 1982. Detoxification enzyme differences between a herbivorous and predatory mite. Science 217:1270-1272.

Nettles, W.C., R.K. Morrison, Z.N. Xie, D. Ball, C.A. Shenkir and S.B. Vinson. 1982. Synergistic action of potassium chloride and magnesium sulfate on parasitoid wasp oviposition. Science 218:164-166.

Newsom, L.D., R.F. Smith and W.H. Whitcomb. 1976. Selective pesticides and selective use of pesticides, pp. 565-591. In: C.B. Huffaker and P.S. Messenger (Eds.), Theory and practice of biological control. Academic Press, N.Y.

Nordlund, D.A., R.L. Jones and W.J. Lewis (Eds.). 1981. Semiochemicals--their role in pest control. John Wiley & Sons, N.Y. 306 pp.

O'Brien, R.D. 1961. Selective toxicity of insecticides. Adv. Pest Control Res. 4:75-116.

O'Brien, R.D. 1967. Insecticides: Action and metabolism. Academic Press, N.Y. 332 pp.

Oonnithan, E.S., and J.E. Casida. 1968. Oxidation of methyl- and dimethyl carbamate insecticide chemicals by microsomal enzymes and anticholinesterase activity of the metabolites. J. Agric. Food Chem. 16:28-44.

Pasteels, J.M. 1978. Apterous and brachypterous coccinellids at the end of the food chain, Cionura erecta (Aslepiadaceae) - Aphis nerii. Entomol. Exp. Appl. 24:379-384.

Plapp, F.W. 1976. Biochemical genetics of insecticide resistance. Annu. Rev. Entomol. 21:179-197.

Plapp, F.W. 1981. Ways and means of avoiding or ameliorating resistance to insecticides, pp. 244-249. In: T. Kommedahl (Ed.), Proceedings of symposia IX international congress of plant protection, Vol. 1. Burgess, Minneapolis, MN.

Plapp, F.W., and D.L. Bull. 1978. Toxicity and selectivity of some insecticides to Chrysopa carnea, a predator of the tobacco budworm. Environ. Entomol. 7: 431-434.

Prestwich, G.D., A.K. Gayen, S. Phirwa and T.B. Kline. 1983. 29-Fluorophytosterols: Novel pro-insecticides which cause death by dealkylation. Biotechnology 1:62-65.

Price, P.W., C.E. Bouton, P. Gross, B.A. McPheron, J.N. Thompson and A.E. Weis. 1980. Interactions among three trophic levels: Influence of plants on interactions between insect herbivores and natural enemies. Annu. Rev. Ecol. Syst. 11:41-65.

Rajakulendran, S.V., and F.W. Plapp. 1982. Comparative toxicities of five synthetic pyrethroids to the tobacco budworm (Lepidoptera: Noctuidae), an ichneumonid parasite, Campoletis sonorensis, and a predator, Chrysopa carnea. J. Econ. Entomol. 75:769-772.

Ripper, W.E., R.M. Greenslade and G.S. Hartley. 1951. Selective insecticides and biological control. J. Econ. Entomol. 44:448-459.

Rodriguez, E., and D.A. Levin. 1976. Biochemical parallelisms of repellents and attractants in higher plants and arthropods, pp. 214-270. In: J.W. Wallace and R. L. Mansell (Eds.), Recent advances in phytochemistry, Vol. 10. Plenum Press, N.Y.

Roelofs, W.L. 1981. Pheromones, plateaus, and platitudes. Bull. Entomol. Soc. Amer. 27:3-8.

Rosenthal, G.A., and D.H. Janzen (Eds.). 1979. Herbivores, their interaction with secondary plant metabolites. Academic Press, N.Y. 718 pp.

Ross, D.C., and T.M. Brown. 1982. Inhibition of larval growth in Spodoptera frugiperda by sublethal dietary concentrations of insecticides. J. Agric. Food Chem. 30:193-196.

Roush, R.T., and F.W. Plapp. 1982. Biochemical genetics of resistance to aryl carbamate insecticides in the predaceous mite, Metaseiulus occidentalis . J. Econ. Entomol. 75:304-307.

Schoonhoven, L.M. 1982. Biological aspects of antifeedants. Entomol. Exp. Appl. 31:57-69.

Schultz, J.C. 1983. Impact of variable plant defensive chemistry on susceptibility of insects to natural enemies, pp. 37-54. In: P.A. Hedin (Ed.), Plant resistance to insects. Symposium series no. 208. Amer. Chem. Soc., Washington, D.C.

Scott, J.G., B.A. Croft and S.W. Wagner. 1983. Studies on the mechanism of permethrin resistance in Amblyseius fallacis (Acarina: Phytoseiidae) relative to previous insecticide use on apple. J. Econ. Entomol. 76:6-10.

Sechser, B. 1981. An approach to integrated pest management from the chemical industry. Acta Phytopath. Acad. Sci. Hung. 16:239-243.

Silverstein, R.M. 1981. Pheromones: Background and potential for use in insect pest control. Science 213:1326-1332.

Slansky, F., and J.M. Scriber. 1982. Selected bibliography and summary of quantitative food utilization by immature insects. Bull. Entomol. Soc. Amer. 28: 43-55.

Smilowitz, Z. 1973. Electrophoretic patterns in hemolymph protein of cabbage looper during development of the parasitoid Hyposoter exiguae. Ann. Entomol. Soc. Amer. 66:93-99.

Smilowitz, Z., and C.L. Smith. 1977. Hemoloymph proteins of developing Pieris rapae larvae parasitized by Apanteles glomeratus. Ann. Entomol. Soc. Amer. 7-0: 447-454.

Smilowitz, Z., C.A. Martinka and E.A. Jowyk. 1976. The influence of a juvenile hormone mimic (JHM) on the growth and development of the cabbage looper, Trichoplusia ni (Lepidoptera: Noctuidae) and the endoparasite, Hyposoter exiguae (Hymenoptera: Ichneumonidae). Environ. Entomol. 5:1178-1182.

Smissaert, H.R., S. Voerman, L. Oostenbrugge and N. Renooy. 1970. Acetyl cholinesterases of organophosphate-susceptible and -resistant spider mites. J. Agric. Food Chem. 18:66-75.

Soderhall, K. 1982. Prophenoloxidase activating system and melanization--a recognition mechanism of arthropods? A review. Developm. Comp. Immunol. 6:601-611.

Soderlund, D.M., J.R. Sanborn and P.W. Lee. 1983. Metabolism of pyrethrins and pyrethroids in insects. Prog. Pestic. Biochem. Toxicol. 3:401-435. In: D.H. Hutson and T.R. Roberts (Eds.), Progress in pesticide biochemistry and toxicology, Vol. 3. John Wiley & Sons, N.Y.

Sparks, T.C., and B.D. Hammock. 1983. Insect growth regulators: Resistance and the future, pp. 615-668. In: G.P. Georghiou and T. Saito (Eds.), Pest resistance to pesticides. Plenum Press, N.Y.

Sroka, P., and S.B. Vinson. 1978. Phenoloxidase activity in the haemolymph of parasitized and unparasitized Heliothis virescens. Insect Biochem. 8:399-402.

Stanley-Samuelson, D.W., and R.H. Dadd. 1983. Long-chain polyunsaturated fatty acids: Patterns of occurrence in insects. Insect Biochem. 13:549-558.

Starks, K.J., R. Muniappan, and R.D. Eikenbary. 1972. Interaction between plant resistance and parasitism against the greenbug on barley and sorghum. Ann. Entomol. Soc. Amer. 65:650-655.

Svoboda, J.A., and W.E. Robbins. 1979. Comparison of sterols from a phytophagous and predacious species of the family Coccinellidae. Experientia 35:186-187.

Svoboda, J.A., M.J. Thompson, W.E. Robbins and J.N. Kaplanis. 1978. Insect steroid metabolism. Lipids 13:742-753.

Tanaka, K., and Y. Ito. 1982. Different response in respiration between predaceous and phytophagous lady beetles (Coleoptera: Coccinellidae) to starvation. Res. Popul. Ecol. 24:132-141.

Thompson, S.N. 1973. A review and comparative characterization of the fatty acid compositions of seven insect orders. Comp. Biochem. Physiol. 74B: 183-211.

Thompson, S.N. 1981. Effects of dietary carbohydrate and lipid on nutrition and metabolism of metazoan parasites with special reference to parasitic Hymenoptera, pp. 215-252. In: G. Bhaskaran, S. Friedman and J.G. Rodriguez (Eds.), Current topics in insect endocrinology and nutrition. Plenum Press, N.Y.

Thompson, S.N. 1983. Biochemical and physiological effects of metazoan endoparasites on their host species. Comp. Biochem. Physiol. 74B:183-211.

Thompson, S.N., and J.S. Barlow. 1974. The fatty acid composition of parasitic Hymenoptera and its possible biological significance. Ann. Entomol. Soc. Amer. 67:627-632.

Thurling, D.J. 1980. Metabolic rate and life stage of the mite Tetranychus cinnabarinus Boisd. (Prostigmata) and Phytoseiulus persimilis A-H. (Mesostigmata). Oecologia 46:391-396.

Vinson, S.B., and D.J. Barras. 1970. Effects of the parasitoid, Cardiochiles nigriceps, on the growth, development, and tissues of Heliothis virescens. J. Insect Physiol. 16:1329-1338.

Vinson, S.B., and G.F. Iwantsch. 1980a. Host suitability for insect parasitoids. Annu. Rev. Entomol. 25:397-419.

Vinson, S.B., and G.F. Iwantsch. 1980b. Host regulation by insect parasitoids. Quart. Rev. Biol. 55:143-165.

Voss, G., and F. Matsumura. 1965. Biochemical studies on a modified and normal cholinesterase found in the Leverkusen strains of the two-spotted spider mite Tetranychus urticae. Can. J. Biochem. 43:63-72.

Whalon, M.E., B.A. Croft and T.M. Mowry. 1982. Introduction and survival of susceptible and pyrethroid-resistant strains of *Amblyseius fallacis* (Acari: Phytoseiidae) in a Michigan apple orchard. Environ. Entomol. 11:1096-1099.

Wilkinson, C.F. 1976. Insecticide synergism, pp. 195-222. *In*: R.L. Metcalf and J.J. McKelvey (Eds.), The future for insecticides: Needs and prospects. John Wiley & Sons, N.Y.

Wilkinson, C.F., K. Hetnarski, M.S. Denison and F.P. Guengerich. 1983. Selectivity of 1-phenylimidazole as a ligand for cytochrome P-450 and as an inhibitor of microsomal oxidation. Biochem. Pharmacol. 32:997-1003.

Williams, M.C., and L.F. James. 1983. Effects of herbicides on the concentration of poisonous compounds in plants: A review. Amer. J. Vet. Res. 44:2420-2422.

Winteringham, F.P.W. 1969. Mechanisms of selective insecticidal action. Annu. Rev. Entomol. 14:409-442.

Yu, S.J. 1983. Induction of detoxifying enzymes by allelochemicals and host plants in the fall armyworm. Pestic. Biochem. Physiol. 19:330-336.

Yu, S.J., and L.C. Terriere. 1978. Metabolism of juvenile hormone 1 by microsomal oxidase, esterase, and epoxide hydrase of *Musca domestica* and some comparisons with *Phormia regina* and *Sarcophaga bullata.* Pestic. Biochem. Physiol. 9:237-246.

Yu, S.J., R.E. Berry and L.C. Terriere. 1979. Host plant stimulation of detoxifying enzymes in a phytophagous insect. Pestic. Biochem. Physiol. 12:280-284.

Zohdy, N.Z.M. 1976. On the effect of the food of *Myzus persicae* Sulz. on the hymenopterous parasite *Aphelinus asychis* Walker. Oecologia 26:185-191.

DISCUSSION: An update on development of selective pesticides favoring arthropod natural enemies.

DISCUSSION LEADER: R.E. Stinner

Although the data presented show a remarkable similarity in enzyme systems for natural enemies, with distinct differences from herbivores, the question was raised as to whether or not this is in fact a general phenomenon. For example, the selectivity for herbivores over natural enemies apparently exhibited by the synthetic pyrethroids does not seem to exist for at least some predaceous mites. It was also noted that many of our most important predators in field crops (e.g., hemipterans, thrips) are also obligatory plant feeders and thus may be much more similar in enzyme systems to phytophagous insects than are the natural enemies so far examined. Also previous studies have been concerned with detoxification systems; comparative studies of target site activities, penetration, and behavioral avoidance of pesticides should be conducted with representative species of major natural enemy groups.

IMPROVING ESTABLISHMENT OF ARTHROPOD NATURAL ENEMIES

Marjorie A. Hoy

For nearly 100 years entomologists have moved parasites, predators, or pathogens from one part of the world to another. Many of these natural enemies, once established in the new environment, have provided partial to complete control of noxious pests (Bartlett & van den Bosch 1964, DeBach 1971, Laing & Hamai 1976). As Rosen & Huffaker (1983) noted, the most important attributes desired in an effective natural enemy are searching capacity, prey (host) specificity, power of increase, and adaptability; high searching capacity commonly is regarded as the most important of these attributes, followed by specificity. Classical biological control with such effective natural enemies has been shown to be cost-effective in providing long term solutions to pest problems in agriculture, rangelands, forests, and greenhouses (Batra 1982, Caltagirone 1981, DeBach 1971, Ehler & Miller 1978). These successes have stimulated the very natural desire to apply this approach to other pests, both exotic and endemic (Carl 1982, Ehler & Andres 1983, Pimentel 1963).

Classical biological control programs usually require that the natural enemy undergo both laboratory and field colonization (DeBach & Bartlett 1964). This discussion will focus on field establishment but will include a discussion of laboratory or insectary colonizations since exotic species are often reared for at least one generation in a quarantine facility in order to assure identity of the species and exclude pathogens or secondary parasites.

Why do classical biological control programs continue to be necessary? There are at least three answers to that question. First, exotic pests continue to enter the United States on a regular basis (Sailer 1978, 1983), despite the enactment of a Plant Quarantine Act in 1912. Between 1910-1920, at least 187 immigrant species established, and between 1920 and 1980, at least 837 additional species became established. Approximately 10% of these (80) subsequently became important pests, including the pink bollworm (*Pectinophora gossypiella*), citrus blackfly (*Aleurocanthus woglumi*), Egyptian alfalfa weevil (*Hypera brunneipennis*), face fly (*Musca autumnalis*), and the cereal leaf beetle (*Oulema melanopus*). Each year approximately 14 species

BIOLOGICAL CONTROL
IN AGRICULTURAL IPM SYSTEMS

ISBN 0-12-357030-1

are added to our fauna, with three beneficials introduced intentionally. Of the 11 species accidentally introduced, seven are likely to be injurious to some degree. About every third year, a pest of major importance is likely to establish (Sailer 1983). Sailer (1978) estimated that foreign species are responsible for about 50% of the crop losses due to insects and mites in California agriculture, with approximately 38% of all losses attributed to species of unquestioned foreign origin. Sailer (1983) further concluded that there is "a very large reservoir of foreign species of insects and mites that are potentially injurious to American agriculture." The number capable of establishing in the United States cannot be predicted because it is difficult to predict whether a species will establish or be a pest in a new environment. However, it is clear that invading foreign insects and mites remain a threat to American agriculture, with Eastern Palearctic and Oriental pests providing an increasing danger to major crops in the U.S. as commerce and travel increases between the U.S. and these areas. Because these exotic pests typically invade without their natural enemies, classical biological control offers an opportunity for cost effective, permanent control (Caltagirone 1981, DeBach 1971, Ehler & Andres 1983, Laing & Hamai 1976), and should be considered if eradication is not achieved or attempted.

A second reason for an enduring interest in classical biological control is that not all previous projects were successful (Beirne 1975; DeBach 1965, 1971; Hoy 1976; Laing & Hamai 1976; Messenger 1971). Renewed efforts in many classical biological control projects are clearly justified. In the intervening years, new information on systematics, biology, host preferences, and geographic range of natural enemies could have provided key information that may result in additional establishments and subsequent effective biological control.

Finally, exotic natural enemies could be sought to control native pests (Carl 1982). Pimentel (1963) considered this approach to be useful with 39% of 66 pest species examined being controlled by an introduced parasite or predator of an allied species or genus.

WHAT IS THE ESTABLISHMENT RATE?

Of the immigrant arthropods established in the U.S., 271 species are considered to be beneficial, either as parasites, predators, or pollinators (Sailer 1978). It is provocative to learn that of these 271 species, 118 species were accidentally introduced into the U.S. Nearly half of the successful introductions involving beneficial species occurred without the deliberate assistance of man!

The overall establishment rate for natural enemies of insects and arachnids worldwide was estimated to be 34%, with establishment rates for predators and parasites being identical at 34%

(Hall & Ehler 1979). Exotic natural enemies introduced against exotic pests established at a higher rate than did exotic natural enemies introduced against native ones, but the establishment rate of 25% for exotic natural enemies of native pests is sufficiently high to encourage continued effort with this approach.

Hall & Ehler (1979) evaluated habitat stability as a factor in establishment and found natural enemies released into unstable habitats, such as annual vegetable or field crops, had an establishment rate of 28%. Natural enemies released into habitats with an intermediate level of stability, such as orchards had an establishment rate of 32%. Natural enemies released into stable habitats, such as forests and rangelands, had the highest rate, 36%. Beirne (1975) evaluated Canadian biological control programs and found that natural enemies released into annual crops had an establishment rate of 16%, while natural enemies released into orchards and ornamental shrubs had a 34% rate of establishment.

Establishment of natural enemies does not, of course, guarantee that the natural enemy will be effective in controlling the pest. But establishment is a prerequisite to successful classical control. Some authors believe that a worldwide average establishment rate of 34% is poor, while others believe this to be a high rate of achievement. Whether this rate is good or poor is determined by the eyes of the beholder. Nevertheless, if we can increase the rates, it is likely that effective biological control of noxious pests will increase as well.

If we can detect a pattern of circumstances that has apparently resulted in lowered establishment rates, we may be able to change procedures favorably in future projects. Actually, it is nearly impossible to determine the precise reason or reasons a natural enemy failed to establish: we can only identify reasons that are sufficient to explain why a species would be expected to fail to establish. Of the hundreds of insects introduced into various countries as biological control agents, few have been studied carefully after introduction, and the outcome of the introduction is often not understood (Price 1972). At the moment, establishing exotic natural enemies in a new environment is largely an empirical process, and will remain so until we can test hypotheses about the colonization of arthropod natural enemies in at least a few selected cases.

REASONS SUFFICIENT FOR FAILURE

Natural enemies probably fail to establish in new environments for many reasons. I have categorized the suspected deterrents as due to biological and ecological, genetical, and procedural problems, but it is possible to arrange the proposed reasons in other categories as they are often interrelated.

Biological and Ecological Impediments to Establishment

Many biological control specialists agree that the selection of inappropriate races, strains, or biotypes is a significant factor precluding establishment and efficacy (Caltagirone, this volume; Kelleher 1968; Knutson 1981; Messenger & van den Bosch 1971; Messenger et al. 1976; van den Bosch 1968; Zwolfer et al. 1976). Thus, the strain may not be well adapted to the new climate or host race of the pest. In some cases, the target host may be different in some critical way from the host in the collection site (Beirne 1975, DeBach & Bartlett 1964, Hoy 1976, van den Bosch 1968, Volker & Simpson 1975). Enhanced knowledge of the geographic distribution and biosystematics of biological control agents would alleviate many of these problems (Delucchi et al. 1976, Rosen 1978, Rosen & DeBach 1973). Unfortunately, support for biosystematics studies of biological control agents has never been sufficient; this should be remedied.

The phenology of the host or prey in the new environment may be inappropriate so that adequate overwintering populations are not produced (Beirne 1975, Ehler 1982, Hoy 1976, Messenger & van den Bosch 1971, Miller 1983, Price 1972, Stehr 1974, Zwolfer et al. 1976). The unavailability of alternate hosts or prey may prevent the establishment of certain species (Hoy 1976, Messenger & van den Bosch 1971, van den Bosch 1968). A lack of detailed knowledge about the fundamental biology and ecology of natural enemies means that we can't remedy some of these difficulties. Clearly, fundamental biological studies remain a critical necessity.

On occasion, natural enemies (hyperparasites, predators, or pathogens) of natural enemies could prevent colonization or efficacy subsequent to establishment (Carl 1982, Ehler & Andres 1983, Myers & Sabath 1981). If these natural enemies are endemic in the new environment, little can be done to combat them. Quarantine procedures have been very successful in screening out most exotic natural enemies of parasites and predators (Beirne 1975). It is difficult to speculate what more can be done to alleviate the effects of endemic hyperparasites, generalist predators, or pathogens.

Competitive exclusion has been suggested to occasionally prevent establishment (Beirne 1975; DeBach 1965; DeBach & Bartlett 1964; DeBach & Hagen 1964; Ehler 1982; Ehler & Hall 1982, 1984; Messenger & van den Bosch 1971; Stehr 1974; Turnbull & Chant 1961). Ehler & Hall (1982) suggested that the rates of establishment of exotic natural enemies introduced against exotic homopteran, lepidopteran, and coleopteran pests were inversely related to the number of species released at a given time and place and the number of exotic incumbent species of natural enemies present. This latter fact suggested competitive exclusion of introduced natural enemies has "probably" occurred (Ehler & Hall 1982) although they "do not believe that competitive exclusion has occurred frequently in the past " (Ehler & Hall 1984).

The issue of competitive exclusion, because of its past admitted rarity (Ehler & Hall 1984), seems to be of lesser importance in establishment rates in agricultural systems than habitat stability, pesticide interference, genetic problems or inadequate numbers released (see below), and concern over competitive exclusion should not delay additional importations.

Habitat stability has already been alluded to as an impediment to establishment, with lower establishment rates occurring in the more temporary sites such as field crops (Beirne 1975, Ehler & Miller 1978). Habitat instability could make it difficult for natural enemies to establish due to limited or insufficient requisites such as alternate hosts or prey, nectar or pollen, or overwintering sites. Monoculture, with its simplification of the habitat, may make establishment of natural enemies in field crops particularly difficult. Special efforts in cultural practices and habitat management may be required to achieve establishment and enhance efficacy of arthropod natural enemies in such systems (see Herzog & Funderburk; Hull & Beers; Lewis & Nordlund, this volume).

Pesticide applications are more often made in annual field crops than in forests and this may further reduce establishment rates and efficacy. While biological control specialists make every effort to release natural enemies into sites where they are less likely to be killed by pesticides, nonselective pesticides could deter both short and long term colonization by natural enemies in field crops unless significant amounts of untreated refuges exist. Integration of biological control agents into agricultural IPM systems can't be achieved unless the natural enemy can survive the pesticides <u>necessary</u> in that crop system. Most crops have recalcitrant key pests for which there is currently no effective biological, cultural, or other non-pesticidal control. Laboratory and subsequent field screening of herbicides, fungicides, insecticides, acaricides or nematicides could provide useful guidelines for their use in IPM programs, as responses of natural enemies to pesticides vary (Croft & Brown 1975). If selective materials were identified and used, additional natural enemies might establish and persist in crop systems. I believe that the extensive use of natural enemies in many agricultural IPM systems will be determined by how well we learn to use pesticides selectively--either through the choice of material, their placement, or rate and timing of application (see Hull & Beers; Mullin & Croft, this volume).

Alternatively, pesticide-resistant natural enemies can be developed through artificial selection in the laboratory, or potentially through the development of genetic engineering techniques (Beckendorf & Hoy, this volume; Hoy 1985). If genetic engineering techniques are adapted to arthropod natural enemies in the next few years, it may become possible to develop pesticide-resistant natural enemies within months (Beckendorf & Hoy, this volume). This time scale could make the approach economically feasible. Pesticide-resistant strains developed in the field can also be identified through surveys.

Procedural Problems Inhibiting Establishment

Various technical problems may interfere with the establishment of natural enemies (DeBach & Barlett 1964, Stehr 1974, van den Bosch 1968). Many parasitoids have complex life cycles, making rearing so difficult that initial laboratory colonization may fail (DeBach & Bartlett 1964, van den Bosch 1968). In some cases natural enemies may not mate prior to their field release, or they may be released at the wrong time of day or in the wrong site.

van den Bosch (1968) was concerned that some failures have been due to a project's organization and recommended that the same biological control specialists carry a project through the quarantine, propagation, release, and evaluation phases so that adequate information is transferred and motivation maintained. He suggested indifference as a contributing factor in some failures. One individual reportedly released parasites into the hot, dry San Joaquin Valley of California while driving "slowly" by the orchard. This is surely an extreme case, but more subtle procedural inadequacies could influence project success.

A compelling, albeit circumstantial, case has been made regarding the numbers (and quality) of the individuals released (Beirne 1975, DeBach & Bartlett 1964, Ehler & Hall 1982, Hoy 1976, Stehr 1974, van den Bosch 1968). Beirne's (1975) analysis of 159 Canadian cases suggeted that the number of individuals released influenced the results. The median number of individuals released for all species was approximately 5000. However, for the 44 species that became established, an average of 31,200 individuals were released. If fewer than 5000 were released, only 10% of the species became established. If more than 31,200 individuals were released, 78% of the 28 species became established.

The numbers released at a site also appear to influence establishment (Beirne 1975). If more than 800 individuals were released at each Canadian site, about 60% of the species established. If fewer than 800 individuals were released, only about 15% of the species established. Beirne (1975) attributed the increased establishment rate to several factors, including a feedback process that reinforced additional releases of "early colonizers" and resulted in termination of releases of the apparently less successful colonizers. He also attributed the success of the larger releases to "the greater genetic diversity of the colony." Beirne (1975) suggested that "adequate genetic diversity is needed to produce the new genetic combinations that are essential to enable the colony to thrive." No evidence was provided to support this genetic hypothesis, however.

While it is true that some natural enemies have established after the release of only a few individuals (DeBach 1971), the above correlation between establishment success and large releases deserves serious consideration because it suggests how additional establishments could be achieved. It seems likely that different arthropods will have very different population structures; some are likely to be better colonizers than others by virtue of their reproductive rates, mating behavior, dispersal propensities, and ability to tolerate inbreeding. Some very effective natural enemies may not be "weedy" colonizers. In fact, those species that are most effective at low host (prey) densities may be least "weedy". Therefore, those species for which we have no sufficient explanation(s) for failure to establish (such as lack of alternate hosts in the new environment, a poor climatic match, or asynchrony with the host or prey) should be collected and released after quarantine processing in as large as numbers as possible. Whether large numbers (more than 800/release site) obtained through mass propagation will improve establishment is another question, as it is not clear whether the "numbers problem" is due to a lack of genetic variability upon which selection can operate after release, or whether the natural enemy species requires a minimal number released to ensure successful colonization. In species which disperse before mating, there could be a threshold number below which the exotic species could not find each other for mating; conversely, those species which commonly sib mate prior to dispersal are less likely to require large releases. In cases where discovery for mating is critical, genetic diversity may be of lesser importance than absolute numbers. On the other hand, it may be that post release genetic adaptation must occur to achieve establishment and a broad genetic pool is necessary to ensure variability. Mass propagation of a small founder colony then would not be appropriate unless all necessary genetic variability could be maintained. It is difficult to discriminate between these situations from most records (and they probably occur in different natural enemy taxonomic groups) because, while the number of individuals received in quarantine is recorded, people rarely record the number of individuals that actually produce progeny. The proportion that reproduce in a new laboratory colony is often a very small fraction of those collected (Bartlett 1984, Unruh et al. 1983). It is highly desirable that detailed documentation be made of colonization attempts of representative natural enemy groups. A better knowledge of the breeding systems and population structures of different natural enemy groups may result that will allow more effective colonization attempts. For example, Unruh et al. (1983) showed that allozyme variability declined in seven laboratory populations of <u>Aphidius ervi</u> (Aphidiidae) and attributed this to the fact that fewer females actually reproduced than expected, thus leading to a rapid loss of average heterozygosity. If such heterozygosity is important in obtaining establishment, possible solutions might include reduction of bottlenecks to enhance the effective population size, and maintenance of multiple inbred lines with migration (Unruh et al. 1983).

Genetics as a Factor

Genetic questions concern the optimal foreigh exploration sites to obtain the appropriate biotypes (Delucchi et al. 1976, Knutson 1981, Rosen 1978, Rosen & DeBach 1973), the value of hybridization and artificial selection (Ashley et al. 1974; Beckendorf & Hoy, this volume; Collins 1984; Hoy 1979a, b; Legner 1972; Roush 1979), and the desirability of genetic variability in establishment and post-release adaptation (Hoy 1979a, b; Levins 1969; Myers & Sabath 1981; Unruh et al. 1985). Concern is voiced over loss of fitness due to inadvertent laboratory selection, reduced genetic variability through bottlenecks, founder effects, and inbreeding during laboratory colonization (Bartlett 1984, Mackauer 1976, Messenger et al. 1976, Unruh et al. 1983). Genetic deterioration of colonized natural enemies is thought to have prevented the establishment of some natural enemies (Beirne 1975; DeBach 1965; Hoy 1976, 1978; Joslyn 1984; Legner 1979; Mackauer 1976; Messenger & van den Bosch 1971; Myers & Sabath 1981; Stehr 1974).

Force (1967) and Messenger et al. (1976) voiced concern that coadapted gene complexes would be disrupted if different strains were hybridized during quarantine procedures, or during release, and suggested that different strains should be kept separate and released at different sites. Biemont & Bouletreau (1980) tested the effects of hybridization and inbreeding on the haplo-diploid parasitoid Cothonaspis boulardi and found that in two strains inbreeding had no effect on productivity, but sex ratio was altered in one strain perhaps due to interference with the sex determining mechanism. Other authors suggested that hybridization might ensure establishment by enhancing vigor through heterosis (Ashley et al. 1974, Hoy 1975a, Legner 1972). Others have stressed the need to provide maximum genetic variability so that selection can operate after release to yield a population adapted to the new environment (Bartlett & van den Bosch 1964; DeBach & Bartlett 1964; Force 1967; Hoy 1976, 1979b; Messenger & van den Bosch 1971; Messenger et al. 1976; Simmonds 1976; Wilson 1965), although Myers & Sabath (1981) found that successful colonizers have had both high and low genetic variability.

Several have suggested that natural enemy species undergo genetic adaptations after release into a new environment (Messenger & van den Bosch 1971, Messenger et al. 1976, Myers 1978), but little data exists to document that it occurs. Myers (1978) showed that post-establishment changes in the cinnabar moth occurred within 17 years after release in British Columbia. Larval dispersal tendency, wing length, pupal size, and temperature threshold for adult emergence were evaluated, and the latter trait most clearly had responded to selection, perhaps within eight generations. The frequency of post-release adaptation is unknown since it had rarely been sought. If post release adaptation commonly occurs, the need for maintaining appropriate genetic variability in founder colonies is reinforced. Additional studies should be conducted to determine if it is a general phenomenon or not. I suspect that it is; it is highly unlikely that an exotic species would be so well preadapted to a new environment that no genetic fine tuning was needed.

The original source and size of the colony concerns many biological control specialists. Thus, a foreign explorer attempts to collect natural enemies in a climatic region comparable to the release site (Messenger 1971). In addition, the genetic nature of the population has been perceived to be important. Marginal populations are proposed to be more homozygous and to have a lower genetic load while central populations are expected to be highly variable, with a high genetic load. Remington (1969) believed marginal populations were good source populations for biological control agents while Lucas (1969) believed them to be poor. Myers & Sabath (1981) reviewed variability in single loci as measured with electrophoresis. They found tremendous variation from species to species. Some isolated populations were low in genetic variability while others were high, and they concluded that variation among central and marginal populations is too great to allow generalizations, and that "practical constraints on how and where insects are collected for biological control releases probably make further consideration of this aspect of biological control unprofitable" (Myers & Sabath 1981).

Insectary rearing of biological control agents is a critical component of classical biological control. Most natural enemy species are reared at least one generation in quarantine facility to ensure that secondary parasites and pathogens are excluded. Problems in the insectary arise from genetic and other causes (Bartlett 1984; Boller & Chambers 1977; Force 1967; Hoy 1975b, 1979a; Huettel 1976; Joslyn 1984; Legner 1979; Mackauer 1976; Messenger & van den Bosch 1971; Messenger et al. 1976; Myers & Sabath 1981; Unruh et al. 1983). Diseases and nongenetic behavioral alterations can influence the quality of natural enemies reared in the insectary, but genetic bottlenecks, inbreeding, drift, and inadvertant selection are often suspected to be deterrents to the establishment of natural enemies. Solutions to these problems are not easy, cheap, or readily apparent (Bartlett 1984, Mackauer 1976).

Few of the solutions proposed from studies of Drosophila have been tested using arthropod natural enemies. It is critical that they be tested with a representative group of natural enemies for several reasons. Firstly, many natural enemies are haplo-diploid rather than diplo-diploid; some are thelytokous, having females only. Thus, population genetics theory developed for diplo-diploid species may not be applicable (Crozier 1970a, b). Secondly, the genetic architecture of arthropod species varies dramatically (White 1973). For example, some species are well adapted to close inbreeding via brother-sister mating, while other species develop aberrent sex ratios after inbreeding. Of the approximately 100,000 known parasitic Hymenoptera, only a few species have been studied genetically (Cassidy 1975). Thus, it seems premature to apply observations from Drosophila to arthropod natural enemies.

The following genetic questions seem particularly relevant:

1. How much genetic variation, measured in several ways, is lost during the colonization of typical natural enemy species in the quarantine facility and insectary?
2. Does post-release genetic adaptation commonly occur in arthropod natural enemies?
3. Are hybrid strains more likely to establish in the field than pure bred strains? Does hybridization disrupt coadapted gene complexes or does it provide heterotic vigor? Which, if any, natural enemy taxonomic groups respond well to hybridization?
4. Do variable environments in the insectary (cyclic temperatures, natural daylengths, variable relative humidities, variable host/prey availabilities) improve colonization?
5. How many colonized individuals actually contribute progeny to the colony? Are genetic bottlenecks common? Is there a relationship between the number of insects colonized in the insectary and the likelihood of field establishment?
6. Can artificial selection of pesticide-resistant natural enemies yield enhanced establishment of natural enemies in agricultural crop systems?

Until some of these questions are answered with carefully designed experiments with taxonomically diverse natural enemies, we may not make significant progress in enhancing establishment rates. Price (1972) noted that "virtually nothing is known of the genetic requirements necessary to insure the best possible chance for establishment of a founder colony of parasites or predators." This statement is still true. Much has been written about genetics and biological control, but very little research has been conducted to test these genetic questions. This does not mean however, that releases should be curtailed while these studies go on; rather, these studies may ultimately provide information to increase the establishment rate.

RECOMMENDATIONS

1. Basic studies on the biosystematics of arthropod natural enemies of key established pests are of utmost importance and need increased emphasis and support. Current knowledge of the natural enemies of exotic pests that might become established in North America also is inadequate. Identification of geographic ranges of pests and the location of potential natural enemy biotypes is important, as specific natural

enemy races or strains may vary in their ability to establish and be effective in North America.

2. Fundamental knowledge of the biology and ecology of biological control agents is needed. Too often we don't know whether exotic natural enemies require an alternate host, whether they have a diapause, or whether their phenology matches that of the target pest in the new environment. Such studies are critically important in establishing optimal release sites and timing, and should lead to improved establishment rates.
3. In most agricultural crop systems, pesticides will be required for the forseeable future to control pests for which there is no other effective control method. When an exotic natural enemy is released into an agricultural crop system we should know whether these pesticides are harmful to the natural enemy. Such screening will allow more realistic expectations of the possibility of establishing the natural enemy; the availability and adoption of selective pesticides will determine the establishment and ultimate efficacy in the specific cropping system.
4. Evidence suggests that in many unsuccessful establishment attempts inadequate numbers of exotic natural enemies were released. Previous projects should be evaluated and, if no other sufficient reason is found to explain the failure to establish, larger numbers of the most promising species should be collected from the foreign site, quarantined, and released. It is preferable to collect large numbers for release than to mass propagate a small founding colony.
5. Some exotic natural enemies imported to control native pests have successfully established and been effective. Projects should be supported to identify and establish exotic natural enemies for serious native crop pests lacking effective native biological control agents.
6. Genetic selection (and perhaps engineering) of arthropod natural enemies to develop pesticide resistances offers promise as a way of integrating biological control agents into agricultural IPM systems. Research should be supported to determine how widely these techniques can be applied to arthropod natural enemies other than phytoseiid predators of spider mites.
7. Fundamental long term genetic studies on arthropod natural enemies should be supported. Artificial selection, hybridization, and genetic engineering techniques should be investigated with systematically diverse natural enemy species to determine if these genetic techniques can alleviate genetic deterioration in the laboratory colonization phase or improve specific attributes. The desirability of hybridizing strains should be tested. The importance of genetic variability in obtaining establishment should be investigated.
8. Attempts to establish exotic arthropod natural enemies should be documented more extensively. It is difficult to publish

the sometimes fragmentary information obtained during quarantine, rearing, and release of exotic species. Standardized documentation forms should be developed that will allow future studies of colonization attempts. Attributes that would be useful to record include: The number of colonizers received in quarantine; the number that actually produce progeny; the numbers released into the field with each colonization attempt and their locations; information on development rates, sex ratios, mating behavior, alternate prey requirements, diapause, pesticide susceptibility or tolerances; host or prey stage preferred; temperature tolerances; geographic distribution in the native range; and host range. This data base should computerized and made available to all researchers working in biological control. Currently, a documentation system is being developed by the USDA-ARS Beneficial Insect Introduction Laboratory (J. Coulson, personal communication) that incorporates most of these suggestions. Such a documentation system, if supported by biological control workers, could provide a useful historical data base for further correlative analyses.

Nearly everyone agrees that biological control by importation of exotic natural enemies is a good idea. However, the idea isn't being adequately pursued (DeBach 1971). The task of importing useful species is less than 20% completed, and present efforts could not finish the job in the next 100 years (Sailer 1983). Identification, location, importation, colonization, dispersion, and evaluation of introduced species must be enhanced.

ACKNOWLEDGEMENTS

This work was supported in part by Experiment Station Project 3522-H, and Regional Project W-84. I thank the following people who commented on earlier versions of the manuscript: J. Coulson, F. Legner, D. Gonzalez, C. Huffaker, J. Rosenheim, D. Rosen, and R. Roush.

REFERENCES

Ashley, T.R., D. Gonzalez and T.F. Leigh. 1974. Selection and hybridization of Trichogramma. Environ. Entomol. 3:43-48.

Bartlett, A.C. 1984. Genetic changes during insect domestication, pp. 2-8. In: E.G. King and N.C. Leppla (Eds.), Advances and challenges in insect rearing. USDA-ARS, Southern Region, New Orleans, LA.

Bartlett, B.R., and R. van den Bosch. 1964. Foreign exploration for beneficial organisms, pp. 283-304. In: P. DeBach (Ed.), Biological control of insect pests and weeds. Chapman & Hall, London.

Batra, S.W.T. 1982. Biological control in agroecosystems. Science 215:134-139.

Beirne, B.P. 1975. Biological control attempts by introductions against pest insects in the field in Canada. Can. Entomol. 107:225-236.

Biemont, C., and M. Bouletreau. 1980. Hybridization and inbreeding effects on genome coadaptation in a haplodiploid Hymenoptera: Cothonaspis boulardi (Eucoilidae). Experientia 36:45-46.

Boller, E.F., and D.L. Chambers (Eds.). 1977. Quality control, an idea book for fruit fly workers. IOBC SROP WPRS Bull. 1977/5, 162 pp.

Caltagirone, L.E. 1981. Landmark examples in classical biological control. Annu. Rev. Entomol. 26:213-232.

Carl, C.P. 1982. Biological control of native pests by introduced natural enemies. Commonw. Inst. Biol. Contr. Biocontr. News Inform. 3(3):191-200.

Cassidy, J.D. 1975. The parasitoid wasps, Habrobracon and Mormoniella, pp. 173-203. In: R.C. King (Ed.), Handbook of genetics, Vol. 3, Invertebrates of genetic interest. Plenum Press, N.Y.

Collins, A.M. 1984. Artificial selection of desired characteristics, pp. 9-19. In: E.G. King and N.C. Leppla (Eds.), Advances and challenges in insect rearing. USDA-ARS, Southern Region, New Orleans, LA.

Croft, B.A., and A.W.A. Brown. 1975. Responses of arthropod natural enemies to insecticides. Annu. Rev. Entomol. 20:285-335.

Crozier, R.H. 1970a. Coefficients of relationship and the identity of genes by descent in the Hymenoptera. Amer. Nat. 104:216-217.

Crozier, R.H. 1970b. On the potential for genetic variability in haplo-diploidy. Genetica 41:551-556.

DeBach, P. 1965. Some biological and ecological phenomena associated with colonizing entomophagous insects, pp. 287-306. In: H.G. Baker and G.L. Stebbins (Eds.), The genetics of colonizing species. Academic Press, N.Y.

DeBach, P. 1971. The use of imported natural enemies in insect pest management ecology. Proc. Tall Timbers Conf. Ecol. Anim. Contr. Habitat Mgt. 3:211-233.

DeBach, P., and B.R. Bartlett. 1964. Methods of colonization, recovery, and evaluation, pp. 402-426. In: P. DeBach (Ed.), Biological control of insect pests and weeds. Chapman & Hall, London.

DeBach, P., and K.S. Hagen. 1964. Manipulation of entomophagous species, pp. 429-458. In: P. DeBach (Ed.), Biological control of insect pests and weeds. Chapman & Hall, London.

Delucchi, V., D. Rosen and E.I. Schlinger. 1976. Relationship of systematics to biological control, pp. 81-91. In: C.B. Huffaker and P.S. Messenger (Eds.), Theory and practice of biological control. Academic Press, N.Y.

Ehler, L.E. 1982. Foreign exploration in California. Environ. Entomol. 11:525-530.

Ehler, L.E,. and L.A. Andres. 1983. Biological control: Exotic natural enemies to control exotic pests, pp. 295-418. In: C.L. Wilson and C.L. Graham (Eds.), Exotic plant pests and North American agriculture. Academic Press, N.Y.

Ehler, L.E., and R.W. Hall. 1982. Evidence for competitive exclusion of introduced natural enemies in biological control. Environ. Entomol. 11:1-4.

Ehler, L.E., and R.W. Hall. 1984. Evidence for competitive exclusion of introduced natural enemies in biological control: An addendum. Environ. Entomol. 13:v-vii.

Ehler, L.E., and J.C. Miller. 1978. Biological control in temporary agroecosystems. Entomophaga 23:207-212.

Force, D.C. 1967. Genetics in the colonization of natural enemies for biological control. Ann. Entomol. Soc. Amer. 60:722-799.

Hall, R.W., and L.E. Ehler. 1979. Rate of establishment of natural enemies in classical biological control. Bull. Entomol. Soc. Amer. 25:280-282.

Hoy, M.A. 1975a. Forest and laboratory evaluations of hybridized Apanteles melanoscelus (Hym.: Braconidae), a parasitoid of Porthetria dispar (Lep.: Lymantriidae). Entomophaga 20:261-268.

Hoy, M.A. 1975b. Hybridization of strains of the gypsy moth parasitoid, Apanteles melanoscelus, and its influence on diapause. Ann. Entomol. Soc. Amer. 68: 261-264.

Hoy, M.A. 1976. Establishment of gypsy moth parasitoids in North America: An evaluation of possible reasons for establishment or nonestablishment, pp. 215-232. In: J.F. Anderson and H.K. Kaya (Eds.), Perspectives in forest entomology. Academic Press, N.Y.

Hoy, M.A. 1978. Variability in diapause attributes of insects and mites: Some evolutionary and practical implications, pp. 101-126. In: H. Dingle (Ed.), Evolution of insect migration and diapause. Springer-Verlag, N.Y.

Hoy, M.A. 1979a. The genetics of biological control agents, pp. 235-238. In: T. Kommedahl (Ed.), Proc. Symp. IX Intern. Congr. Plant Prot., Washington, D.C., Vol. I.

Hoy, M.A. 1979b. The potental for genetic improvement of predators for pest management programs, pp. 106-115. In: M.A. Hoy and J.J. McKelvey (Eds.), Genetics in relation to insect management. Rockefeller Found. Press, N.Y.

Hoy, M.A. 1985. Recent advances in genetics and genetic improvement of the Phytoseiidae. Annu. Rev. Entomol. 30:345-370.

Huettel, M.D. 1976. Monitoring the quality of laboratory-reared insects: A biological and behavioral perspective. Environ. Entomol. 5:807-814.

Joslyn, D.J. 1984. Maintenance of genetic variability in reared insects, pp. 20-29. In: E.G. King and N.C. Leppla (Eds.), Advances and challenges in insect rearing. USDA-ARS, Southern Region, New Orleans, LA.

Kelleher, J.S. 1968. Introduction practices--past and present. Bull. Entomol. Soc. Amer. 15:236-236.

Knutson, L. 1981. Symbiosis of biosystematics and biological control, pp. 61-78. In: G.C. Papavizas (Ed.), Biological control in crop production. BARC Symp. No. 5, Allanheld Press, Osmun, Ottowa.

Laing, J.E., and J. Hamai. 1976. Biological control of insect pests and weeds by imported parasites, predators, and pathogens, pp. 685-743. In: C.B. Huffaker and P.S. Messenger (Eds.), Theory and practice of biological control. Academic Press, N.Y.

Legner, E.F. 1972. Observations on hybridization and heterosis in parasitoids of synanthropic flies. Ann. Entomol. Soc. Amer. 65:254-263.

Legner, E.F. 1979. Prolonged culture and inbreeding effects on reproductive rates of two pteromalid parasites of muscoid flies. Ann. Entomol. Soc. Amer. 72: 114-118.

Levins, R. 1969. Some demographic and genetic consequences of environmental heterogeneity for biological control. Bull. Entomol. Soc. Amer. 15:237-240.

Lucas, A.M. 1969 The effect of population structure on the success of insect introductions. Heredity 24:151-157.

Mackauer, M. 1976. Genetic problems in the production of biological control agents. Annu. Rev. Entomol. 21:369-385.

Messenger, P.S. 1971. Climatic limitations to biological controls. Proc. Tall Timbers Conf. Ecol. Animal Contr. Habitat Mgt. 3:97-114.

Messenger, P.S., and R. van den Bosch. 1971. The adaptability of introduced biological control agents, pp. 68-92. In: C.B. Huffaker (Ed.), Biological control. Plenum Press, N.Y.

Messenger, P.S., F. Wilson and M.J. Whitten. 1976. Variability, fitness, and adaptability of natural enemies, pp. 209-231. In: C.B. Huffaker and P.S. Messenger (Eds.), Theory and practice of biological control. Academic Press, N.Y.

Miller, J.C. 1983. Ecological relationships among parasites and the practice of biological control. Environ. Entomol. 12:620-624.

Myers, J.H. 1978. Biocontrol introductions as grandiose field experiments: Adaptations of the cinnabar moth to new surroundings, pp. 181-188. In: T.E. Freeman (Ed.), Proc. 4th Intern. Symp. Biol. Contr. Weeds.

Myers, J.H., and M.D. Sabath. 1981. Genetic and phenotypic variability, genetic variance, and the success of establishment of insect introductions for the biological control of weeds, pp. 91-102. In: Proc. 5th Intern. Symp. Biol. Contr. Weeds Brisbane, Australia, 1980.

Pimentel, D. 1963. Introducing parasites and predators to control native pests. Can. Entomol. 95:785-792.

Price, P.W. 1972. Methods of sampling and analysis for predictive results in the introduction of entomophagous insects. Entomophaga 17:211-222.

Remington, C.L. 1968. The population genetics of insect introduction. Annu. Rev. Entomol. 13:415-426.

Rosen, D. 1978. The importance of cryptic species and specific identifications a s related to biological control, pp. 23-35. In: Beltsville Symp. Agric. Res. No. 2, Biosystematics in agriculture. Allanheld, Osmun & Co.

Rosen, D., and P. DeBach. 1973. Systematics, morphology and biological control. Entomophaga 18:215-22.

Rosen, D., and C.B. Huffaker. 1983. An overview of desired attributes of effective biological control agents, with particular emphasis on mites, pp. 2-11. In: M.A. Hoy, G.L. Cunningham, and L. Knutson (Eds.), Biological control of pests by mites. Univ. California, Div. Agric. Nat. Res. Spec. Publ. 3304.

Roush, R.T. 1979. Genetic improvement of parasites, pp. 97-105. In: M.A. Hoy & J.J. McKelvey (Eds.), Genetics in relation to insect management. Rockefeller Found. Press, N.Y.

Sailer, R.I. 1978. Our immigrant insect fauna. Bull. Entomol. Soc. Amer. 24:3-11.

Sailer, R.I. 1983. History of insect introductions, pp. 15-38. In: C.L. Wilson and C.L. Graham (Eds.), Exotic plant pests and North American agriculture. Academic Press, N.Y.

Simmonds, F.J. 1976. Some recent puzzles in biological control. Entomophaga 21: 327-332.

Stehr, F.W. 1974. Release, establishment and evaluation of parasites and predators, pp. 124-136. In: F.G. Maxwell and F.A. Harris (Eds.), Proc. Summer Inst. Biol. Contr. Plant Insects and Diseases. Univ. Mississippi Press, Jackson.

Turnbull, A.L., and D.A. Chant. 1961. The practice and theory of biological control of insects in Canada. Can. J. Zool. 39:627-753.

Unruh, T.R., W. White, D. Gonzalez, G. Gordh and R.F. Luck. 1983. Heterozygosity and effective size in laboratory populations of *Aphidius ervi* (Hym.: Aphidiidae). Entomophaga 28:245-258.

van den Bosch, R. 1968. Comments on population dynamics of exotic insects. Bull. Entomol. Soc. Amer. 14:112-115.

Volker, K.C., and R.G. Simpson. 1975. Behavior of alfalfa weevil larvae affecting the establishment of Tetrastichus incertus in Colorado. Environ. Entomol. 4: 742-744.

White, M.J.D. 1973. Animal cytology and evolution, 3rd Ed., Cambridge Univ. Press, Cambridge. 961 pp.

Wilson, F. 1965. Biological control and the genetics of colonizing species, pp. 307-329. In: H.G. Baker and G.L. Stebbins (Eds), Genetics of colonizing species. Academic Press, N.Y.

Zwolfer, H., M.A. Ghani and V.P. Rao. 1976. Foreign exploration and importation of natural enemies, pp. 189-207. In: C.B. Huffaker and P.S. Messenger (Eds.), Theory and practice of biological control. Academic Press, N.Y.

DISCUSSION: Improving establishment of arthropod natural enemies.

DISCUSSION LEADER: R.I. Sailer

To what extent is it possible to transform biological control through importation of natural enemies into a rigorous science? Data show that there is a 20-30% success rate of establishment of intentionally-introduced species. How can this rate be improved? Assuming that success could be improved, if not assured, would the results be cost effective? Examples were cited of species repeatedly introduced over extended periods of time that failed to establish and of others that, while established, were of little economic consequence. Other examples were cited of highly successful species that would never have been introduced had pre-introduction studies been required; they might not have appeared to be good natural enemies. One example was cited of a highly successful species that was held in the laboratory for several years for studies intended to determine whether an economically injurious effect would result if it were released and entered into competition with an already-established but relatively ineffective natural enemy. Once released, the species, Aphytis melinus, did competitively displace Aphytis lingnanensis and also provided effective control of the California red scale. In the meantime, growers had suffered unnecessary losses.

There was general consensus that more pre-introduction studies are desirable, but difficult to conduct under the constrained budgets available. All agreed that studies of host range, phenology and behavior should be conducted in order to facilitate developing an optimal release strategy for the natural enemy. To do such studies, additional resources are necessary and should be sought. However, such studies should not be made a mandatory prerequisite for introduction.

There was general agreement on the need for post colonization and terminal evaluation studies, as these provide basic information needed to guide future introduction programs, as well as the information needed to assess economic benefits of completed projects.

GENETIC IMPROVEMENT OF ARTHROPOD NATURAL ENEMIES THROUGH SELECTION, HYBRIDIZATION OR GENETIC ENGINEERING TECHNIQUES

Steven K. Beckendorf and Marjorie A. Hoy

The term "genetic improvement" refers to man's attempts to alter the genetic characteristics of a species to suit his needs. Genetic improvement is regularly achieved with domestic animals and crops, using standard breeding and hybridization (Allard 1960, Johansson & Rendel 1968). More recently, genetic engineering techniques have been developed with Drosophila, microorganisms, and plants (Bender et al. 1983, Horsch et al. 1984, Rubin & Spradling 1982, Watson et al. 1983). Genetic improvement projects are predicated on the belief that man can identify desirable attributes and, through some type of genetic manipulation, improve the usefulness of the candidate species.

Genetic improvement projects with arthropod natural enemies have, until recently, received little attention because of the concern that genetic manipulation and prolonged rearing under artificial conditions would necessarily result in laboratory-adapted strains that would perform poorly in the field (Boller 1972, Hoy 1976, Mackauer 1976, Messenger & van den Bosch 1971, Roush 1979). However, the successful genetic improvement of phytoseiid predators of spider mites has suggested that this approach is practical (Hoy 1982b; Hoy et al. 1982a, 1983, 1984; Roush & Hoy 1981a, b; Strickler & Croft 1982; Whalon et al. 1982).

Genetic improvement of arthropod natural enemies could be achieved by employing artificial selection, hybridization and genetic engineering. This paper will review the status of these three approaches, assess their usefulness for improving arthropod natural enemies for agricultural crop systems, and outline specific recommendations regarding the directions additional research might take.

BIOLOGICAL CONTROL
IN AGRICULTURAL IPM SYSTEMS

ISBN 0-12-357030-1

SELECTION AND HYBRIDIZATION

History and Examples

Genetic improvement of parasites and predators is an old concept (DeBach 1958, Malley 1916, Sailer 1961), and insects that serve as biological control agents of arthropod pests have undergone laboratory selection for improved climatic tolerance (White et al. 1970, Wilkes 1942), improved sex ratio (Simmonds 1947, Wilkes 1947), host finding ability (Landaluze 1950, Urquijo 1946), host preference (Allen 1954, Box 1956), insecticide resistance (Pielou & Glasser 1952, Robertson 1957), and increased fecundity (Ram & Sharma 1977, Szmidt 1972, Wilkes 1947). Evidence that an artificially-selected biological control agent has been improved in its field effectiveness was provided only recently, with phytoseiid predators of spider mites (Hoy 1982b, Roush & Hoy 1981a, Whalon et al. 1982).

In addition to the use of artificial selection, hybridization has been attempted as a method for increasing vigor or fitness through heterosis. Ashley et al. (1974) reported generally negative results with hybridization and selection of *Trichogramma*, while Hoy (1975a, b) found that a hybrid strain of the gypsy moth parasite *Apanteles melanoscelus* was easier to mass rear than the three strains from which it was derived. Its development rate, diapause attributes, sex ratio, and parasitization rate in the insectary was improved. However, the limited evaluation conducted with the hybrid strain did not indicate that it was more effective than the Connecticut strain under forest conditions (Hoy 1975a). Legner (1972) found that intraspecific hybrids of strains of *Muscidifurax raptor*, *Spalangia endius* and *S. cameroni* exhibited hybrid vigor in the laboratory and suggested that production of superior parasitoids through heterosis should be considered. Interspecific hybridization has also been suggested as a method for introducing desirable genes into closely related insect species (Sailer 1954, 1961). In those cases where the hybrids are viable and fertile, a primitive form of genetic engineering might be accomplished.

Genetic Improvement of Phytoseiid Mites

Genetic improvement of the phytoseiid mite *Metaseiulus occidentalis* has been documented in field trials. *M. occidentalis* is the most important predator of spider mites in deciduous orchards and vineyards in western North America (Flaherty & Huffaker 1970, Huffaker et al. 1970). It acquired resistance to organophosphorus (OP) insecticides such as azinphosmethyl, diazinon, and phosmet through selection in orchards and vineyards (Hoy & Knop 1979, Hoy et al. 1979, Hoyt 1969). Resistance to sulfur was also discovered in native California vineyard populations (Hoy & Standow 1982). To augment these naturally-acquired resistances, *M. occidentalis* was artificially selected in the laboratory with permethrin and carbaryl and crosses and selection were conducted to obtain multi-resistant strains.

Laboratory-selected strains of M. occidentalis have successfully met the criteria for a field success. Strains selected for resistance to permethrin (Hoy & Knop 1979, 1981) and carbaryl (Roush & Hoy 1981a) retained their resistance to OP insecticides despite the fact that they were not subjected to selection with OP pesticides during the projects. The carbaryl-OP and permethrin-OP strains have established in apple, pear, and almond orchards in California, Oregon, and Washington. The resistant strains have spread within the release trees, multiplied, overwintered, survived pesticide applications in the field, and had measurable impacts on spider mite populations (Hoy 1982a; Hoy & Knop 1981, Hoy et al. 1982a, 1983a; Roush & Hoy 1981b). The carbaryl-OP-permethrin and carbaryl-OP-sulfur resistant strains were developed through laboratory crosses and selection and these have also performed well in field trials (Hoy 1984, Hoy et al. 1984).

The genetic improvement project with M. occidentalis took several years to accomplish, and there were a number of steps involved (Hoy 1979, 1982b). First, M. occidentalis had been determined to be an economically-important predator, in part because it acquired resistance to OP insecticides in the field. As a result, it could control spider mites in deciduous orchards and vineyards even though OP insecticides had to be used to control insect pests. Because the target insects may not be controlled with OP insecticides perpetually due to the development of resistance, it was easy to determine a trait to improve--namely, increasing the number of pesticides to which M. occidentalis is resistant. A survey of California populations of M. occidentalis during 1977-78 indicated that native M. occidentalis were susceptible to carbaryl and to permethrin and that substantial variability in OP resistance levels existed in California orchards and vineyards (Hoy & Knop 1979, Hoy et al. 1979, Roush & Hoy 1981a). M. occidentalis is easily reared and mass selection in the laboratory of adult females or immature stages was conducted using these two pesticides.

Genetic analyses indicated that the carbaryl resistance obtained is conferred by a major semidominant gene (Roush & Hoy 1981a), while permethrin resistance is polygenically determined (Hoy & Knop 1981). Laboratory and greenhouse evaluations showed that the resistant strains had normal longevity, fecundity, and diapause capacity and these results led to a decision to release both strains into small field plots beginning in 1979 (Hoy 1984, Hoy & Knop 1981, Hoy et al. 1982a, Roush & Hoy 1981b).

Release strategy is dependent on the genetic basis of the resistance. Because the permethrin resistance is polygenically determined, and is lost if the strain interbreeds with susceptible natives, strains with this resistance must be released after a permethrin application has decimated the native population, or where no natives exist. The release strategy with the carbaryl resistant strain involves field selection with carbaryl to favor the resistance gene.

Both the carbaryl and permethrin resistant strains performed well in small field plots, and the carbaryl-OP-sulfur resistant strain is now being reared and released into commercial almond orchards and vineyards in California on a large scale (Hoy et al. 1982a, 1984). Several companies began selling these resistant predators and management advice in 1984. Two mass rearing methods were developed that should allow nearly unlimited production of the resistant strains and thousands of acres of deciduous orchards and vineyards could receive predators in the next few years (Hoy et al. 1982b). Mass releases of the carbaryl-OP resistant strain were made during 1982 in pear orchards in Australia (R. Field, personal communication).

Permanent establishment of the resistant predator strains in orchards and vineyards should reduce the number of acaricide applications needed, or allow a reduction in the rates of acaricides applied. If so, a favorable cost:benefit ratio for this program should result. Crops in which this predator is known to be important include apples, pears, peaches, almonds, grapes, and walnuts in western North America and Australia.

Other genetic selection projects have been, or are being, conducted with phytoseiid mites (Schulten & van de Klashorst 1974, Strickler & Croft 1982, Voroshilov 1979, Voroshilov & Kolmakova 1977). Attributes selected include pesticide resistance, increased high temperature tolerance, non-diapause, and increased fecundity. A non-diapause strain of M. occidentalis is also resistant to carbaryl and OP insecticides (Hoy 1984). Field (1981) evaluated this carbaryl-OP-nondiapause strain as a biological control agent of the two-spotted spider mite (Tetranychus urticae) on roses. Data obtained from small plots in a University of California greenhouse were promising and suggest that larger-scale tests with this improved predator strain should be conducted (Field 1981, Field & Hoy 1984). Permethrin-resistant strains of Amblyseius fallacis were released into a Michigan apple orchard and successfully established, survived pyrethroid applications, and overwintered (Whalon et al. 1982). The successful genetic improvement of these phytoseiids could serve as a model for other taxonomic groups of natural enemies, including parasites and pathogens of pest arthropods, as well as predators.

GENETIC ENGINEERING

Techniques of genetic engineering offer the possibility of introducing a desirable trait directly into a species chosen for genetic improvement. This may be done by transferring the gene controlling this trait from one species to another, thereby avoiding normal boundaries that prevent interspecific hybridization, or by modifying a known gene in vitro and returning it to the species of origin. Such experiments require not only that a desirable trait and the gene or genes controlling it be identified, but also that methods for purification and isolation of these genes and for gene transfer be available. Currently such methods have been worked out for just one insect, *Drosophila melanogaster*. For this reason and because so much information has been accumulated about the *Drosophila* genome, we propose that *Drosophila* genes can be the starting point for several strategies designed to improve other species. To illustrate how such an approach could work, we first describe the basic methods for preparing and screening genomic libraries that contain cloned copies of all the DNA in the genome of a particular species. We then describe several refinements of the basic library construction methods which allow one to go after and isolate almost any desired *Drosophila* gene. Since the introduction of an isolated gene into a new host species is crucial for genetic improvement strategies, we next describe a method of gene transfer that has recently been worked out in *Drosophila*. Finally we suggest a few experimental strategies that, starting with cloned *Drosophila* sequences, could lead to the modification of an arthropod natural enemy. Our suggestions are based directly on current technology and knowledge of gene organization and expression. Additional opportunities will certainly appear as new techniques and results become available.

We will not cover basic recombinant DNA techniques in this section. General discussions of the use of enzymes to cleave, polymerize, or ligate DNA, of DNA hybridization to identify specific sequences, of gel electrophoresis to separate DNAs and RNAs, and of other relevant techniques can be found in recent texts (for example, Alberts et al. 1983, Lewin 1983) and in more detail in Watson et al. (1983). For specific experimental details we strongly recommend Maniatis et al. (1982). The protocols in this book are not only complete and easily followed; they have all been tested and found to work in the authors' laboratories. More extensive reviews of particular techniques and the results obtained with them can be found in the series *Genetic Engineering* (edited by Williamson 1981-1983) and in volumes 68, 100 & 101 of *Methods in Enzymology* (Colowick & Kaplan 1979-1983).

Construction of Recombinant DNA Libraries

Basic steps in the isolation of genes from any species are indicated in Fig. 1. To construct a library of clones containing all the sequences in the DNA of a particular organism, genomic DNA is fragmented by shearing or by partial digestion with a restriction enzyme. The resulting pieces are then inserted into a specially designed bacteriophage (a bacterial virus) or bacterial plasmid (a circular DNA molecule that can replicate independently of the bacterial chromosome). Convenient restriction sites and genetic markers have been built into these phage or plasmid vectors so it is easy to insert foreign DNA, so that they are easily propagated, and so that detection of inserted DNA is simplified. For

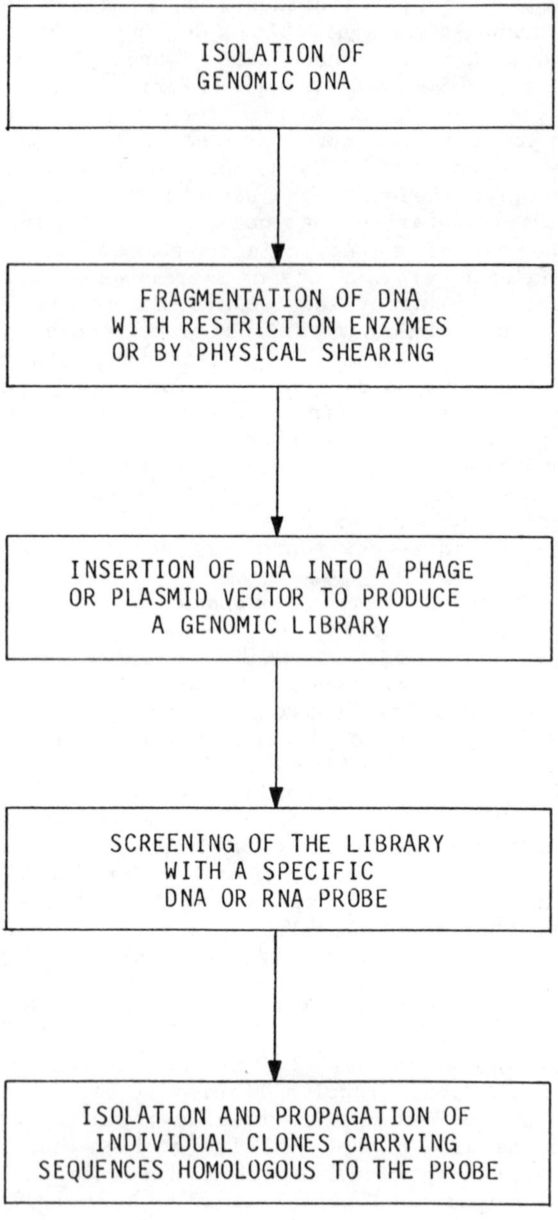

Fig. 1. Flow chart for construction of a genomic library and isolation of clones from it.

properties of common vectors used for library construction see Karn et al. (1983), Maniatis et al. (1983), or Messing (1983). Once the genomic fragments are inserted, the entire population of phage or plasmids can be propagated by growth in bacteria (usually E. coli). The aim is to construct and propagate a sufficient number of recombinant molecules so that the probability is high that any particular genomic sequence will be represented in the library.

The library must then be screened to pick out individual clones containing the gene or genes of interest. An insect genomic DNA library will contain between 10,000 and 1,000,000 distinct molecules, depending on the length of the inserted DNA and the total genome size. Screening techniques must, therefore, allow discrimination of the desired clone from a large background of unrelated clones. The most widely used techniques employ radioactively labeled DNA (or RNA) of a specific sequence as a hybridization probe to identify an homologous sequence in the library. Filter hybridization techniques have been developed for simultaneously screening large numbers of clones from plasmid or bacteriophage libraries (Benton & Davis 1977, Grunstein & Hogness 1975). Briefly, cells or bacteriophage particles containing several thousand recombinant molecules from a library are spread on a petri plate and allowed to grow into colonies (or plaques). A nitrocellulose filter is placed on the surface of the plate, and some of the cells from each colony (or bacteriophage from each plaque) adhere to it. The filter is peeled off, treated to disrupt the cells (or bacteriophage), to denature the DNA, and to fix the DNA irreversibly to the nitrocellulose giving a DNA print of the original plate. Such filter-bound DNA is capable of hybridizing with radioactively labeled DNA of homologous sequence (the probe) and the colonies (or plaques) which hybridize can be identified by autoradiography.

Upon isolation, the cloned fragment can be propagated as an abundant source of a specific sequence that is not contaminated by any other sequences from the starting organism. One can go back to the library again and again to isolate additional clones carrying DNA from other parts of the genome.

Identification and Isolation of Drosophila Genes

Construction of genomic libraries is a general technique which makes all the genes of an organism accessible for isolation as long as a specific probe is available for screening the library. But of course a specific probe for the gene of interest is rarely available before the gene is cloned. In Drosophila a number of approaches have been developed to focus the search for a particular gene. These include the use of semi-specific probes such as hybridization between the complementary DNA copy (cDNA) of an abundantly-transcribed RNA and a cloned sequence (Livak et al. 1978, Spradling et al. 1980) and cross hybridization between a gene previously cloned from another species and its Drosophila homologue (Tobin et al. 1980). Three other approaches which allow one to concentrate on a subset of the sequences in the genome are discussed in more detail below because of their generality. They make most, if not all, *Drosophila* genes available for molecular analysis.

Perhaps the most direct method involves *microdissection* of chromosomal DNA from a region known to contain the target gene. Using glass needles attached to micromanipulators it is possible to cut squashed chromosomes at defined positions and recover the region between the cuts. The technique has been most precisely used to excise one or a few bands from salivary gland polytene chromosomes. The DNA from a few dissected bands is then inserted into a plasmid or phage vector to produce a micro library similar, except in scale, to the genomic libraries just described. This technique has recently been used to isolate clones for several *Drosophila* genes, among them *white* (Scalenghe et al. 1981) and *Antennapedia* (Scott et al. 1983).

The second approach is known as *chromosomal walking*. Since genomic libraries are usually constructed by randomly fragmenting the genomic DNA, any particular sequence can be found in the library in several overlapping fragments. Starting with an initial cloned segment, one can initiate a "walk" along the chromosome by identifying the restriction fragments at the ends of the initial clone (see Maniatis et al. 1982, for methods of mapping restriction fragments) and using them to isolate additional fragments that overlap the ends of the initial one. The secondary clones can be aligned by restriction mapping and the most distal fragment used to isolate a third set of clones that extend further along the chromosome. Each successive step will extend the walk 10-15 kb, and the process is rapid enough that several walks of 200-500 kb have been reported (Bender et al. 1983, Scott et al. 1983). Much larger distances can be crossed if the end points of chromosomal rearrangements (deletions, inversions, translocations) lie within the region cloned during the walk. One can then "jump" to the other end of the rearrangement, whether that be thousands of kb away or even on another chromosome (Bender et al. 1983). This technique is very useful in *Drosophila* because many rearrangements have been characterized and because more than 300 regions have been cloned and their chromosomal positions determined by hybridizing radioactive DNA to polytene chromosomes in situ. These sequences are distributed throughout the euchromatic regions of all *Drosophila* chromosomes at intervals of approximately 500 kb. Thus many, if not most, *Drosophila* genes can be approached by walking or jumping from a known cloned sequence. For example, this approach was used to isolate the *bithorax* complex, the *rosy* gene and the gene encoding acetylcholinesterase (*Ace*) in one walk on the third chromosome (Bender et al. 1983).

The third method, known as *transposon tagging*, is best used when mutants of a particular gene are known but its chromosomal position has not been precisely determined. This method takes advantage of the observation that transposable elements can inactivate a gene by inserting into it. The transposable element can then act as a label to identify the gene of interest in a library made from the mutant strain. In Drosophila this method is most easily applied by using the transposable P element whose movement can be controlled. In strains that have many copies of the P element, it transposes very infrequently, apparently because it encodes a repressor that prevents transposition. However, when sperm from a fly that carries multiple copies of the P element (a P strain) fertilizes an egg from a strain with no P elements (an M strain), the paternal chromosomes encounter no repressor and P elements within them are induced to transpose rapidly (Engels 1979). As a result, mutations caused by the insertion of the P element occur frequently. Once a mutation has been identified in the Drosophila gene of interest, a genomic library can be constructed from the mutant line and clones containing the desired sequence identified by hybridization to the inserted P element. The mutant clone can then be used as a hybridization probe to identify in a wild type genomic library clones containing the nonmutant gene. This method was first used to clone the RNA polymerase II gene (Searles et al. 1982) and has since been applied to several other loci.

Gene Transfer

If we are to use genetic engineering to improve arthropod natural enemies, we need methods not only to isolate genes but also to put isolated genes into new hosts. At the moment Drosophila is the only organism into which a cloned gene can be reintroduced so that it is properly expressed. The method uses the transposable P element because, as just described, its transposition can be genetically manipulated. If a cloned P element is injected into an early Drosophila embryo which came from an M strain, P will transpose at high frequency into the resident chromosomal DNA (Spradling & Rubin 1983). When other cloned sequences are placed between the ends of the P element, they are inserted into the chromosomal DNA along with P (Rubin & Spradling 1983). In almost all cases tested, the developmental regulation of genes introduced by P-element transformation is normal and unaffected by the position of insertion into the chromosome (Goldberg et al. 1983, Rubin & Spradling 1982, Scholnick et al. 1983, Spradling & Rubin 1983). This method of gene transfer provides, therefore, a precise and reliable method for reintroducing cloned sequences into the Drosophila genome so that their functional capabilities can be monitored.

While P element mediated transformation has only been demonstrated in Drosophila, it may be possible to apply the technique to other species. Since the P element apparently encodes at least some of the functions required for its transposition (O'Hare & Rubin 1983, Rubin & Spradling 1982), its host range may extend well outside D. melanogaster.

If the host range of the P element turns out to be too restricted, other vectors will need to be developed. One of the most likely alternates would be a transposon indigenous to the species slated for improvement. Since transposons seem to be widely dispersed in all eukaryotes that have been examined, they should be readily identified in the target species as moderately repeated sequences that reside in different stocks of the species. By analogy with the known *Drosophila* transposable elements, it is unlikely that the movement of many will be easily controlled. Perhaps most promising would be an element found, as is P, to be present only in some individuals of a species or in one species but not its close relatives. It might then be induced to transpose when introduced into an individual lacking the element.

Insect viruses might offer another alternative. A virus which was able to infect the target organism and integrate its genome into the chromosome but could not replicate autonomously, might act as a vector for additional sequences. Unfortunately, no suitable insect viruses have yet been identified (but see Faulkner & Boucias, this volume).

One might imagine that injection of the desired sequence directly into developing embryos without a vector might provide a generally applicable method of gene transfer since it would not require the identification and characterization of specific vectors. Unfortunately the experience with *Drosophila* is not encouraging. Such experiments were tried for years in *Drosophila* with no reproducible results. However, recent unpublished results on the nematode *Caernorhabditis elegans* suggest that with strong selective markers such techniques can work at low frequencies to give at least limited expression of the injected genes. The progress of work with this organism should be followed closely to determine whether technical innovations are introduced which can be applied to other species.

Strategies for Genetic Improvement

In this section we propose three ways in which the techniques of gene isolation and gene transfer outlined above could be used to improve arthropod natural enemies. Gene transfer seems limited at this time to traits governed by one or a small number of genes that could be included within the same P element. While no definite upper limit has been established for the amount of DNA which can be inserted into a P element and then transferred, the frequency of transfer seems to decrease rapidly when the insert exceeds 15-20 kb (G.M. Rubin, personal communication). Thus it seems unlikely that genes responsible for polygenic traits could be successfully transferred with current techniques. Unfortunately many of the traits that would be desirable in an arthropod natural enemy are, or are likely to be, under polygenic control. These include increases in overall viability or fecundity, changes in host or habitat selection, and changes in dispersal or diapause characteristics.

What are we left with? The most obvious target for gene transfer, and the one used in the examples below, is insecticide resistance. In many cases it is controlled by a single gene (Tsukamoto 1983), and it has the added advantage that successfully transformed individuals can be identified by treatment with an insecticide.

Each of the examples described below starts with genes from D. melanogaster, thereby taking advantage of the genetic and molecular information and methodology available for this organism. The one current drawback is the small number of investigations on pesticide resistance in Drosophila (Tsukamoto 1983). Prior analysis to define additional pesticide resistant genes would facilitate the strategies outlined below.

The first example assumes that there exist alleles of a previously isolated gene which confer insecticide resistance on Drosophila. Such a gene might be the acetylcholinesterase gene (Ace) which was isolated during a chromosome walk on the third chromosome (Bender et al. 1983). In other insects changes in acetylcholinesterase often lead to resistance to DDT and OP insecticides (J. Casida, personal communication; Tsukamoto 1983). If this result were duplicated in Drosophila, the normal Ace gene could be used to identify the resistance gene in a library made from resistant Drosophila. The resistance gene could then be inserted into a P element vector and, if the P element functioned in species other than Drosophila, could be transferred to a suitable recipient species. Transformants would be identified by resistance to the insecticide.

In the second example, insecticide resistance might reside in mutants of a previously unknown but genetically mappable gene. An example is the DDT resistance in Drosophila described by Tsukamoto & Ogaki (1953). By recombination analysis this has been approximately located at genetic map position 65 on the second chromosome, a site which corresponds to regions 47 or 48 on the polytene chromosome map but does not coincide with a previously characterized gene. If one wished to isolate this gene, its cytological location could be determined more precisely with the aid of previously characterized translocations, at least five of which have breakpoints in this region (Lindsley & Grell 1968). One could then isolate the gene either by microdissection or by a chromosome walk from one of the seven previously identified cloned sequences that come from this region. Once the gene had been isolated, it could be inserted into a P element vector and tested for the ability to confer resistance on Drosophila and on the species to be improved.

The third strategy relies on the evolutionary conservation of genes present in both Drosophila and other insects. Proteins responsible for insecticide resistance have been identified in several other insects. If the gene for such a protein has been identified in Drosophila, it may be possible to isolate the corresponding gene from the resistant insect by DNA sequence cross homology. Again we can use the example of acetylcholinesterase, changes in which are responsible for several types of pesticide resistance in houseflies and other insects (Tsukamoto 1983). Using the Drosophila Ace gene as a hybridization probe, homologous sequences could be isolated from a genomic library of housefly DNA. If the genes have retained sufficient homology, the housefly Ace gene should be among the sequences isolated in this way. Since many structural genes, among them actin, tubulin and histones show cross hybridization across much wider evolutionary distances (for example see Tobin et al. 1980), this seems like a promising approach. Once the Ace gene were isolated from a pesticide resistant housefly, it could be tested to be sure that the resistance could be transferred into Drosophila along with the gene. If so, it could then be transferred to another species.

CONSTRAINTS UPON GENETIC IMPROVEMENT PROGRAMS

Why have biological control workers been slow to exploit artificial selection and hybridization techniques? [Genetic engineering is so new that we can't say there has been any delay in its use.] There may be several reasons. 1) Entomologists are aware that laboratory-reared insects often become laboratory-adapted and their field efficacy is reduced either from genetic or behavioral changes in the colonies being reared (Boller 1972, Bush et al. 1976, Huettel 1976, Mackauer 1976, Messenger & van den Bosch 1971). The concern over these problems has served as one deterrent to genetic improvement projects. 2) There may be an information gap. Most biological control workers are trained in entomology and ecology, but rarely have training in genetics and are often unaware of genetic techniques that might solve rearing problems. Conversely, few geneticists are trained in biological control and rarely understand the needs of such applied projects. They are most often interested in advancing fundamental genetic knowledge and less interested in applying genetics as a tool to solve entomological problems. There is an array of fundamental genetic theory that should be tested to determine if the quality of laboratory reared biological control agents can be improved. 3) The failure to document that the improved strain is successful in the field has given genetic improvement a negative image. This image is not warranted because few of the "improved" natural enemies were ever field tested. Thus, we don't know if the projects were a failure or not. The success of genetic improvement projects currently under development in China, New Zealand, Israel, and the United States will determine the acceptance of this tool. 4) The logistics of maintaining cultures of host/prey species and natural enemies in the laboratory is a deterrent in certain cases.

Genetic improvement projects need not always produce a biological control agent that retains all of the attributes of a "wild" natural enemy. Classical biological control releases require the species to establish, survive over many seasons, and control the target pest. Thus inoculative (classical) releases with genetically-improved biological agents will be the most difficult to achieve, as the improved agent should be as good as, or better than, a "wild" insect in the field. Despite these constraints, resistant strains of M. occidentalis have persisted for at least five years in the field (Hoy 1982a; Hoy & Knop 1981; Hoy et al. 1982a, b, 1983; Roush & Hoy 1981b), and permethrin-resistant strains of A. fallacis have overwintered (Whalon et al. 1982).

Inoculative release strategy using genetically-improved biological control agents will depend upon the genetic basis of the traits involved. If the improved attribute is determined by a single (or very few) gene(s), and if the releases are made where the species is endemic, then the desired gene(s) can be introduced into the established, presumably well-adapted native population, particularly if a pesticide can be applied as a selective agent. If the trait that has been modified is monogenic, then crosses with wild stocks prior to release could increase the fitness of the released individuals. An alternative method involves replacing the native population with the improved strain; again the availability of a selective pesticide can be an effective management tool. If the native species is absent, the released strain should establish and competition with the native strain is not an issue. Whether native and released strains could ever coexist in the same general area is unknown. Strains with polygenic traits can be established if the native strain can be decimated or eliminated so that a high ratio of released to native natural enemies exists. Subsequent selection in the field with an appropriate selective pesticide will help to maintain them.

Inundative releases typically involve large numbers and the released natural enemy is expected to persist only a relatively short time, often in a temporary crop system. The genetic basis of the improved trait (polygenic vs. major gene) and the reproductive isolation of the strain may be much less important than they are in inoculative releases.

Is the field success achieved with phytoseiids likely to be repeated with other biological control agents? Only time will tell. These predators have desirable attributes that contribute to their successful manipulation. The field effectiveness of the laboratory-selected strains of M. occidentalis may be due to the fact that: a) we reared the colonies under cyclic temperatures, which may have helped to retain genetic variability; b) we provided them natural prey, which varied in availability (sometimes the predators were stressed and this may have helped to retain colony vigor); c) predators were required to search in a moderately large arena, which may have ensured that they retain normal prey location behavior; and/or d) colonies were initiated with 40-100 females, and the selected lines were never reduced below 100 females so that the influences of founder effects and genetic drift may have been reduced.

Selection programs with species destined for inundative releases are most likely to be successful since it is not necessary to retain all traits of a "normal" natural enemy, particularly if the biological control agent is being released periodically into temporary crop systems. Under these conditions, it is not necessary to retain normal diapause responses or dispersal behavior or habitat selection ability since the biological control agents can be mass-reared and delivered to the target site. Fecundity, sex ratio, and viability should be maintained through adequate quality controls and even may be improved.

Production of "standard strains" through the use of single or double hybrids offers another potentially effective method for improving efficacy of biological control agents as it does with silkworm breeding (Hoy 1976). Hybridization has probably been undervalued as a colony management tool.

Genetic improvement of biological control agents is time consuming, expensive compared to many classical importation projects, and requires a substantial amount of biological and ecological knowledge about the natural enemy under consideration. Thus, it is not a panacea. Our knowledge about the genetics of biological control agents is poor, and our ability to determine which traits need improving is a major limiting factor. The availability of genetic variability upon which selection can operate is not likely to be a limiting factor, since most arthropods (and pathogens of arthropods) are quite variable for a wide array of attributes. However, rarely do we know why an arthropod natural enemy fails to establish or to control a pest unless substantial research efforts have been made. Thus, laboratory selection may remain a special tool for improving limited numbers of natural enemies about which we know a great deal. This means that the pest must be serious, probably widely distributed, and may occur in a number of crops. Despite these constraints, biological control agents other than phytoseiids should be considered suitable candidates for genetic improvement projects.

RECOMMENDATIONS

Hybridization and Selection Research

1. Genetic improvement of arthropod natural enemies is a useful technique for enhancing the efficacy of biological control agents in agricultural crop systems and is worthy of additional research support. Research projects should be funded for at least five years as feasibility projects since little fundamental genetic information is currently available on arthropod natural enemies.
2. Key natural enemies in major agricultural crop systems should be identified as potential candidates for genetic improvement

projects using the following criteria: a) ability to identify characters needing improvement, b) ability to rear the natural enemy under laboratory conditions, and c) ability to choose an appropriate improvement approach, such as genetic engineering, hybridization, or artificial selection.

3. Genetically-improved natural enemies can be released as though they were exotic species being established in a new environment; if the improved traits are quantitatively determined, then some form of reproductive isolation is required if the native strains are present or the traits will be lost after interbreeding unless strong selection is exerted to favor the released strain. Research on methods to enhance reproductive isolation of improved strains to be released sympatrically with the native strains is necessary.
4. Maintaining quality of mass-reared natural enemies is necessary if periodic or inundative releases are to be feasible. Genetic studies should complement research on improved artificial diets, preventing or treating arthropod diseases, and improving the abiotic conditions of the environment so that inadvertent selection is reduced. Strategies for maintaining quality might include maintenance of different lines which could be hybridized prior to releases, but several different approaches should be evaluated. Such research requires long term funding, but the solutions obtained will be of value for genetic control, as well as for biological control programs.
5. We rarely know why a natural enemy is not as effective as desired in agricultural crops. Studies of natural enemy behavior, biology, and ecology in agricultural crop systems should receive high priority so that candidates can be identified for additional genetic improvement efforts. Genetic improvement should be attempted ONLY when naturally-occurring biotypes do not appear to exist that would solve the particular problem.
6. Development of pesticide resistance in arthropod natural enemies in agricultural crops is desirable as certain key pests will need to be managed with pesticides until cultural, pheromonal, biological, or other control methods can be devised for them. Pesticide-resistant natural enemies can play a role in many agricultural IPM systems. Research should be funded that will determine if it is possible to select additional resistant natural enemy strains other than phytoseiids.

Genetic Engineering Research

1. The host range of the P element should be examined to determine whether it will be a widely useable vector. To carry out such studies a marker gene that will confer an easily recognizable phenotype on transformed individuals should be incorporated into the P element vector. Because genetic markers are unlikely to exist in the hosts which will be

examined, we suggest that a gene specifying pesticide resistance be used as this marker.

2. While the host range of the P element may be limited, there may well be transposons in the target species which would work equally well as vectors for gene transfer. Therefore, if P will not work, we recomend that transposons in the target species or its near relatives be characterized. It may be possible to identify one or more transposons which would work in a manner analogous to the P element. An ideal candidate would be a transposon which is present in some, but not all, strains of the target species or in a near relative but not in the target species itself.
3. Similarly, insect viruses should be surveyed to see whether they have the appropriate properties to serve as infective or injectable carriers of desirable genes. These properties include integration in a fairly stable way into the host genome under conditions that do not kill the host cell but may allow the expression of a gene that has been inserted into the viral genome. Vertebrate retroviruses display these properties and a possible insect analog has been identified in the Copia-like elements of *Drosophila*.
4. The strategies we outlined above rely heavily on the use of *Drosophila* genes associated with pesticide resistance. However, little work has been done to identify such genes. Reports of loci conferring resistance to DDT, BHC, parathion, and carbaryl seem to have identified a single gene or a complex of closely linked genes on the second chromosomes (Kikkawa 1964a, b; Tsukamoto & Ogaki 1954). It seems likely that more extensive analysis would identify several other pesticide resistance loci, similar to those found in the housefly (Tsukamoto 1983). After characterization, one or more of these loci could be cloned and used in gene transfer experiments.

ACKNOWLEDGMENT

This work was supported in part by grant number 12866 from the National Institute of Child Health and Human Development to SKB and by the Almond Board of California, Statewide UC/IPM, California Department of Agriculture, Experiment Station Project 3522-H, Western Regional Project W-84, and Pear Zone to MAH. We thank Jay Rosenheim, Reece Sailer and Susan McNabb for comments on the manuscript, and Terri DeLuca for assistance in preparing the manuscript.

REFERENCES

Alberts, B., D. Bray, J. Lewis, M. Rolf, K. Roberts and J.D. Watson. 1983. Molecular biology of the cell. Garland Publ., N.Y. 1181 pp.

Allard, R.W 1960. Principles of plant breeding. John Wiley & Sons, N.Y.

Allen, H.W. 1954. Propagation of Horogenes molestae, an Asiatic parasite of the Oriental fruit moth, on the potato tuberworm. J. Econ. Entomol. 47: 278-281.

Ashley, T.R., D. Gonzalez and T.F. Leigh. 1974. Selection and hybridization of Trichogramma. Environ. Entomol. 3:43-48.

Bender, W., P. Spierer and D.S. Hogness. 1983. Chromosomal walking and jumping to isolate DNA from the Ace and rosy loci and the bithorax complex in Drosophila melanogaster. J. Mol. Biol. 168:17-33.

Benton, W.D., and R.W. Davis. 1977. Screening qt recombination clones by hybridization to single plaques in situ. Science 196:180.

Boller, E. 1972. Behavioral aspects of mass-rearing insects. Entomophaga 17:9-25.

Box, H.E. 1956. Battle against Venezuela's cane borer. I. Preliminary investigations and the launching of a general campaign. Sugar 25-27, 30:45.

Bush, G.L., R. Neck and G.B. Kitto. 1976. Screwworm eradication; inadvertent selection for non-competitive ecotypes during mass rearing. Science 193:491-493.

Colowick, S.P., and N.O. Kaplan (Eds.). 1979-1983. Methods in enzymology, Academic Press, N.Y., Vols. 68, 100, 101.

DeBach, P. 1958. Selective breeding to improve adaptations of parasitic insects. Proc. Tenth Intern. Congr. Entomol. 4:759-768.

Engels, W.R. 1979. Hybrid dysgenesis in Drosophila melanogaster: Rules of inheritance of female sterility. Genet. Res. 33:219-236.

Field, R.P. 1981. Evaluation of genetically-improved strains of Metaseiulus occidentalis (Nesbitt) (Acarina: Phytoseiidae) for integrated control of spider mites on roses in greenhouses. Ph.D. thesis, Univ. of California, Berkeley.

Field, R.P., and M.A. Hoy. 1984. Biological control of spider mites on greenhouse roses. Calif. Agric. 38(3/4):29-32.

Flaherty, D.L., and C.B. Huffaker. 1970. Biological control of Pacific mites and Willamette mites in San Joaquin Valley vineyards. I. Role of Metaseiulus occidentalis. II. Influence of dispersion patterns of Metaseiulus occidentalis. Hilgardia 40:267-330.

Goldberg, D.A., J.W. Posakony and T. Maniatis. 1983. Correct developmental expression of a cloned alcohol dehydrogenase gene transduced into the Drosophila germ line. Cell 34:59-73.

Grunstein, M., and D. Hogness. 1975. Colony hybridization: A method for the isolation of cloned DNAs that contain a specific gene. Proc. U.S. Natl. Acad. Sci. 72:3961.

Horsch, R.B., R.T. Fraley, S.G. Rogers, P.R. Sanders, A. Lloyd and N. Hoffman. 1984. Inheritance of functional foreign genes in plants. Science 223:496-498.

Hoy, M.A. 1975a. Forest and laboratory evaluations of hybridized Apanteles melanoscelus (Hym.: Braconidae), a parasitoid of Porthetria dispar (Lep.: Lymantriidae). Entomophaga 20:261-268.

Hoy, M.A. 1975b. Hybridization of strains of the gypsy moth parasitoid, Apanteles melanoscelus, its influence upon diapause. Ann. Entomol. Soc. Amer. 68:261-264.

Hoy, M.A. 1976. Genetic improvement of insects: Fact or fantasy. Environ. Entomol. 5:833-839.

Hoy, M.A. 1979. The potential for genetic improvement of predators for pest management programs, pp. 106-115. In: M.A. Hoy and J.J. McKelvey (Eds.), Genetics in relation to insect management. Rockefeller Found. Press, N.Y.

Hoy, M.A. 1982a. Aerial dispersal and field efficacy of a genetically-improved strain of the spider mite predator Metaseiulus occidentalis. Entomol. Expt. Appl. 32:205-212.

Hoy, M.A. 1982b. Genetics and genetic improvement of the Phytoseiidae, pp. 72-89. In: M.A. Hoy (Ed.), Recent advances in knowledge of the Phytoseiidae. Univ. California, Div. Agric. Sci., Berkeley Publ. 3284.

Hoy, M.A. 1984. Genetic improvement of a biological control agent: Multiple pesticide resistances and nondiapause in Metaseiulus occidentalis (Nesbitt) (Phytoseiidae). Proc. VI Intern. Congr. Acarology, Edinburgh, August 1982. (In Press).

Hoy, M.A., W.W. Barnett, L.C. Hendricks, D. Castro, D. Cahn and W.J. Bentley. 1984. Managing spider mites in almonds with pesticide-resistant predators. Calif. Agric. 38(7/8):18-20.

Hoy, M.A., W.W. Barnett, W.D. Reil, D. Castro, D. Cahn, L.C. Hendricks, R. Coviello and W.J. Bentley. 1982a. Large-scale releases of pesticide-resistant spider mite predators. Calif. Agric. 36(1/2):8-10.

Hoy, M.A., D. Castro and D. Cahn. 1982b. Two methods for large-scale production of pesticide-resistant strains of the spider mite predator Metaseiulus occidentalis (Nesbitt) (Acarina: Phytoseiidae). Zeitsch. Angew. Entomol. 94:1-9.

Hoy, M.A., D. Flaherty, W. Peacock and D. Culver. 1979. Vineyard and laboratory evaluations of methomyl, dimethoate, and permethrin for a grape pest management program in the San Joaquin Valley of California. J. Econ. Entomol. 72: 250-255.

Hoy, M.A., and N.F. Knop. 1979. Studies on pesticide resistance in the phytoseiid Metaseiulus occidentalis in California. Rec. Adv. Acarol. 1:89-94.

Hoy, M.A., and N.F. Knop. 1981. Selection for and genetic analysis of permethrin resistance in Metaseiulus occidentalis: Genetic improvement of a biological control agent. Entomol. Exp. Appl. 30:10-18.

Hoy, M.A., and K.A. Standow. 1982. Inheritance of resistance to sulfur in the spider mite predator Metaseiulus occidentalis. Entomol. Exp. Appl. 31:316-323.

Hoy, M.A., P.H. Westigard and S.C. Hoyt. 1983. Release and evaluation of a laboratory-selected, pyrethroid resistant strain of the predaceous mite Metaseiulus occidentalis (Acari: Phytoseiidae) in southern Oregon pear orchards and a Washington apple orchard. J. Econ. Entomol. 76:383-388.

Hoyt, S.C. 1969. Integrated chemical control of insects and biological control of spider mites on apple in Washington. J. Econ. Entomol. 62:74-86.

Huettel, M.D. 1976. Monitoring the quality of laboratory-reared insects; a biological and behavioral perspective. Environ. Entomol. 5:807-814.

Huffaker, C.B., M. van de Vrie and J.A. McMurty. 1970. Ecology of tetranychid mites and their natural enemies: A review. II. Tetranychid populations and their possible control by predators: An evaluation. Hilgardia 40:391-458.

Johansson, I., and J. Rendel. 1968. Genetics and animal breeding. Oliver & Boyd, Edinburgh.

Karn, J., S. Brenner and L. Barnett. New bacteriophage lambda vectors with positive selection for cloned inserts, pp. 3-19. In: S.P. Colowick and N.O. Kaplan (Eds.), Methods in enzymology. Academic Press, N.Y. Vol. 101.

Kikkawa, H. 1964a. Genetical studies on the resistance to parathion in Drosophila melanogaster. II. Induction of a resistance gene from its susceptible allele. Botyi-Kagaku 29:37.

Kikkawa, H. 1964b. Genetical studies on the resistance to Sevin in Drosophila melanogaster. Botyu-Kagaku 29:42.

Landaluze, P.U. 1950. Aplicacion de la genetia al aumento de la eficacia del Trichogramma minutum en la lucha biologica. Bolet. Patol. Vegetal Entomol. Agric. 18(1/2):1-12.

Legner, E.G. 1972. Observations on hybrdization and heterosis in parasitoids of synanthropic flies. Ann. Entomol. Soc. Amer. 65:254-263.

Lewin, B. 1983. Genes. John Wiley & Sons, N.Y. 715 pp.

Lindsley, D.L., and E.H. Grell. 1968. Genetic variations of Drosophila melanogaster. Carnegie Instit., Washington, D.C. 472 pp.

Livak, K.J., R. Freund, M. Schweber, P.C. Wensiuk and M. Meselson. 1978. Sequence organization and transcription at two heat shock loci in Drosophila. Proc. Nat. Acad. Sci. USA 75:5613-5617.

Mackauer, M. 1976. Genetic problems in the production of biological control agents. Annu. Rev. Entomol. 21:369-385.

Mally, C.W. 1916. On the selection and breeding of desirable strains of beneficial insects. S. Agr. J. Sci. 13:191-195.

Maniatis, T., E.F. Fritsch and J. Sambrook. 1982. Molecular cloning: A laboratory manual. Cold Spring Harbor Lab., Cold Spring Harbor, N.Y.

Messenger, P.A., and R. van den Bosch. 1971. The adaptability of introduced biological agents, pp. 69-92. In: C.B. Huffaker (Ed.), Biological control. Plenum Press, N.Y.

Messing, J. 1983. New M13 vectors for cloning, pp. 20-78. In: S.P. Colowick and N.O. Kaplan (Eds.), Method in enzymology. Academic Press, N.Y. Vol. 101.

O'Hare, K., and G.M. Rubin. 1983. Structures of P transposable elements and their sites of insertion and excision in the Drosopila melanogaster genome. Cell 34:25-35.

Pielou, D.P., and R.F. Glasser. 1952. Selection for DDT resistance in a beneficial insect parasite. Science 115:117.

Ram, A., and A.K. Sharma. 1977. Selective breeding for improving the fecundity and sex ratio of Trichogramma fasciatum (Perkins) (Trichogrammatidae: Hymenoptera), an egg parasite of lepidopterous hosts. Entomologist 2:133-137.

Robertson, J.G. 1957. Changes in resistance to DDT in Macrocentrus ancylivorus Rohw. (Hymenoptera: Braconidae). Canad. J. Zool. 35:629-633.

Roush, R.T. 1979. Genetic improvement of parasites, pp. 97-105. In: M.A. Hoy and J.J. McKelvey (Eds.), Genetics in relation to insect management. Rockefeller Found. Press, N.Y.

Roush, R.T., and M.A. Hoy. 1981a. Genetic improvement of Metaseiulus occidentalis: Selection with methomyl, dimethoate, and carbaryl and genetic analysis of carbaryl resistance. J. Econ. Entomol. 74:138-141.

Roush, R.T., and M.A. Hoy. 1981b. Laboratory, glasshouse, and field studies of artificially selected carbaryl resistance in Metaseiulus occidentalis. J. Econ. Entomol. 74:142-147.

Rubin, G.M., and A.C. Spradling. 1982. Genetic transformation of Drosophila with transposable element vectors. Science 218:348-353.

Sailer, R.I. 1954. Interspecific hybridization among insects with a report on crossbreeding experiments with stink bugs. J. Econ. Entomol. 47:377-383.

Sailer, R.I. 1961. Possibilities for genetic improvement of beneficial insects, pp. 295-303. In: Germplasm resources. Amer. Assoc. Adv. Sci., N.Y.

Scalenghe, F., E. Turco, J.E. Edstrom, V. Pixotta and M. Melli. 1981. Microdissection and cloning of DNA from a specific region of Drosophila melanogaster polytene chromosomes. Chromosoma 82:205-216.

Scholnick, S.B., B.A. Morgan and J. Hirsh. 1983. The cloned dopa decarboxylase gene is developmentally regulated when reintegrated into the Drosophila genome. Cell 34:37-45.

Schulten, G.G.M. and G. van de Klashorst. 1974. Genetics of resistance to parathion and demeton-s-methyl in Phytoseiulus persimilis A.H. (Acari: Phytoseiidae). Proc. 4th Intern. Congr. Acarology, pp. 519-524.

Scott, M.P., A.J. Weiner, T.I. Hazelrigg, B.A. Polisky, V. Pirotta, F. Scalenghe and T.C. Kaufman. 1983. The molecular organization of the Antennapedia locus of Drosophila. Cell 35:763-776.

Searles, L.L., R.S. Jokerst, P.M. Bingham, R.A. Voekler and A.L. Greenleaf. 1982. Molecular cloning of sequences from a Drosophila RNA polymerase II locus by P element transposon tagging. Cell 31:585-592.

Simmonds, F.J. 1947. Improvement of the sex ratio of a parasite by selection. Can. Entomol. 79:41-44.

Spradling, A.C., M.R. Digan, A.P. Mahowald, M. Scott and E.A. Craig. 1980. Two clusters of genes for major chorion proteins of Drosophila melanogaster. Cell 19:905-914.

Spradling, A.C., and G.M. Rubin. 1983. The effect of chromosomal position on the expression of the Drosophila xanthine dehydrogenase gene. Cell 34:47-57.

Strickler, K.A., and B.A. Croft. 1982. Selection for permethrin resistance in the predatory mite Amblyseius fallacis Garman (Acarina: Phytoseiidae). Entomol. Exp. Appl. 31:339-345.

Szimdt, A. 1972. Studies on the efficiency of various strains of the parasite Dahlbominus fuscipennis (Zett.) (Hymenoptera, Chalcidoidea) under natural conditions. Ekol. Pol. 20:299-313.

Tobin, S.L., E. Zulauf, F. Sanchez, E.A. Craig and B.J. McCarthy. 1980. Multiple actin-related sequences in the Drosophila melanogaster genome. Cell 19:121-131.

Tsukamoto, M. 1983. Methods of genetic analysis of insecticide resistance. In: G.P. Georghiou and T. Saito (Eds.), Pest resistance to pesticides, Plenum Press, N.Y.

Tsukamoto, M., and M. Ogaki. 1953. Inheritance of resistance to DDT in Drosophila melanogaster. Botyu-Kagaku 18:39.

Urquijo, P. 1946. Seleccion des estirpes de Trichogramma minutum Riley de maxima effectividad parasitaria. Boln. Patol. Veg. Ent. Agric. 14:199-216.

Voroshilov, H.V. 1979. Heat-resistant lines of the mite Phytoseiulus persimilis A.-H. Genetika 15:70-76.

Voroshilov, H.V., and L.I. Kolmakova. 1977. Heritability of fertility in a hybrid population of Phytoseiulus. Genetika 13:1496-1497.

Watson, J.D., J. Tooze and D.T. Kung. 1983. Recombinant DNA: A short course. W.A. Freeman and Co., N.Y.

Whalon, M.E., B.A. Croft and T.M. Mowry. 1982. Introduction and survival of susceptible and pyrethroid-resistant strains of Amblyseius fallacis (Acari: Phytoseiidae) in a Michigan apple orchard. Environ. Entomol. 11:1096-1099.

White, E.B., P. DeBach and M.J. Garber. 1970. Artificial selection for genetic adaptation to temperature extremes in Aphytis lingnanensis Compere (Hymenoptera: Aphelinidae). Hilgardia 40:161-192.

Wilkes, A. 1942. The influence of selection on the preferendum of a chalcid, (Microplectron fuscipennis Zett.) and its significance in the biological control of an insect pest. Proc. Roy. Soc. London Ser. B 130:400-415.

Wilkes, A. 1947. The effects of selective breeding on the laboratory propagation of insect parasites. Proc. Roy. Soc. London Ser. B 134:227-245.

Williamson, R. (Ed.). 1981-1983. Genetic engineering. Academic Press, London, Vols. 1-4.

DISCUSSION: Genetic improvement of arthropod natural enemies through selection, hybridization or genetic engineering techniques.

DISCUSSION LEADER: R.I. Sailer

The availability of marker genes to facilitate selection for desired traits was questioned and it was noted that lack of genetic knowledge is a problem in work with arthropod natural enemies. Use of mutagens as a means of increasing heritable variation from which desired traits might be obtained through selection may rarely be required since the intrinsic variability in arthropods appears high.

The patentability of genetically-improved natural enemies was questioned. Problems in obtaining patents might arise from multiple sources of funding and wide dispersal of information and of cultures. However, it might be desirable to patent improved strains so that quality control can be maintained during commercial production of such strains.

Genetic engineering techniques may become useful in producing superior strains of natural enemy species. However, the point was made that it is most sensible to first search for pre-adapted biotypes. If none are found, then prospects for developing improved strains should be explored. One problem confronting both genetic selection and genetic engineering is the quantitative nature of so many desirable behavioral and physiological traits. Such quantitative traits are difficult to stabilize or otherwise work with.

In considering the use of genetic engineering for developing pesticide-resistant strains of arthropod natural enemies, the possibility of producing new strains more quickly than through standard artificial selection projects is particularly intriguing. Thus, a process that would normally require years could be reduced to weeks; this assumes that a major gene can be identified, reliable vectors can be identified, once the gene is in the new host that it is stable and is expressed in an appropriate manner. These assumptions are stringent, but if such a scenario works, it should prove a highly cost effective way to develop improved natural enemies for agricultural IPM systems.

IDENTIFYING AND DISCRIMINATING AMONG BIOTYPES OF PARASITES AND PREDATORS

Leopoldo E. Caltagirone

Identification and discrimination of biotypes of parasites and predators would significantly increase the probability of success in biological control programs. Unfortunately, there is a paucity of information on the attributes that make a parasite or a predator an effective natural enemy, i.e., one that regulates its host/prey populations at desired density levels. Less is known of the genetic makeup of effective natural enemies that make the expression of the desirable attributes possible. I am not, then, presenting here a set of guidelines to identify and discriminate biotypes. Nevertheless, I decided to maintain the title of this chapter as given to me because it clearly indicates what is required to procure effective biotypes that exist as natural populations, and eventually to improve others that appear not so effective. I hope that the summary discussions of a few biological control cases in which biotypes were recognized as deciding factors in the outcome will stimulate researchers to seek the information needed.

The performance (phenotype) of any organism is determined by the genotype and by the environment to which it is exposed. A genotype expresses itself differently, even if ever so slightly, when in different environments. This is the norm of reaction of a genotype, which can be defined as the "correspondence between the diffenent possible environments and the phenotypes that would result" (Lewontin 1982). Conversely, different genotypes express themselves differently when in the same environment. A species exists as a group of populations within each of which there is genotypic variation, even in thelytokus species. So, different phenotypes may be the result of different genotypes, different environments, or both. When the difference is such that a population can be consistently singled out from other conspecific populations, that population is called a biotype. Diehl & Bush (1984) define biotypes as conspecific populations that differ among themselves in some biological trait. They classify biotypes in several groups: 1) nongenetic polyphenisms (ecomorphs), 2) polymorphic or polygenic variations, 3) geographic races, 4) host races, or 5) species.

ISBN 0-12-357030-1

Nongenetic polyphenisms are biotypes that result from their exposure to different environmental conditions. They are the expression of the norm of reaction of their genotypes. Examples of polyphenisms include the solitary and gregarious "phases" of locusts. Factors such as temperature, photoperiod, and feeding on different plants or different parts of the same plant influence the production of alate or apterous forms in aphids. Increasingly crowded conditions determine whether the proportion of macropterous female plant hoppers increase. Salt (1937) showed that the antennal structure and wing veination differed in individuals of *Trichogramma semblidis* reared on different hosts. In certain species of *Torymus*, the ovipositor is longer in the spring generation. These individuals are able to reach hosts in the larger galls that occur at that time (Askew 1965).

As defined by Diehl & Bush (1984), "genetic polymorphism consists of discontinuous phenotypes among individuals within a freely interbreeding population that are the result of allelic variation of a frequency higher than can be maintained by recurrent mutations." Genetic polymorphism is exhibited by the Hessian fly, *Mayetiola destructor*, a major pest of wheat. Wheat varieties that are resistant to this fly have been selected. Resistance to the fly is conferred by dominant, nonallelic genes. However, there are biotypes of *M. destructor* that are able to overcome this resistance. This counteradaptation is provided by a series of recessive, nonallelic genes. Genetic analysis of the fly biotypes indicates that this "is the one definite case in insects of a 'gene-for-gene' relationship" (Futuyma 1983). Virulence loci in the fly correspond to resistance loci in the plant.

A geographic race is a biotype only if it is conspecific with similar other races. Geographic separation is a prerequisite for a mode of speciation (allopatric) (Mayr 1942) and within this concept, geographic races could be considered as early steps in the process of speciation. Biotypes resulting from geographic variation and sibling species in entomophagous insects have a great significance in biological control. A few cases will be presented later.

Host races are conspecific populations that are partially reproductively isolated from each other as a result of adaptation to specific hosts. It is difficult to decide whether allopatric, conspecific populations are host races because the differences observed when compared to each other may not indicate that the necessary level of reproductive isolation will be reached if the populations become sympatric.

Finally, there is the category of sibling species, which are morphologically identical but reproductively isolated populations. This category has only a temporary status as a biotype, until reproductive isolation is confirmed. Gonzalez et al. (1979) have proposed a scheme for identifying infraspecific populations.

BIOTYPES: EXAMPLES OF THEIR ROLE IN BIOLOGICAL CONTROL PROGRAMS

There are a number of cases in classical biological control in which biotypes played an important role in the outcome of the projects. In most, if not all cases, the ultimately efficient biotype was not specifically searched for. In fact, as far as I know, the "desirable" traits were not even defined. These populations were found as a result of the prevailing procedure of searching for and collecting natural enemies in as many areas and situations as possible when the first material colonized was not effective. In some cases, especially when the addition of other natural enemies improved the overall control, biotypes were identified and studied. One can suspect that in biological control programs that were considered failures, a number of the natural enemies that did not become established or failed to provide adequate control may have been biotypes, but they were not identified as such, perhaps because it is generally deemed fiscally unsound to spend resources studying failures. If this has been the prevailing attitude, I strongly disagree with it. I believe that if these cases are carefully studied, it would be easier to find additional, more effective biotypes, or eventually to generate them.

The following are examples of biological control cases in which biotypes played a significant role.

Aphytis spp. (Hymenoptera: Aphelinidae)

Aphytis is a very important genus of ectoparasites of diaspine (armored) scales. Ninety species are recognized by Rosen & DeBach (1979). According to these authors, the various species groups include "numerous complexes of cryptic or sibling species, as well as various biological races and other infraspecific entities exhibiting differing degrees of morphological and biological distinctness. In fact, practically every one of the originally described, more common species of *Aphytis* is now known to include a complex of cryptic forms."

An Iranian population of *Aphytis paramaculicornis*, which is primarily a parasite of the olive scale, *Parlatoria oleae*, develops readily on the oleander scale, *Aspidiotus nerii*, which is an unacceptable host to allopatric populations of this parasite. Similarly, in Greece, there are two sympatric populations of *Aphytis mytilaspidis*, an important parasite of the oystershell scale, *Lepidosaphes ulmi*; one of these populations develops on latania scale, *Hemiberlesia lataniae*, while the other does not.

The designations "biotype," "race," "strain" or "form" may be more the result of inadequate knowledge of the systematics of a group than of the presence of intraspecific variability. In 1952, the then-called "Persian strain" of *Aphytis maculicornis* was colonized in California. This and other natural enemies were imported in an attempt to control olive scale. Materials from Afghanistan, Cyprus, Egypt, India, Iran, Iraq, Israel, Lebanon,

Pakistan, Spain and Syria yielded several parasites, among which there were biparental and uniparental forms that were considered strains of *A. maculicornis*. At the time these "strains" were morphologically indistinguishable from each other, but they were obviously different in certain biological attributes (Doutt 1954, Hafez & Doutt 1954). In 1976, the Persian strain of *A. maculicornis* was recognized as a separate species: *A. paramaculicornis* (Rosen & DeBach 1979). *A. paramaculicornis* and another aphlinid, *Coccophagoides utilis*, complement each other and have successfully controlled the olive scale in California.

Trichogramma spp. (Hymenoptera: Trichogrammatidae)

This is a genus that contains species in which environmental factors influence morphological and physiological characteristics in several of the species included, to the point that Quednau (1960) included in his diagnostic key to the species characters such as coloration, developmental time in days and hours, and length of wing fringe hairs with the condition that they were valid only when the specimens had been reared in Mediterranean flour moth eggs at 30°C. This suggests that the norms of reaction of the different genotypes are very variable. The problem of the lack of discreetness of these characters has been solved to a point by using electrophoretic techniques (Pintureau & Babault 1981, 1982).

Various species of *Trichogramma* are extensively used in inundative releases against a number of pests in several countries. China, Mexico, and USSR have very large, intensive release programs. Races, mainly geographic, are selected naturally and adopted in suitable areas. In Russia, *T. evanescens* is the species used against several field crop pests. This species has many local or specialized races, four of which have been distinguished as being adapted to several groups of hosts: noctuids, tortricids, and the European corn borer (Shchepetilnikova et al. 1975)

Li (1982) reports that forms of *T. dendrolimi* from different parts of China have different responses to temperature. Crossing of these strains indicates that there is no reproductive isolation.

Comperiella bifasciata (Hymenoptera: Encyrtidae)

Three distinct biotypes of this parasite of diaspine scales are recognized (Rosen & DeBach 1978). Although morphologically indistinguishable and apparently able to interbreed in the laboratory, they differ in their host relations. The South African strain parasitizes the Florida red scale, *Chrysomphalus aonidum*, which is not parasitized by either of the other two strains. The Japanese strain parasitizes *Chrysomphalus bifasciculatus*; the Asiatic red scale, *Aonidiella taxus*; and the yellow scale, *A. citrina*. The Chinese strain parasitizes yellow scale and California red scale, A. aurantii. The Japanese and Chinese strains were introduced into California where they live sympatrically. Little is known of the distribution and host relationships of these two "races" in their areas of indigeneity. It is not known whether they are sympatric or whether the biotypes originated sympatrically or after geographic separation.

Trioxys pallidus (Hymenoptera: Aphidiidae)

This aphidiid is an endoparasite of the walnut aphid, Chromaphis juglandicola. It was introduced twice into California: In 1959 from France and 1n 1968 from Iran. The performances of these two populations were markedly different from each other. The French material performed very well in the coastal plains of southern California. However, in spite of numerous releases in central California, it failed to get a permanent foothold there. When Iranian material was colonized in several places in central California in 1968, recoveries were made even in sites with very low aphid densities. The parasite also dispersed very rapidly. The two populations are considered to be different races of T. pallidus, as no morphological differences between the two have been detected. However, van den Bosch et al. (1970) state that the two strains do not interbreed. His would suggest that the so-called strains may be sibling species.

Hambletonia pseudococcina (Hymenoptera: Encyrtidae)

This species is parasitic on the pineapple mealybug, Dismicoccus brevipes, and was introduced into Hawaii for the control of this pest. There are two races of H. pseudococcina: A thelytokous (uniparental) one from Colombia and Venezuela, and an arrhenotokous (biparental) one from Brazil. In Hawaii, the uniparental race was more effective (Bartlett 1978).

Ceranisus brui (Hymenoptera: Eulophidae)

This chalcid is parasitic on the onion thrips, Thrips tabaci. It occurs as a biparental species in Japan, where it exhibits a sex ratio of three females to two males, while in Europe it is reported as uniparental (Clausen 1978b).

Chrysopa carnea (Neuroptera: Chrysopidae)

Tauber & Tauber (1973) present morphological and biological evidence indicating that Chrysopa carnea consists of two biotypes: C. carnea strain carnea from eastern and midwestern North America and C. carnea strain mohave from California. The carnea strain was secondarily introduced into California where interstate crossing resulted in genetically varied populations with carnea characteristics predominating. As a result of further biological studies, Tauber & Tauber (1975) are of the opinion that the pure biotype carnea, originating from eastern and midwestern North America," should be used in biological control programs involving this predator in agricultural situations in western North America. The biotype mohave may be more suitable to control their prey in native, non-agricultural vegetation (oak, laurel) because they are adapted to situations in which prey available is highly variable in summer.

Bathyplectes curculionis (Hymenoptera: Ichneumonidae)

The alfalfa weevil, Hypera postica, is native to the Old World. It reached the United States (Utah) by the turn of the century, and dispersed to most of the western U.S. by 1933. Another invasion occurred in 1940-50. In 1955, it was found in Maryland, whence it spread to other eastern states.

The ichneumonid, Bathyplectes curculionis, is one of the natural enemies that have been imported in attempt to control the weevil. It was brought from Italy and Switzerland into Utah, from where it spread into Idaho and Wyoming. In 1933-34, it was brought from Utah into central California, where it became established.

In 1939, the Egyptian alfalfa weevil, H. brunneipennis, a species closely related to H. postica, was discovered in Yuma, Arizona, whence it spread into the Imperial and Coachella Valleys of southern California. By 1950 it had reached San Diego. Bathyplectes curculionis from Utah was colonized in southern Arizona in 1942. Years later it was found attacking the weevil in San Diego (Dietrick & van den Bosch 1953). B. curculionis, primarily a parasite of H. postica suffered 35-40% encapsulation by H. brunneipennis in San Diego by 1950 (Messenger & van den Bosch 1971). When the populations were examined again in 1966, the level of encapsulation was found to be only 5% (Salt & van den Bosch 1967). In California there are two biotypes of B. curculionis: One in the north, parasitizing the alfalfa weevil; and one in the south, parasitizing the Egyptian alfalfa weevil. As far as it is known, all B. curculionis originated from material collected in Utah (Salt & van den Bosch 1967). This case suggests that under certain circumstances, in this case closely related hosts, parasites can develop better adapted biotypes in a surprisingly short time.

Mesoleius tenthredinis (Hymenoptera: Ichneumonidae)

Another case that illustrates the important role played by biotypes in the outcome of biological control programs involves the larch sawfly, Pristiphora erichsoni, and one of its parasites, the ichneumonid Mesoleius tenthredinis. Biological control of the sawfly in North America was satisfactory until the sawfly became resistant to M. tenthredinis in part of the area of distribution. The effectiveness of this parasite was reduced in central and eastern North America as a result of mortality of the parasite by encapsulation (Clausen 1978a, Messenger & van den Bosch 1971, Turnock et al. 1976).

In this system there are interactions the details of which are not completely understood. One of the possibilities to explain the development of a resistant larch sawfly strain is that the original populations consisted of two strains (biotypes): One susceptible to M. tenthredinis, and the other virtually resistant to it. In the absence of the parasite, the susceptible strain was dominant. When the parasite was introduced, it decimated the susceptible strain, making it possible for the resistant strain to dominate (Clausen 1978a). Another possibility is that the development of a resistant strain was due to an increase of alleles for resistance as a result of selective pressure by the parasite.

Introduction from Bavaria of a strain of M. tenthredinis that is not encapsulated, and the establishment of the European ichneumonid Olesicampe benefactor has improved the control of the sawfly (Turnock et al. 1976).

BIOTYPE: IS THE CONCEPT USEFUL IN BIOLOGICAL CONTROL?

According to Diehl & Bush (1984) the term biotype has "very little descriptive power" because it has been "too loosely applied." They urge that the term be used only temporarily until "the underlying nature" of the variation observed in the population in consideration is known. At that time, the proper terminology should be used (nongenetic polyphenisms, polymorphic variations, geographic races, etc.) and "further use of the ambiguous and all-encompassing term 'biotype' should be discontinued." I could not agree more with the lofty goal of these authors, but in reality, at least for biological control agents, I venture to predict that the term biotype will be in use for a long time, first because the number of biotypes so far detected is relatively small, and second because "the genetic basis and the evolutionary status of the differences" in biotypes of biological control agents have not yet been ascertained for the large majority of those so far detected. acknowledging its limited descriptiveness, the term is irreplaceable in theoretical and practical biological control. nonetheless, the appropriate, more specific term should be adopted as soon as the class to which the biotype belongs is known.

PROCUREMENT OF BIOTYPES

When searching for candidate natural enemies in poorly known systems, where the area of distribution of the target species (host, prey, weed) is poorly known, or not known at all, the explorer should assume that there are biotypes present, and consequently the material collected from different sites should be kept separate in order to preserve their integrity. In the mountain valleys of the Zagros, Iran, where alfalfa seems to be native, evidence is accumulating that would demonstrate the existence there of alfalfa and alfalfa weevil biotypes (M. Esmaili, personal communication). If these biotypes actually exist, it is conceivable that several biotypes of natural enemies of the alfalfa weevil are also present in the area. Similar situations may be common elsewhere.

Explorers for natural enemies may be able to identify biotypes based mostly on their field experience. They may identify situations (microclimates, host plants, etc.) suggestive of the presence of biotypes. However, quarantine operators are the first to be in a position to determine experimentally whether the material being processed consists of biotypes by detecting differences in attributes such as fecundity, longevity, diapause, sex ratio or life-cycle. Other characteristics such as temperature tolerance, dispersal, alternative host/prey, intrinsic and extrinsic competition should be detected during subsequent laboratory and field studies.

COMMENTS

The few examples of classical biological control programs in which it has been demonstrated that "the right" or "a better" biotype of natural enemy determined a successful outcome point to the importance of devoting efforts to identify, procure and evaluate these forms. It is essential to know what are the different traits that characterize biotypes and the genetic basis of these traits if we want to preserve or augment their efficacy as biological control agents. Studies leading to the generation of this information should be done whether a project is in progress or has terminated because of either success or failure. Failure in a project may be the result of natural enemies failing to become established in the new environment (biological failure), or, although well established, they may not regulate their host/prey at a population level below the economic injury level (economic failure). I believe that if we know why natural enemies fail, our capability to successfully manipulate biological control agents would increase considerably.

One very important fact should be always borne in mind when studying the efficacy, or lack thereof, of biological control agents. Biological control agents do not exist in order to regulate pests below economic injury levels. The successful cases of biological control are simply due to the fact that, under certain circumstances, certain natural enemies reduce populations of pests below the levels set by man as being economically injurious. In these cases, we will not reap all the benefits of such endeavors until the question "What makes a successful biological control agent?" is answered.

The characteristics of entomophagous insects have been reviewed by Doutt (1959, 1964), Rosen & Huffaker (1983) and Vinson & Iwantsch (1980). For an individual parasite to develop to the reproductive stage, several critical steps must be successfully completed: 1) a host must be reached, either as a result of the female's activity, or through other mechanisms (host contacting the parasite, phoretic agent carrying the parasite to the host); 2) the host must be nutritionally and physiologically suitable; 3) if exposed to intrinsic competition, either intra- or inter-specific, the parasite must be the victor; 4) at the population level, parasites must have adequate searching capacity, fecundity, general dispersal ability and, in the case of biparental populations, adequate sex ratio; and 5) they compete also at the extrinsic level. Based on these characteristics, we could construct a profile for the ideal natural enemy, and, on that basis, search for the natural enemy that most closely resembles the model, or conceivably select for one. However, we still do not know that specific characteristics are the ones needed to make a natural enemy an effective one in a specific situation. Cases in which a natural enemy that was not considered to be an effective candidate nevertheless did regulate a pest population at or below a tolerable density make predictability in biological control a vexing proposition.

One assumption is that a biotype from an area climatically similar to that of the target area has greater probability of becoming established than one from a climatically different place (Messenger & van den Bosch 1971). One of the criteria for procuring biotypes should be based on this assumption.

In systems where effective natural enemies maintain potential pests at low population levels, disruption can occur as a result of the adoption of some cultural practice, such as the use of chemicals to control pests, decimating the natural enemies and rendering them unable to continue controlling the potential pest. In such cases the limiting characteristic (susceptibility to a chemical) of the natural enemies is known, consequently the approach to manipulate natural enemies (rendering it resistant or tolerant to the chemical if other strategies are not available) offers possibilities. The potential for artificial improvement (selection, genetic manipulation) of natural enemies has been comprehensively discussed by Hoy (1979, 1982) and demonstrated by Hoy & Knop (1981) and Roush & Hoy (1981). Beckendorf & Hoy (this volume) discuss the potential of genetic engineering of arthropod natural enemies.

RECOMMENDATIONS

1. Intensify studies on the biosystematics of natural enemies, especially of those considered key, and of their host/prey. The aim should be at producing monographs similar to that on *Aphytis* by Rosen & DeBach (1979).

2. During procurement of exotic natural enemies, materials from different geographic areas, or from different host plants in the same area, should be kept separate. Comparative laboratory and field studies should be conducted. Special attention should be paid to materials that include biparental and uniparental biotypes. The controversy over whether a successful genotype is more faithfully transmitted to progeny in uniparental, closely inbred or outbred populations is still quite alive (Shields 1982).
3. Newly imported material should be routinely tested to detect biotypes. High resolution techniques, such as electrophoresis, should be used in cases where traditional ones, such as morphology and biology (fecundity, developmental rate, sex ratio) prove insufficient.
4. Detailed studies on the biology and phenology of the host plant, the host, and the natural enemy should be conducted in all biological control programs, whether successes or failures, so as to identify the possible trait or combination of traits that may account for the outcome.

ACKNOWLEDGMENTS

I thank D.C. Herzog and M.A. Hoy for inviting me to participate in the Conference on Biological Control in Agricultural IPM Systems and for their kindness before, during and after the Conference. I am thankful to my fellow participants for their enlightening discussions and for their comradeship during the meetings. I am grateful to D. Gonzalez, M.A. Hoy, C.B. Huffaker, R.I. Sailer, M.J. Tauber and K.V. Yeargan for their detailed review of all or parts of my manuscript. Whatever is of value in this paper is, at least in part, the result of their criticism and suggestions. All errors, either of omission or commission, are entirely my responsibility.

REFERENCES

Askew, R.R. 1965. The biology of the British species of Torymus Dalman (Hymenoptera: Torymidae) associated with galls of Cynipidae (Hymenoptera) on oak, with special reference to alternation of forms. Trans. Soc. Brit. Entomol. 16:217-232.

Bartlett, B.R. 1978. Pseudococcidae, pp. 137-170. In: C.P. Clausen (Ed.), Introduced parasites and predators of arthropod pests and weeds: A world review. USDA Agric. Handb. 480.

Clausen, C.P. 1978a. Tenthredinidae, pp. 308-313. In: C.P. Clausen (Ed.), Introduced parasites and predators of arthropod pests and weeds: A world review. USDA Agric. Handb. 480.

Clausen, C.P. 1978b. Thripidae, pp. 19-21. In: C.P. Clausen (Ed.), Introduced parasites and predators of arthropod pests and weeds: A world review. USDA Agric. Handb. 480.

Diehl, S.R., and G.L. Bush. 1984. An evolutionary and applied perspective of insect biotypes. Annu. Rev. Entomol. 29:471-504.

Dietrick, E.J., and R. van den Bosch. 1953. Further notes on Hypera bruneipennis and its parasite Bathyplectes curculionis. J. Econ. Entomol. 46:1114.

Doutt, R.L. 1954. An evaluation of some natural enemies of the olive scale. J. Econ. Entomol. 47:39-43.

Doutt, R.L. 1959. The biology of parasitic Hymenoptera. Annu. Rev. Entomol. 4: 161-182.

Doutt, R.L. 1964. Biological characteristics of entomophagous adults, pp. 145-167. In: P. DeBach (Ed.), Biological control of insect pests and weeds. Chapman & Hall, London.

Futuyma, D.J. 1983. Evolutionary interactions among herbivorous insects and plants, pp. 207-231. In: D.J. Futuyma and M. Slatkin (Eds.), Coevolution. Sinauer, N.Y.

Gonzalez, D., G. Gordh, S.N. Thompson and J. Adler. 1979. Biotype discrimination and its importance to biological control, pp. 129-136. In: M.A. Hoy and J.J. McKelvey (Eds.), Genetics in relation to insect management. Rockefeller Found. Press, N.Y.

Hafez, M., and R.L. Doutt. 1954. Biological evidence of sibling species in Aphytis maculicornis (Masi) (Hymenoptera, Aphelinidae). Can. Entomol. 86:90-96.

Hoy, M.A. 1979. The potential for genetic improvement of predators for pest management programs, pp. 106-115. In: M.A. Hoy and J.J. McKelvey (Eds.), Genetics in relation to insect management. Rockefeller Found. Press, N.Y.

Hoy, M.A. 1982. Genetics and genetic improvement of the Phytoseiidae, pp. 72-89. In: M.A. Hoy (Ed.), Recent advances in knowledge of the Phytoseiidae. Univ. Calif., Div. Agr. Sci Publ 3284.

Hoy, M.A., and N.F. Knop. 1981. Selection for and genetic analysis of permethrin resistance in Metaseiulus occidentalis: Genetic improvement of a biological control agent. Entomol. Exp. Appl. 30:10-18.

Lewontin, R. 1982. Human diversity. Sci. Amer. Books, N.Y. 179 pp.

Li, L.-Y. 1982. Trichogramma sp. and their utilization in People's Republic of China. Colloq de l'INRA 9:23-29.

Mayr, E. 1942. Systematics and the origin of species. Columbia Univ. Press, N.Y. 334 pp.

Messenger, P.S., and R. van den Bosch. 1971. The adaptability of introduced biological control agents, pp. 68-92. In: C.B. Huffaker (Ed.), Biological control. Plenum Press, N.Y.

Pintureau, B., and M. Babault. 1981. Caracterisation enzymatique de Trichogramma evanescens et de T. maidis (Hym.: Trichogrammatidae); etude des hybrides. Entomophaga 26:11-22.

Pintureau, B., and M. Babault. 1982. Comparaisons des inzymes chez 10 souches des Trichogramma (Hym. Trichogrammatidae). Colloq. de l'INRA 9:32-44.

Quednau, W. 1960. Uber die Identitat der Trichogramma--Arten und einiger ihrer Okotypen (Hymenoptera, Chalcidoidea, Trichogrammatidae). Bitt. Biol. Bundes. Land. Forstwirtsh. 100:11-50.

Rosen, D., and P. DeBach. 1978. Diaspididae, pp. 78-128. In: C.P. Clausen (Ed.), Introduced parasites and predators of arthropod pests and weeds: A world review. USDA Agric. Handb. 480.

Rosen, D., and P. DeBach. 1979. Species of Aphytis of the world (Hymenoptera: Aphelinidae). Israel Univ. Press, Jerusalem. 801 pp.

Rosen, D., and C.B. Huffaker. 1983. An overview of desired attributes of effective biological control agents, with particular emphasis on mites, pp. 2-11. In: M.A. Hoy, G.L. Cunningham and L. Knutson (Eds.), Biological control of pests by mites. Univ. Calif., Div. Agric. Spec Publ. 3304.

Roush, R.T., and M.A. Hoy. 1981. Genetic improvement of Metaselulus occidentalis: Selection with methomyl, dimethoate, and carbaryl and genetic analysis of carbaryl resistance. J. Econ. Entomol. 74:138-141.

Salt, G. 1937. The egg parasite of Sialis lutaria: A study of the influence of the host upon a dimorphic parasite. Parasitology 29:539-553.

Salt, G., and R. van den Bosch. 1967. The defense reaction of three species of Hypera (Coleoptera, Curculionidae) to an ichneumon wasp. J. Invertebr. Pathol. 9:164-177.

Shchepetilnikova, V.A., S.S. Murashevskaya, O.V. Kaputina and G.I. Kalashnikova. 1975. Bases of regional distribution of the common Trichogramma. Proc. All-Union Sci. Res. Inst. Plant Prot., Leningrad 44:5-32.

Shields, W.M. 1982. Phylopatry, inbreeding, and the evolution of sex. State Univ. New York Press, Albany. 245 pp.

Tauber, C.A., and M.J. Tauber. 1973. Diversification and secondary intergradation of two Chrysopa carnea strains (Neuroptera: Chrysopidae). Can. Entomol. 105: 1153-1167.

Tauber, M.J., and C.A. Tauber. 1975. Criteria for selecting Chrysopa carnea biotypes for biological control: Adult dietary requirements. Can. Entomol. 107: 589-595.

Turnock, W.J., K.L. Taylor, D. Schrader and D.L. Dahlsten. 1976. Biological control of pests of coniferous forests, pp. 289-336. In: C.B. Huffaker and P.S. Messenger (Eds.), Theory and practice of biological control. Academic Press, N.Y.

van den Bosch, R., B.D. Frazer, C.S. Davis, P.S. Messenger and R. Hom. 1970. An effective walnut aphid parasite from Iran. Calif. Agric. 24(11):8-10.

Vinson, S.B., and G.F. Iwantsch. 1980. Host suitability for insect parasitoids. Annu. Rev. Entomol. 25:397-419.

DISCUSSION: Identifying and discriminating among biotypes of parasites and predators.

DISCUSSION LEADER: R.I. Sailer

The group agreed that many natural enemy introduction failures could be the consequence of selecting an ill-adapted or inappropriate biotype. The importance of biotypes is widely recognized and concern focused on methods for rapid and accurate identification of biotypes for specific purposes. Characterizing biotypes by behavioral differences is particularly difficult (although potentially extremely desirable) because behavior is so often modified by environment. In some cases, behavioral differences have been found to correlate with isozyme differences; however, isozymes may be influenced by factors such as diet and age. The value of using restriction enzymes to detect distinctive DNA biotypes was discussed as a potential new method for use in biological control projects. Identification and characterization of biotypes should be a high priority in biological control projects utilizing arthropods, nematodes, antagonists of plant pathogens, pathogens of arthropods, weeds and nematodes.

ANALYSIS OF SYSTEMS FOR BIOLOGICAL CONTROL OF CROP ARTHROPOD PESTS IN THE U.S. BY AUGMENTATION OF PREDATORS AND PARASITES

Edgar G. King, Keith R. Hopper and Janine E. Powell

Suppression of arthropod pests by naturally occurring predators and parasites (entomophagous arthropods) is well recognized (Anonymous 1978, DeBach 1964, Huffaker & Messenger 1976). However, these entomophages may fail to maintain pests below acceptable population levels. Three strategies are often identified for encouraging and using predators and parasites. First, exotic species of predators and parasites may be introduced for establishment (classical biological control). Second, where a pest is separated in time or space from its natural enemies because of ecosystem instability, means may be developed to protect and improve the effectiveness of naturally occurring entomophages (conservation). Finally, entomophage numbers may be increased within a defined area (biological control by augmentation). Classical biological control is examined elsewhere in these proceedings as is the practice of conserving entomophages and control by augmentation using behavior-modifying chemicals.

Knipling (1979) suggested that augmentation of host specific parasites and predators can provide a solution to some of the world's major insect problems based on his efforts for using theoretical simulation models to estimate the number of parasites required in host habitats to achieve various levels of parasitism. Ridgway & Vinson (1977) reviewed control by augmentation of predators and parasites worldwide, including theoretical aspects and implementation technology. Other reviews include Ables & Ridgway (1981), DeBach & Hagen (1964), Rabb et al. (1976), and Stinner (1977). Reviews specific to the Western Hemisphere and U.S. include King et al. (1984c) and Ridgway et al. (1977). These researchers often described augmentative releases based on the purpose intended using terms such as inoculative releases, seasonal colonization, supplemental releases, strategic releases, programmed releases, inundative releases, and compensatory releases. In this paper, we summarize the state-of-the-art of propagation and releasing entomophagous arthropods that attack arthropod pests of crops in the U.S., analyze the

ISBN 0-12-357030-1

results of augmentative releases to control a range of crop pests, and recommend directions for future research.

AUGMENTATION IN THE U.S.

Predators and parasites tested for biological control of crop arthropod pests by augmentative releases in the U.S. are listed in Table 1. Pest populations and/or their damage have often been supressed by released entomophages, yet economic feasibility has seldom been demonstrated. Reichelderfer (1981) defined technical feasibility as the capability of a control method to exhibit a decrease in pest numbers and/or pest effects and economic feasibility as the probable capability of the technology to yield positive monetary returns to agricultural production that are greater than or equal to those realized from alternative technologies or no control.

There are 62 different firms which supply predators and parasites to the U.S. public (Brunetti 1981, Ridgway & Vinson 1977). Many retailers don't manage their own insectaries, thus they receive their entomophages from larger outlets.

ANALYSES OF SOME AUGMENTATIVE RELEASES

The efficacy of experimental releases as well as the extent and effectiveness of commercial use of selected entomophages is discussed here. Certain attributes of the crop, pest, entomophage, and agroecosystem for each case are discussed briefly while those contributing to success of the releases are emphasized.

Case Studies

Phytoseiids The use of phytoseiids in biological control has been reviewed by McMurtry (1982) and van Lenteren et al. (1980). Although over 30 species of predatory mites have been listed as important in biological control, only five have been evaluated in augmentative releases in the U.S. on crops (see Table 1).

Use of *Phytoseiulus persimilis* is greatest in Great Britain and the Netherlands for control of *Tetranychus urticae* in greenhouses. Studies conducted in the U.S. for control of phytophagous mites attacking strawberry and apple have demonstrated the effectiveness of augmentative releases of phytoseiids. McMurtry (1982) provided information on use of phytoseiids in the greenhouse and field and commented on future prospects for use in supplementary releases. Inoculative releases of pesticide-resistant *Metaseiulus occidentalis* for establishment are not considered augmentative releases because the mite is considered a new biotype (Hoy et al. 1982a). However effective control could be accelerated by release of higher numbers and a need may exist for

Table 1. Some entomophagous arthropods evaluated in augmentative releases to control agricultural arthropod pests in the USA.[a/]

Entomophage	Target Pest	Production Host	Crop	Reference
Amblyseius californicus	Two-spotted spider mite	Not given	Strawberry	Oatman et al. 1977
Aphidius smithi	Pea aphid	Pea aphid	Pea	Halfhill & Featherston 1973
Aphytis spp.	California red scale	Not given	Citrus	DeBach et al. 1950
Bracon spp.	Pink bollworm	Beet armyworm	Cotton	Bryan et al. 1973
		Potato tuberworm		Legner & Medved 1979
Agathis (Bassus) diversus	Oriental fruit moth	Oriental fruit moth	Peach	Brunson & Allen 1944
Chelonus spp.	Pink bollworm	Pink bollworm	Cotton	Bryan et al. 1976
		Potato tuberworm		Legner & Medved 1979
Chrysopa spp.	Bollworm, Tobacco budworm	Angoumois grain moth	Cotton	Ridgway & Jones 1969
	Green peach aphid	Not given	Snapdragon	Harbaugh & Mattson 1973
	Grape mealybug	Potato tuberworm	Pear	Doutt & Hagen 1950
	Potato aphid, Green peach aphid	Not given	Potato	Shands et al. 1972
	Mealybug	Potato tuberworm	Gardenia	Doutt 1951
Coccinella septempunctata	Potato aphid, Green peach aphid	Aphid	Potato	Shands et al. 1972

(continued)

Table 1. (Continued).

Entomophage	Target Pest	Production Host	Crop	Reference
Cryptolaemus montrouzieri	Mealybugs	Mealybug	Citrus	Armitage 1929
Diaretiella rapae	Aphids	Green peach aphid	Potato	Shands et al. 1975
Encarsia formosa	Greenhouse whitefly	Not given	Poinsettia	Helgesen & Tauber 1974
Hippodamia convergens	Aphids	- - - - - - [b/]	Pea	DeBach & Hagen 1964
				Cooke 1963
	Greenbug		Grain sorghum	Starks et al. 1975
Jalysus spinosus	Tobacco budworm,	Tobacco budworm	Tobacco	Elsey 1975
	Tobacco hornworm	Tobacco hornworm		
Leptomastix dactylopii	Mealybugs	Citrus mealybug	Citrus	Fisher 1963
Lixophaga diatraeae	Sugarcane borer	Sugarcane borer	Sugarcane	Summers et al. 1976
		Greater wax moth		King et al. 1981
Lysiphlebus testaceipes	Greenbug	Greenbug	Sorghum	Starks et al. 1976
Macrocentrus ancylivorus	Oriental fruit moth	Oriental fruit moth	Peach	Brunson & Allen 1944
Microterys flavus	Brown soft scale	Brown soft scale	Citrus	Hart 1972
Pediobius foveolatus	Mexican bean beetle	Mexican bean beetle	Soybean	Stevens et al. 1975a
Phytoseiulus persimilis	Two-spotted spider mite	Not given	Strawberry	Oatman et al. 1968, 1976, 1977
				McMurtry et al. 1978
Polistes spp.	Tobacco budworm	- - - - - - - [b/]	Tobacco	Lawson et al. 1961
	Lepidopterous larvae	- - - - - - - [b/]	Cotton	Fye 1972

(continued)

Table 1. (Continued).

Entomophage	Target Pest	Production Host	Crop	Reference
Praon spp.	Aphids	Green peach aphid	Potato	Shands et al. 1975
Stethorus picipes	Avocado brown mite	Not given	Avocado	McMurtry et al. 1969
Trichogramma pretiosum	Bollworm, Tobacco budworm	Angoumois grain moth	Cotton	Stinner et al. 1974 Ables et al. 1979 King et al. 1984a
Trichogramma minutum	Tomato fruitworm, Cabbage looper, Tomato hornworm	Cabbage looper	Tomato	Oatman & Platner 1978
	Imported cabbageworm, Cabbage looper	Angoumois grain moth Cabbage looper	Cabbage	Parker 1971
	Sugarcane borer	Angoumois grain moth	Sugarcane	Jaynes & Bynum 1941
Trichogramma cacoeciae	Codling moth	Codling moth	Apple	Dolphin et al. 1972
Trichogramma nubilale	European corn borer	European corn borer	Corn	Kanour & Burbutis 1984
Typhlodromus spp.	Cyclamen mite	Not given	Strawberry	Huffaker & Kennet 1956
	McDaniel spider mite	Not given	Apple	Croft & McMurtry 1972

a/ Evaluations in laboratory and field cages are not included.
b/ Insects obtained through field collections.

periodic release thereafter to maintain an acceptable predator-prey ratio.

Unfortunately, phytoseiids are expensive to rear, in part, because satisfactory artificial diets have not been developed (Kennett & Hamai 1980). Hoy et al. (1982b) reported methods for large-scale rearing of *M. occidentalis* in the greenhouse and field. Presently, at least six species of phytoseiids are commercially available in the U.S. from at least six companies here and one in the Netherlands. Prices vary from $12.50 to $50/1000 predators depending on quantity purchased, species, and vendor. An effective field release rate of 312,500 *P. persimilis*/ha (Oatman et al. 1976) priced at $25/1000 predators would cost $7800/ha to control *T. urticae* in strawberry. Current markets for phytoseiid use are on relatively high value greenhouse crops such as cucumber and tomato (van Lenteren et al. 1980).

Spider mites reduce yield indirectly by damaging foliage, and where foliage appearance is important (ornamentals) growers have been slow to adopt biological control. Besides being expensive to rear, release of the predators is labor intensive due to low dispersal rates within and between generations. Therefore, phytoseiids must be delivered near their prey to effect rapid biological control. However, low dispersal rates coupled with a tendency to remain in an area despite low prey density results in the predators staying in the area where control is desired. Unfortunately, this tendency to stay in areas of low prey density results in failure to encounter and eliminate patches of high pest density and may lead to unstable predator/prey population interactions.

Little has been published regarding the economics of phytoseiid augmentative release vs. alternative controls. Implementation of the augmentative release strategy has been due largely to resistance in tetranychid mites to available acaricides (Hussey et al. 1965). Nevertheless, elegant methods have been developed for rearing, distributing, and releasing the predators which have effectively controlled tetranychid pests.

Encarsia formosa *E. formosa* is an aphelinid parasite of greenhouse whitefly, *Trialeurodes vaporariorum*. Biology, use, and rearing of *E. formosa* have been reviewed by King & Morrison (1984), van Lenteren et al. (1980) and Vet et al. (1980). This parasite is used extensively in greenhouses for augmentative control in tomato and sweet pepper in Great Britain and the Netherlands. *E. formosa* is used, in part, because *P. persimilis* is used for control of tetranychids in greenhouses and insecticide use for control of *T. vaporariorum* would be detrimental to the predatory mite (Parr et al. 1976).

Effectiveness of releasing the parasite for control of *T. vaporariorum* in greenhouses has been documented by Vet et al. (1980). Scopes (1969) claimed that 100,000 parasites/ha would control the greenhouse whitefly in cucumber for as low as $40/ha in Great Britain. However, Helgesen & Tauber (1974) suggested that this estimate may have been overly optimistic because the practice of reducing greenhouse temperatures to lessen energy costs may adversely affect the parasite's rate of increase. Moreover, some host plants (e.g., cucumber) apparently are not suitable for the parasite to achieve adequate biological control.

Several firms distribute E. formosa in the U.S. (Brunetti 1981), but extent of use is low relative to that in western Europe. According to one research entomologist (L.A. Osborne, personal communication), E. formosa might be used more extensively in ornamental stock production if they were available. Currently, E. formosa is used mostly on tomato, pepper, and egg plant, which can withstand high levels of infestation, but not on ornamentals because of the low threshold for aesthetic damage.

Chrysopa carnea Rearing of C. carnea was reviewed by King & Morrison (1984). Doutt & Hagen (1950) achieved maximum control of mealybug on pear by three successive releases of the predator during the prey's first generation; and the trees remained relatively free of mealybugs for two years. Thus, in the absence of insecticides, biological control was effective. Lingren et al. (1968) and Ridgway & Jones (1968, 1969) demonstrated the effectiveness of C. carnea for supressing Heliothis spp. in cotton. However, numbers of C. carnea larvae required to supress Heliothis spp. larvae below the economic injury level were near 1 million/ha. These tests culminated in a series of releases in small field plots at rates of 227 to 494 thousand predator larvae/ha where a peak Heliothis spp. larval population (44,460/ha) was reduced 96% and cotton yield increased three-fold over the untreated check. They conclude from other small plot studies that release of 123,500 C. carnea/ha might provide effective control of Heliothis spp. in cotton. Unfortunately, at a cost of $12.50/4000 predator larvae (E.J. Dietrick, personal communication) each release would cost $385/ha. Harbaugh & Mattson (1973) reported that it might be feasible to integrate chemical control and release of C. carnea on snapdragon for aphid control after releasing two to six larvae/plant and reducing aphid populations up to 98%. Thus, it might be feasible to use this predator in augmentative releases for control of aphids in greenhouse crops if pests such as greenhouse whitefly and two spotted spider mite are controlled biologically.

Trichogramma spp. Trichogramma spp. are presently the most widely augmented entomophagous arthropod in the world. They are used particularly in the USSR (8 million ha) and PRC (2 million ha) (Li 1984, Voronin & Grinberg 1981). Ridgway et al. (1981) reported that Trichogramma spp. may be released over as many as 200,000 ha in the U.S. Rearing of Trichogramma spp. has been reviewed by King & Morrison (1984).

Several U.S. commercial insectaries produce Trichogramma spp. with as many as 50 to 100 million wasps produced daily by one company, Rincon-Vitova Insectaries, Inc. (E.J. Dietrick, personal communication). This company supplies *T. platneri*, *T. minutum* and *T. pretiosum* for control of caterpillar pests infesting cotton, corn, apple, spruce, and avocado.

Many research reports and review articles cite increased egg parasitism in response to release of *Trichogramma* (Ridgway et al. 1977, 1981). However, relatively few reports correlate increased parasitism with a decline in the pest population, damage due to the pest, and/or yield response as compared to a nonrelease area. Data reported by Oatman & Platner (1971, 1978) demonstrated the technical feasibility of augmenting *T. pretiosum* populations to reduce damage in tomato caused by the tomato fruitworm, *Heliothis zea*, cabbage looper, *Trichoplusia ni*, and *Manduca* spp. However, the need for chemically controlling other pests not attacked by *T. pretiosum* was also expressed. Stinner et al. (1974) demonstrated the technical feasibility of reducing *Heliothis* spp. larval populations in cotton by releasing *T. pretiosum*, though parasite release rates were high (up to 957,000/ha) and neither damage supression nor yield response were shown. King et al. (1984b) could not, after three years of tests in large plots, correlate increased *T. pretiosum* parasitism of *Heliothis* eggs with reduced larval populations or their damage. During the third year of tests, however, parasite release plots yielded more than untreated plots, but only 77% as much lint as insecticide-treated plots.

Tests with other *Trichogramma* spp. have also been variable. Parker (1971) summarized research where the host *Pieris rapae*, a larval parasite *Cotesia* (*Apanteles*) *rubecula*, and *T. evanescens* were released in cabbage. Parasitization of *P. rapae* eggs and larvae averaged 95 and 80%, respectively. Consequently, pest larval populations and damage were greatly reduced. In another study, Dolphin et al. (1972) demonstrated that it is technically feasible to reduce codling moth, *Cydia* (*Laspeyresia*) *pomonella*, larvae in fruit by release of *T. minutum*, but release numbers were about 0.5 million parasites per tree. And in sugarcane, there is a succession of studies attempting to augment *Trichogramma* populations to control the sugarcane borer; however, no studies in the U.S. have conclusively demonstrated control (Metcalfe & Breniere 1969).

It is obvious that under optimum conditions in most crops, it is technically feasible to control selected caterpillar pests by augmentation of *Trichogramma*. However, the need for releasing high numbers of parasites at two to four day intervals limits economic feasibility of the approach. An exception to this might be the use of *T. nubilale* to control the European corn borer in corn and sweet pepper. Kanour & Burbutis (1984) calculated that release of ca. 2400 and 12,000 female *T. nubilale*/day/ha on corn would be needed to achieve 80% parasitism of F. and F. egg populations respectively, of the European corn borer. Nevertheless, they estimated control costs of $60/ha based on a cost of $0.12/1000 *T. nubilale* reared on European corn borer eggs (Burbutis & Goldstein 1983).

Lysiphlebus testaceipes This native braconid has often been reported to heavily parasitize field populations of the greenbug, Schizaphis graminum, with a resultant collapse of the greenbug population (Sekhar 1957, Wood & Chada 1969). However, Starks et al. (1976) noted that L. testaceipes' effectiveness as a natural control agent was limited since the greenbug reproduces at lower temperatures than the parasite. Also, the greenbug's life cycle (6-10 days) is shorter than that of the parasite's (10-14 days), resulting in high numbers of greenbugs before the buildup of parasite populations. Moreover, hyperparasites attacking L. testaceipes may become numerous and suppress the primary parasite population.

Starks et al. (1976) reported on the feasibility of early, inoculative release of L. testaceipes for control of the greenbug on sorghum in Texas. Significantly more parasitized greenbugs per plant were observed in fields where parasites were released at 36,600 or 69,200/ha than in nonrelease fields. Moreover, grain yield was increased over nonrelease fields. No significant difference in yield occurred between release fields and insecticide treated fields. Nevertheless, Starks et al. (1976) concluded that plant resistance was a more desirable control strategy than augmentative release of L. testaceipes or insecticide control. Augmentation of the parasite was expensive and it seemed likely that the greenbug would gain resistance to commonly used insecticides. A new greenbug biotype (Biotype E) has developed which successfully attacks Biotype C-resistant sorghum hybrids, necessitating incorporation of new sources of resistance. Greenbug control with insecticides continues despite use of resistant varieties. Thus, biological control of greenbugs in sorghum with augmentative releases of L. testaceipes during early season remains a viable alternative control strategy to insecticides or host plant resistance.

Pediobius foveolatus This parasite is native to the Far East and was imported into the U.S. in 1966. Host range studies revealed that it reproduced readily on third and fourth instar Mexican bean beetle (MBB) (Epilachna varivestis) larvae (Angalet et al. 1968). The MBB has become an important foliage-feeding pest of soybean in the eastern U.S. (Turnipseed & Shepard 1980). The MBB overwinters as an adult; but P. foveolatus cannot overwinter and must be reinoculated annually (Schroder 1981). Stevens et al. (1975b) described rearing and inoculative release procedures for this parasite.

Stevens et al. (1975a) demonstrated the technical feasibility of inoculating MBB populations by infesting snapbean nurse plots with P. foveolatus; afterward the parasitized larvae were collected and distributed into surrounding soybean fields. Parasitism of MBB larvae by P. foveolatus approached 100%, resulting in a four-fold reduction in insecticide usage where the parasite was released in soybean.

A simulation model of soybean pest control and production to compare net benefits obtained by various control methods of MBB in Maryland revealed that annual release of P. foveolatus had the highest return per dollar spent (Reichelderfer & Bender 1979). Reichelderfer (1979) conducted a benefit-cost analysis of controlling MBB in soybean and reported that use of this parasite was economically feasible in areas where MBB was the major pest. In 1976, the Maryland Department of Agriculture began rearing and distributing P. foveolatus over 12,146 ha of soybean at $3.46/ha. Additionally, snapbean nurse plots (ca. one per ha of soybean) cost about $50 each.

Lixophaga diatraeae This parasite is native to Cuba and has been reared and augmentatively released there as well as in other Caribbean and South American countries for control of sugarcane borer, Diatraea saccharalis, in sugarcane (King et al. 1981). Unfortunately, experimental data demonstrating the effectiveness of augmentative release of this tachinid were equivocal (Bennett 1969).

The U.S. Department of Agriculture undertook a study in 1973 to evaluate the technical feasibility of using L. diatraeae in augmentative releases for suppressing sugarcane borer populations in Louisiana and Florida. Mass production of the parasite on an unnatural host, the greater wax moth, Galleria mellonella, cost about $10/1000 parasites, which was about 1/5 the cost of rearing them on sugarcane borer (King et al. 1979). Parasitism by L. diatraeae increased significantly after parasite releases in Louisiana (King et al. 1981) and Florida (Summers et al. 1976), with suppression apparent in some fields. Although reduced feeding by parasitized larvae apparently occurred (Brewer & King 1978), the economic injury level was exceeded. It was concluded that early season augmentative releases of L. diatraeae may be valuable for population suppression (King et al. 1981), but implementation will require developing an inexpensive method of mass rearing a quality parasite, limiting parasite dispersal from release areas, and establishing the relationship between numbers of L. diatraeae released, their impact on borer populations, subsequent damage, yield, and net profit of the crop.

Macrocentrus ancylivorus This parasite is native to North America, but attacked the oriental fruit moth (OFM), Grapholitha molesta, after this pest became established. Brunson & Allen (1944) reported on observations made in 145 ha of peach orchards where M. ancyclivorus, as well as another braconid, Agathis diversus, were released at the rate of 3.3 females per tree (815/ha). They concluded that the released parasites, particularly M. ancyclivorus, significantly increased parasitization of second brood OFM larvae, decreased the number of moths emerging per tree by half, and reduced fruit injury by half. Subsequently, a method for mass producing M. ancyclivorus on an unnatural host, the potato tuberworm, Phthorimaea operculella, was developed (Finney et al. 1947).

When DDT became available (ca. 1948), use of it and other insecticides for control of OFM largely displaced M. ancyclivorus. Nevertheless, Brunson & Allen (1954) summarized a five year study where M. ancyclivorus releases (<5 female parasites/tree) were integrated with insecticides and showed that the number of insecticide applications could be reduced. In fact, it was recommended that three female M. ancyclivorus females should be released per peach tree at four day intervals, five times, against first and second brood larvae in areas (Massachusetts to Michigan, southward to eastern Missouri, and Arkansas and northern Georgia) not equipped to spray for OFM (Snapp 1954).

Presently, M. ancyclivorus is released for augmentative purposes only in Colorado during the spring and early summer to aid in maintaining OFM below economic injury levels (Anonymous 1983). These releases are integrated with insecticide applications for the peach twig borer, Anarsia lineatella. Parasite pupae at the rate of 2470/ha are released two to three times. The parasites cost $13/1000 and release of the parasite can essentially replace two insecticide applications (A.J. Merlino, personal communication).

PROPAGATION OF ENTOMOPHAGOUS ARTHROPODS

Methods for rearing selected parasites and predators have been most recently reviewed by King & Morrison (1984). Present propagation methods are largely restricted to rearing the predator or parasite on live hosts (often unnatural) in an insectary. However, Aphidius spp. and P. foveolatus can be mass produced on hosts reared in the field under controlled conditions (Halfhill & Featherston 1973, Stevens et al. 1975a). Also, Hoy et al. (1982b) produced the phytoseiid M. occidentalis in the greenhouse and field.

Unnatural Hosts

Rearing parasites or predators on unnatural hosts may result in a change of host preference, consequently rendering them less effective when released for control of the natural host (National Academy of Science 1969). However, this phenomenon has not been observed under experimental conditions (Arthur 1965, Legner & Thompson 1977), though rearing a parasite on an unnatural host may increase its acceptance of that host (Thorpe & Jones 1937). Obviously, selection of genotypes responding more rapidly to unnatural hosts and surviving better on it may occur. King & Morrison (1984) concluded that entomophagous arthropods reared for a short time on an unnatural host will probably respond as strongly to the natural host as those reared on it.

Cost of Entomophagous Arthropods

The cost of mass producing entomophagous arthropods coupled with the need to release large numbers at periodic intervals greatly limits the economic feasibility of the augmentative approach. Using some of the production costs cited in selected papers it is apparent that the cost per ha for using predators and parasites at effective rates usually greatly exceeds the cost using other control strategies. For example, control of *Heliothis* spp. in cotton with *T. pretiosum* required a total of at least 192,000 parasites/ha per release, and parasites were released at two to four day intervals (Stinner et al. 1974). Assuming the cost of parasites at $0.04/1000 (B. Maedgen, personal communication) then the cost would be $7.68/ha per application, which compares favorably with cost of a commonly used synthetic pyrethroid, fenvalerate, at $16.18/ha (1983 costs for material plus aerial application) when applied at the rate 0.11 kg/ha. However, the pyrethroid need be applied only once for every two to three parasite applications. Also, Biogenesis, Inc. prices for *Trichogramma* are 1/7 to 1/9 those quoted by other companies. Certainly, the *Trichogramma* releases made by Oatman & Platner (1971, 1978) seem to be comparable in cost to insecticide control, but other pests requiring insecticide treatments were present.

Phytoseiids are not used more extensively in the U.S. for augmentative releases since costs could amount to $7800/ha if they are released at an effective rate (see case study above). Likewise, the use of *C. carnea* to control *Heliothis* spp. in cotton is simply not economically feasible at effective release rates of 247-494 thousand/ha (see case study above). Unfortunately, most commercial insectaries compensate for the high cost of their product by either making no release rate recommendations, or the release rates recommended are so low as to be ineffective. Typically, the only guarantee is delivery of live insects. The insectary, Biogenesis, Inc., may be an exception because they are recommending releases of *T. pretiosum* at rates ranging from 123,500 to 185,250 parasites/ha per release for control of *Heliothis* spp. in cotton and southwestern corn borer, *Diatraea grandiosella*, control in corn in the Texas High Plains (B. Maedgen, personal communication).

Nutrition and Artificial Diets

Large-scale commercialization of the augmentative approach in the U.S. will be practical only for selected organisms and is largely dependent upon the development of suitable diets and methods for rearing and storing predators and parasites. Media costs for rearing entomophagous arthropods would be minimal compared to costs for rearing them on hosts. Thus, mass production would be less expensive, and potentially a more consistently high quality organism could be produced.

Research on rearing predators and parasites on artificial diet may be divided into 3 areas: collection of eggs or larvae; formulation of nutritious and economical artificial diets; and development of suitable holding conditions and efficient handling procedures. L. diatraeae maggots were extracted from adults using a food blender (King et al. 1979); however, Nettles & Burks (1975) induced Archytas marmoratus to larviposit on acidic solutions of a protein extracted from Heliothis virescens frass. Nettles et al. (1982) demonstrated a $KCl-Mg_2SO_4$ solution to be an excellent ovipositional stimulant for T. pretiosum, and eggs collected from the solution and transferred to an artificial diet later produced adults. King et al. (1984c) enumerated 22 parasite species reared in vitro.

An oligidic diet for Coleomegilla maculata (Attalah & Newsom 1966) and a meridic diet for Geocoris punctipes (Cohen 1981) have been developed. Significant developments in the U.S. include an artificial diet for rearing C. carnea larvae (Vanderzant 1973), encapsulation of the larval diet (Martin et al. 1978), and formulation of a nutritious adult diet (Hagen & Tassan 1970).

Maintaining Control Characteristics

Emphasis has been placed on the maintenance of genetic variability in insect rearing programs since it is important in the production of insects which will interact effectively with the wild host. However, equally important is the production of nutritionally fit insects free of debilitating pathogenic organisms. Also, the physical conditions surrounding production and use of the entomophage may affect its fitness.

Several researchers have divided overall quality of an organism into various components and then attached relative values to the components based on the proposed use of the entomophage (Boller 1972, Hoy 1976, Huettel 1976). We have modified their categorization by dividing overall quality into four components: adaptability, sexual activity, host selection, and mobility. Fitness traits or attributes can be assigned to each of these components and Huettel (1976) lists a number of monitoring methods for measuring these traits. Unfortunately most routine measurements of traits are production oriented, e.g., percentage parasitization, adult emergence, and egg production. Traits more directly related to behavioral performance such as diurnal rhythmicity, flight propensity, genetic variation, and host finding ability are not routinely measured.

Insects typically undergo a low level of reproduction during laboratory colonization, though reproduction usually increases after five to seven generations. Apparently, selection for individuals suitable to laboratory conditions has taken place (Boller 1972). Consequently, a decline in genetic variation occurs concurrent with the organism's ability to interact with its host. Insectary managers often compensate for this loss of genetic variation by initiating new colonies each season with feral stock. Nevertheless, it is possible to select for desirable traits in the laboratory. Perhaps the most outstanding example is work by Hoy et al. (1982b) in selecting for insecticide resistance in the predaceous mite M. occidentalis for use in inoculative releases to control spider mites.

Desirable characteristics in an entomophagous arthropod chosen for augmentation would include: 1) adaptation to the biotic environment, 2) good searching ability, 3) appropriate prey or host range, 4) synchrony with host or prey life cycle, 5) easily and economically mass reared, and 6) pesticide tolerance. These characteristics may be selected for, and certainly must be preserved during the rearing process.

Maintaining genetic variation in laboratory colonies will remain a serious problem, but with proper management these problems can be minimized. Pre-colonization measures may include pooling large, multiple founder colonies and afterwards maintaining several colonies under varying environmental conditions. Genetic changes must be characterized for each species to determine effective colony size and retain desirable characteristics in the colony. Token stimuli may be necessary in the rearing procedure for maintaining characteristics such as oviposition site preferences.

Storage of Entomophagous Arthropods

Shelf life of parasites and predators is brief compared to that of inorganic insecticides and miticides. Dietrick (1981) stated that "other pest control products can be kept in readiness for problems to develop, but the insectary must arrange for an orderly delivery of its products as they emerge daily from the production cage". The demand for entomophages is seasonal and there may be only brief periods in selected crops when insect or mite pest control is required. However, the insectary must maintain colonies throughout the year, often discarding much of their product. This necessity inflates production cost, thereby limiting the economic feasibility of augmentation.

Research must be directed toward developing methods for long-term storage of selected natural enemies and their insectary host without loss of viability and effectiveness. Li (1984) reported that Antheraea pernyi pupae can be stored at 3-5°C for 3-5 months, and eggs parasitized by T. dendrolimi can be stored at 3-5°C for 2 months. King & Morrison (1984) reported on a reduced temperature storage procedure for T. pretiosum whereby parasitized host eggs are held at 26.7°C for 6 days then at 16.7°C for 6 days. This system synchronized emergence of the parasites (75% within 4 h after application to field) and permitted flexibility of release date since parasite pupae could be held up to 10 days at 16.7°C before adult emergence occurred in storage.

Since storage methods usually involve retarding development with reduced temperatures, reduction in viability and effectiveness may occur (King et al. 1979, Scopes & Biggerstaff 1971,

Starks et al. 1976). Moreover, effects of storage may not be apparent under laboratory conditions, but differences between stored and nonstored organisms may occur when they are stressed by field conditions.

RELEASE METHODS FOR ENTOMOPHAGOUS ARTHROPODS

Field release of parasites and predators in the U.S. ranges from labor-intensive operations to broadcast aerial release methods. Entomophages with poor dispersal capability, e.g., phytoseiids, must be placed in close proximity to their prey. Others such as *L. testaceipes* (see case history above) may be point released. Perhaps the most dramatic example of dispersal ability and biological control is use of *P. foveolatus* to control the Mexican bean beetle (see case history above). This parasite can be reared on its host in snapbean nurse plots near soybean and disperse up to 64 km; thus, broadcast release would be unnecessary for this parasite. Reeves (1975) reported distribution of 2-3 day old *C. carnea* larvae in a sawdust medium dispensed from a back-pack release unit, a 4-row motorized unit, or aircraft. Aircraft release of *L. diatraeae* utilized a refrigerated box housing a mechanism to dispense parasites at desired rates (King et al. 1984c).

King et al. (1984b) reviewed advances in release technology for *Trichogramma* spp. This parasite has been typically point released either from cards suspended from the plant or distributed at random in the field. In the People's Republic of China (Coulson et al. 1982) parasitized and unparasitized eggs are suspended from trees in plastic capsules (with ports for parasite escape), parasitized eggs on cards are pinned to the underside of corn leaf sheaths, or the sugarcane leaf sheath is doubled over the card of parasitized eggs. Distribution of *Trichogramma* has been automated in the U.S. Morrison et al. (1978) developed a system which yields individual parasitized host eggs unattached to a substrate. Bouse et al. (1981) and King et al. (1984b) developed a thermoelectrically cooled device for broadcast aerial application of *Trichogramma* pupae contained in the free-form Angoumois grain moth egg.

Significant advances have been made on the mechanized releases of entomophagous arthropods, most notably *Trichogramma*. However, dispersal of the organism after release must be known to develop the best release system. Methods will necessarily be unique to each entomophage to ensure maximum viability and effectiveness.

MODELING IN AUGMENTATION

Predictive models are required for rational use of augmentative releases. Timing of release and number of predators and/or parasites necessary for release to achieve the desired level of biological control must be known. One approach to obtaining this information is to test the entomophages at various timing and rate regimes under various field conditions. Then crop yield could be regressed on pest density, entomophage release rate, and other covariates. Scouting and weather data could be used in the regressions for decision making on timing and release rate. Unfortunately, regression models may be inaccurate when applied to conditions other than those under which they were generated and this trial-and-error approach provides little information on the underlying causes of success or failure of augmentation.

A mechanistic approach to predict effects of augmentations involves building mathematical models of population dynamics of the crop, pest, and natural enemies. The models can be applied in a variety of field conditions since they incorporate the underlying mechanisms of the crop-pest-entomophage interaction. Moreover, these models can reveal why a particular release rate of entomophage succeeds or fails in achieving adequate levels of biological control. However, building these models requires extensive knowledge of: 1) realized fecundity of the pest and its natural enemies in the field; 2) developmental rate; 3) mortality and density dependence of the mortality; 4) functional and aggregative response of natural enemies; 5) host or prey preference including stage preference; 6) effect of entomophage attack on host or prey behavior; 7) effect of pest activity on crop yield; and 8) dispersal rate of pest and entomophage(s). Besides requiring extensive data, mechanistic mathematical models are difficult to test. This problem might be solved by combining the empirical and mechanistic approaches.

DISCUSSION

In October 1978, the National Program Staff, Agricultural Research Service, U.S. Department of Agriculture issued a statement entitled "Status and Potential for Production and Use of Biological Agents for Pest Control". We paraphrase here the section relevant to augmentation of predators and parasites:

> Careful consideration of a number of complex factors is essential in determining whether or not adequate arthropod pest control can be obtained with the use of commercially available entomophagous arthropods. Reliance upon use of entomophagous arthropods to control pests should be limited to those situations where scientifically, environmentally, and economically sound procedures are available.

The authors support the above statement and cannot condone the use of predators and parasites as placebos such as the release of *Trichogramma* in West Texas to "lessen the farmers' worry when cotton bollworm moths were reported to be flying...and provide the pest control advisor [with] a logical alternative product to counteract the fear generated in the farmers by the unrealistic economic thresholds set for bollworms" (Dietrick 1981).

We appreciate the apparent economic success of grower cooperation such as the Fillmore Citrus District, Ventura County, CA (DeBach 1974, Reichelderfer 1981, Ridgway et al. 1977) and their integrated approach to pest control including augmentation of entomophages. However, there is a need to quantify the impact of each management tactic, including the augmentations. For example, DeBach et al. (1950) first reported that release of 100,000 Aphytis chrysomphali or A. lingnanensis/acre would keep citrus orchards "commercially clean" of California red scale. DeBach et al. (1955) revised the release rate to 500,000/acre. Now A. melinus is produced commercially by several companies, yet little research has been published demonstrating the technical or economic feasibility of the approach. In the case of C. montrouzieri, recommended release rates vary from 10 to 50 adults/tree depending on the level of infestation and mealybug species (Armitage 1929, Bartlett & Clancy 1972, Fisher 1963). However, no recent research has been published on the economic feasibility of using this predator in augmentative releases.

Commercialization of the augmentative approach in the U.S. will increase in part due to problems associated with pesticide residues, resistance and impact on non-target organisms. In other cases, it will increase because it is simply economically more feasible than alternative tactics, e.g., P. foveolatus to control Mexican bean beetle. The cost of producing entomophages will be reduced as artificial diets and storage techniques are perfected. The quality of mass reared entomophages will improve as our knowledge of nutritional requirements increases, including development of artificial diets, characterization of genetic changes during colonization, and development of means for retaining essential control characteristics. Finally, integrated pest management systems are being developed that emphasize host plant resistance, cultural control, and biological control with a consequent reduction in pesticide use. Knowledge of the biology and behavior of the organisms involved, coupled with modeling, will facilitate decisions in these complex systems.

RECOMMENDATIONS

Two biological control workshops have been conducted within the past year; each included a section on technology for propagation and release of entomophagous arthropods for augmentation purposes. The first workshop, "National Interdisciplinary Biological Control Conference," developed a list of 28 research areas in the section "Propagation and Release" within "Increasing Abundance and Effectiveness" (Anonymous 1983a). Of these 28 ideas, six were given priority. The second meeting was the "ARS

National Biological Control Planning Conference/ Workshop." The discussion group for "Technology for Propagation and Release (Augmentation): Arthropods to Control Arthropod and Weed Pests" developed a list of nine research objectives and 11 research approaches (Anonymous 1984b). Three areas of research were considered major obstacles in advancement of the augmentative approach. Other reviews that list priority research areas and provide a perspective for future prospects in biological control by augmentative release of entomophages include King et al. (1984c), Rabb et al. (1976), Ridgway et al. (1977) and Stinner (1977). Additionally, recommendations on insect rearing and quality control germane to augmentation were developed as an outgrowth of the conference "Advances and Challenges in Insect Rearing," Atlanta, GA, March 4-6, 1980 (King & Leppla 1984a). The recommendations below are listed in priority (1 = highest priority) based on the authors' experiences and interpretation of recommendations and perspective from above workshops and reviews.

1. Inability to efficiently mass rear entomophagous arthropods of a consistently high quality will continually limit the augmentative approach to a few specialized situations. The necessity of rearing host insects is expensive and often complex. The cost of media for rearing predators and parasites would probably be minimal compared to rearing them on natural or unnatural hosts, and dietary ingredients could be monitored to assure that nutrition was adequate.
   RECOMMENDATION: Interdisciplinary teams, each including an insect physiologist, biochemist, and nutritionist should be formed to establish the nutritional requirements for selected entomophages and to develop artificial diets and in vitro rearing procedures where appropriate.
2. In order to develop rational augmentation programs the relationship between numbers of entomophages released and their impact on the pest population and the commodity protected must be assessed. Dynamic models based on an understanding of the population processes underlying these interactions will be required for these assessments. This quantitative information will aid in defining benefits of the augmentation.
   RECOMMENDATION: Fundamental studies must be conducted to elucidate processes underlying interactions between the host or prey, the augmented entomophage, and their biotic and abiotic environment. Modeling each system will serve as a guide toward developing an understanding of these complex interactions.
3. Biotypes of the entomophagous arthropod and host or prey influence the potential for biological control. Biotypes of entomophages may have differences in host specificity, and subtle differences in behavior or other attributes may affect the efficacy of biological control. Genetic improvement of entomophages is possible through standard breeding techniques

and perhaps high technology gene transfer thereby creating a new, more effective biotype.

RECOMMENDATION: Comprehensive biosystematic studies, including development of criteria for biotype selection, are needed to ensure that the most suitable species or biotype is selected for each specific augmentative use. Genetic improvement should be attempted for key attributes.

4. Dispersal of the pest and entomophage continues to hamper attempts to design, conduct, and evaluate experiments on the augmentative approach. Minimal size of the release area is often unknown and movement of the natural enemy from the release area limits its effectiveness.

   RECOMMENDATION: For each augmentation system the optimal size of the release area should be determined based on dispersal attributes of the pest and natural enemy as well as spatial heterogeneity of the agroecosystem. This determination should be made prior to, or be a part of, augmentation experiments.

5. Insecticide interference often is a major factor limiting the use or evaluation of an augmentation. For example, broad-spectrum insecticides or miticides should not be used at a time when they will substantially reduce released and naturally-occurring entomophages. Cooperation between growers will be required to limit insecticide interference, and improved decision making will require indepth knowledge of the complex ecosystem.

   RECOMMENDATION: Develop pest management systems that eliminate or limit insecticide interference of released and naturally occurring entomophages.

6. Laboratory domestication, the restriction of population size, and use of unnatural diets inevitably lead to differences in gene frequency between native and colonized populations. These changes may reduce the effectiveness of the entomophage.

   RECOMMENDATION: Establish methods for the selection and maintenance of the best genotypic constitution of selected mass reared entomophagous arthropods, and methods for monitoring genotypic changes as quality control procedures.

7. The cost of mass producing entomophagous arthropods often limits the economic feasibility of the approach. Moreover, the perishability of the product and its need for a narrow range of temperature, humidity, and light condition increases its cost. Other augmentation requirements include packaging, shipment, and application of the entomophages.

   RECOMMENDATION: Efficient production and use of entomophagous arthropods for augmentative purposes will require engineering expertise. Environmental parameters and engineering specifications for storage, shipment, and release of entomophages must be defined.

ACKNOWLEDGMENT

The authors thank Mr. R.J. Coleman, Biological Technician, in the Southern Field Crop Insect Management Laboratory, for his assistance in the collection and cataloguing of literature in this paper. Also, we thank Drs. J.D. Lopez, College Station, TX; E.F. Knipling, Beltsville, MD; and S.B. Stark, Stoneville, MS, for their critical review of the manuscript.

REFERENCES CITED

Ables, J.R., S.L. Jones, R.K. Morrison, V.S. House, D.L. Bull, L.F. Bouse and J.B. Carlton. 1979. New developments in the use of Trichogramma to control lepidopteran pests of cotton, pp. 125-127. In: Proc. 1979 Cotton Prod. Res. Conf. Memphis, TN.

Ables, J.R., and R.L. Ridgway. 1981. Augmentation of entomophagous arthropods to control pest insects and mites, pp. 273-303. In: G.C. Papavizas (Ed.), Biological control in crop production. BARC Symposium No. 5. Allanheld, Osmun & Co. Totowa, NJ.

Angalet, G.W., L.W. Coles and J.A. Stewart. 1968. Two potential parasites of the Mexican bean beetle from India. J. Econ. Entomol. 61:1073-1075.

Anonymous. 1978. Biological agents for pest control: Status and prospects. U.S. Government Printing Office, Washington, D.C. 138 pp.

Anonymous. 1983. Colorado Dept. Agric., Div. Plant Industry Ann. Rept. 1982-83, 44 pp.

Anonymous. 1983a. Practice and implementation of biological control, pp. 49-58. In: Proc. nat. interdisc. biological control conf. Feb 15-17, Las Vegas, NV. USDA/CSRS, Washington.

Anonymous. 1984. Research planning conference on biological control. March 20-22, Laurel, MD.

Armitage, H.M. 1929. Timing field liberations of Cryptolaemus in the control of the citrophilus mealybug in the infested citrus orchards of southern California. J. Econ. Entomol. 22:910-915.

Arthur, A.P. 1965. Absence of preimaginal conditioning in Itoplectis conquisitor (Say) (Hymenoptera: Ichneumonidae). Can. Entomol. 97:1000-1001.

Attalah, Y.H., and L.D. Newsom. 1966. Ecological and nutritional studies on Coleomegilla maculata DeGeer (Coleoptera: Coccinellidae). I. The development of an artificial diet and a laboratory rearing technique. J. Econ. Entomol. 59:1173-1179.

Bartlett, B.R., and D.W. Clancy. 1972. The Comstock mealybug in California and observations on some of its natural enemies. J. Econ. Entomol. 65: 1329-1332.

Bennett, F.D. 1969. Tachinid flies as biological control agents for sugarcane moth borers, pp. 117-148. In: J.R. Williams, J.R. Metcalfe, R.W. Montgomery and R. Mathes (Eds.), Pests of sugar cane. Elsevier Publ. Co., N.Y.

Boller, E.F. 1972. Behavioral aspects of mass rearing of insects. Entomophaga 17:9-25.

Bouse, L.F., J.B. Carlton and R.K. Morrison. 1981. Aerial application of insect egg parasites. Trans. Amer. Soc. Agric. Eng. 24:1093-1098.

Brewer, F.D., and E.G. King. 1978. Effects of parasitism by a tachinid, Lixophaga diatraeae, on growth and food consumption of sugarcane borer larvae. Ann. Entomol. Soc. Amer. 71:19-22.

Brunetti, K.M. 1981. Suppliers of beneficial organisms in North America. Calif. Dept. of Food and Agric., Biological Control Services Program, Sacramento, CA., 4 pp.

Brunson, M.H., and A.W. Allen. 1944. Mass liberation of parasites for immediate reduction of oriental fruit moth injury to ripe peaches. J. Econ. Entomol. 37:411-416.

Brunson, M.H., and A.W. Allen. 1954. Joint use of parasites and insecticides for control of the oriental fruit moth. J. Econ. Entomol. 47:147-152.

Bryan, D.E., R.E. Fye, C.G. Jackson and R. Patana. 1973. Release of parasites for suppression of pink bollworms in Arizona. USDA-ARS W-7, 8 pp.

Burbutis, P.P., and L.F. Goldstein. 1983. Mass rearing of Trichogramma nubilale on European corn borer, its natural host. Prot. Ecol. 5:269-275.

Cohen, A.C. 1981. An artificial diet for Geocoris punctipes (Say). S.W. Entomol. 6:109-113.

Cooke, W.C. 1963. Ecology of the pea aphid in the Blue Mountain area of eastern Washington and Oregon. USDA Tech. Bull. 1287, 48 pp.

Coulson, J.R., W. Klassen, R.J. Cook, E.G. King, H.C. Chiang, K.S. Hagen and W.G. Yendol. 1982. Notes on biological control of pests in China, 1979, pp. 1-192. In: Biological control of pests in China. USDA (Unnumbered Series).

Croft, B.A., and J.A. McMurtry. 1972. Minimum releases of Typhlodromus occidentalis to control Tetranychus mcdanieli on apple. J. Econ. Entomol. 65:188-191.

DeBach, P. (ed.). 1964. Biological control of insect pests and weeds. Chapman & Hall, London. 844 pp.

DeBach, P. 1974. Biological control by natural enemies. Cambridge Univ. Press, N.Y. 323 pp.

DeBach, P., E.J. Dietrick, C.A. Fleschner and L.W. Fisher. 1950. Periodic colonization of Aphytis for control of the California red scale. Preliminary tests, 1949. J. Econ. Entomol. 43:783-802.

DeBach, P., and K.S. Hagen. 1964. Manipulation of entomophagous species, pp. 429458. In: P. DeBach (Ed.), Biological control of insect pests and weeds. Chapman & Hall, London.

Dietrick, E.J. 1981. Commercial production of entomophagous insects and their successful use in agriculture, pp. 151-160. In: G.C. Papavizas (Ed.), Biological control in crop production. Beltsville Symposia in Agriculture Research. Allanheld, Osmun Publ., Towata, N.J.

Dolphin, R.E., M.L. Cleveland, L.E. Mouzin and R.K. Morrison. 1972. Releases of Trichogramma minutum and T. cacoeciae in an apple orchard and the effects on populations of codling moths. Environ. Entomol. 1:481-484.

Doutt, R.L. 1951. Biological control of mealybugs infesting commercial greenhouse gardenias. J. Econ. Entomol. 44:37-40.

Doutt, R.L., and K.S. Hagen. 1950. Biological control measures applied against Pseudococcus maritimus on pears. J. Econ. Entomol. 43:94-96.

Elsey, K.D. 1975. Jalysus spinosus: Increased numbers produced on tobacco by early-season releases. Tobacco Sci. 19:13-15.

Finney, G.L., S.E. Flanders and H.S. Smith. 1947. Mass culture of Macrocentrus ancylivorus and its host, the potato tuber moth. Hilgardia 17: 437-483.

Fisher, L.W. 1963. Mass culture of Cryptolaemus and Leptomastix--natural enemies of citrus mealybug. Calif. Agric. Exp. Sta. Bull. 797, 39 pp.

Fye, R.E. 1972. Manipulation of Polistes exclamans arizonensis. Environ. Entomol. 1:55-57.

Hagen, K.S., and R.L. Tassan. 1970. The influence of food wheast and related Saccaramyces fragilis yeast products on the fecundity of Chrysopa carnea. Can. Entomol. 102:806-811.

Halfhill, J.E., and P.E. Featherston. 1973. Inundative releases of Aphidius smithi against Acyrthosiphon pisum. Environ. Entomol. 2:469-472.

Harbaugh, B.K., and R.H. Mattson. 1973. Lacewing larvae control aphids on greenhouse snapdragons. J. Amer. Soc. Hort. Sci. 98:306-309.

Hart, R.E. 1972. Compensatory releases of Microterys flavus as a biological control agent against brown soft scale. Environ. Entomol. 1:414-419.

Hassan, S.A. 1982. Mass production and utilization of Trichogramma: 3. Results of some research projects related to the practical use in the Federal Republic of Germany, pp. 213-218. In: Les trichogrammes, Antibes (France), April 20-23.

Helgesen, R.G., and M.J. Tauber. 1974. Biological control of greenhouse whitefly, Trialeurodes vaporariorum (Aleyrodidae: Homoptera), on short-term crops by manipulating biotic and abiotic factors. Can. Entomol. 106: 1175-1188.

Hoy, M.A. 1976. Genetic improvement of insects: Fact or fantasy. Environ. Entomol. 5:833-839.

Hoy, M.A., W.W. Barnett, W.O. Reil, D. Castro, D. Cahn, L.C. Hendricks, R. Coviello and W.J. Bentley. 1982a. Large-scale releases of pesticide resistant spider mite predators. Calif. Agric. 36(1/2):8-10.

Hoy, M.A., D. Castro and C. Cahn. 1982b. Two methods for large scale productions of pesticide-resistant strains of the spider mite predator Metaseiulus occidentalis (Nesbitt) (Acarina, Phytoseiidae). Z. angew. Entomol. 94:1-9.

Huettel, M.D. 1976. Monitoring the quality of laboratory-reared insects: A biological and behavioral perspective. Environ. Entomol. 5:807-814.

Huffaker, C.B., and C.E. Kennett. 1956. Experimental studies on predation: Predation and cyclamen mite populations on strawberries in California. Hilgardia 26:191-222.

Huffaker, C.B., and P.S. Messenger (Eds.). 1976. Theory and practice of biological control. Academic Press, N.Y. 788 pp.

Hussey, N.W., W.J. Parr and H.J. Gould. 1965. Observations on the control of Tetranychus urticae Koch on cucumbers by the predatory mite Phytoseiulus riegeli Dosse. Entomol. Exp. Appl. 8:271-281.

Jaynes, H.A., and E.K. Bynum. 1941. Experiments with Trichogramma minutum Riley as a control of the sugarcane borer in Louisiana. USDA Tech. Bull. 743, 43 pp.

Kanour, W.W., and P. Burbutis. 1984. Trichogramma nubilale (Hymenoptera: Trichogrammatidae) field releases in corn and a hypothetical model for control of European corn borer (Lepidoptera: Pyralidae). J. Econ. Entomol. 77:103-107.

Kennett, C.E., and J. Hamai. 1980. Oviposition and development in predaceous mites fed with artificial and natural diets (Acari: Phytoseiidae). Entomol. Exp. Appl. 28:116-122.

King, E.G., L.F. Bouse, D.L. Bull, W.A. Dickerson, W.J. Lewis, P. Liapis, J.D. Lopez, R.K. Morrison and J.R. Phillips. 1984b. Potential for management of Heliothis spp. in cotton by augmentative releases of Trichogramma pretiosum, pp. 232-236. In: Proc. Beltwide Cotton Prod. Res. Conf. National Cotton Council, Memphis, TN.

King, E.G., G.G. Hartley, D.F. Martin, J.W. Smith, T.E. Summers and R.D. Jackson. 1979. Production of the tachinid Lixophaga diatraeae on its natural host, the sugarcane borer, and on an unnatural host, the greater wax moth. USDA, SEA, Adv. Agric. Technol., Southern Ser. No. 3, 16 pp.

King, E.G., and N.C. Leppla (Eds.). 1984a. Advances and challenges in insect rearing. USDA-ARS, Southern Region, New Orleans, LA. 306 pp.

King, E.G., and R.K. Morrison. 1984. Some systems for production of eight entomophagous arthropods, pp. 206-222. In: E.G. King and N.C. Leppla (Eds.), Advances and challenges in insect rearing. USDA-ARS, Southern Region, New Orleans, LA.

King, E.G., R.L. Ridgway and A.W. Hartstack. 1984c. Propagation and release of entomophagous arthropods for control by augmentation. In: Ma Shijun and P.L. Adkisson (Eds.), Joint U.S./China Symposium on the biological control of insects. Sci. Sinica (In Press).

King, E.G., J. Sanford, J.W. Smith and D.F. Martin. 1981. Augmentative releases of Lixophaga diatraeae (Dip.: Tachinidae) for suppression of early-season sugarcane borer populations in Louisiana. Entomophaga 26:59-69.

Knipling, E.F. 1979. The basic principles of insect population suppression and management. USDA Agric. Handb. 512, 623 pp.

Lawson, F.R., R.L. Rabb, F.E. Guthrie and L.G. Boweny. 1961. Studies of an integrated control system for hornworms on tobacco. J. Econ. Entomol. 540:93-97.

Legner, E.F., and R.A. Medved. 1979. Influence of parasitic Hymenoptera on the regulation of pink bollworm, Pectinophora gossypiella, on cotton in the lower Colorado desert. Environ. Entomol. 8:922-930.

Legner, E.F., and S.N. Thompson. 1977. Effects of the parental host on host selection, reproductive potential, survival and fecundity of the egg-larval parasitoid, Chelonus sp. near curvimaculatus, reared on Pectinophora gossypiella and Phthorimaea operculiella. Entomophaga 22:75-84.

Li, L.-Y. 1984. Research and utilization of Trichogramma in China. In: Ma Shijun and P.L. Adkisson (Eds.), Joint U.S./China Symposium on the biological control of insects. Sci. Sinica (In Press).

Lingren, P.D., R.L. Ridgway and S.L. Jones. 1968. Consumption by several common arthropod predators of eggs and larvae of two Heliothis species that attack cotton. Ann. Entomol. Soc. Amer. 61:613-618.

Martin, P.B., R.L. Ridgway and C.E. Schultze. 1978. Physical and biological evaluation of an encapsulated diet for rearing Chrysopa carnea. Fla. Entomol. 61: 145-152.

McMurtry J.A. 1982. The use of phytoseiids for biological control: Progress and future prospects, pp. 23-48. In: M.A. Hoy (Ed.), Recent advances in knowledge of the Phytoseiidae. Univ. California, Agric. Sci. Publ. No. 3284.

McMurtry, J.A., H.G. Johnson and G.T. Scriver. 1969. Experiments to determine effects of mass release of Stethorus picipes on the level of infestations of the avocado brown mite. J. Econ. Entomol. 62:1216-1221.

McMurtry, J.A., E.R. Oatman, P.A. Phillips and C.W. Wood. 1978. Establishment of Phytoseiulus persimilis [Acari: Phytoseiidae] in southern California. Entomophaga 23:175.

Metcalfe, J.R., and J. Breniere. 1969. Egg parasites (Trichogramma spp.) for control of sugarcane moth borers, pp. 81-115. In: J.R. Williams, J.R. Metcalfe, R.W. Montgomery and R. Mathes (Eds.), Pests of sugar cane. Elsevier Publ. Co., N.Y.

Morrison, R.K., S.L. Jones and J.D. Lopez. 1978. A unified system for the production and preparation of Trichogramma pretiosum for field release. S.W. Entomol. 3:62-68.

National Academy of Sciences. 1969. Control by parasites, predators, and competitors, pp. 100-164. In: Insect management and control. Washington, D.C.

Nettles, W.C., and M. L. Burks. 1975. A substance from Heliothis virescens larvae stimulating larviposition by females of the tachinid Archytus marmoratus. J. Insect Physiol. 21:965-978.

Nettles, W.C., R.K. Morrison, Z.N. Zie, D.L. Bull, C.A. Shenkir and S.B. Vinson. 1982. Synergistic action of potassium chloride and magnesium sulfate on parasitoid wasp oviposition. Science 218:164-166.

Oatman, E.R., F.E. Gilstrap and V. Voth. 1976. Effect of different release rates of Phytoseiulus persimilis [Acarina: Phytoseiidae] on the twospotted spider mite on strawberry in southern California. Entomophaga 21:269-273.

Oatman, E.R., J.A. McMurtry, F.E. Gilstrap and V. Voth. 1977. Effect of releases of Amblyseius californica, Phytoseiulus persimilis, and Typhlodromus occidentalis on the twospotted spider mite on strawberry in southern California. J. Econ. Entomol. 70:45-47.

Oatman, E.R., J.A. McMurtry and V. Voth. 1968. Suppression of the twospotted spider mite on strawberry with mass releases of Phytoseiulus persimilis. J. Econ. Entomol. 61:1517-1521.

Oatman, E.R., and G.R. Platner. 1971. Biological coptrol of the tomato fruitworm, cabbage looper, and hornworms on processing tomatoes in southern California, using mass releases of Trichogramma pretiosum. J. Econ. Entomol. 64: 501-506.

Oatman, E.R., and G.R. Platner. 1978. Effect of mass releases of Trichogramma pretiosum against lepidopterous pests on processing tomatoes in southern California, with notes on host egg population trends. J. Econ. Entomol. 71:896-900.

Parker, F.D. 1971. Manipulation of pest populations by manipulating densities of both hosts and parasites through periodic releases, pp. 365-376. In: C.B. Huffaker (Ed.), Biological control. Plenum Press, N.Y.

Parr, W.J., A.J. Gould, N.H. Jessop and F.A.B. Ludlam. 1976. Progress towards a biological control programme for glasshouse whitefly (Trialeurodes vaporariorum) on tomatoes. Ann. Appl. Biol. 83:349-363.

Rabb, R.L., R.E. Stinner and R. van den Bosch. 1976. Conservation and augmentation of natural enemies, pp. 233-254. In: C.B. Huffaker and P.S. Messenger (Eds.), Theory and practice of biological control. Academic Press, N.Y.

Reeves, B.G. 1975. Design and evaluation of facilities and equipment for mass production and field release of an insect parasite and an insect predator. Ph.D. Thesis, Texas A&M University, College Station.

Reichelderfer, K.H. 1979. Economic feasibility of a biological control technology: Using a parasitic wasp, Pediobius foveolatus, to manage Mexican bean beetle on soybeans. USDA/ESCS, Agric. Econ. Rept. 430, 20 pp.

Reichelderfer, K.H. 1981. Economic feasibility of biological control of crop pests, pp. 403-417. In: G.C. Papavizas (Ed.), Biological control in crop production. BARC Symposium No. 5. Allanheld, Osmun Co., Totowa.

Reichelderfer, K.H., and F.E. Bender. 1979. Applications of a simulative approach to evaluating alternative methods for the control of agricultural pests. Amer. J. Agric. Econ. 61:258-267.

Ridgway, R.L., J.R. Ables, C. Goodpasture and A.W. Hartstack. 1981. Trichogramma and its utilization for crop protection in the United States, pp. 41-48. In: J.R. Coulson (Ed.), Proc. Joint American-Soviet Conference on use of beneficial organisms in the control of crop pests. Entomol. Soc. Amer., College Park, MD.

Ridgway, R.L., and S.L. Jones. 1968. Field-cage releases of Chrysopa carnea for suppression of populations of the bollworm and tobacco budworm on cotton. J. Econ. Entomol. 61:895-898.

Ridgway, R.L., and S.L. Jones. 1969. Inundative releases of Chrysopa carnea for control of Heliothis on cotton. J. Econ. Entomol. 62:177-180.

Ridgway, R.L., E.G. King and J.L. Carrillo. 1977. Augmentation of natural enemies for control of plant pests in the Western Hemisphere, pp. 379-416. In: R.L. Ridgway and S.B. Vinson (Eds.), Biological control by augmentation of natural enemies. Plenum Press, N.Y.

Ridgway, R.L., and S.B. Vinson (Eds.). 1977. Biological control by augmentation of natural enemies. Plenum Press, N.Y. 480 pp.

Schroder, R.F.W. 1981. Biological control of the Mexican bean beetle, Epilachna varivestis Mulsant, in the United States, pp. 351-360. In: G.C. Papavizas (Ed.), Biological control in crop production. BARC Symposium No. 5. Allanheld, Osmun Co. Totowa.

Scopes, N.E.A. 1969. The economics of mass rearing Encarsia formosa, a parasite of the whitefly Trialeurodes vaporariorum, for use in commercial horticulture. Plant Pathol. 18:130-132.

Scopes, N.E.A., and S.M. Biggerstaff. 1971. The production, handling, and distribution of the whitefly Trialeurodes vaporariorum and its parasite Encarsia formosa for use in biological control programmes in glasshouses. Plant Pathol. 20:111-116.

Sekhar, P.S. 1957. Mating, oviposition and discrimination of hosts by Aphidius testaceipes (Cres.) and Praon aguite Smithi, primary parasites of aphids. Ann. Entomol. Soc. Amer. 50:370-375.

Shands, N.A., G.W. Simpson and C.C. Gordon. 1972. Insect predators for controlling aphids on potatoes. 5. Numbers of eggs and schedules for introducing them in large field cages. J. Econ. Entomol. 65:810-817.

Shands, N.A., G.W. Simpson and B.A. Simpson. 1975. Evaluations of field introductions of two insect parasites (Hymenoptera: Braconidae) for controlling potato-infesting aphids. Environ. Entomol. 4:499-503.

Snapp, O.I. 1954. Insect pests of the peach in the eastern states. USDA Farm. Bull. 1861, 32 pp.

Starks, K.J., R.L. Burton, G.L. Teetes and E.A. Wood. 1976. Release of parasitoids to control greenbugs on sorghum. USDA ARS-S-91, 12 pp.

Starks, K.J., E.A. Wood, R.L. Burton and H.W. Somsen. 1975. Behavior of convergent lady beetles in relation to greenbug control in sorghum: Observations and preliminary tests. USDA ARS-S-53, 10 pp.

Stevens, L.M., A.L. Steinhauer and J.R. Coulson. 1975a. Suppression of Mexican bean beetle on soybeans with annual inoculative releases of Pediobius foveolatus. Environ. Entomol. 4:947-952.

Stevens, L.M., A.L. Steinhauer and T.C. Elden. 1975b. Laboratory rearing of the Mexican bean beetle and the parasite, Pediobius foveolatus, with emphasis on parasite longevity and host-parasite ratios. Environ. Entomol. 4:953-957.

Stinner, R.E. 1977. Efficacy of inundative releases. Annu. Rev. Entomol. 22: 515-531.

Stinner, R.E., R.L. Ridgway, J.R. Coppedge, R.K. Morrison and W.A. Dickerson. 1974. Parasitism of Heliothis eggs after field releases of Trichogramma pretiosum in cotton. Environ. Entomol. 3:492-500.

Summers, T.E., E.G. King, D.F. Martin and R.D. Jackson. 1976. Biological control of Diatraea saccharalis in Florida by periodic releases of Lixophaga diatraeae. Entomophaga 4:359-366.

Thorpe, W.H., and F.G.W. Jones. 1937. Olfactory conditioning in a parasitic insect and its relation to the problem of host selection. Proc. Roy. Soc. London [Ser. B] 124:56-81.

Turnipseed, S.G., and M. Shepard. 1980. Sampling Mexican bean beetle on soybean, pp. 189-200. In: M. Kogan and D.C. Herzog (Eds.), Sampling methods in soybean entomology. Springer-Verlag, N.Y.

Vanderzant, E.G. 1973. Improvements in the rearing diet for Chrysopa carnea and the amino acid requirements for growth. J. Econ. Entomol. 66:336-338.

van Lenteren, J.C., P.M.J. Ramakers, and J. Woets. 1980. World situation of biological control in greenhouses, with special attention to factors limiting application. Med. Fac. Landbouww. Rijksuniv. Gent. 45:537-544.

Vet, L.E.M., J.C. van Lenteren, and J. Woets. 1980. The parasite-host relationship between Encarsia formosa (Hymenoptera: Aphelinidae) and Trialeurodes vaporariorum (Homoptera: Aleyrodidae). IX. A review of the biological control of the greenhouse whitefly with suggestions for future research. Z. angew. Entomol. 90:26-51.

Voronin, K.E., and A.M. Grinberg. 1981. The current status and prospects of Trichogramma utilization in the USSR, pp. 49-51. In: J.R. Coulson (Ed.), Proc. Joint American-Soviet Conference on use of beneficial organisms in the control of crop pests. Entomol. Soc. Amer. Publ., College Park, MD.

Wood, E.A., and H.L. Chada. 1969. Chemical control of the sorghum "greenbug," Guymon, Oklahoma 1968. Okla. Agric. Exp. Sta. Prog. Rep. P-616, 4 pp.

DISCUSSION: Analysis of systems for biological control of crop arthropod pests in the U.S. by augmentation of predators and parasites.

DISCUSSION LEADER: R.I. Sailer

In view of the pessimistic picture with respect to large-scale application of augmentation, questions were raised concerning augmentation in other countries. Augmentation may be practiced in other areas with very different agricultural practices, different standards, and an absence of the alternatives available in the U.S.

Some interest was expressed in augmentation in small farm situations as opposed to large farms in the U.S. Relatively high natural enemy populations normally present under small farm conditions should reduce the number of released entomophages required to achieve control, which could enhance the economic feasibility. Also, monitoring procedures characteristic of good IPM programs may often allow release of entomophages into relatively small areas within large fields as needed to reduce localized infestations. Such a "spot treatment" approach should also increase economic feasibility. There was general agreement on the need for more basic research relating to systematics, nutritional requirements of natural enemies, as well as genetics, behavior, dispersal and population dynamics of both the pest and natural enemies--a very large order!

DEVELOPMENT AND USE OF MICROBIAL INSECTICIDES

Louis A. Falcon

Microbial agents are an important natural resource that contribute in various ways to population regulation of plant and animal species, both in wild and people-managed ecosystems. Those that are sought for use in agricultural IPM systems as microbial insecticides usually have certain special features. One of the most important is safety. This means that the microbial agent should be harmless to nontarget organisms and not pollute or damage the environment. As a result, microbial agents used in this way may operate harmoniously with other biological control agents and 1) complement their effects or 2) fill in during periods when they are scarce. Some other important features for microbial agents used as insecticides in agricultural IPM systems include: They may be industrialized, which means they can be mass-cultured, formulated, packaged, stored and marketed like chemical insecticides; they may be applied with conventional spray equipment normally employed in agricultural operations or a variety of other ways; once applied, some are short-lived and require repeated applications like chemical insecticides while others may persist at effective levels for long periods of time and behave in a density-dependent manner (Ignoffo, this volume). In the U.S. and many other countries of the world, the transformation of microbial agents to insecticides requires that each be researched, developed, shown to be non-hazardous, registered, industrialized and marketed. The procedures and rules employed are very similar to those used for chemical pesticides. Today, microbial agents represent an important source of raw materials for biotechnology and, especially, genetic engineering (Faulkner & Boucias, this volume). In the future many new pesticides may be derived from this source.

In my opinion, microbial agents represent an ideal form of pest control for use in agricultural IPM systems. Despite all of their positive features, microbial agents are little used, and research and development have been poorly supported to date. There are several reasons for this situation. In this review I shall briefly report on what is available and how they are used. Also, I shall discuss research needs for the development and use of microbial agents for pest control in agricultural IPM systems.

ISBN 0-12-357030-1

The subjects of insect pathology and microbial control have been extensively reviewed in the past 20 years. The Appendix to this Chapter contains a list of the major reviews produced during that time.

WHAT IS AVAILABLE

While nature has provided a large pool of microbial agents that perform biological control of all kinds of pests, I believe this great natural resource has been only slightly exploited to date. So far, over a 1000 microbial agents have been isolated from insects and mites. Most have been found by chance, very few by organized surveys. Several hundred of the isolated microbial agents are viruses. The next largest group are the fungi, followed by the protozoa, bacteria, and rickettsiae (Poinar & Thomas 1978).

At least one agent from each microbial group has been registered and industrialized in the U.S. As of early 1983, 10 microbial insecticides had been registered with the United States Environmental Protection Agency (EPA) (Table 1). Four contain bacteria, one a fungus, another a protozoan and four are viruses. The oldest registration (1948) belongs to the product Doom®. Next came products containing the bacterium Bacillus thuringiensis variety Kurstaki (Btk) registered in 1961. Fifteen years later, Elcar®, the first product containing an insect virus was registered. Since then several microbial agents have been registered as microbial insecticides. I expect this activity to continue and increase during the next decade.

A few years ago the EPA began to refer to microbial pesticides as, "biorational pesticides" (EPA 1980). In this category they also included biochemical substances such as insect hormones and pheromones. Biorational pesticides are derived from naturally occurring substances or, if synthetic, must be structurally identical to a naturally-occurring chemical. Their mode of action is other than direct toxicity. Some of the characteristics of biorational agents are their unique mode of action, generally low mammalian toxicity, low use volume, target species specificity and natural occurrence. All biorational pesticides must be registered with the EPA before they can be sold commercially. The estimated cost to complete the major required tests is $141,000 (Table 2) (Markle 1983). This is a small amount of money compared to the millions of dollars required to register a chemical pesticide. Naturally-occurring and strain-improved microbial agents are handled in this manner. Genetically-engineered microbial agents will be examined on a case by case basis. After a pesticide is registered with the EPA it may be sold in all states except California. A separate registration is needed in California and, in addition to the safety data required by the EPA, the California Department of Food and Agriculture (CDFA)

Table 1. Microbial insecticides registered in the U.S.

Microbial Agent (Trade Name)	Registrant (Year)	Uses
Bacillus thuringiensis var. Kurstaki (Btk) (Thuricide®, Dipel®)	Sandoz, Abbott, others (1961)	Against lepidopterous larvae on many agricultural crops and sites including: forest, ornamental, and shade trees; [Exempt from tolerance for all raw agricultural commodities.]
Bacillus popilliae and Bacillus lentimorbus (Doom®)	Fairfax, Hyponex, Reuter (1948)	Japanese beetle larvae on turf.
Heliothis NPV (inclusion bodies) (Elcar®)	Sandoz (1975)	Bollworm and tobacco budworm on cotton. [Exempt from tolerance for all raw agricultural commodities.]
Douglas fir tussock moth NPV (inclusion bodies) (Biocontrol®)	USDA Forest Service (1976)	Douglas fir tussock moth on Douglas fir.
Gypsy Moth NPV (inclusion bodies) (Gypcheck®)	USDA Forest Service (1978)	Gypsy moth on forest, shade and ornamental trees.
Nosema locustae (Noloc®)	Sandoz (1980)	Rangeland grasshoppers.
Hirsutella thompsoni (Mycar®)	Abbott (1981)	Citrus red mite.
Bacillus thuringiensis var. israelensis (Bti) (Teknar®)	Sandoz (1981)	Larvae of mosquitoes and blackflies.
Bacillus thuringiensis (Certan®)	Sandoz (1981)	Wax moth larvae in honeycombs.
Neodiprion sertifer NPV (Neocheck®)	USDA Forest Service (1983)	Pine sawfly.

Table 2. Microbial pesticides non-target organism and environmental fate data requirements for tier I tests only.

Type of Test	Test Substance[a] T	M	F	Notes[b]	Cost (X $1000)	Duration Days
Avian Injection and Acute Oral Tests	X					
Wild Mammalian Injection Test	X			A		
Fresh Water Fish Test	X					
Fresh Water Invertebrate Test	X					
Estuarine and Marine Animal Testing	X			A		
Plant Studies	X			A		
Honeybee	X					
Acute Oral	X	X	X			
Acute Dermal	X	X	X		45	14-21
Acute Inhalation	X	X	X			
Primary Eye Irritation		X	X		5	3-skin
Primary Dermal Irritation		X	X			7-eye
Hypersensitivity Study		X	X		6	38
Hypersensitivity Incidents				A		
Cellular Immune Response	X				5	14-21
Intravenous Injection	X			B	20	14
Intracerebral Injection	X			C	20	14
Intraperitoneal Injection	X			D	20	14
Tissue Culture	X			E	20	45
				Total	141	

[a] T = Technical; M = Manufacturing use product; F = Formulation.
[b] A = May not be required; B = Bacteria and virus only; [c] C = Protozoa and virus only; D = Protozoa and fungus only; and E = Virus only.

Table 3. Candidate entomopathogens under development in the U.S.

Bacteria:	*Bacillus thuringiensis thuringiensis* (δ-Exotoxin) (Lepidoptera, Diptera, Hymenoptera, Coleoptera)
Baculoviruses:	*Anticarsia gemmatalis* NPV (velvetbean caterpillar)
	Autographa californica NPV (loopers, armyworms, bollworm)
	Laspeyresia pomonella GV (codling moth, oriental fruit moth, pine tip moth)
	Phthorimea operculella GV (potato tuberworm)
	Pieris rapae GV (cabbage butterfly)
	Pseudoplusia includens NPV (soybean looper)
	Spodoptera spp. NPV (armyworms)
	Trichoplusia ni NPV (cabbage looper)
Fungi:	*Beauveria bassiana*; *Coelomomyces* spp.; *Entomophthora* spp.; *Erynia phytonomii*; *Lagenidium gigantium*; *Metarhizium anisopliae*; *Nomuraea rileyi*; *Paecilomyces farinosus*; *Verticillium lecanii* (Host range includes Coleoptera, Diptera, Hemiptera, Homoptera, Lepidoptera).
Protozoa:	*Nosema* spp. (Lepidoptera, Orthoptera)
	Vairimorpha necatrix (Lepidoptera)

requires efficacy data (see Food and Agricultural Code, State of California, Chapter 2).

There are several entomopathogens under development as microbial insecticides in the U.S. (Table 3). Representatives from all microbial groups are included. The bulk of the research is being done by State Agricultural Experiment Station and USDA Laboratories. A Regional Research Project (S-135) entitled, "Development of Microbial Agents for Use in Integrated Pest Management Systems," serves to coordinate much of the research. This project began in 1978 and involves about 20 State Agricultural Experiment Station groups and 14 USDA Laboratories. In addition, there are cooperating scientists from Canada and Newfoundland.

The Interregional Research Project 4 (IR-4) is assisting with the development and registration of biorational pesticides. IR-4 was established by the State Agricultural Experiment Stations and the U.S. Department of Agriculture, Cooperative State Research Service in 1963 to provide national leadership, coordination and a focal point for obtaining clearances for pesticides used on so-called minor or specialty crops. The traditional objectives of the IR-4 Program have been expanded to provide for the registration of biological materials for pest management programs (Guest 1983a, b).

Microbial agents have been industrialized as microbial insecticides in other parts of the world. Some have been registered following procedures similar to those employed in the U.S. Also, cottage industry production projects have been attempted in various countries. In the U.S., Trichoplusia ni NPV has been produced in this way in California since the 1950's (Falcon 1971). The first American entomology team that visited the People's Republic of China (PRC) in the early 1970's came back with reports of a Btk producing cottage industry at the commune level (CSCPRC 1977). By the time I visited in 1982 this activity had virtually ceased. A lack of quality control which resulted in a loss of the effective Btk strains during production appeared to be the major cause for the loss of interest in Btk.

HOW CAN THEY BE USED?

Microbial insecticides may be employed to control a specific pest such as the use of a Btk product for the control of alfalfa caterpillar (Colias eurytheme) or Certan for the control of wax moth (Galleria mellonella) larvae in honeycombs. They may be mixed with chemical pesticides to control a particular life stage of an insect such as the combining of Heliothis NPV with chlordimeform for the control of bollworm (H. zea) on cotton (Yearian et al. 1980). The chemical is used to destroy the egg and the NPV kills the larvae. A microbial insecticide may be used at harvest when it is too late to employ a chemical pesticide because of the toxic residues it will deposit. Btk is employed in this manner to control the cabbage looper (T. ni) and other caterpillars at harvest on lettuce and crucifers in California (L.A.Falcon, unpublished data). Some workers have combined a pathogen with a chemical pesticide and looked for synergistic effects (Abrahamson et al. 1982, Onsanger et al. 1981). Two or more pathogens may be combined (Fuxa 1979, Luttrell et al. 1982). A predator may be used to disseminate the pathogen (Biever et al. 1982). The product Doom, which contains the bacteria B. popilliae and B. lentimorbus is mixed with soil for the control of Japanese beetle larvae (P. japonica) (Klein 1981). Btk is added to wheat and other stored grains to protect them from damage by lepidopterous larvae (McGaughey 1980). The protozoan Nosema locustae is combined with a feeding bait in the product Noloc. This material is dispersed by airplane over range lands infested by grasshoppers. The protozoan is ingested by the grasshoppers when they feed on the bait. The protozoan does not kill the hoppers quickly, but rather debilitates them and interferes with reproduction. The hoppers eat less, produce fewer young, and cause less damage then they would otherwise be capable of doing (Henry & Onsager 1982, Onsager et al. 1981). B.t. israeliensis (Bti) is sprayed or dumped into water where it effectively controls the larvae of disease-bearing mosquitoes and black flies (Gaugler & Finney 1982, Hall et al. 1977). This product may be useful in agricultural areas where mosquitoes and black flies are resistant to chemical pesticides. The effect of mosquito and black fly feeding on humans, and the diseases they transmit may have a debilitating effect on farmworkers.

Although microbial insecticides may be applied with the same kinds of equipment used for chemical pesticides, there are some important differences (Smith 1981). The most important differences are in the modes of entry and action. Most chemical insecticides used since World War II are synthetic organic substances. They are mostly nerve poisons and enter the insect in various ways (by contact through the pulvilli on the tarsi, as a gas through the spiracles, by ingestion). In many cases all stages of an insect species are susceptible (Ware 1983). As a result, thorough spray coverage may not be necessary for a chemical insecticide to do its job. By contrast, microbial insecticides enter the insect primarily by ingestion (all groups) or contact (fungi only). In most cases only one stage (usually larva/nymph) is susceptible. For many plant protection situations, the microbial insecticide must be placed exactly where the susceptible target stage will consume it.

Another important difference is dosage. Chemical insecticides are effective at very low dosages whereas microbial insecticides may require significantly greater amounts as the target larval or nymphal stage increases in size. Chemical insecticides may have a long effective residual life on the plant surface (several days to 4 weeks), most microbial insecticides so not. Commercial preparations of *Heliothis* NPV, for example, may have a half life of only 12 hours (Bullock 1967). The need to control the larva/nymph at the youngest age plus the short residual life of microbial insecticides requires that timing of application be more precise than with chemical insecticides.

Let us look at the control of bollworm (*H. zea*) on cotton. The aerial spray application of an effective chemical insecticide (methyl parathion, carbaryl or permethrin) results in deposits mostly on the upper surfaces of the uppermost leaves (Falcon et al. 1974). Many egg-laying moths entering a newly-treated area may be quickly killed by the toxic fumes produced by the chemical pesticide. Those that escape the fumes may be killed by contacting the treated plant surfaces. If any survive to oviposit, the eggs may be destroyed by contact action. If larvae succeed in hatching, they may be poisoned by walking across treated plant surfaces. If all of these modes of entry fail, then the final assault is through the mouth as larvae bite into treated plant surfaces. The residual effectiveness of many chemicals lasts for several days. A similar aerial spray application of *Heliothis* NPV is also mostly deposited on the upper surfaces of the uppermost leaves. For maximum effectiveness the application must coincide with egg hatch. The virus must be deposited in sufficient quantity on or around the eggs so that when the larvae begin to chew their way out of the eggs, or feed on the foliage around the eggs, a lethal dose is ingested. After ingesting a lethal dose, a larva may continue to feed for several hours and live for three or more days before death occurs. The addition of feeding or gustatory stimulants such as Coax® or Gustol® may enhance the effects of Heliothis NPV (Bell & Romine 1980). Also, the use of an ultraviolet-absorbing material (cottonseed oil; A. Berlowitz, personal communication) may prolong its residual life. But even with all of these additions, Heliothis NPV still cannot readily compete with an effective chemical insecticide. The virus works best as part of an IPM program.

To be used most successfully for the control of bollworm on cotton, Heliothis NPV must be applied frequently (daily) during the period of egg hatch in the presence of bollworm egg predators. Another approach is to apply it following treatment with a chemical insecticide used for the control of another pest and which also suppresses predators. In California, dimethoate is used for the supression of lygus bug (Lygus hesperus) on cotton in the San Joaquin Valley. Dimethoate destroys the predators which normally consume over 90% of the bollworm eggs and hatching larvae. The use of Heliothis NPV serves to replace the predators until their return. This normally takes about two weeks following a single application of dimethoate (1.1 kg ai/ha) (Falcon et al. 1974).

Other methods for dispersing microbial insecticides have been studied. Agudelo & Falcon (1983) and Gard & Falcon (1978) used light traps to attract nocturnal insects. In the trap mechanism the trapped insects are coated with a microbial insecticide-laden dust and released. Upon mating with an untreated mate in the field the pathogen may be transferred to it much in the same way as a venereal disease. When alighting upon a leaf surface or other plant parts the treated insects deposit some of the microbial insecticide. If an egg is deposited at the same time, the microbial agent is in the proper place (on the chorion and the area surrounding the egg) to effect control as soon as the larva begins to feed. Both host and non-host insects pass through the trap and may pick up the microbial insecticide dust preparation and help to disseminate it throughout the target area.

WHY ARE THEY NOT USED?

The best available information indicates that microbial insecticide sales represent less than 1% of the total yearly insecticide sales in the U.S. In the last 20 years more companies have given up development and marketing of microbial insecticides than are currently doing so. The immediate future does not look bright and unless biotechnological research makes some major breakthroughs, there may be no future at all for microbial insecticides as commercial products.

One major problem is that microbial insecticides must compete with chemical insecticides in an arena where the latter dominate. Most published information giving pest control advice and recommendations to farmers deals primarily with the use of chemical pesticides. The majority of people giving farmers advice on crop protection are chemical pesticide-oriented. The chemical insecticide companies (manufacturers and retailers) are very aggressive and competitive in selling their products. Farmers have come to expect a pest control product to control many pests, very rapidly and at low cost. When an investment is at stake there is little concern for environmental problems and deleterious side-effects caused by the use of toxic chemical pesticides.

Even if the above were not true, microbial insecticides would still probably not be a high-volume sales item. A major deficiency is that few strategies utilizing microbial and other biological control agents as the main weapons to suppress major (key) pests are available to agricultural producers. Normally this is accomplished with chemical pesticides (90+% of the time) which also 1) destroy existing natural enemies (e.g., fungicides may kill entomogenous fungi), 2) deposit toxic residues (delaying the natural enemies' reestablishment), and 3) condition the plant making it more attractive to and/or better-suited for the pest (hormoligosis). In cotton, for example, key pests such as pink bollworm, bollweevil and lygus bug are commonly controlled with chemical pesticides. This predisposes the cotton ecosystem to pest control by chemical pesticides (the pesticide treadmill effect). Secondary pest outbreaks are common (bollworms, loopers, armyworms) and it is here that currently-available microbial insecticides may be used. Heliothis NPV is a useful tool for the control of bollworms. Btk controls loopers and may supress armyworms. Commonly though, several species of caterpillars occur simultaneously and their numbers may increase rapidly. As stated earlier, the farmer is conditioned into wanting rapid and inexpensive control measures. The chemical insecticides fit these requirements more readily than the available microbial insecticides.

The situation described above primarily determines the market potential of a product. Market potential is what drives industry in the western world: No market potential, no development. For this reason, private industry has made only the feeblest attempt to develop microbial insecticides to date. The bottom line is profit, something the microbial insecticides have not provided so far, and most market analysts do not believe they will provide it in the future.

WHAT IS NEEDED?

We need to develop effective strategies for using microbiological and macrobiological agents in agricultural IPM systems. Then we need to do a good job of educating the public about what can be done. There are several ways to do this. One is to develop a strategy around an existing microbial agent. Another, is to search for useful microbial agents around which to develop strategies. In the former case, one example is the codling moth granulosis virus. In the latter situation, Bti seems to fit.

The codling moth is a key pest of pome fruit and Persian walnut in most parts of the world. Methods of codling moth control usually predetermine other insect and mite problems. Codling moth research workers interested in biological control have long searched for selective and effective ways to control this insect. Several microbial agents have been examined for this purpose. The discovery through foreign exploration of a codling moth granulosis virus (CMGV) in 1963 opened a new chapter (Tanada 1964). Preliminary field studies in California (Falcon et al. 1968) and in other parts of the world produced promising results (Huber & Dickler 1975, Keller 1973, Morris 1972). But not until 1980 was there a sufficient quantity of virus to conduct the research needed to develop effective strategies. Prior to that viral quantities were limited to what each researcher could produce in their own laboratory. This situation made it difficult to conduct large scale field studies and impossible to compare results. In 1980, Sandoz Inc. began to produce CMGV as an experimental product labelled SAN 406. Although production ceased in 1982, a sufficient quantity of standardized product was made availible for large-scale testing and some strategies were developed. The strategy we developed in California consists of several parts including pest monitoring, accurate timing of application, proper application methods to achieve maximum coverage, orchard sanitation, fertilizer, irrigation, pruning and cover crop management (L.A. Falcon, unpublished data). Other strategies were developed in Europe where an active research and development program has been underway since 1980 (Huber 1982). There activities have been curtailed since 1983, again because of a lack of standardized product. The availability of large quantities of an effective standardized product is the key to developing strategies.

The development of Bti was driven by the necessity to find new ways to control chemical insecticide-resistant disease-bearing mosquitoes. In this case, an organized effort led to the discovery of a bacterial strain that produces a lethal toxin for many species of mosquito and black fly larvae (de Barjac 1978). The need was there, a highly effective agent was found and the necessary research funds were made available, mainly through the World Health Organization. This microbial agent received the fastest development of any to date. The discovery was made in 1977 and the first products were registered in 1980. This shows that where an effective microbial agent is available to provide relief for a difficult problem, it can be developed rapidly into a useful product.

RECOMMENDATIONS

1. Encourage the sale and use of registered microbial insecticide products in the following ways:

   a. Price commercial microbial insecticide products so they can compete successfully with chemical insecticides. In this way every grower who wishes to use them may do so.
   b. Disseminate information about the use of microbial insecticides. Include in pest control information available to farmers. Educate extension workers and farmers in the use of microbial insecticides.
   c. Concentrate on the development of IPM strategies in which micro- and macrobiological agents play a significant role.

2. Provide the resources and support needed to develop IPM strategies in which micro- and macrobiological agents play a significant role. Some of these are:

   a. Large quantities of standardized, effective preparations of promising candidate microbial insecticides must be made available for field studies and for the development of pest control strategies.
   b. Funding is needed to increase the number of scientists working with the research and development of strategies employing microbial insecticides and to provide adequate support budgets and facilities.
   c. Funds are needed to provide crop insurance for growers cooperating with the field development of microbial insecticides.

3. Field development of microbial insecticides requires special considerations and should only be undertaken by researchers understanding the importance of biological control in pest management.

REFERENCES

Abrahamson, L.P., J.D. Harper, I.R. Ragenovich and J.R. Hyland. 1982. Pilot control project using trichlorfon and Bacillus thuringiensis against forest tent caterpillar in southwest Alabama. USDA Forest Serv., State and Private Forestry Tech. Publ. SA-TP 15:29.

Agudelo, F., and L.A. Falcon. 1983. Mass production, infectivity, and field application studies with the entomogenous fungus, Paecilomyces farinosus. J. Invertebr. Pathol. 42:124-132.

Bell, M.R., and C.L. Romine. 1980. Tobacco budworm field evaluation of microbial control using Bacillus thuringiensis and a nuclear polyhedrosis virus with a feeding adjuvant. J. Econ. Entomol. 73:427-430.

Biever, K.D., P.L. Andrews and P.A. Andrews. 1982. Use of a predator, Podisus maculiventris, to distribute virus and initiate epizootics. J. Econ. Entomol. 75:150-152.

Bullock, H.R. 1967. Persistence of Heliothis nuclear-polyhedrosis virus on cotton foliage. J. Invertebr. Pathol. 9:434-436.

CSCPRC. 1977. Insect control in the People's Republic of China. National Academy of Sciences, Washington D.C. 218 pp.

de Barjac, H. 1978. Une nouvelle variete de Bacillus thuringiensis tres toxique pour les moustique: Bacillus thuringiensis var. israelensis serotype 14. C.R. Hebd. Seanc. Acad. Sci. Ser. D, Sci. Nat. 286:797-800.

EPA. 1980. Guidelines for registering pesticides in the U.S., Subpart M, Data requirements for biorational pesticides, preamble and guidelines (draft). (40 CFR Parts 163) (OPP-300; FRL), Pesticide Programs. NTIS. Springfield, VA.

Falcon, L.A. 1971. Microbial control as a tool in integrated control programs, pp. 346-364. In: C.B. Huffaker (Ed.), Biological control. Plenum Press, N.Y.

Falcon, L.A., W.R. Kane and R.S. Bethell. 1968. Preliminary evaluation of a granulosis virus for control of the codling moth. J. Econ. Entomol. 61:1208-1213.

Falcon, L.A., A. Sorensen and N.B. Akerson. 1974. Aerosol application of insect pathogens. Calif. Agric. 28:1-13.

Fuxa, J.R. 1979. Interactions of the microsporidium Vairimorpha necatrix with a bacterium, virus and fungus in Heliothis zea. J. Invertebr. Pathol. 33:16-23.

Gard, I.E., and L.A. Falcon. 1978. Autodissemination of entomopathogens: Virus, pp. 46-54. In: G.E. Allen, C.M. Ignoffo and R.E. Jaques (Eds.), Microbial control of insect pests: Future strategies in pest management systems. Proc. NSF-USDA-Univ. Florida Workshop.

Gaugler, R., and J.R. Finney. 1982. A review of Bacillus thuringiensis var. israelensis (serotype 14) as a biological control agent of black flies. (Simuliidae). Misc. Publ. Entomol. Soc. Amer. 12:1-17.

Guest, R.T. 1983a. Registration of biorationals. Interregional Research Project No. 4, pp. 65-67. In: G.E. Allen and M.R. Nelson (Conf. Organizers), Proc. national interdisciplinary biological control conference. USDA, CSRS. July 1983.

Guest, R.T. 1983b. Report of the IR-4 biorationals workshop, St. Louis, MO, Mar. 1-3, 1983. Office of IR-4, Cook College, Rutgers Univ., New Brunswick, NJ.

Hall, I.M., K.Y. Arakawa, H.T. Dulmage and J.A. Correa. 1977. The pathogenicity of strains of Bacillus thuringiensis to larvae of Aedes and Culex mosquitoes. Mosq. News 37:246-251.

Henry, J.E., and J.A. Onsager. 1982. Large-scale test of control of grasshoppers on rangeland with Nosema locustae. J. Econ. Entomol. 75:31-35.

Huber, J. 1982. Report of meeting of experts on "Virus production and specific control techniques in orchards", Darmstadt, Federal Republic of Germany, Dec. 2-3, 1982. Commission of European Communities. Integrated and Biological Control Programme, Joint Research Centre, I-21020 Ispra (Va).

Huber, J., and E. Dickler. 1975. Freilandversuche zur bekampfung des apfelwicklers, Laspeyresia pomonella (L.), mit granuloseviren. Z. Pflanzen-Krankh. Pflanzensch. 82:540-546.

Ignoffo, C.M., C. Garcia, D.L. Hostetter and R.E. Pinnel. 1980. Transplanting: A method of introducing an insect virus in an ecosystem. Environ. Entomol. 9: 153-154.

Keller, S. 1973. Microbiologische bekampfung des apfelwicklers Laspeyresia pomonella (L.) mit spezifischem granulosisvirus. Z. angew. Entomol. 73:137-181.

Klein, M.G. 1981. Advances in the use of Bacillus popilliae for pest control, pp. 183-192. In: H.D. Burges (Ed.), Microbial control of pests and plant diseases 1970-1980. Academic Press, N.Y.

Luttrell, R.G., W.C. Yearian, S.Y. Young and D.L. Horton. 1982. Evaluation of Bacillus thuringiensis--spray adjuvant viral insecticide combinations against Heliothis spp. Environ. Entomol. 11:783-787.

McGaughey, W.H. 1980. Bacillus thuringiensis for moth control in stored wheat. Can. Entomol. 112:327-331.

Markle, G.M. 1983. Registration of biorationals. Interregional Research Project No. 4, p. 106. In: G.E. Allen and M.R. Nelson (Conf. Organizers), Proc. of the national interdisciplinary biological control conference. USDA, CSRS. July, 1983.

Morris, D.S. 1972. A cooperative programme of research into the management of pomefruit pests in south eastern Australia: IV. Evaluation of a nuclear granulosis virus for control of codling moth, p. 238. In: Abstr. 14th Intern. Congr. Entomol., Canberra, Australia.

Onsager, J.A., N.E. Rees, J.E. Henry and R.N. Foster. 1981. Integration of bait formulations of Nosema locustae and carbaryl for control of rangeland grasshoppers. J. Econ. Entomol. 74:183-187.

Poinar, G.O., and G.M. Thomas. 1978. Diagnostic manual for the identification of insect pathogens. Plenum Press, N.Y. 218 pp.

Smith, D.B. 1981. Application technology with microbial insecticides, pp. 91-98. In: Summary proceedings of cotton biological control conference. Sponsered by Sandoz, Inc. & EPA. Dallas, Texas. Jan. 15-16.

Tanada, Y. 1964. A granulosis virus of the codling moth, Carpocapsa pomonella. J. Insect Pathol. 6:378-380.

Ware, G.W. 1983. Pesticides, theory and application. W.H. Freeman, San Francisco. 308 pp.

Yearian, W.C., R.G. Luttrell, A.L. Stacey and S.Y. Young. 1980. Efficacy of Bacillus thuringiensis and Baculovirus heliothis chlordimeform spray mixtures against Heliothis spp. on cotton. J. Ga. Entomol. Soc. 15:260-271.

APPENDIX

Selected References - Insect Pathology and Microbial Control

Steinhaus, E.A. 1946. Insect microbiology. Comstock Publ. Co., Ithaca, N.Y. 763 pp. (Reprinted by Hafner Publ. Co.)

Steinhaus, E.A. 1949. Principles of insect pathology. McGraw-Hill, N.Y. 757 pp. (Reprinted by Hafner Publ. Co.)

Franz, J.M. 1961. Biologische Schadlingsbekampfund, pp. 1-302. In: H. Richter (Ed.), Handbuch der Pflanzenkrankheiten. 2nd Ed., Vol. 6, Paul Parey, Berlin.

Krieg, A. 1961. Grundlagen der Insektenpathologie, -Viran -Rickettsien und Bakterien-infektionen. Steinhopff, Darmstadt. 304 pp.

Steinhaus, E.A. (Ed.). 1963. Insect pathology, an advanced treatise. Vol. 1, 661 pp.; Vol. 2., 689 pp. Academic Press, N.Y.

DeBach, P., and E.I. Schlinger. 1964. Biological control of insect pests and weeds. Chapman & Hall, Reinhold Book Division, N.Y. 844 pp.

Kilgore, W.W., and R.L. Doutt (Eds.). 1967. Pest control, biological, physical and selected chemical methods. Academic Press, N.Y. 477 pp.

Weiser, J. 1969. An atlas of insect diseases. Academia, Prague. 292 pp.

Burges, H.D., and N.W. Hussey. 1971. Microbial control of insects and mites. Academic Press, N.Y. 861 pp.

Huffaker, C.B. 1971. Biological control. Plenum Press, N.Y. 511 pp.

Aruga, H. 1973. General insect pathology. Yokendo, Tokyo. 674 pp. (In Japanese).

Cantwell, G.E. (Ed.). 1974. Insect diseases. Vols. 1 and 2. Marcel Dekker, N.Y.

Evalakhova, A.A. 1974. Entomogenous fungi. Classification, biology, practical significance. Nauka, Leningrad. 260 pp.

Maxwell, F.G., and F.A. Harris (Eds.). 1974. Proceedings of the summer institute on biological control of plant insects and diseases. Univ. Press of Miss., Jackson. 647 pp.

Summers, M., R. Engler, L.A. Falcon and P.V. Vail. 1975. Baculoviruses for insect pest control: Safety considerations. Amer. Soc. Microbiol., Washington, D.C. 186 pp.

Weiser, J. 1977. An atlas of insect diseases. Academia, Prague, 2nd revised ed. 240 pp.

Burges, H.D. (Ed.). 1981. Microbial control of pests and plant diseases 1970-1980. Academic Press, N.Y. 949 pp.

Kurstak, E. (Ed.). 1981. Microbial and viral pesticides. Marcel Dekker, N.Y.

MANIPULATING ENZOOTIC-EPIZOOTIC DISEASES OF ARTHROPODS

Carlo M. Ignoffo

Natural enzootics have been recorded for representatives within all major groupings of insect pathogens (protozoa, fungi, viruses, bacteria) but epizootics are more commonly observed in fungi and viruses. If disease(s) naturally occurs at low incidence (an enzootic) effective control may never be attained unless the incidence can be raised to a level above the economic threshold of the arthropod pest or if the enzootic becomes epizootic. Thus, I have assumed that if an epizootic is observed then at one time it was an enzootic and that some limiting host, pathogen or environmental factor was released to permit its expression. Since there are many examples of epizootics among arthropod pests the presumption is that we may somehow manipulate those factors that induce enzootic diseases to become epizootic.

A triad of overlapping factors are required for the existence of an enzootic and its development into an epizootic (Steinhaus 1954). These are: 1) presence of the host; 2) presence of the pathogen, and; 3) proper environmental conditions. These triad factors can act independently but also are woven together with many interactions between them. For ease of presentation I have individually focused on each of the triad factors and, when possible, discussed how they might be manipulated to initiate, orchestrate or optimize enzootics and epizootics of arthropod pests.

More detailed information of the use of microbials in IPM systems is available in the National Science Foundation proceedings, "Microbial Control of Insect Pests: Future Strategies in Pest Management Systems" edited by Allen et al. (1978). Additional studies as well as ideas, recommendations and attempts to manipulate enzootics and epizootics also are summarized in Ignoffo (1978) and Ignoffo & Hostetter (1977).

MANIPULATING THE PATHOGEN

In an enzootic the pathogen inducing the disease, by definition, is already present. However, although present, the inoculum may be too low to express an epizootic that will economically control

ISBN 0-12-357030-1

the arthropod pest. If the inoculum is absent, low or limiting we might increase it by either colonization, augmentation, or conservation.

Colonization of Inoculum

Although colonization, introduction of inoculum to establish a new enzootic disease or to attain a higher prevalence of an enzootic disease, has been rarely used it has been highly successful. Bacterial spores of Bacillus popilliae (a pathogen of Japanese beetle grubs, Popilliae japonica) were introduced to become established (White & Dutky 1940). In another example of colonization a nuclear polyhedrosis virus (NPV) of the spruce sawfly, Gilpinia hercyniae (accidentally introduced into North America), was established and today is the major, natural controlling factor (Balch & Bird 1944). This sucessful, accidental introduction stimulated the purposeful establishment of another sawfly virus, the NPV of Neodiprion sertifer (Bird 1961). Colonization also has been sucessfully used to establish fungi in aquatic and terrestrial environments (Laird 1967, Soper 1984).

Efforts to use colonization to establish enzootic diseases should be further researched, not only as a means of providing direct control but also to reduce arthropod pest populations so that they might be more easily controlled by other methods. The colonization of enzootic, new entomopathogens into domestic agroecosystems as well as evaluating the strategy of establishing domestic entomopathogens into new ecosystems where they do not normally occur should also be researched. As an example, the Heliothis NPV normally infects Heliothis zea on corn; it is rarely isolated in the United States from in cotton; and occurs only rarely in soybeans (Ignoffo 1965). Might this domestic entomopathogen be enzootically established in alien agroecosystems in the United States (cotton, soybeans, peanuts, vegetables) where it is only rarely found?

Augmentation of Inoculum

Augmentation can follow two approaches, distribution of inoculum as a microbial insecticide or release of inoculum to induce an epizootic. As a microbial insecticide the inoculum is used to suppress or control an incipient pest population. Suppression or control is generally sought for a short period, within a field season or at most for an entire field season. Many examples of the strategic use of pathogens as microbial insecticides are cited in Burges (1981), Burges & Hussey (1971) and Steinhaus (1963).

In the second augmentative approach the inoculum is introduced before damaging pests are anticipated so that it may build-up to controlling levels (Table 1). This strategy was successfully used to induce early epizootics of the fungus, Nomuraea rileyi, in caterpillar pests of soybeans (Ignoffo et al. 1976a,

Table 1. Selected examples of augmentative releases of entomopathogens to induce epizootics.

Pathogen	Ecosystem	Reference
BACTERIA		
Coccabacillus acridiorum	range-land	d'Herelle 1912
FUNGI		
Entomophthora sphaerosperma	cabbage field	Jaques & Patterson 1962
Hirsutella thompsonii	citrus orchard	McCoy et al. 1976
Nomuraea rileyi	soybean field	Sprenkel & Brooks 1975, Ignoffo et al. 1976a
PROTOZOANS		
Mattesia grandis	cotton field	McLaughlin et al. 1969
Mattesia trogodermae	stored products	Shapas et al. 1977
Nosema locustae	range-land	Henry et al. 1973
Thelohania hyphantriae	fruit orchard	Weiser & Veber 1957
VIRUSES[a]		
GIV of Pieris rapae	cabbage field	Wilson 1960, Jaques 1977a
NPV of Gilpinia hercyniae	spruce-pine forest	Bird & Burk 1961, Smirnoff 1972
NPV of Orgyia pseudosugata	fir forest	Morris 1963
NPV of Malacosoma disstria	aspen forest	Stairs 1965
NPV of Trichoplusia ni	cabbage field	Thomas et al. 1972

[a] GIV = granulosis inclusion virus; NPV = nuclear polyhedrosis viruses.

Sprenkel & Brooks 1975). Although a one-time release of a large amount of inoculum (10^{13} conidia/0.4 hectare) was employed (Ignoffo et al. 1976a) to prove the concept it may not be necessary to do this. N. rileyi incidence of only 10% in a population of 100,000 susceptible caterpillars (per 0.4 hectare) can provide the same inoculum as the release of 100 grams of conidia. The economic defoliation level for soybeans is about 110,000 mature larvae per 0.4 hectare (Thomas et al. 1974, 1976, 1978). Six weekly applications (prior to the anticipated economic pest levels) of only 1 gram per hectare of *N. rileyi* conidia against non-economic populations of susceptible first and second instar larvae may be more effective than one heavy application of inoculum. Nor are we restricted to utilizing only one strategy. Combinations of strategies also may be possible. As an example, host-manipulation and pathogen-manipulation were combined to induce an epizootic of *N. rileyi* of green cloverworms (Ignoffo et al. 1976a). Epizootics were induced about two weeks before the natural epizootic after an alternate susceptible host (*Trichoplusia ni*) and conidia were both released into soybean fields. This approach also was successful with viral, protozoan and bacterial diseases. Baculovirus enzootics and epizootics have been induced in species of *Neodiprion*, (Bird & Burk 1961), *Orygia pseudotsugata* (Morris 1963), *Malacosoma disstria* (Stairs 1965) and in *Pieris rapae* (Jaques 1977a, Wilson 1960). A bacterium, *Bacillus thuringiensis dendrolimus*, released into a forest ecosystem, became established and eventually supressed populations of *Dendrolimus sibiricus* (Talalev 1958). Protozoa have been introduced to colonize row crops, pastures, rangeland, and stored grain (Decker 1960, Hall & Dunn 1959, Henry et al. 1973, McLaughlin et al. 1969, Shapas et al. 1977, Weiser & Veber 1957).

Conservation of Inoculum

Conservation focuses on protecting or preserving inoculum (spores of bacteria, fungi, and protozoa or inclusion bodies of viruses) that is produced following natural enzootics or epizootics. Can the massive quantity of inoculum generated from epizootics be conserved?

Although there are many reports of attempts to preserve the activity of inoculum after it has been artificially applied I know of no instance where we have attempted to conserve the inoculum produced from a natural epizootic. As an example of what is theoretically possible, a full-blown viral or fungal epizootic can provide as much as 10^{13} to 10^{16} polyhedral inclusion bodies (PIB) or conidia per 0.4 hectare, respectively. What happens to this inoculum? Is it lost? How much of it is lost? I estimate that as much as 99.9999% of the inoculum provided from an epizootic is not available to initiate an epizootic in the next field season. Can this loss be conserved? If reduced to only 99% loss would sufficient inoculum be conserved and accumulate each year to higher levels of incidence? What effects would this yearly accumulation have on prevailing enzootic levels and on initiating earlier-than-normal epizootics each year? Can the enzootic level of incidence be eventually raised to provide economic control each year via an epizootic?

MANIPULATING THE ARTHROPOD HOST

How might we manage host insects in order to initiate and regulate enzootic and epizootic diseases of pest arthropods? Two management approaches immediately come to mind. We might manipulate the number of available susceptible hosts or manipulate the species composition of hosts. Both techniques are designed to provide living substrates for a more rapid increase and dispersal of inoculum.

Manipulating Host Number

Using the change-in-host-number approach susceptible hosts would be introduced earlier than their normal occurence, in larger numbers than their normal occurrence, and using susceptible stages that would not cause economic damage. Using Nomuraea rileyi as a model, we could release as many as ten susceptible first-instar larvae per 0.3 meter-row into soybeans at an early, vegetative growth-stage without any loss in yield or quality. Death of only 10% of these larvae would provide about 40 grams of conidia/0.4 hectare, and provide it at the best time and place for subsequent dispersal.

Another variation of this approach is release of living, diseased hosts preprogrammed to die at a stage that would provide the greatest quantity of inoculum, before serious damage occurred, and at sites frequented by healthy hosts. This approach was successful in field-cage studies of Heliothis zea on soybean (Ignoffo et al. 1978) and citrus red mite populations in orchards (Shaw et al. 1968).

When impossible to artificially rear the natural host then field-collected larvae could be infected and subsequently released in the field. In those instances where neither the host nor the pathogen can be propagated then it may be possible to field-collect early stages of the susceptible host from enzootic sites, concentrate the host under ideal conditions that encourage growth and development of the pathogen, and later release these infected hosts to establish foci for subsequent rapid dispersal and spread to non-infected field populations of the pest. This strategy might be employed to induce early epizootics of the enzootic Erynia (Zoophthora) nr. phytonomi in populations of the alfalfa weevil, Hypera postica, recently modeled by Nordin et al. (1983).

Manipulating Species Composition

Using the change-in-species approach highly susceptible, noneconomic alternate hosts could be introduced (instead of the primary pest) or encouraged to develop via some manipulative activity. These alternate hosts would not necessarily have to be those naturally present in the ecosystem under study. As in the previous example the introduced alternate host would be the substrate for a more rapid build-up and dispersal of large quantities of inoculum.

Behavioral Patterns

We should not limit our conceptualization to simply manipulate host number or species composition in order to establish enzootics or induce epizootics of arthropod pests. Behavioral characteristics, i.e., transovarial or venereal transmission, reported for many protozoans and some viral diseases, might be the vehicle for introduction and dispersion of inoculum (Bird 1961, Brooks 1968, Kellen & Lindegren 1971, Nordin 1975, Schwalbe et al. 1974, Shapas et al. 1977, Zimmack & Brindley 1957). Living insects, both the natural host (Elmore & Howland 1964, Gard & Falcon 1979, Martignioni & Milstead 1962) as well as parasites and predators might be used to introduce and disperse inoculum (Brooks 1973, Weiser 1961). By-products of arthropods, i.e., frass, meconium, and regurgitants, often contaminated with sufficient inoculum (especially protozoa and virus) might be manipulated to infect many healthy individuals. Indirect transmission via cannabalism or feeding on diseased cadavers or contaminated eggs has been suggested as the means of dispersal of viral and protozoan diseases of arthropods (caterpillars, grasshoppers, stored grain pests). In fact ovarial transmission, byproducts contamination and transmission by feeding may be the major ways in which a disease is maintained as an enzootic in the environment.

MANIPULATING THE ENVIRONMENT

Environmental conditions resulting from weather are probably the most difficult to manipulate. Yet, any condition, such as drought, can severely limit the existence of an enzootic or the eventual expression of an epizootic. We are aware that sunlight and high temperatures will inactivate inoculum, that inoculum can be dispersed by wind and water, and that lack of water may inhibit the development of many pathogens. But how can these be used to enhance enzootics and epizootics. In this strategy we must try to create or manipulate sunlight, temperature, water or wind to initiate or optimize an epizootic. We also must consider manipulation of environmental conditions other than weather such as: the substrate (soil or foliage); cultural crop practices (plant variety, density, planting date, tillage); as well as the chemical substrate, (pH, pesticides, adjuvants). The effects of both weather and non-weather conditions on microbials, reviewed by several authors, has been summarized in Ignoffo & Hostetter (1977).

Solar Radiation

Natural sunlight (radiation UV between 300 and 320 nm) is one of the most destructive environmental factors affecting survival and persistence of inoculum (Ignoffo et al. 1977). The inoculum half-life in the environment is estimated at less than one hour for the most sensitive pathogen to about 80 hours for the most resistant pathogen (Table 2). The half-life of viruses exposed to sunlight is less than three days. Unprotected spores of protozoa (Nosema) generally have a half-life of less than eight

Table 2. Estimated sunlight half-life and temperature longevity for selected representative species of entomopathogenic bacteria, fungi, protozoa and viruses.[a]

Pathogen	Sunlight Half-life (Hours)	Longevity at Temperature °C (days)		
		5°-10°	20°-25°	45°-50°
BACTERIA				
Bacillus thuringiensis spores	24	> 5,000	> 300	100
Bacillus thuringiensis crystal toxin	72	> 3,000	90	> 10
Bacillus popilliae	--	> 9,001		< 1
FUNGI				
Beauveria bassiana	--	900	85	< 1
Metarrhizium anisiopliae	8	455	85	< 1
Nomuraea rileyi	60	120	30	< 1
PROTOZOA				
Nosema necatrix	<1	--	300	< 1
Nosema trichoplusiae	8	--	--	--
Nosema plodiae	--	120	270	--
VIRUSES				
Heliothis NPV	24	>10,000	>10,000	< 100
Trichoplusia NPV	48	> 1,500	> 1,000	< 100
Pieris GV	42	> 1,500	> 365	< 5

[a] Abstracted from Ignoffo (1964), Ignoffo & Garcia (unpublished data), Ignoffo et al. (1977), Jaques (1977b), Maddox (1977), Pinnock et al. (1977), Roberts & Campbell (1977), West et al. (1984).

hours in natural sunlight (Maddox 1977). The half-life of unprotected bacterial spores (*Bacillus*) exposed to direct sunlight is less than 1 day, and the crystal-endotoxin of *B. thuringiensis* has a half-life of about three days (Ignoffo et al. 1977, Pinnock et al. 1977). Generally, conidia of entomopathogenic fungi such as *Beauveria*, *Metarrhizium*, and *Nomuraea* are more resistant to sunlight than protozoan spores of *Nosema* species but less resistant than bacterial spores of *Bacillus* species (Ignoffo et al. 1977, Maddox 1977, Pinnock et al. 1977, Roberts & Campbell 1977).

Most manipulative activities, to minimize the detrimental effects of sunlight-UV, have focused on formulation protectants. The use of UV-protectants has ranged from: tank additives (UV-protectants added to the spray tank prior to application); to formulation mixtures of a UV-protectant and inoculum; and to microencapsulation of intimate mixtures of the protectant and the inoculum (Couch & Ignoffo 1981). Although the use of sunlight protectant has been moderately successful (3-6 fold increase in field half-life) additional protection is needed under conditions where: populations of the pest extend over a time longer than the half-life; the plant substrate is rapidly growing and thus diluting the inoculum; continuous cycling of heavy populations of pest insects are encountered; and when even minimal damage by the pest cannot be tolerated.

Restricting the dispersal of inoculum to the dusk-to-dawn period to escape exposure to sunlight, as well as applications of inoculum to the underside of foliage or to protected plant parts has been recommended in an effort to minimize the effects of sunlight. As an example, the persistence of *Heliothis* NPV was 10X greater on the calyx and inner surface of cotton bracts than on cotton leaves (Yearian & Young 1974). The half-life on leaves, protected bracts, and calyxes was less than 24 hours, 24-48 hours, and greater than 96 hours, respectively.

Additional research on use of sunlight protectants should be considered since this method has provided the best manipulative activity to date. Researching the actual mode of inactivation by sunlight rather than the empirical approach used in the past will probably provide more insights on how to provide greater sunlight protection to natural and artificially introduced inoculum.

Temperature

Temperature during the growing season in most agroecosystems ranges from about 0 to 50°C. The optimum range for infection, growth and development of most entomopathogens, however, lies between 10 and 30°C. In general, temperature exposure within the 10 to 30°C range for less than 30 days does not affect the stability of the entomopathogens (Table 2). Deleterious effects at temperatures less than 10°C or greater than 30°C however, might be obtained when stressed by water, sunlight, wind, substrate (soil, foliage), or chemicals. In general, temperatures above 30°C inhibit development of entomopathogens, and viability or activity of inoculum is markedly decreased as temperatures approach 50°C (Table 2). Low temperatures (less than 10°C) also inhibit growth and development but may increase persistence of the resting-stage inoculum (spores, inclusion bodies). As an example (Table 2), viral activity of *Heliothis* NPV was still present after more than 25 years storage at 5°C; but, was completely inactivated after only 100 days at 50°C. Similar relationships were obtained for: spores of *Bacillus thuringiensis*; spores of *Nosema* species; and conidia of *Nomuraea rileyi*. The upper extreme of field temperatures (30 to 50°C) is most detrimental to species of entomopathogenic protozoa and fungi (Table 2). The half-life of *Nosema* spores exposed at 50°C is less than 8 hours but greater than 30 days at 10°C. Species of *Beauveria* and *Metarrhizium* have a thermal death point of less than 15 min at 50°C. The half-life of dry spores of the bacterium *B. thuringiensis* exposed at 50°C is greater than 100 days while wet spores have a half-life of less than 60 days (Table 2). Spores of *B. popilliae* exposed at 60°C are inactivated after an exposure of 15 minutes. Although occluded NPV are not inactivated at low temperatures, the inactivation profile at 50°C (for *Heliothis* NPV) may indicate that long exposure at 30°C could reduce viral activity (Ignoffo & Couch 1981).

At first, it might seem impossible to manipulate temperature to enhance enzootics or epizootics. Temperature increases, however, induced by solar heat, might be used to stimulate an enzootic to develop into an earlier-than-normal localized epizootic. The temperature-induced localized epizootics then would be the foci for an earlier-than-normal dispersal of inoculum to the entire field. Concentration of solar heat was successfully used to control weeds and sterilize cotton fields (Katan et al. 1983). Other methods of directing or concentrating solar heat to increase temperature might include: use of parabolic, directional heat reflectors; absorbent or transparent netting or covering placed over plants; or spraying plants with a solar heat-absorbing substance (carbon) or with heat-dissipating substances.

Water

In general, the presence of natural water, other than as a dispersal and diluting vehicle or in combinations with other factors (pH, temperature, sunlight), is not a major limiting factor to the establishment and persistence of entomopathogens. The absence of free water, however, can limit the germination and subsequent infection of arthropods by species of fungi and protozoans. Contrastingly, dew combined with mineral deposits or foliage swell PIB and if followed by additional sequential wetting and drying may hasten loss of viral activity (Allen et al. 1978, Andrews & Sikorowski 1973, David et al. 1971, Jaques 1972, McLoed et al. 1977, Young et al. 1977).

Results summarized in Ignoffo & Hostetter (1977) indicated that the half-life of bacterial spores (*Bacillus thuringiensis*) and inclusion bodies (NPV) stored in water below 10°C is greater than 300 days. Spores of some Nosema survive best in a water suspension: however, other species still are infective after dry-storage for several months (Maddox 1977). In contrast drying completely destroyed the infectivity of spores of another Nosema species (N. algerae). Moisture, generally as free water or high relative humidity, also is needed for sporulation of many species of fungi (Beauveria, Entomophthora, Metarrhizium, and Nomuraea).

Water as falling rain or streaming flows can disperse the inoculum of many soil and plant dwelling arthropod pathogens (Entomophthora, Paecilomyces, Cordyceps, Isaria, Hymenostilbe, Beauveria, Metarrhizium, Nomuraea). Dispersal via falling or flowing water is generally beneficial for soil-borne inoculum and detrimental for air-borne inoculum. As an example, heavy rains will wash the air of wind-dispersed conidia of Nomuraea rileyi and thus may inhibit the rapid development of an epizootic (Kish & Allen 1978).

Efforts to increase water in the environment have generally focused on the use of overhead and ground irrigation, modifying cultural conditions to produce an overlapping foliage canopy, and changing planting date to coincide with optimum dew development at the desired time. The overlapping canopy technique was successfully employed by both Burleigh (1975) in cotton and Sprenkel et al. (1979) in soybeans to increase effectiveness of the fungus Nomuraea rileyi. Although never attempted, some method of formulating a timed, atmospheric-water-absorbing material with a free-water-requiring spore might provide sufficient water under drought conditions to permit spore germination and subsequent infection of arthropod pests.

Wind

Air movement also can be a limiting factor: in releasing inoculum, dispersing inoculum, and determining how long the inoculum is airborne. Wind-generated, driving rains will wash the air of air-borne inoculum and thus inhibit development of an epizootic of N. rileyi (Kish & Allen 1978). However, air movement as low as 2-3 km per hour will release spores of fungi from infected cadavers to surrounding foliage and thus aid in the more rapid expression of an epizootic (Garcia & Ignoffo 1977). Low atmospheric humidity or exposure to sunlight-UV may completely inactivate inoculum that is airborne for a long time.

I do not know of attempts to manipulate air movements to enhance the effectiveness of an enzootic or epizootic. It is conceivable, however, that foci of inoculum might be established at the border or in selected sites in order to take advantage of prevailing wind patterns and thus maximize the dispersal of inoculum. Also, some type of solar- or wind-generated device might be developed to specifically direct air movements from regions of high inoculum to regions of low inoculum.

Non-Weather Factors

Soil and Foliage Substrate Since soil is the natural reservoir of inoculum, it is important to know how soil conditions might impact on the extent of enzootic and the development of epizootic diseases. An indication of the significance of the combined affects of sun and substrate was reported by Maddox (1977). Viability of spores of Nosema necatrix exposed to sunlight sequentially decreased as the substrate was changed. Their viability on glass, corn leaf, artificial diet, and soil was 3, 4.5, 9, and greater than 28 hours, respectively.

Natural soil microorganisms may produce metabolites or induce pH changes that prevent inoculum build-up or inactivate inoculum activity. As examples, viruses are more stable in neutral soil than in either highly alkaline or acidic soils while fungi develop better under slightly acidic conditions (Ignoffo & Hostetter 1977). The half-life of baculoviruses in soil is estimated at greater than 600 days and although spores of Bacillus popilliae have been recovered from soil 45 years after their initial introduction (Hutton & Burbutis 1974) their half-life in soil, devoid of the host, is probably less than that of the baculoviruses (Jaques 1964). About 10% of the introduced spores of Bacillus thuringiensis could be recovered from loam soil after four months (Ignoffo 1964, Ignoffo & Hostetter 1977).

Persistence of pathogens on foliage can be affected by the kind of leaf, leaf texture, leaf exudate or dew. As an example, Young & Yearian (1976) demonstrated that the Heliothis NPV persisted longer on tomato and soybean leaves than on cotton leaves. The greater stability on tomatoes was attributed to mechanical screening of the virus by the numerous leaf hairs or leaf curling.

The pH of dew on foliage may inactivate viruses as well as induce premature polar-filament extrusion of protozoan spores, thus resulting in a loss of spore infectivity. Although solutions of high alkaline and low acidic pH will inactivate the Heliothis NPV, the effects of alkaline dew on leaf surfaces is less understood. Buffered viral suspensions were more effective on cotton in California (Falcon 1969) than in Arkansas (Young & Yearian 1976). Andrews & Sikorowski (1973) concluded, however, that repeated drying and wetting at alkaline pH of dew was what inactivated the Heliothis NPV deposited on foliage. Young et al. (1977) did not observe any inactivation when the Heliothis virus was directly placed in dew collected from cotton (7.4 to 8.8 pH) or soybean (7.2 to 7.8 pH) leaves. Although entomopathogenic fungi generally can tolerate a wide range of hydrogen ion concentrations, germination, mycelial growth, and sporulation is generally restricted to near neutrality. There are, of course, some notable exceptions (e.g., the optimum pH for Beauveria bassiana is 4.5 while the pH range for Entomophthora coronata is 3 to 10). The pH and soil type (muck at 5.6; silty-clay loam at 6.3-7.3;

silt loams at 4.8-5.6) had little effect on survival of spores of Bacillus thuringiensis (Saleh et al. 1969).

Foliage substrates also contain phagostimulants, phagodeterrents, or antibacterial substances that may either enhance or inhibit the activity of entomopathogens (Kushner & Harvey 1962; Morris 1969, 1972; Maksymiuk 1970; Smirnoff 1972; Smirnoff & Hutchinson 1965). Pinnock et al. (1977) reported that the physical and chemical characteristics of foliage might inactivate Bacillus thuringiensis spores with the rate of inactivation dependent upon the foliage substrate: live oak, red bud, walnut or eucalyptus.

Efforts to manipulate the substrates have included the mechanical dispersal of inoculum from contaminated soil to plants and neutralizing and buffering foliage substrate via formulation additives. Other methods that also have been recommended, but rarely adopted, are minimum tillage to reduce dilution of soil inoculum and neutralizing and buffering soil to increase persistence of inoculum.

Chemicals In general, pesticides, as well as their additives or adjuvants, do not adversely affect microbial inoculum. However, there are some notable exceptions to this generalization. As examples, methyl parathion, an organophosphate insecticide, inactivates the Heliothis NPV (Ignoffo & Couch 1981). The insecticides methyl parathion, monocrotophos, and phenthoate interfere as much as a fungicide with the germination and growth of Nomuraea rileyi (Ignoffo 1981). Another organophosphate insecticide, malathion, completely destroyed the viability of Nosema spores after 30 minutes of exposure (Maddox 1977) and there are indications that chlorinated hydrocarbons, carbamates and organophosphate insecticides may reduce the viability of spores or growth of Bacillus thuringiensis (Chen et al. 1975, Dougherty et al. 1971, Sutter et al. 1971).

Severe effects of insecticides on entomopathogenic fungi were reported using BHC, carbaryl, carbophenothion, chlorpyrifos, DBCP, DDT, demeton, diazinon, ethion, ekatin, endosulfan, endrin, fenitrothion, heptachlor, heptophor, malathion, methyl parathion, monocrotophos, parathion, phenthoate, supracide, toxophene, and trithion (Cadatal & Gabriel 1970, Dirimanov & Angelova 1962, Hall & Dunn 1959, Ignoffo et al. 1975Olmert & Kenneth 1974, Ramaraje et al. 1967, Soper et al. 1974, Wang & Leu 1972). Fungicides, when used to control plant diseases, also can reduce either the viability or infectivity of entomopathogenic fungi. Those fungicides with the greatest spectrum of activity against entomopathogenic fungi include benomyl, captan, ferbam and maneb (Catadal & Gabriel 1970, Hall & Dunn 1959, Ignoffo et al. 1975, Machrowicz 1967, Olmert & Kenneth 1974, Soper et al. 1974, Yendol 1968).

Of all the environmental factors, manipulation of enzootics or epizootics by use of chemicals has been tried the most. Adjuvants to increase persistence have, as examples, included: sunlight protectants (Allen et al. 1978, Couch & Ignoffo 1981);

evaporation retardants (Ignoffo et al. 1976b); and organic and non-organic buffers (Couch & Ignoffo 1981). Additives to enhance activity of inoculum have included feeding stimulants (Montoya & Ignoffo 1966), infection aids such as chitinase (Smirnoff 1971, 1973), and even the use of combinations of microbials (fungus + bacteria, virus + bacteria).

Combinations of synthetic insecticides and microbial insecticides have been used to stress the host thus making them more susceptible to disease (Roberts & Campbell 1977). Epizootics also may be induced by use of stressors or incitants as was demonstrated by McCoy et al. (1976) in their use of oils to induce late summer epizootics of *Hirsutella thompsonii* in populations of the citrus rust mite, *Phyllocoptruta oleivora*. Fungicides, normally used to control plant pathogens, have been prudently and selectively used to permit the natural expression of epizootics of *Entomophthora sphaerosperma* and *Nomuraea rileyi* (Ignoffo 1981, Jaques & Patterson 1962, Johnson et al. 1976).

CONCLUSIONS AND RECOMMENDATIONS

Most of our current knowledge on strategic uses of entomopathogens is based upon only a few representative species. Data on viruses, for example, comes largely from the baculoviruses, and more specifically, the NPV of *Heliothis* spp., *Trichoplusia*, *Pieris*, *Orgyia*, *Lymantria*, and *Neodiprion*. Data on bacteria are predominantly from studies on *Bacillus thuringiensis* and *Bacillus popilliae*, while data on protozoa are primarily studies of microsporidians, more specifically species of *Nosema*. Most data on fungal entomopathogens come from studies of species of *Beauveria*, *Entomophthora*, *Erynia*, *Metarrhizium*, *Nomuraea* and *Verticillium*. Consequently, any interpretation that is presented herein should be viewed as a generalization based on about one percent of the described arthropod pathogens.

The following are selected comments on areas that might be researched to productively manipulate enzootic and epizootic diseases of arthropod pests. This list is not meant to be an all inclusive or prioritized list although all subjects have been discussed in the text.

1. Develop epizootiological models for representative species of entomopathogens (bacteria, fungi, protozoa, viruses) in model ecosystems (row-crops, forage, truck crops, forests, ornamentals).
2. Develop quantitative ecological sub-models of the pathogen insect complex and integrate these into pest management models.
3. Evaluate the concept of induction of earlier than normal epizootics by manipulation of the pathogen, the insect and the environment.

4. Evaluate the impact of introductions and colonizations of new pathogens into domestic ecosystems and domestic pathogens into new ecosystems.
5. Determine the fate of inoculum produced from natural epizootics and determine how this inoculum could be conserved.
6. Evaluate the impact of release of susceptible healthy insects and diseased insects (preprogrammed to die) on the initiation and development of epizootics.
7. Explore the use of behavioral patterns of the insects to introduce, maintain enzootics, and maximize dispersal of entomopathogens.
8. Determine the mechanism and mode of inactivation of entomopathogens by sunlight and develop sunlight protectants that provide at least 50% protection of the inoculum up to the next regularly scheduled application.
9. Evaluate the use of solar-induced heat to initiate earlier than normal localized epizootics that serve as a foci for the rapid build-up and dispersal of inoculum.
10. Develop efficient, economical ways to manipulate air movements so as to maximize dispersal of inoculum.
11. Determine and overcome those factors that limit the effectiveness of augmentative uses of entomopathogens as microbial insecticides.
12. Provide formulation microclimates (containing free water, gustatory stimulants, sunlight protectants) and modify those cultural practices that would enhance the effectiveness of entomopathogens.

As previously concluded (Ignoffo 1978), it may be that none of the suggested approaches or strategies will provide levels of control equal to or better than that currently obtained with the microbial insecticide strategy. However, these differing approaches (that use the most important attributes of a pathogen, i.e., their ability to establish, disperse, multiply and persist) should be further researched to evaluate their potential in specific agricultural ecosystems. At least some of these approaches can be immediately applied; undoubtedly more will be utilized in the future. We who are interested in advancing this field should become knowledgeable about older strategies as well as developing newer, innovative strategies. We should broaden our horizons, use our imagination, and devise, develop, and eventually implement other ways to more effectively utilize/manipulate enzootic and epizootic diseases of arthropod pests.

REFERENCES

Allen, G.E., C.M. Ignoffo and R.P. Jaques. 1978. Proceedings of microbial control of insect pests: Future strategies in pest management systems. NSF-USDA, Univ. of Florida Workshop. 290 pp.

Andrews, G.L., and P.P. Sikorowski. 1973. Effects of cotton leaf surface on the nuclear-polyhedrosis virus of Heliothis zea and Heliothis virescens (Lepidoptera: Noctuidae). J. Invertebr. Pathol. 22:290-291.

Balch, R.E., and F.T. Bird. 1944. A disease of the European spruce sawfly, Gilpinia hercyniae (Htg.), and its place in natural control. Sci. Agric. 25:65-80.

Bird, F.T. 1961. Transmission of some insect viruses with particular reference to ovarial transmission and its importance in the development of epizootics. J. Insect Pathol. 3:352-380.

Bird, F.T., and J.M. Burk. 1961. Artificially disseminated virus as a factor controlling the European spruce sawfly, Diprion hercyniae (Htg.) in the absence of introduced parasites. Can. Entomol. 93:228-238.

Brooks, W.A. 1968. Transovarial transmission of Nosema heliothides in the corn earworm, Heliothis zea. J. Invertebr. Pathol. 11:511-512.

Brooks, W.A. 1973. Protozoa: Host-parasite-pathogen-interrelationships. Misc. Publ. Entomol. Soc. Amer. 9:105-111.

Burges, H.D. 1981. Microbial control of pests and plant diseases, 1970-1980. Academic Press, N.Y. 949 pp.

Burges, H.D., and N.W. Hussey. 1971. Microbial control of insects and mites. Academic Press, N.Y. 861 pp.

Burleigh, J.G. 1975. Comparison of Heliothis spp. larval parasitism and Spicaria infection in closed and open canopy cotton varieties. Environ. Entomol. 4: 574-576.

Cadatal, T.D., and B.D. Gabriel. 1970. Effect of chemical pesticides on the development of fungi pathogenic to some rice insects. Phillipp. Entomol. 1:379-395.

Chen, K., B.R. Funke, F.L. Shultz, R.B. Carlson and F.I. Proshold. 1974. Effects of certain organophosphate and carbamate insecticides on Bacillus thuringiensis. J. Econ. Entomol. 67:471-473.

Couch, T.L., and C.M. Ignoffo. 1981. Formulation of insect pathogens, pp. 621-634. In: H.D. Burges (Ed.), Microbial control of pests and plant diseases 1970-1980. Academic Press, N.Y.

d'Herelle, F. 1912. Sur la propagation, dans la Republique Argentine, de l'epizootie des sauterelles du Mexique. Compt. Rend. Acad. Sci., Paris, 154:623-625.

David, W.A.L., S.J. Ellaby and G. Taylor. 1971. The stability of a purified granulosis virus of the European cabbageworm, Pieris brassicae, in dry deposits of intact capsules. J. Invertebr. Pathol. 17:228-233.

Decker, G.C. 1960. Microbial insecticides and their future. Agric. Chem. 15(1): 30-33.

Dirimanov, M., and R. Angelova. 1962. Rasit Zasht. 10:63-67. Cited in Benz, G. 1971. Synergism of micro-organisms and chemical insecticides, pp. 327-355. In: H.D. Burges and N.W. Hussey (Eds.), Microbial control of insects and mites. Academic Press, N.Y.

Dougherty, E.M., C.R. Reichelderfer and R.M. Faust. 1971. Sensitivity of Bacillus thuringiensis var. thuringiensis to various insecticides and herbicides. J. Invertebr. Pathol. 17:292-293.

Elmore, J.C., and A.F. Howland. 1964. Natural versus artificial dissemination of nuclear polyhedrosis virus by contaminated adult cabbage loopers. J. Invertebr. Pathol. 6:430.

Falcon, L.A. 1969. Microbial control as a tool in integrated control programs, pp. 346-364. *In*: C.B. Huffaker (Ed.), Biological control. Plenum/Rosetta Ed., N.Y.

Garcia, C., and C.M. Ignoffo. 1977. Dislodgment of conidia of *Nomuraea rileyi* from cadavers of cabbage looper, *Trichoplusia ni*. J. Invertebr. Pathol. 30: 114-116.

Gard, I.M., and L.A. Falcon. 1979. Auto-dissemination of entomopathogens: Viruses, pp. 46-54. *In*: G.E. Allen, C.M. Ignoffo and R.P. Jaques (Eds.), Microbial control of insect pests: Future strategies in pest management systems. NSF-USDA, Univ. of Florida Workshop.

Hall, I.M., and P.H. Dunn. 1959. The effect of certain insecticides and fungicides on fungi pathogenic to the spotted alfalfa aphid. J. Econ. Entomol. 52: 28-29.

Henry, J.E., K. Tiahrt and E.A. Oma. 1973. Importance of timing, spore concentrations, and levels of spore carrier in applications of *Nosema locustae* (Microsporidia: Nosematidae) for control of grasshoppers. J. Invertebr. Pathol. 21: 263-272.

Hutton, P.O., and P.B. Burbutis. 1974. Milky disease and Japanese beetle in Delaware. J. Econ. Entomol. 67:247-248.

Ignoffo, C.M. 1964. Effects of temperature and water on viability and virulence of *Bacillus thuringiensis* var. *thuringiensis* Berliner spores. Entomophaga Mem. 2:293-298.

Ignoffo, C.M. 1965. The nuclear-polyhedrosis virus of *Heliothis zea* (Boddie) and *Heliothis virescens* (Fabricus). I. Virus propagation and its virulence. J. Invertebr. Pathol. 7:209-216.

Ignoffo, C.M. 1978. Strategies to increase the use of entomopathogens. J. Invertebr. Pathol. 31:1-3.

Ignoffo, C.M. 1981. The fungus *Nomuraea rileyi* as a microbial insecticide, pp. 513-538. *In*: H.D. Burges (Ed.), Microbial control of pests and plant diseases. Academic Press, N.Y.

Ignoffo, C.M., and T.L. Couch. 1981. The nucleopolyhedrosis virus of *Heliothis* species as a microbial insecticide, pp. 329-362. *In*: H.D. Burges (Ed.), Microbial control of pests and plant diseases 1970-1980. Academic Press, N.Y.

Ignoffo, C.M., and D.L. Hostetter (Eds.). 1977. Environmental stability of microbial insecticides. Misc. Publ. Entomol. Soc. Amer. 10, 119 pp.

Ignoffo, C.M., D.L. Hostetter, K.D. Biever, C. Garcia, G.D. Thomas, W.A. Dickerson and R. Pinnell. 1978. Evaluation of an entomopathogenic bacterium, fungus, and virus for control of *Heliothis zea* on soybeans. J. Econ. Entomol. 71:165-168.

Ignoffo, C.M., D.L. Hostetter, C. Garcia and R.E. Pinnell. 1975. Sensitivity of the entomopathogenic fungus *Nomuraea rileyi* to chemical pesticides used on soybeans. Environ. Entomol. 4:765-768.

Ignoffo, C.M., D.L. Hostetter, P.P. Sikrowski, G. Sutter and W.M. Brooks. 1977. Inactivation of representative species of entomopathogenic viruses, a bacterium, fungus, and protozoan by an ultraviolet light source. Environ. Entomol. 6:411-415.

Ignoffo, C.M., D.L. Hostetter and D.B. Smith. 1976b. Gustatory stimulant, sunlight protectant, evaporation retardant: Three characteristics of a microbial insecticidal adjuvant. J. Econ. Entomol. 69:207-210.

Ignoffo, C.M., N.L. Marston, D.L. Hostetter, B. Puttler and J.V. Bell. 1976a. Natural and induced epizootics of Nomuraea rileyi in soybean caterpillars. J. Invertebr. Pathol. 27:191-198.

Jaques, R.P. 1964. The persistence of a nuclear-polyhedrosis virus in soil. J. Insect Pathol. 6:251-254.

Jaques, R.P. 1972. The inactivation of foliar deposits of viruses of Trichoplusia ni (Lepidoptera: Noctuidae) and Pieris rapae (Lepidoptera: Pieridae) and tests on protectant additives. Can. Entomol. 104:1985-1994.

Jaques, R.P. 1977a. Field efficacy of viruses infectious to the cabbage looper and imported cabbageworm on late cabbage. J. Econ. Entomol. 70:111-118.

Jaques, R.P. 1977b. Stability of entomopathogenic viruses. In: C.M. Ignoffo and D.L. Hostetter (Eds.), Environmental stability of microbial insecticides. Misc. Publ. Entomol. Soc. Amer. 10:99-116.

Jaques, R.P., and N. Patterson. 1962. Control of the apple sucker, Psylla mali Schmidb., by the fungus Entomophthora sphaerosperma (Fresenius). Can. Entomol. 94:818-825.

Johnson, D.W., L.P. Kish and G.E. Allen. 1976. Field evaluation of selected pesticides on the natural development of the entomopathogen, Nomuraea rileyi, on the velvetbean caterpillar in soybeans. Environ. Entomol. 5:964-966.

Katan, I., G. Fishler and A. Grinstein. 1983. Short- and long-term effects of soil solarization and crop sequence on Fusarium wilt and yield of cotton in Israel. Phytopathology 73:1215-1219.

Kellen, W.R., and Lindegren. 1971. Modes of transmission of Nosema plodiae Kellen and Lindegren, a pathogen of Plodia interpunctella (Hubner). J. Stored Prod. Res. 7:31-34.

Kish, L.P., and G.E. Allen. 1978. The biology and ecology of Nomuraea rileyi and a program for predicting its incidence on Anticarsia gemmatalis in soybean. Fla. Agr. Exp. Sta. Tech. Bull. 795, 48 pp.

Kushner, D.J., and G.T. Harvey. 1962. Antibacterial substances in leaves: Their possible role in insect resistance to disease. J. Insect Pathol. 4:155-184.

Laird, M. 1967. A coral island experiment: A new approach to mosquito control. World Health Organiz. Chron. 21, 18-26.

Machrowicz, I. 1967. Effect of some plant protection chemicals on the development of some fungi in pure culture. Zesz. nauk. wyzsz. Szk. Voln. Szczec. 24: 179-184. Abstr. in 1969 Rev. Appl. Mycol. 48:189.

Maddox, J.V. 1977. Stability of entomopathogenic protozoa. In: C.M. Ignoffo and D.L. Hostetter (Eds.), Environmental stability of microbial insecticides. Misc. Publ. Entomol. Soc. Amer. 10:3-18.

Maksymiuk, B. 1970. Occurrence and nature of antibacterial substances in plants affecting Bacillus thuringiensis and other entomogenous bacteria. J. Invertebr. Pathol. 15:356-371.

Martignoni, M.E., and J.E. Milstead. 1962. Transovum transmission of the nuclear polyhedral virus of Colias eurytheme Boisduval through contamination of the female genitalia. J. Insect Pathol. 4:113-121.

McCoy, C.W., R.F. Brooks, J.C. Allen and A.G. Selhime. 1976. Management of arthropod pests and plant diseases in citrus agroecosystem. Proc. Tall Timbers Conf. Ecol. Animal Control by Habitat Mgt. 6:1-17.

McIntosh, A.H., and C.M. Ignoffo. 1982. Notes on a low-cost medium for growth of insect cell lines and replication of insect viruses. J. Kansas Entomol. Soc. 55:354-356.

McLaughlin, R.E., T.C. Cleveland, R.J. Daum and M.R. Bell. 1969. Development of the bait principle for boll weevil control. IV. Field tests with a bait containing a feeding stimulant and the sporozoans Glugea gasti and Mattesia grandis. J. Invertebr. Pathol. 13:429-441.

McLeod, P.J., W.C. Yearian and S.Y. Young. 1977. Inactivation of Baculovirus heliothis by ultraviolet irradiation, dew, and temperature. J. Invertebr. Pathol. 30:237-241.

Monotoya, E.L., and C.M. Ignoffo. 1966. Laboratory technique and apparatus for testing viruses applied as dusts. J. Invertebr. Pathol. 8:251-254.

Morris, O.N. 1963. The natural and artificial control of the Douglas-fir tussock moth, Orgyia pseudosugata McDunnough, by a nuclearpolyhedrisis virus. J. Insect Pathol. 5:401-414.

Morris, O.N. 1969. Susceptibility of several forest insects of British Columbia to commercially produced Bacillus thuringiensis. I. Studies on the physiological properties of some commercial products. J. Invertebr. Pathol. 13:134-146.

Morris, O.N. 1972. Inhibitory effects of foliage extracts of some forest trees on commercial Bacillus thuringiensis. Can. Entomol. 104:1357-1361.

Nordin, G.L. 1975. Transovarial transmission of a Nosema sp. infecting Malacosoma americanum. J. Invertebr. Pathol. 25:221-228.

Nordin, G.L., G.C. Brown and J.A. Millstein. 1983. Epizootic phenology of Erynia disease of the alfalfa weevil, Hypera postica (Gyllenhal) (Coleoptera: Curculionidae), in central Kentucky. Environ. Entomol. 12:1350-1355.

Olmert, I., and R.G. Kenneth. 1974. Sensitivity of the entomopathogenic fungi, Beauveria bassiana, Verticillium lecani and Verticillium sp. to fungicides and insecticides. Environ. Entomol. 3:33-38.

Pinnock, D.E., J.E. Milstead, M.E. Kirby and B.J. Nelson. 1977. Stability of entomopathogenic bacteria. In: C.M. Ignoffo and D.L. Hostetter (Eds.), Environmental stability of microbial insecticides. Misc. Publ. Entomol. Soc. Amer. 10:77-98.

Ramaraje U.R.S., N.V., H.C. Govindu and K.S. Shivashankara Shastry. 1967. The effect of certain insecticides on the entomogenous fungi Beauveria bassiana and Metarrhizium anisopliae. J. Invertebr. Pathol. 9:398-403.

Roberts, D.W., and A.S. Campbell. 1977. Stability of entomopathogenic fungi. In: C.M. Ignoffo and D.L. Hostetter (Eds.), Environmental stability of microbial insecticides. Misc. Publ. Entomol. Soc. Amer. 10:19-76.

Saleh, S.M., R.F. Harris and O.N. Allen. 1969. Method for determining Bacillus thuringiensis var. thuringiensis Berliner in soil. Can. J. Microbiol. 15:1101-1104.

Schwalbe, C.P., W.E. Burkholder and G.M. Bousch. 1974. Mattesia trogodermae infection rates as influenced by mode of transmission, dosage, and host species. J. Stored Prod. Res. 10:161-166.

Shapas, J., W.E. Burkholder and G.M. Bousch. 1977. Population supression of Trogoderma glabrum by using pheromone luring for protozoan pathogen dissemination. J. Econ. Entomol. 70:469-474.

Shaw, J.G., D.L. Chambers and H. Tashlor. 1968. Introducing and establishing the noninclusion virus of the citrus red mite in citrus groves. J. Econ. Entomol. 61:1352-1355.

Smirnoff, W.A. 1971. Effect of chitinase on the action of Bacillus thuringiensis. Can. Entomol. 103:1829-1831.

Smirnoff, W.A. 1972. Effects of volatile substances released by foliage of Abies balsamea. J. Invertebr. Pathol. 19:32-35.

Smirnoff, W.A. 1973. Results of tests with Bacillus thuringiensis and chitinase on larvae of the spruce budworm. J. Invertebr. Pathol. 21:116-118.

Smirnoff, W.A., and P.M. Hutchinson. 1965. Bacteriostatic and bacteriocidal effects of extracts of foliage from various plant species on Bacillus thuringiensis var. thuringiensis Berliner. J. Invertebr. Pathol. 7:273-280.

Soper, R.S. 1984. Current research in Entomophthorales. In: Program schedule, p. 10. Paper presented at North Central Branch Entomol. Soc. Amer. Wichita, KS., March 1984.

Soper, R.S., F.R. Holbrook and C.C. Gordon. 1974. Comparative pesticide effects on Entomophthora and the phytopathogen Alternaria solani. Environ. Entomol. 3:560-562.

Sprenkel, R.K., and W.M. Brooks. 1975. Artificial dissemination and epizootic initiation of Nomuraea rileyi, an entomogenous fungus of lepidopterous pests in soybeans. J. Econ. Entomol. 68:847-851.

Sprenkel, R.K., W.M. Brooks, J.W. Van Duyn and L.L. Dietz. 1979. The effects of three cultural variables on the incidence of Nomuraea rileyi, phytophagous Lepidoptera, and their predators on soybeans. Environ. Entomol. 8:334-339.

Stairs, G.R. 1965. Artificial initiation of virus epizootics in forest tent caterpillar populations. Can. Entomol. 97:1059-1062.

Steinhaus, E.A. 1954. The effects of disease on insect populations. Hilgardia 23, 261 pp.

Steinhaus, E.A. 1963. Insect pathology an advanced treatise. Vols. 1 and 2, Academic Press, N.Y., 661 pp. and 689 pp., respectively.

Sutter, G.R., M.D. Abrahamson, E.W. Hamilton and I.D. Vick. 1971. Compatability of Bacillus thuringiensis var. thuringiensis and chemical insecticides. I. Effect of insecticide doses on bacterial replication rate. J. Econ. Entomol. 64:1348-1350.

Talalev, E.V. 1958. Introduction of epizootic septicemia in the caterpillars of Siberian silkworm moth, Dendrolimus sibericus Tschtv. (Lepidoptera, Lasiocampidae). Entomol. Rev. USSR 37:562-641.

Thomas, G.D., C.M. Ignoffo, K.D. Biever and D.B. Smith. 1974. Influence of defoliation and depodding on yield of soybeans. J. Econ. Entomol. 67:683-685.

Thomas, G.D., C.M. Ignoffo and D.B. Smith. 1976. Influence of defoliation and depodding on quality of soybeans. J. Econ. Entomol. 69:737-740.

Thomas, G.D., C.M. Ignoffo, D.B. Smith and C.E. Morgan. 1978. Effects of single and sequential defoliations on yield and quality of soybeans. J. Econ. Entomol. 71:871-874.

Thomas, G.D., C.F. Reichelderfer and A.M. Heimpel. 1972. Accumulation and persistence of a nuclear polyhedrosis virus of the cabbage looper in the field. J. Invertebr. Pathol. 20:157-164.

Wang, P., and L. Leu. 1972. Effects of insecticides and herbicides to the fungi parasitic on nymph of grass cicada, Mogannia hebes Walker. Rept. Taiwan Sugar Exp. Stn. 55:103-109.

Weiser, J. 1961. A new microsporidian from the bark beetle, Pityokteines curvidens Germar (Coleoptera, Scolytidae) in Czechoslovakia. J. Insect Pathol. 3: 324-329.

Weiser, J., and J. Veber. 1957. Die Mikrosporidie Thelohania hyphantriae Weiser des weissen Barenspinners and anderer Mitglieder seiner Bioconose. Z. angew. Entomol. 40:55-70.

West, A.W., N.E. Crook and H.D. Burges. 1984. Detection of Bacillus thuringiensis in soil by immunofluorescence. J. Invertebr. Pathol. 43:150-155.

White, R.T., and S.R. Dutky. 1940. Effect of introduction of milky disease on populations of Japanese beetle larvae. J. Econ. Entomol. 33:306-309.

Wilson, F. 1960. The effectiveness of a granulosis virus applied to field populations of Pieris rapae (Lepidoptera). Australian J. Agric. Res. 2:485-497.

Yearian, W.C., and S.Y. Young. 1974. Persistence of Heliothis nuclear polyhedrosis virus on cotton plant parts. Environ. Entomol. 3:1031-1036.

Yendol, W.G. 1968. Factors affecting germination of Entomophthora conidia. J. Invertebr. Pathol. 10:116-121.

Young, S.Y., and W.C. Yearian. 1976. Influence of buffers on pH of cotton leaf surfaces and activity of a Heliothis nuclear polyhedrosis virus. J. Georgia Entomol. Soc. 11:277-282.

Young, S.Y., W.C. Yearian and K.S. Kim. 1977. Effect of dew from cotton and soybean foliage on activity of Heliothis nuclear polyhedrosis virus. J. Invertebr. Pathol. 29:105-111.

Zimmack, H.L., and T.A. Brindley. 1957. The effect of the protozoan parasite Perezia pyraustae Paillot on the European corn borer. J. Econ. Entomol. 50: 637-640.

GENETIC IMPROVEMENT OF INSECT PATHOGENS: EMPHASIS ON THE USE OF BACULOVIRUSES

Peter Faulkner and Drion G. Boucias

The use of integrated chemical and biological procedures for integrated pest management (IPM) appears to be gaining general acceptance as a realistic goal, and has stimulated research aimed at understanding the role played by microbial pathogens in affecting insect populations. Drawbacks inherent in the application of chemical pesticides alone, particularly the broad ranging toxicity to species other than the target insect and the development of resistance by insects, can be avoided by judicious use of protocols in which microbial pathogens supplement or replace the chemical. Owing to the complex nature of parameters influencing pest insect populations the roles played by microorganisms as components of IPM are not always analyzed. Few persons working in field situations now expect applications of pathogenic microorganisms alone to stabilize an insect population at a low level.

In biological terms, factors influencing population levels of insects in nature depend on interactions between the environment (climate, geography, food supply), genetic vigor of the species, and assault on the target by predators, parasites and microorganisms. Modification of one component is scarcely likely to obliterate the species, but a sustained alteration in a parameter could result in observable shifts in the genetic constitution and numbers of a pest. Insects probably completed their evolution as a class and found their biological niche many millions of years ago; hence it is likely that their associations with microorganisms are also of long standing and often chronic. Acute disease is only likely to be observed when the stress level on the pest is raised by some natural or external pressure. This review will focus on viruses from the family Baculoviridae and the bacterial pathogen Bacillus thuringiensis.

The insect host range of a baculovirus species and and of B. thuringiensis is restricted to a small group of species or genera, hence, when used as pesticidal control agents, they do not cause disease in natural predators of the target insect or in other beneficial insect, plant or animal species (Payne 1982, Tanada 1984). Another factor that makes these pathogens attractive as microbial insecticides is that there are no clear cut

ISBN 0-12-357030-1

instances of selection for resistance among insect populations under field conditions (Anderson 1982). There is a long history of study of the general nature of baculovirus infections and *B. thuringiensis* in the annals of insect pathology but recent research has focused on the biochemistry and molecular genetics of these organisms. Many current studies use recombinant DNA technology to investigate virus gene expression and the molecular basis of biological properties such as host range and virulence. In addition the feasibility of using baculoviruses as vectors for the expression of 'foreign' bacterial and eukaryotic genes has been demonstrated (Miller 1982, Pennock et al. 1984, Smith et al. 1983a). Similar studies have been directed at studying the chemistry of the proteinaceous endotoxin of *B. thuringiensis*, the isolation of the gene(s) for endotoxin synthesis and the development of genetic exchange systems to analyze and manipulate the expression of the endotoxin gene.

PATHOGENIC VIRUSES OF INSECTS

Seven families of animal viruses produce disease in insects. These are the Baculoviridae (genome type: circular ds DNA); Reoviridae (segmented ds RNA); Poxviridae (linear ds DNA); Iridoviridae (linear ds DNA); Parvoviridae (linear ss DNA); Rhabdoviridae (linear ss negative stranded RNA) and Picornaviridae (linear ss RNA). The insect baculoviruses, poxviruses and reoviruses have been utilized most often as microbial control agents principally because the virus particles become embedded within protective occlusion bodies (OB) (Burges 1981, Kurstak 1982, Payne 1982, Tinsley 1979). The acquistion of an OB has contributed to the stability of many of the viruses used for control of insect pests in forestry (Cunningham 1982) and in agriculture (Yearian & Young 1982). Although in this review we will be mostly considering the genetics of one class of these occlusion body-bearing viruses, the NPV, it should be borne in mind that viruses derived from all seven families can cause disease and have potential as microbial insecticides. Discovery of many species of insect viruses occurred when attempts were being made to rear field collected insects under laboratory conditions.

Properties of the Baculoviruses

The family Baculoviridae has been divided into three subgroups (Matthews 1982). Subgroup A contains the nuclear polyhedrosis viruses (NPV) and is the most studied. They were the first to be discovered because cells in tissues of infected insects become filled with (OB) up to 15 um in size and were easily seen in the light microscope. The OB of the NPV encapsulate many enveloped virus nucleocapsids which are dispersed throughout the OB either as separate enveloped nucleocapsids (SNPV) or as groups (often

1-7) of nucleocapsids within a common envelope (MNPV). Subgroup B are the granulosis viruses (GV); the OB are much smaller (about 0.5 um) and occlude one virus particle. The type virus in subgroup C is Oryctes virus which infects the rhinoceros beetle. The rod-shaped enveloped nucleocapsid does not become occluded and the virus is transmitted as the free particle.

Two forms of the NPV virion are replicated when lepidopteran larvae become infected. Following ingestion, the OB dissolve and release virus particles known as occluded virus (OV) in the vicinity of midgut epithelial cells. After attachment and penetration a round of virus replication occurs and enveloped nucleocapsids known as non-occluded virus (NOV) are released into the hemocoel; these spread within the insect, and cause a secondary infection of most tissues. More NOV are released from the secondary sites but OV are also assembled within the nuclei. Eventually occlusion and OB morphogenesis is observed (Faulkner 1981, Granados 1980). Infection is mostly confined to midgut cells when sawflies (Hymenoptera) are infected by an NPV (Granados 1980). In tissue culture the infection process is initiated by NOV and is clearly biphasic. Progeny NOV are released in the early stages of infection in vitro but later production of NOV is curtailed and OB are assembled in the nucleus (Volkman et al. 1976).

GENETIC STUDIES WITH BACULOVIRUSES

A primary objective with genetic studies with baculoviruses is to relate the genetic components in the DNA sequence of the genome with the overall expression of viral pathogenesis. In the case of the baculovirus such an objective implies an understanding of the many stages in the infection process: 1) the means by which the virus is transmitted to a susceptible insect; 2) its fate upon ingestion; 3) factors affecting the dissolution of OB and release of virus particles within the mid-gut; 4) mechanisms by which primary and secondary target cells within the insect are recognized by the virus; 5) factors permitting the adsorption, entry and uncoating of the virus particle; 6) an understanding of the molecular aspects of gene regulation, and expression of virus polypeptides and their processing; 7) the control of virus assembly into progeny particles; 8) assembly of the particles into OB; and 9) an understanding of the mechanisms by which viruses kill the host cell and are released from the infected insect. In addition, an awareness that intergenomic reactions may take place between the genomes of the virus and its insect host needs to be borne in mind (Fraser et al. 1983, Miller & Miller 1982). In particular, knowledge is sought on possible persistence of the virus genome or portions of it by integration into the host cell material. The in-depth analysis of virus pathogenesis is complex as it is likely that virus survival in nature may depend on the persistence of the genome within the host insect at some stages in its life history.

The genome of NPV is a closed circular DNA molecule of MW 80-150X10^6. Physical maps for four NPV have been published (A. californica MNPV, Vlak and Smith 1982; S. frugiperda MNPV, Loh et al. 1981; Choristoneura fumiferana MNPV, Arif et al. 1984; Heliothis zea SNPV, Knell and Summers 1984); each virus is a candidate virus insecticide. Biochemical studies have shown that the genomes of OV and NOV have identical DNA in nucleotide sequences (Cochran et al. 1982) and that the two forms of the virion share some common polypeptides, but each form also has some unique polypeptides (Volkman 1983). Control of gene expression thus occurs throughout the replication cycle and many contemporary studies are focused on understanding the underlying mechanisms.

Most genetic studies with baculoviruses include cloning of the wild type (wt) virus, mutation, and screening of strains; this has been done with a few NPV (Autographa californica NPV, Spodoptera frugiperda NPV, Trichoplusia ni NPV, and Galleria mellonella NPV) which grow in permissive cell lines of Lepidoptera.

Occurrence of Variants of Baculoviruses

Inheritable variations in virulence, host range, and phenotype of the occlusion bodies arise spontaneously in nature and field isolates collected in diverse geographic locations are often genetically heterogeneous, but they can often be distinguished on the basis of restriction endonuclease (REN) analysis of the viral DNA (Knell & Summers 1981, Lee & Miller 1978, Smith & Summers 1979). In some instances some variants have an altered virulence. Harvey & Volkman (1983) compared the virulence of three strains of codling moth GV and noted differences in REN patterns. A segment of genomic DNA and two small structural polypeptides were absent in the strain of lowest virulence. Variants of baculoviruses also arise in laboratory studies when a virus infects an alternate host. An isolate of C. fumiferana (spruce budworm) NPV was fed to neonate larvae of the cabbage looper and the wax moth. After passage in the alternate hosts the isolated virus became increasingly virulent for new insect hosts but also retained infectivity for the budworm but with an altered expression of pathogenicity (Stairs et al. 1981). Unfortunately, the REN of the passaged virus was not compared to the original material and no serological comparisons were made. In vivo recombination has been carried out by mixedly infecting Galleria larvae with two closely related baculoviruses (AcMNPV and GmMNPV). REN analysis of pooled virus DNA extracted from OB harvested after five in vivo passages in the wax moth showed that recombinants were present (Croizier & Quiot 1981). The degree of genetic recombination between baculoviruses seems to depend upon their DNA sequence homology. When the SMPV and MNPV strains of Orgyia pseudotsugata NPV were examined for in vivo genetic recombination, the two strains of virus apparently remained distinct (Rohrmann et al. 1978).

Genetic recombination between viruses with similar but not identical restriction endonuclease patterns occurs in vitro and gene reassortment can be demonstrated. The first report on the location of the polyhedrin gene on the physical map of AcMNPV was done by infecting cells with a mixture of AcMNPV and *Rachiplusia ou* NPV. Analysis of recombinants showed that the distinctive polyhedrin genes derived from each of the viruses had crossed over in some instances (Summers et al. 1980).

A conclusion that may be drawn based on studies on recombination of closely related baculoviruses is that novel genes present in a genetically engineered baculovirus would probably become distributed into the gene pool of field isolates.

Mutants of Baculoviruses

Several classes of mutants have been identified in laboratory studies with baculoviruses. Usually the mutants have been generated in tissue culture by growing the wild type virus in the presence of a chemical mutagen.

Mutants with enhanced virulence One of the more compelling needs in IPM is to have available strains of microorganisms with enhanced virulence. A direct application of genetics to viral insecticides is thus to generate stable variants of baculoviruses which are more effective in the field. Wood et al. (1981) describe the isolation of a mutant strain of AcMNPV, designate HOB, which produces large numbers of OB in infected cells and is of higher virulence in insects than the parent strain. Over 600 plaque isolates were screened in the study. HOB was passed in insects and in culture which suggests that the genome of HOB was modified extensively. In another study the spruce budworm virus (CfMNPV) was grown in the presence of the mutagen nitrosomethyl guanidine and the surviving virus was cloned (Ireland 1982). Of the 34 plaque isolates examined, one, $Cf_{NTG}29$, had an LT_{50} of 99.3 hours against neonate budworm larvae compared with a standard LT_{50} of 113.6 hours. This shotgun approach to generate mutants is highly labor intensive using existing methodology and has other drawbacks, especially with regard to mapping the genetic alterations. In the future it is likely that gene modification will be undertaken using recombinant DNA techniques to alter specific regions on the genome.

Temperature sensitive (ts) mutants Using ts mutants, it has been possible to carry out complementation analysis to define genetic functional groups. Genetic recombination occurs at high frequency between many ts mutants and some have been used to develop a genetic map based on recombination frequencies (Brown & Faulkner 1980) or by marker rescue (Miller 1981). Studies on ts mutants exhibiting OB^- phenotype (absence of occlusion bodies) have demonstrated that up to seven complementation groups affect OB morphogenesis (Brown et al. 1979, Duncan & Faulkner 1982, Faulkner et al. 1980, Lee & Miller 1979).

Non-temperature sensitive (non-ts) mutants A class of non-ts morphology mutants (morph) have also been isolated on the basis of gross defects in OB morphogenesis (Brown et al. 1980). Infection with the morph mutants, m-5 or m-6, produces a single giant OB in infected cell nuclei. Thin sectioning of the OB from the mutants has revealed the normal crystalline lattice structure but the mutants failed to occlude complete virus nucleocapsids. Carstens (1982) made a detailed genetic analysis of the m-5 mutant and showed that virus contains a mixture of two types of genomic DNA. In the type 1 molecule, 400 base pair inserts were present at two locations, but there was a major deletion of over 40% of the genome in the type two molecule. Another morph mutant, m-29, forms an amorphous mass of noncrystalline protein in place of discrete OB (Duncan et al. 1983). In this mutant a single restriction site was eliminated within the region of the polyhedrin gene. Studies with the morph mutants suggest that minor changes in the amino sequence of polyhedrin may account for major phenotypic differences. Since the OB is the principal vehicle of virus transmission, modifications of OB structure may greatly influence the virulence of NPV and GV.

FP Mutants Recent studies with few polyhedra (FP) mutants have indicated that some strains may arise by inserting a transposon-like element of insect host DNA into a specific locus of the baculovirus genome. FP strains arise when a cloned standard strain of baculovirus is passed serially in tissue culture or in insects (Potter et al. 1976, 1978). One such mutant, FP-D, was isolated by plaque purification after AcMNPV was passed 25 times in tissue cell culture (Potter & Miller 1980). The mutant contained an insertion of host cell DNA 7.3 kilobases in size and located at 85.1-87.2 map units on the genome (Miller & Miller 1982). The insertion was considered to be a transposable element derived from Trichoplusia ni DNA. The transposable element was present in multiple copies in T. ni and consisted of a 5 kilobase segment of host DNA flanked by 0.27 kilobase identical sequences. Another set of FP mutants has also been analyzed (Fraser et al. 1983). These were generated by passing GmMNPV or AcMNPV for 3 passages in the T. ni cell line. The insertions 0.8-2.8 kilobase in size were in a specific region between 35.0 and 35.7 map units of the virus genome, and consisted of DNA homologous to a 1280 bp Alu I fragment of the host cellular DNA. The studies also showed that the FP mutants failed to synthesize a 25,000 MW polypeptide normally detected in cells infected with standard virus. The standard occlusion body phenotype and production of a 25,000 MW peptide was regenerated when FP DNA was rescued by transfecting S. frugiperda cells with Hind III-I DNA, thus identifying a portion of the AcMNPV genome important in OB morphogenesis. The studies with FP mutants have also shown clearly that host DNA sequences may become incorporated into the baculovirus genome. Present information is that it is unlikely that FP mutants carrying an insert could be propagated in nature since the OB are either not infectious or are greatly attenuated (Frazer & Hink 1982, Potter et al. 1978, Wood 1980).

CONSTRUCTION OF SITE-SPECIFIC MUTATIONS OF BACULOVIRUSES

Deletion of DNA sequences at specific locations within the genome of a virus can provide important information on the control of gene expression. For example, by deletion one can determine which are essential or non-essential genes for baculovirus replication and for OB morphogenesis. Genes may be excised from the genome, amplified in a plasmid, modified in vitro, then religated into a plasmid and transfected back into the baculovirus. Such studies could be undertaken to modify genes affecting virus host range or virulence. In addition, techniques could be worked out to insert foreign genes into the baculovirus genome at specific sites and use the baculovirus as an expression vector to produce novel proteins in insect cells.

Expression of Eucaryotic and Procaryotic Genes

Smith et al. (1983b) constructed deletion mutations in the protein coding sequences of AcMNPV polyhedrin gene and selected for OB^- mutants. The study was done in 3 stages. First, deletions were introduced into a cloned fragment (EcoRI-I) of the virus genome that contained the polyhedrin gene. Plasmid DNA was linearized by cutting it at a Kpn restriction site lying within the polyhedrin coding region and nibbling back from the site using an exonuclease. In the second stage the plasmid was religated, mixed with intact AcMNPV DNA, and transfected into *S. frugiperda* cells. Finally, recombinant viruses carrying the defective polyhedrin genes were selected by screening plaques that did not produce OB. A 300 bp deletion was present at the expected location in separate clones picked following transfection with one of the plasmids. The study showed that the complete polyhedrin gene was not essential for the production of infectious virus and that deletion of some sequences did not alter the control or decrease the level of expression of the modified polyhedrin.

Two recent reports have demonstrated that AcMNPV can be used to construct expression vectors for both bacterial and eucaryotic genes by inserting them downsteam from strong promotor sequences. Smith et al. (1983a) inserted a human beta-interferon gene within the baculovirus polyhedrin transcription unit. Interferon activity was detected at high titres in the culture media when the gene was inserted near the transcription initiation site and at lower levels when the insert was in frame with the peptide coding region. This study demonstrated that a eucaryotic gene could be expressed in an insect cell at high titre and that post-translational modification (glycosylation) occurred. In another study the E. coli beta-galactosidase gene was inserted in frame with the poyhedrin gene at the 58th codon. Beta-galactosidase was synthesized as a polyhedrin fusion product and its synthesis was temporally regulated in a manner similar to polyhedrin (Pennock et al. 1984). This class of constructs is of value as expression vectors of foreign genes since the beta-galactosidase gene can act as a selectable marker for the presence or absence of a specific inserted gene.

These studies have demonstrated the feasibility of generating new strains of baculoviruses and they indicate a direction along which research is likely to develop over the next period. It seems that contraction or expansion of the length of the baculovirus genome is possible; elements of the genome may be deleted or homologous and heterologous genes may be inserted at specific site without altering the fundamental properties of the virus.

BACILLUS THURINGIENSIS

The gram-positive spore-forming bacterium, Bacillus thuringiensis, has been the subject of extensive research and development by both insect pathologists and applied microbiologists. Several strains of B. thuringiensis possessing insecticidal activity, have been commercially produced and marketed as control agents against both lepidopteran and dipteran pests. This bacterium, having little or no activity against non-target organisms (parasites, predators, vertebrates) yet being highly toxic to certain pests and compatible with current agronomic practices, is considered an ideal candidate for implementation into various IPM programs (Tanada 1984). Current acceptance of B. thuringiensis formulations for pest management in agricultural systems has been hindered by the availability of less expensive broad-spectrum chemical pesticides.

Since the initial identification of the type species B. thuringiensis var. thuringiensis by Berliner, numerous strains have been isolated and identified to possess insecticidal activity. Presently, approximately 800 isolates of B. thuringiensis identified according to serovar and crystal type are being maintained in a culture collection by the U.S. Department of Agriculture at laboratories in Brownsville, Texas (Dulmage & Beegle 1982). Comparative studies have demonstrated that the activity spectrum of B. thuringiensis is highly variable (Burgerjorn & Dulmage 1977, Dulmage 1981, Rogoff et al. 1969). Certain strains having little or no toxicity against one lepidopterous pest may be highly efficacious in controlling other defoliators. Potential improvement of B. thuringiensis as a commercial product may be achieved by hybridizing B. thuringiensis strains and selecting strains having desired host spectra activities.

The insecticidal activity of B. thuringiensis against lepidopterous pests is related to the presence of a proteinaceous toxin which is produced and packaged as a parasporal crystal by sporulating cells. Recent studies have associated the production of this -endotoxin with genes located both on extrachromosomal plasmids and/or within the bacterial chromosome. Potential manipulation of genes responsible for toxin synthesis and for crystal formation can be expected to enhance the development of cost effective B. thuringiensis formulations which can compete successfully with conventional chemical agents.

Bacillus thuringiensis produced via large scale submerged fermentation methods is formulated and applied to crops as conventional pesticides. Like the baculoviruses, lethal amounts of B. thuringiensis formulations must be ingested by target insects to induce mortality. Application of B. thuringiensis for control against defoliators which mine or tunnel plant parts would not be expected to be effective. Unlike the baculoviruses, this bacterial pathogen does not undergo extensive biomagnification in host insects, which presently precludes the establishment of B. thuringiensis as a long-term control agent. During the past several years various investigators, recognizing the potential of genetic engineering methods, have proposed possible strategies for the incorporation and the expression of genes responsible for endotoxin production into the microflora associated with plant surfaces (providing it biomagnification and dissemination capabilities) or into plants (conferring a general resistance to insect herbivory).

During the past decade, research on B. thuringiensis has greatly extended our understanding of the factors responsible for its insecticidal activity. Chemical studies have elucidated the nature of the crystal toxin, molecular studies have resulted in the identification of the genes responsible for toxin synthesis, and genetic research has provided the methods by which these genes may be manipulated. The information derived from these studies has provided a foundation from which a new era of research aimed at improving B. thuringiensis as an effective biorational insecticide may be pursued.

Composition and Mode of Action of δ-Endotoxin

B. thuringiensis strains are characterized by the presence of a parasporal crystal which is produced adjacent to the bacterial spore during post exponential cellular development. Parasporal crystal preparations dissociated with mild alkali (Bulla et al. 1981, Nishiitsutsujii-Uwo et al. 1977) or a mild alkali and 2-mercaptoethanol solution (Yamamoto & Iizuka 1983) will liberate the protoxin (MW 130,000). This alkaline-liberated protoxin may be activated by gut juice proteases producing active toxin having a molecular weight of 68,000 daltons (Bulla et al. 1981). Recently, a toxin proteinase-resistant core (MW 62,000) from several B. thuringiensis strains has been isolated (Yamamoto &

Iizuka 1983). Comparative analysis (peptide mapping) indicated that slight differences in the structure of this proteinase-resistant core protein may cause differences in the activity spectrum.

The primary target of the activated toxin is the larval midgut epithelium. Ingested parasporal crystals disrupted by the alkaline gut juices release the activated toxin which binds to the plasma membrane of the microvilli of the midgut cells. Knowles et al. (1984) have reported that preincubation of the activated toxin with the monosaccharides N-acetyl-D-galactosamine and N-acetylnueraminic acid will neutralize toxic activity. These results suggest that the active toxin recognizes specific glycoconjugates present on the plasma membrane. Treatment of susceptible hosts with active toxin results in cytolysis of midgut epithelium. The molecular mode of the toxin on host cells has been proposed to involve the uncoupling of oxidative phosphorylation in mitochondria (Travers et al. 1976), the disruption of potassium regulation (Fast & Morrison 1972), or possibly directly on the cell membrane. Treated insects, upon exposure to the δ-endotoxin will stop feeding (gut paralysis). Lethal dosages will lead to a general body paralysis and host death. At sublethal concentrations, regenerative nidi will replace damaged midgut cells with host recovery occurring in as little as 2-3 days.

GENETIC STUDIES ON δ-ENDOTOXIN PRODUCTION

Early studies demonstrated that certain B. thuringiensis strains would irreversibly lose their capability to produce parasporal crystals (Norris 1970, Vankova 1957) suggesting the involvement of an extrachromosomal plasmid. Initial attempts to correlate the presence of a plasmid(s) to the production of parasporal crystals were inconclusive. For example, Debabov et al. (1977) and Stahly et al. (1978) reported acrystalliferous mutants lost the plasmids normally present in their respective wild type strains. However, Gonzalez & Carlton (1980) and Miteva (1978) were not able to correlate the presence of specific plasmids to parasporal crystal formation. Comparative investigations of plasmids present in crystalliferous (cry^+) and their respective acrystalliferous (cry^-) mutant strains using more sensitive extraction and electrophoretic techniques have provided evidence that the gene for δ-endotoxin activity is related to the presence of specific plasmids (Gonzalez et al. 1981). Interestingly, this research also indicated that the size of specific plasmids involved in δ-endotoxin production will vary among strains. For example, loss of a 75 megadalton (Md) plasmid was associated with the loss of crystal production in strain HD-2 whereas crystal formation in strain HD-8 was associated with a larger (130 Md) plasmid. Hybridization studies employing a cloned crystal gene (Wong et al. 1983) as a probe revealed that the crystal protein gene may be on one or more plasmids in a particular *B. thuringiensis* strain (Kronstad et al. 1983). In addition to being associated with the extrachromosomal plasmid the protein crystal gene has been associated with the bacterial chromosome (Held et al. 1982; Klier et al. 1982, 1983a).

Currently several methods, protoplast transformation, transduction, and conjugation, are available for vectoring genetic material into *B. thuringiensis*. Transformation, a process involving the introduction of DNA into bacterial protoplasts, has been utilized to transfer plasmid DNA into various *B. thuringiensis* strains (Martin et al. 1981, Miteva et al. 1981). This technique has also been utilized in the cloning of various *B. thuringiensis* genes in other bacterial systems. Several bacteriophages (CP51, TP13) are capable of generalized transduction in *B. thuringiensis* (Perlak et al. 1979, Ruhfel et al. 1984, Thorne 1978). The large TP13 bacterophage has been reported to be capable of converting at high frequency an oligosporogenic, acrystalliferous mutant of *B. thuringiensis* to a crystal^{+} spore^{+}. The bacteriophage CP 51 has been utilized to transfer plasmid-borne antibiotic resistance genes among various *B. thuringiensis*, *B. cereus* and *B. anthracis* species.

The conjugation-like mating system, involving growing strains in mixed culture, has been shown to be a highly effective means for transferring plasmids between *B. thuringiensis* strains (Gonzalez et al. 1982, Klier et al. 1983a). Using this technique it is possible to transfer at high frequencies the plasmid(s) containing the crystal gene of donor Cry^{+} strains to both Cry^{-} *B. thuringiensis* and *B. cereus* recipient strains. Similarly, a plasmid (pAMβ1) coding for erythromycin and lincomycin was transferred from *Streptococcus faecalis* to several *B. thuringiensis* strains using the filter mating process (Lereclus et al. 1983). Resulting transconjugant strains of *B. thuringiensis* were capable of transferring this plasmid (pAMβ1) and a hybrid plasmid into recipient *B. thuringiensis* strains. Utilizing the conjugation method it has been possible to modify the host spectrum of certain *B. thuringiensis* strains. For example, the crystal gene from *B. thuringiensis* strain *berliner* (toxic to lepidopteran larvae) has been introduced into the *B. thuringiensis* strain *israelensis* (toxic to dipteran larvae). Resulting transcipient strains produced two types of δ-endotoxin, active against both lepidopteran and dipteran larvae (Klier et al. 1983a).

Expression of the gene responsible for crystal synthesis appears to be under transcriptional control during sporulation. During sporulation *B. thuringiensis* produces RNA polymerase which differs from the RNA polymerase produced during the vegetative cycle (Klier & Lecadet 1976). The sporulation specific RNA-polymerase of *B. thuringiensis* actively transcribes a relatively stable mRNA coding for the crystal toxin (Klier et al. 1983b). Transcription of the cloned chromosomal crystal gene of *B. thuringiensis* var *berliner* has been demonstrated under both *in vivo* and in vitro conditions to require the RNA polymerase from sporulating cells (Klier et al. 1982, 1983a, b). Crystal protein genes from B. thuringiensis plasmids have been successful cloned into and expressed in Escherichia coli (Schnepf & Whiteley 1981). Analysis of the promotor sequence of the cloned crystal gene has revealed the presence of two adjacent start sites with transcription being initiated from one site early in sporulation and from the other in the middle of sporulation (Wong et al. 1983).

RECOMMENDATIONS

These recommendations are made with the objective of enhancing the efficacy of microbial insecticides.

1. It must always be borne in mind that steps from the laboratory to the test plot and field must be taken cautiously and the full range of environmental impact testing should have prominence in any proposal. Teams interested in the full range of activities working in IPM need to be assembled. Molecular genetics and genetic engineering are powerful tools that should not be shunned by mainsteam environmentalists; instead, centers of excellence covering the whole field need to be encouraged.
2. Genes that affect any aspect of virulence need to identified and modified. Obvious candidates for the baculoviruses are those genes that specify the binding of the virus particle to cell receptors. Other genes which influence the expression of the viral genome and efficiency of viral replication should be investigated. Similiarly, modification of the crystal genes for B. thuringiensis should be attempted to increase the efficiency of δ-endotoxin production. Genetic reassortment and gene site-specific modification are now possible by the techniques of genetic engineering. Variants constructed by modifying these genes should be monitored in terms of host spectrum characteristics, virulence or toxicity against target insect and in the abundance of new progeny pathogens available for dispersion.
3. Phenotypic mixing of insect viruses should be attempted. Specifically dual infection of insects with insect picornaviruses and a virus with an OB (NPV, GV, entomopox virus) may lead to encapsulation of the picornavirus within the OB. The effective conjugation exhibited by B. thuringiensis should be utilized to hybridize strains having distinct and/or overlapping host range characteristics. Research will need to be carried out to determine whether such an admixture would enhance virulence and host spectrum characteristics of these pathogens.
4. Construct new baculoviruses containing selected foreign genes which when expressed have a direct effect on the physiology of the insect pest. Likely candidates are genes that disrupt

the feeding behavior of the insect or which specify a fast acting insect neurotoxin. Insertion of the crystal gene of B. thuringiensis into the baculovirus genome will be possible in the near future. Such a construct may express the toxin form within the gut cells and may destroy their permiability from within. Similiarly neurotoxins and hormones which affect feeding behavior or host development have also great potential as pesticides.

5. At present there is little information on the processes controlling the expression of baculovirus genes and the δ-endotoxin gene. For example, the genes encoding virion enveloped polypeptides have not yet been identified; neither is there basic knowledge of mechanisms and control of viral DNA synthesis. Similar understanding of the factors controlling transcription and translation of the δ-endotoxin gene is vital to its proposed incorporation into alternate host systems (plant cells, microflora associated with plant surfaces).
6. A planned approach to genetic reassortment and gene site-specific modification is now possible by the techniques of genetic engineering. It may be possible soon to expand the host range of a given pathogen by introducing genes from a less virulent donor and thus isolate a genetically defined variant. Actual gene manipulation would likely be done using gene fragments isolated from plasmids. The reconstructed gene would be re-inserted into the pathogen at a specific locus on the genome using a vector that already carries a genetic marker at the insertion site.

ACKNOWLEDGMENTS

We thank Dianne Robertson and Anne Keene for typing and assembling this manuscript, and the Medical Research Council of Canada for continued support.

REFERENCES

Anderson, R.M. 1982. Theoretical basis for the use of pathogens as biological control agents of pest species. Parasitology 84:35-77.

Arif, B.M., J. Kuzio, P. Faulkner and W. Doerfler. 1984. The genome of Choristoneura fumiferana nuclear polyhedrosis virus: Molecular mapping of the EcoRI, BamHI, SmaI and BglII restriction sites. Virus Res. (In Press).

Brown, M., A.M. Crawford and P. Faulkner. 1979. Genetic analysis of a baculovirus Autographa californica nuclear polyhedrosis virus. Isolation of temperature sensitive mutants and assortment into complementation group. J. Virol. 31:335-387.

Brown, M., and P. Faulkner. 1980. A partial genetic map of the baculovirus, Autographa californica nuclear polyhedrosis virus, based on recombination studies with ts mutants. J. Gen. Virol. 48:247-251.

Brown, M., P. Faulkner, M.A. Cochran and K.L. Chung. 1980. Characterization of 2 morphology mutants of Autographa californica nuclear polyhedrosis virus with large cuboidal inclusion bodies. J. Gen. Virol. 50:309-316.

Bulla, L.A., K.J. Kramer, D.J. Cox, B.L. Jones, L.I. Davidson and G.L. Lookhart. 1981. Purification and characterization of the entomocidal protoxin of Bacillus thuringiensis. J. Biol. Chem. 256:3000-3004.

Burgerjon, A., and H. Dulmage. 1977. Industrial standardization of microbial pesticides. I. Bacillus thuringiensis. Entomophaga 22:121-129.

Burges, H.D. (Ed.). 1981. Microbial control of pests and plant diseases 1970-1980. Academic Press, N.Y. 949 pp.

Carstens, E.B. 1982. Mapping the mutation site of an Autographa californica nuclear polyhedrosis virus polyhedron morphology mutant. J. Virol. 43: 809-818.

Cochran, M.A., E.B. Carstens, B.T. Eaton and P. Faulkner. 1982. Molecular cloning and physical mapping of restriction endonuclease fragments of Autographa californica nuclear polyhedrosis virus DNA. J. Virol. 41:940-946.

Croizier, G., and J.M. Quiot. 1981. Isolation and analysis of two genetic recombinants of baculoviruses of Lepidoptera, Autographa californica Speyer and Galleria mellonella L. Ann. Virol. 132:3-18.

Cunningham, J.C. 1982. Field trials with baculoviruses: Control of forest insect pests, pp. 335-386. In: E. Kurstak (Ed.), Microbial and viral pesticides. Marcel Dekker, N.Y.

Debavov, V.G., R.R. Azizbekyan, O.I. Kihebalina, V.V. D'Yachenko, F.P. Galushka and R.A. Belykh. 1977. Isolation and preliminary characterization of extra-chromosomal elements of Bacillus thuringiensis DNA. Genetika 13: 496-501.

Dulmage, H.T. 1981. Insecticidal activity of isolates of Bacillus thuringiensis and their potential for pest control, pp. 93-222. In: H.D. Burges (Ed.), Microbial control of insects and plant diseases 1970-1980. Academic Press, N.Y.

Dulmage, H.T., and C.C. Beegle. 1982. Bacillus thuringiensis cultures. Available from the U.S. Department of Agriculture. USDA Agri. Rev. Man. ARMS-S-30/ October, 42 pp.

Duncan, R., K.L. Chung, and P. Faulkner. 1983. Analysis of a mutant of Autographa californica nuclear polyhedrosis virus with a defect in the morphogenesis of the occlusion body macromolecular lattice. J. Gen. Virol. 64:1531-1542.

Duncan, R., and P. Faulkner. 1982. Bromodeoxyuridine-induced mutants of Autographa californica nuclear polyhedrosis virus defective in occlusion body formation. J. Gen. Virol. 62:369-373.

Fast, P., and I. Morrison. 1972. The δ-endotoxin of Bacillus thuringiensis. IV. The effect of δ-endotoxin on ion regulation by midgut tissue of Bombyx mori. J. Invertebr. Pathol. 20:208-211.

Faulkner, P. 1981. Baculoviruses, pp. 3-37. In: E.W. Davidson (Ed.), Pathogenesis of invertebrate microbial diseases. Allanheld, Osmun, Totowa, NJ.

Faulkner, P., M. Brown and A.M. Crawford. 1980. Use of ts mutants to identify complementation group and to develop a genetic map for a baculovirus AcNPV, pp. 451-460. In: E. Kurstak, K. Maramorosch and A. Dubendorfer (Eds.), Invertebrate systems in vitro. North Holland Biochemical Press, Amsterdam.

Fraser, M.J., and W.F. Hink. 1982. The isolation and characterization of the MP and FPK variants of Galleria mellonella nuclear polyhedrosis virus. Virology 117:366-378.

Fraser, M.J., G.E. Smith and M.D. Summers. 1983. Acquisition of host-cell DNA sequences by baculoviruses-relationship between host DNA insertions and FP mutants of Autographa californica and Galleria mellonella nuclear polyhedrosis viruses. J. Virology 47:287-300.

Gonzalez, J.M., B.J. Brown and B.C. Carlton. 1982. Transfer of Bacillus thuringiensis plasmids coding for δ-endotoxin among strains of B. thuringiensis and B. cereus. Proc. Nat. Acad. Sci. 79:6951-6955.

Gonzalez, J.M., and B.C. Carlton. 1980. Patterns of plasmid DNA in crystalliferous and acrystalliferous strains of Bacillus thuringiensis. Plasmid 3:92-98.

Gonzalez, J.M., H.T. Dulmage and B.C. Carlton. 1981. Correlation between specific plasmids and δ-endotoxin production in Bacillus thuringiensis. Plasmid 5: 351-365.

Granados, R.R. 1980. Infectivity and mode of action of baculoviruses. Biotechnol. Bioeng. 22:1377-1405.

Harvey, J.P., and L.E. Volkman. 1983. Biochemical and biological variation of Cydia pomonella (codling moth) granulosis virus. Virology 124:21-34.

Held, G.A., L.A. Bulla, E. Ferrari, J. Hoch, A.I. Aronson and S.A. Munnich. 1982. Cloning and localization of the lepidopteran protoxin gene of Bacillus thuringiensis subsp. kurstaki. Proc. Nat. Acad. Sci. 79:6065-6069.

Ireland, D. 1982. Infectivity of spruce budworm baculovirus variants. M.S. Thesis, Queen's University, Kingston, Ontario, Canada.

Klier, A., C. Bourgouin and G. Rapoport. 1983a. Mating between Bacillus subtilis and Bacillus thuringiensis and transfer of cloned crystal genes. Mol. Gen. Genet. 191:257-262.

Klier, A., F. Fargett, J. Ribier and G. Rapoport. 1982. Cloning and expression of the crystal protein genes from Bacillus thuringiensis strain Berliner 1715. EMBO J. 1:791-799.

Klier, A., and M.M. Lecadet. 1976. Agruments based on hybridization-competition experiments in favor of the in vitro synthesis of sporulation specific mRNAs by the RNA polymerase of Bacillus thuringiensis. Biochem. Biphys. Res. Comm. 73:263-270.

Klier, A., C. Parsot and G. Rapoport. 1983b. In vitro transcription of the cloned chromosomal crystal gene from Bacillus thuringiensis. Nucleic Acids Res. 11 (12):3973-3987.

Knell, J.D., and M.D. Summers. 1981. Investigation of genetic heterogeneity in wild isolates of Spodoptera frugiperda nuclear polyhedrosis virus by restriction endonuclase analysis of plaque purified variants. Virology 112:190-197.

Knell, J.D., and M.D. Summers. 1984. A physical map for the Heliothis zea SNPV genome. J. Gen. Virol. 65:445-450.

Knowles, B.H., W.E. Thomas and D.J. Ellar. 1984. Lectin-like binding of Bacillus thuringiensis var. kurstaki lepidopteran-specific toxin is an initial step in insecticidal action. FEBS Letters 168:197-202.

Kronstad, J.W., H.E. Schnepf and H.R. Whiteley. 1983. Diversity of locations for Bacillus thuringiensis crytal protein genes. J. Bact. 154:419-428.

Kurstak, E. (Ed.). 1982. Microbial and viral pesticides. Marcel Dekker, N.Y. 720 pp.

Lee, H.H., and L.K. Miller. 1979. Isolation, complementation and initial characterization of temperature sensitive mutants of the baculovirus Autographa californica nuclear polyhedrosis virus. J. Virol. 31:240-252.

Lereclus, D., G. Menou and M.M. Lecadet. 1983. Isolation of a DNA sequence related to several plasmids from Bacillus thuringiensis after a mating involving the Streptococcus faecalis plasmid pAM1. Mol. Gen. Genet. 191:307-313.

Loh, L.C., J.J. Hamm and E.S. Huang. 1981. Spodoptera frugiperda nuclear polyhedrosis virus genome: Physical maps for restriction endonucleases BamHI and HindIII. J. Virol. 38:922-931.

Martin, P.A.W., J.R. Lohr and D.H. Dean. 1981. Transformation of Bacillus thuringiensis protoplasts by plasmid deoxyribonucleic acid. J. Bacteriol. 145: 980-983.

Matthews, R.E.F. 1982. Classification and nomenclature of viruses. Intevirology 17:1-199.

Miller, D.W., and L.K. Miller. 1982. A virus mutant with an insertion of a copia-like transposable element. Nature 299:562-564.

Miller, L.K. 1981. Construction of a genetic map of the baculovirus Autographa californica nuclear polyhedrosis virus by marker rescue of temperature sensitive mutants. J. Virol. 39:973-976.

Miller, L.K. 1982. A virus vector for genetic engineering in invertebrates. In: N. Panopoulous (Ed.), Genetic engineering in the plant sciences. Praeger Publishers, N.Y.

Miteva, V.I. 1978. Isolation of plasmid DNA from various strains of Bacillus thuringiensis and Bacillus cereus. C. R. Acad. Bulg. Sci. 31:913-916.

Miteva, V.I., N.I. Shivarova and R.T. Grigorova. 1981. Transformtion of Bacillus thuringiensis protoplasts by plasmid DNA. FEMS Microbiol. Lett. 12:253-256.

Nishiitsutsujii-Uwo, S., A. Ohsawa and M.S. Nishimura. 1977. Factors affecting the insecticidal activity of δ-endotoxin of Bacillus thuringiensis. J. Invertebr. Pathol. 29:162-169.

Norris, J.R. 1970. Spore formers as insecticides. J. Appl. Bacteriol. 33:192-206.

Payne, C.C. 1982. Insect viruses as control agents. Parasitology 84:35-77.

Pennock, G.D., C. Shoemaker and L.K. Miller. 1984. Strong and regulated expression of E. coli B-galactosidase in insect cells using a baculovirus vector. Mol. Cell. Biol. 4:399-406.

Perlak, F.J., C.L. Mendelsohn and C.B. Thorne. 1979. Converting bacteriophage for sporulation and crystal formation in Bacillus thuringiensis. J. Bacteriol. 140:699-706.

Potter, K.N., P. Faulkner and E.A. Mackinnon. 1976. Strain selection during serial passage of Trichoplusia ni nuclear polyhedrosis virus. J. Virol. 18: 1040-1050.

Potter, K.N., R.P. Jaques and P. Faulkner. 1978. Modification of Trichoplusia ni nuclear polyhdrosis virus passaged in vivo. Intervirology 9:76-85.

Potter, K.N., and L.K. Miller. 1980. Correlating genetic mutations of a baculovirus with the physical map of the DNA genome, pp. 71-80. In: B Fields, R. Jaenisch and C.F. Fox (Eds.). Animal virus genetics. Academic Press, N.Y.

Rogoff, M.H., C. Ignoffo, S. Singer, I. Gard and A. Prieto. 1969. Insecticidal activity of thirty-one strains of Bacillus against five insect species. J. Invertebr. Pathol. 14:122-129.

Rohrmann, G.F., R.H. Mcparland, M.E. Martigoni and G.S. Beaudreau. 1978. Genetic relatedness of two nucleopolyhedrosis viruses pathogenic for Orgyia pseudotsugata. Virology 84:213-217.

Ruhfel, R.E., N.J. Robillard and C.B. Thorne. 1984. Interspecies transduction of plasmids among Bacillus anthracis, B. cereus, and B. thuringiensis. J. Bacteriol. 157:708-711.

Schnepf, H.E., and H.R. Whiteley. 1981. Cloning and expression of the Bacillus thuringiensis crystal protein gene in Escherichia coli. Proc. Nat. Acad. Sci. 78:2893-2897.

Smith, G.E., M.J. Fraser and M.D. Summers. 1983b. Molecular engineering of the Autographa californica nuclear polyhedrosis virus genome--deletion mutations within the polyhedrin gene. J. Virol. 46:584-593.

Smith, G.E., and M.D. Summers. 1979. Restriction maps of five Autographa californica MNPV variants, Trichoplusia ni MNPV and Galleria mellonella MNPV DNAs with endonucleases SmaI, KpnI, BamHI, SacI, XhoI, and EcoRI. J. Virol. 30: 828-838.

Smith, G.E., M.D. Summers and M.J. Fraser. 1983a. Production of beta human interferon in insect cells with a baculovirus expression vector. Mol. Cell. Biol. 3:2156-2165.

Stahly, D.P., D.W. Dingman, R.L. Irgens, C.C. Field, M.G. Feiss and G.L. Smith. 1978. Multiple extrachromosomal deoxyribonucleic acid molecules in Bacillus thuringiensis. FEMS Microbiol. Lett. 3:139.

Stairs, G.R., T. Fraser and M. Fraser. 1981. Changes in growth and virulence of a nuclear polyhedrosis virus from Choristoneura fumiferana after passage in Trichoplusia ni and Galleria mellonella. J. Invertebr. Pathol. 38:230-235.

Summers, M.D., G.E. Smith, J.D. Knell and J.P. Burand. 1980. Physical maps of Autographa californica and Rachiplusia ou nuclear polyhedrosis virus recombinants. J. Virol. 34:696-703.

Tanada, Y. 1984. Bacillus thuringiensis: Integrated control--past, present, and future, pp. 59-90. In: T.C. Cheng (Ed.), Comparative pathobiology, Vol. 7, Pathogens of Invertebrates. Plenum Press, N.Y.

Thorne, C.B. 1978. Transduction in Bacillus thuringiensis. Appl. Environ. Microbiol. 34:1109-1115.

Tinsley, T.W. 1979. The potential of insect pathogenic viruses as pesticidal agents. Annu. Rev. Entomol. 24:63-87.

Travers, R.S., R.M. Faust and C.F. Reichelderfer. 1976. Effects of Bacillus thuringiensis var kurstaki δ-endotoxin on isolated lepidoptera mitochondria. J. Invertebr. Pathol. 28:351-356.

Vankova, J. 1957. Study of the effects of Bacillus thuringiensis on insects. Folia Biol. 3:175-182.

Vlak, J.M., and G.E. Smith. 1982. Orientation of the genome of Autographa californica nuclear polyhedrosis virus--a proposal. J. Virol. 41:1118-1121.

Volkman, L.E. 1983. Occluded and budded Autographa californica nuclear polyhedrosis virus--immunological relatedness of structural proteins. J. Virol. 46:221-229.

Volkman, L.E., M.D. Summers and C.H. Hsieh. 1976. Occluded and nonoccluded nuclear polyhedrosis virus grown in Trichoplusia ni: Comparative neutralization, comparative infectivity, and in vitro growth studies. J. Virol. 19:820-832.

Wong, H.C., H.E. Schnepf and H.R. Whiteley. 1983. Transcriptional and translational start sites for the Bacillus thuringiensis crystal protein gene. J. Biol. Chem. 258:1960-1967.

Wood, H.A. 1980. Isolation and replication of an occlusion body-deficient mutant of the Autographa californica nuclear polyhedrosis virus. Virology 105:338-344.

Wood, H.A., P.R. Hughes, L.B. Johnston and W.H. Langridge. 1981. Increased virulence of Autographa californica nuclear polyhedrosis virus by mutagenesis. J. Invertebr. Pathol. 38:236-241.

Yamamato, T., and T. Iizuka. 1983. Two types of entomocidal toxins in the parasporal crystals of Bacillus thuringiensis var kurstaki. Arch. Biochem. Biophy. 227:233-241.

Yearian, W.C., and S.Y. Young. 1982. Control of insect pests of agricultural importance by viral insecticides, pp 357-423. In: E. Kurstak (Ed.), Microbial and viral pesticides. Marcel Dekker, N.Y.

DISCUSSION: Insect pathology: Development, use, manipulation and improvement of insect pathogens.

DISCUSSION LEADER: D.G. Boucias

An array of bacterial, viral, fungal, and protozoan entomopathogens have been identified as potential microbial agents against a variety of agriculturally important pests. Many of these pathogens infect only the target pest and are not detrimental to the naturally-occurring biological control complex. Research over the last 20 years on pathogens such as Bacillus thuringiensis, the Heliothis NPV, and Hirsutella thompsonii has provided the basis for the large-scale production and formulation of microbial insecticides by industrial interests. Such biorational pesticides, in light of their selectivity and safety, have been considered to be ideally suited for implementation into IPM programs. Such was the case with B. thuringiensis and Elcar (Heliothis NPV) for Heliothis management in High Plains cotton fields of Texas until the introduction of the synthetic pyrethroids.

Currently in the U.S., microbial agents comprise less than 1% of the total pesticide market. The inability of microbes to compete successfully against chemical pesticides can be attributed to several factors. Costs of producing and marketing these agents are significantly greater than those for the broad-spectrum pesticides, which can be formulated for use against a range of pest insects affecting a broad spectrum of agricultural commodity systems. Microbial agents effects are currently not as predictable or as fast-acting as conventional chemical controls. In many cases, established pest threshold levels are not suitable for use with microbial pesticide applications. Agricultural producers and extension personnel must be made aware of the mode of action and limitations of using microbials as pest control agents. Generally, microbial agents, although capable of suppressing pest population increases, cannot annihilate the pest species as do certain chemical pesticides.

In order to enhance the utilization of microbial pesticides, it was suggested that research be directed at developing programs suited for the use of microbial agents rather than attempting to fit microbials into existing chemical management programs. This approach, requiring a multiseasonal effort, has been successful in organic pear orchards in California. In these orchards, the codling moth, the major pest insect, has been managed through use of a granulosis virus. The targeting of microbials against insect pests which are not easily controlled by conventional chemicals should also be investigated. For example, soil-inhabiting insects, unlike many phytophagous defoliators, are difficult and expensive pests to control using conventional chemicals. Entomopathogens, developed against such pests can be expected to be more competitive in the market place than those microbes targeted against defoliating pest complexes.

Survival of infectious inoculum in the field is a primary factor which determines the success of microbial agents. Abiotic factors such as temperature, sunlight (ultraviolet light), and pH of the foliage, all have been reported to reduce persistence of entomopathogens. The development of improved strains of entomopathogens via conventional and non-conventional methods were discussed. Improvements in formulation of microbials (microencapsulation, UV-screens) have resulted in dramatic increases in the field half-life of pathogens applied in the field. Research directed at identifying gene products responsible for persistence (UV-resistant proteins) increased virulence (receptor type molecules, toxins), and/or specificity should be pursued and utilized to construct improved strains of entomopathogens.

ENTOMOGENOUS NEMATODES FOR INSECT CONTROL IN IPM SYSTEMS

Harry K. Kaya

Nematodes are morphologically, genetically and ecologically diverse organisms occupying more varied habitats than any other animal group except arthropods. They may occur as free-living organisms or as facultative or obligate parasites. Many species are endoparasites of insects of agricultural, forest, medical, and veterinary importance. Interest in entomogenous nematodes as biological control agents has increased in recent years as amply demonstrated by the number of review articles and books on this subject (Finney 1981; Gaugler 1981; Nickle 1981, 1984; Petersen 1982; Platzer 1981; Poinar 1975, 1979; Webster 1980). Moreover, comprehensive bibliographies on nematodes of arthropods (Shephard 1974) and on the steinernematid nematodes have been compiled (Gaugler & Kaya 1983).

Nematode parasitism of insects may result in sterility, reduced fecundity, delayed development, aberrant behavior or death of the host. A highly desirable attribute of nematodes in control programs is rapid host mortality which prevents or limits the degree of insect damage to crops. However, host mortality per se is not essential for biological control. The nematode, *Deladenus siricidicola* which sterilizes its woodwasp host, *Sirex noctilio*, has been successfully used to reduce woodwasp populations below economic threshold levels in *Pinus radiata* plantations in Australia (Bedding 1979).

Although *Deladenus* has been successful in biological control by inducing host sterility, this is more of an exception than the rule. Accordingly, the major research emphasis is on nematodes that kill their hosts in a relatively short period of time. Nematodes in the families Steinernematidae, Heterorhabditidae, and Mermithidae have this attribute. The first two families are mutualistically associated with bacteria that kill the host quickly by causing septicemia. The action of mermithids is analogous to internal parasitoids because the nematodes obtain nutrients at the expense of their hosts and kill them upon exit. Major research emphasis with mermithids has been against aquatic insects rather than agricultural ones. Consequently, the steinernematids and heterorhabditids which are potentially useful for biological control in agricultural systems will be discussed.

ISBN 0-12-357030-1

SYSTEMATICS AND NATURAL OCCURRENCE OF NEMATODES

Four species are recognized within the family Steinernematidae: *Steinernema kraussei*, *S. glaseri*, *S. bibionis* and *S. feltiae* (= *Neoaplectana carpocapsae*) (Wouts et al. 1982). The family Heterorhabditidae also currently contains four species. They are *Heterorhabditis heliothidis*, *H. bacteriophora*, *H. hoptha*, and *H. hambletoni* (Poinar 1979). *S. bibionis*, *S. feltiae*, *H. heliothidis* and *H. bacteriophora* have received the most attention as potential biological control agents. Depending upon the target host, certain strains of the nematode species are more efficacious than others (Bedding & Miller 1981b, Bedding et al. 1983, Molyneux et al. 1983, Silverman et al. 1982).

These nematodes have been isolated from all continents except Antarctica. Steinernematids have been found in soil-inhabiting insects, in insects on tree trunks and by using trap insects. Bedding & Akhurst's (1975) technique of using *Galleria* larvae for isolating these nematodes from soil is very effective.

Two species of bacteria, *Xenorhabdus nematophilus* and *X. luminescens*, are symbionts of steinernematids and heterorhabditids, respectively.

LIFE CYCLES

Most nematodes have a simple life cycle which includes the egg, four larval (= juvenile) stages and the adult. Heterorhabditids and steinernematids have a resistant stage called the "dauer" meaning durability or permanence. The dauer, the third stage nematode ensheathed in the second stage cuticle, is the infective stage and contains cells of the symbiotic bacteria within the intestinal lumen (Poinar & Leutenegger 1968).

Steinernematids have a simple life cycle. Host finding by the infectives can be an active process in response to physical and chemical host cues. The dauers enter suitable hosts via natural body openings (mouth, anus or spiracles), exsheath, and penetrate mechanically into the hemocoel. The nematodes then release *Xenorhabdus* cells which kill the host within 24 to 48 hours by septicemia. Feeding upon the bacterial cells and host tissues, the nematodes develop rapidly. As resources are depleted and crowding occurs, the progeny of the second or third generation develop into infective dauers which exit from the cadaver and seek new hosts. If no host is found, the infectives can survive for a long time under appropriately humid conditions.

Heterorhabditids have a life cycle similar to steinernematids. The infectives enter their host as described above (Khan et al. 1976, Poinar 1975) or penetrate directly through the host's cuticle (Bedding & Molyneux 1982). After its release the symbiotic bacterium, *X. luminescens*, multiplies rapidly and kills the host within 24 to 48 hours. Unlike the steinernematids which need at least one male and female per host for reproduction, heterorhabditid infectives are hermaphroditic (Khan et al. 1976, Poinar 1975) and therefore, one nematode per host is sufficient for reproduction. The progeny of the hermaphrodite are dioecious and they produce the new hermaphroditic dauers.

HOST RANGE AND SAFETY

Because of their broad host range (Poinar 1979), steinernematids and heterorhabditids have generated great interest as biological control agents. However, tests under laboratory conditions exclude behavioral or ecological barriers to infection. The effective host range in the field is limited by the nematodes' moisture requirement to insects of soil and cryptic habitats. On the other hand, insects in such habitats are difficult to control with chemical pesticides, and these nematodes may provide an effective means of control alone or in an integrated approach.

Since bacteria are not exempted from Environmental Protection Agency (EPA) registration, many tests were conducted to determine the effect of the nematode-bacteria complex on vertebrates (see Obendorf et al. 1983, and references therein). On the basis of such information the EPA has ruled that nematodes vectoring non-exempt biological control agents (bacteria) are also exempt from registration (Gorsuch 1982). EPA did not give blanket approval and will carefully monitor the development and use of these nematodes as biological control agents. If the situation warrants it, the nematode-bacterial complex may be added to the list of non-exempt organisms.

MASS PRODUCTION, STORAGE AND TRANSPORTATION

The utilization of any biological control agent is dependent upon its availability in sufficient quantities at acceptable costs. Steinernematids and heterorhabditids have been mass produced in vivo and in vitro. For these nematodes, almost any available insect can serve as a host for mass propagation; however, such production is not cost effective because it is labor intensive. These nematodes are easily grown in vitro because specialized media are not required (Bedding 1981). Bedding's (1984) method consists of coating shredded polyether-polyurethane sponge with a homogenate of chicken offal (steinernematids) or chicken offal plus 10% beef fat (heterorhabditids), sterilizing the medium in large autoclave bags, and adding the appropriate bacterium and nematode. The estimated expense was less than 1 cent/million. The eventual use of fermentation vats will increase production substantially.

Steinernematid infectives, in general, can be stored for extended periods (5 years) under cool, moist conditions (Bedding 1981, Dutky et al. 1964). For industrial storage, the infectives are placed onto clean, crumbed polyether-polyurethane sponge at rate of 250 million *S. glaseri* or 500 million *S. feltiae* or *S. bibionis* per 100 grams of dry sponge and maintained at 1-2°C in aerated polyethylene tubes (Bedding 1984). For trips less than 12 hours, oxygen is added to the tubes before transport. For longer transport the tubes are aerated with a battery operated air pump. These techniques result in less than 10% mortality of infectives in storage or in transport.

The heterorhabditid infectives are best stored in culture flasks above 12°C (Bedding 1981). Transport of these nematodes will be more difficult and expensive because of the additional weight of the culture flasks.

ENVIRONMENTAL CONSIDERATIONS IN USING NEMATODES IN AGRICULTURE

Entomogenous nematodes have unrealized potential for use in the agroecosystem. At the present time, however, the successful use of these nematodes is in situations that protect them from desiccation, radiation and temperature extremes.

Moisture

Unless nematodes have mechanisms to undergo anhydrobiosis (Crowe & Madin 1975), they desiccate and die. *S. feltiae* is not resistant to rapid drying (Dutky 1959). Even on foliage where the relative humidity (RH) of the microclimate is high, this nematode survives for 90 minutes or less (Moore 1965, Welch & Briand 1961a). Under environmentally controlled conditions at 85% RH, 98% of the infectives died after 102 hours at 30°C and after 36 hours at 5°C (Kamionek et al. 1974a). At 20% RH, 98% of the infectives died after 2.5 hours at 30°C.

The use of antidesiccants improved the survival of *S. feltiae* but not sufficiently to warrant their use against foliage feeding insects (Kaya & Reardon 1982, MacVean et al. 1982, Nash & Fox 1969, Webster & Bronskill 1968, Welch & Briand 1961a). At the present state of the formulation art with nematodes, foliar applications should be discouraged until effective antidesiccants are found.

Sunlight

In addition to antidesiccants, consideration must be given to protecting the nematodes from solar radiation which is detrimental to *S. feltiae* (Gaugler & Boush 1978). The use of ultraviolet protectants extended the longevity of the infectives (Gaugler & Boush 1979).

Soil

The soil environment is a favorable habitat for most nematodes because adequate moisture is usually present. If the soil is gradually allowed to dry, *S. feltiae* infectives can survive. Moore (1965) showed that infectives placed into moist soil survived for 20 days when it was slowly dried at 70% RH. Similarly, 90% of the dauers were alive in the soil after 12 days at 79.5% RH, the permanent wilting point of most plants (Simons & Poinar 1973).

In recent years, the behavior of steinernematids and heterorhabditids in soil has been examined. In moist sandy soil, the majority of *S. feltiae* infectives remained at the point of placement if placed on the surface, although a few dispersed up to 14 cm laterally and 12 cm vertically (Moyle & Kaya 1981). When deposited 15 cm below the surface, the majority remained at that location but dispersal tended to be upwards rather than downwards. Georgis & Poinar (1983a, b) confirmed these results with *S. feltiae* and *S. glaseri*. In addition, they showed that the presence of clay and silt resulted in less dispersal than in sand alone. Moreover, the presence of a host in the soil increased the number of dispersing nematodes, evidence that the infectives could orient to kairomones in the soil. *H. bacteriophora* and *H. heliothidis* showed a greater tendency than steinernematids to disperse vertically in sandy loam soil (Georgis & Poinar 1983c).

Temperature

Extreme temperatures can be an important limiting factor. *S. feltiae* can infect its host at 9°C (Dutky et al. 1964, Gaugler & Molloy 1981), but host mortality may not occur until 312 hours later. In comparison, at 30°C host mortality takes 16 hours (Dutky et al. 1964). Infectives are most active between 22 and 32°C (Schmiege 1963). Temperatures above 35°C are lethal while at the other extreme, 30% of the infectives survived -10°C for 18 hours (Schmiege 1963). Their recovery from temperate soils throughout much of the world also indicates that these nematodes survive low temperatures.

The upper temperature limit for the development of *S. feltiae* is 30°C and the lower limit is about 15°C (Kaya 1977, Pye & Burman 1978). Similar results were obtained for *H. bacteriophora* by Milstead (1981).

The rearing temperature of the nematode can affect the response of the infectives to a thermal gradient. Burman & Pye (1980a) showed that *S. feltiae* cultured at 15, 20 and 25°C tended to aggregate at or near their respective culture temperature. To improve efficacy they suggested that the nematodes should be cultured at the temperature in which the target insects occur.

Oxygen and Chemicals

In laboratory studies, oxygen consumption of *S. feltiae* was strongly temperature dependent; infectives held at a high temperature had greater oxygen consumption (Burman & Pye 1980b). Infectives can survive 43 days with oxygen only 0.5% of saturation at 20°C.

In studies with chemical gradients, Pye & Burman (1981) demonstrated that S. feltiae infectives respond positively to the following ions: Na, Mg, Ca and Cl. The infectives are attracted to basic pH but are repelled by acidic pH at 2.5 and by ammonium. The negative response to ammonium is postulated as a mechanism for dispersal away from expended cadavers. Infectives also respond to various host cues.

COMPATIBILITY WITH PESTICIDES AND OTHER BIOCONTROL AGENTS

Use of entomogenous nematodes in IPM programs has been suggested by Dutky (1974). Ideally, in an IPM approach the chemicals should not be toxic to the nematodes. Entomogenous nematodes may be compatible (Dutky 1974, Fedorko et al. 1977a, Welch 1971) or incompatible (Fedorko et al. 1977b, c; Hara & Kaya 1982, 1983a, b; Prakasa Rao et al. 1975) with chlorinated hydrocarbons, organophosphate insecticides or carbamates in water solution or suspension. Certain fungicides, herbicides, miticides, and nematicides had little or no adverse effects on the infectives (Dutky 1974).

Compatibility must be determined before implementation of an IPM strategy. However, compatibility per se does not necessarily insure effective insect control. The combination should result in additive or supplemental mortality of the pest, reduce the amount of chemical needed for insect control and allow the nematode to recycle. Certainly, an important consideration is the cost/benefit ratio derived from such a strategy.

Generally, studies with hymenopterous parasitoids demonstrated that larval parasitoids within a host or in the process of exiting from a host to pupate were susceptible to nematode infection (Kaya 1978a, b; Kaya & Hotchkin 1981). Once the parasitoids formed cocoons, they were protected by the nonporous silken layer on the inner cocoon surface (Kaya & Hotchkin 1981). Early instars of the tachinid parasitoids, Compsilura concinnata, are adversely affected when the lepidopterous host is infected by S. feltiae (Kaya 1984) because the host dies before the tachinid is sufficiently developed. Older instars of the tachinid do not require a living host, so they develop unaffected even though nematodes may be present. Mracek & Spitzer (1983) obtained no infection of a sawfly tachinid parasitoid with S. kraussei and concluded that the nematode and parasitoid were compatible.

Dipterous predators (Therevidae and Rhagionidae) which feed on sawflies were found to be susceptible to S. kraussei (Mracek & Spitzer 1983). However, this nematode does not infect these predators under natural conditions.

S. feltiae can develop in granulosis-virus-infected hosts. The infectives from such hosts sequester the virus in their intestinal lumen (Kaya & Brayton 1978). The virus retains its infectivity even when the nematodes are treated with antiviral agents (Kaya 1980). Only when a susceptible host consumes macerated nematodes will the virus be transmitted. Similar results have been obtained with infectives from a nuclear polyhedrosis-virus-infected insect (H.K. Kaya, unpublished data).

S. feltiae is compatible with certain entomogenous fungi and bacteria. Infectives applied along with conidia of Beauveria bassiana and Paecilomyces farinosus resulted in higher mortality of Galleria than either the fungi or the nematode alone (Kamionek et al. 1974b, c). Similar results were obtained against diapausing Colorado potato beetle, Leptinotarsa decemlineata (Seryczynska 1975). However, increased mortality with the combination was not achieved with Tribolium (Kamionek et al. 1974b, c). Combining Bacillus thuringiensis with the infectives of S. feltiae in attempts to control the artichoke plume moth, Platyptilia carduidactyla, did not significantly increase control over that achieved by the use of the nematode alone (Bari & Kaya 1984).

S. feltiae is susceptible to infection by a microsporidian of an insect (Veremchuk & Issi 1970). Although not known, the infection of steinernematids by other pathogens is highly probable. For example, Poinar et al. (1980) found an iridescent virus replicating in mermithids.

FIELD APPLICATION OF NEMATODES

Gaugler (1981) and Poinar (1979) have summarized much of the research concerning field applications. Inconsistent results of these field tests were probably related to the poor choice of target habitats (Gaugler 1981). In general, attempts to control foliage insects have been discouraging, with low host mortality, insignificant population reduction, or inadequate crop protection. Successful applications have been made against certain insects in cryptic and soil habitats.

Foliage Applications

In trials against the Colorado potato beetle with S. feltiae, Welch (1958) obtained a significant reduction (14%) in larvae and adults in 1957, but no significant reductions were observed in 1958 and 1959 (Welch & Briand 1961a). These poor results were attributed to the lack of ideal conditions for the nematodes (80% RH or higher, little air movement, and temperature range of 22-28°C). The addition of glycerine, honey, glucose, sorbitol, urea, or agar to the nematode suspension reduced evaporation but only at a high concentration. Moreover, many of the additives were nematicidal or phytotoxic. MacVean et al. (1982) added inert thickening agents to their nematode suspensions and obtained 30 to 60% infection of the Colorado potato beetle larvae compared to 10% with nematodes in water suspension. Applying nematodes at night resulted in higher mortality (15 to 62%) than morning applications (5-51%).

Field trials have been conducted with S. feltiae against several foliage feeding lepidopterous insects. Nematode applications against the imported cabbageworm, Pieris rapae, resulted in 73 to 77% mortality compared to 82 to 84% mortality with the chemical insecticide treatment (Welch & Briand 1961b). The mortality data appeared substantial, but continued observation of the cabbage plants showed that no foliage protection occurred. The reason is related to the nature of the nematode's mortality. The nematode's persistence on foliage is, at best, a few hours; hence, primarily larvae consuming small amounts are less likely to encounter nematodes. In contrast, the chemical insecticides kill small and large larvae providing the necessary foliage protection. Additional tests conducted against the cabbageworm compared S. feltiae with other microbial and chemical insecticides (Fox & Jaques 1966). The nematode was the least effective of the materials tested, but still provided significant mortality, particularly late in the growing season.

Chamberlin & Dutky (1958) showed that application of S. feltiae onto tobacco leaves wet from rains or high humidity and at 30°C caused 80 to 85% larval reduction of the tobacco budworm, Heliothis virescens, within three to four days. However, when temperatures exceeded 30°C and the foliage was dry, no larval mortality occurred. Tests conducted on corn infested with the fall armyworm, Spodoptera frugiperda, resulted in 39% larval reduction at the highest rate (4000 nematodes/plant) compared to the insecticide check with 74% larval reduction (Landazabal et al. 1973). Higher RH conditions resulted in better control with the nematode than low RH. Application of this nematode to control apple defoliators was not successful because of rapid desiccation of nematodes on the leaves (Jaques 1967).

These examples emphasize the importance of high humidity for effective control of insect pests on foliage. The addition of antidesiccants to the nematode suspension increases insect mortality but not to acceptable economic levels. Even if high mortality of the insect occurs with the nematode application, this mortality may not be sufficient to reduce the population levels below the economic threshold level because of differential mortality of larvae (Welch & Briand 1961b).

Cryptic Habitats

The most encouraging use of steinernematids has been against insects in protected habitats, such as galleries in trees, where humidity is high and the infective nematodes are sheltered from the hostile environment (Gaugler 1981). The carpenterworm, Prionoxystus robiniae, was totally supressed in commercial fig orchards by S. feltiae (Lindegren & Barnett 1982, Lindegren et al. 1981b). Similarly, this nematode killed 85-90% of the Zeuzera pyrina larvae, a pest of orchard trees in Italy (Deseo 1982, Foschi & Deseo 1983).

In Australia, large scale use of S. bibionis has developed to control a sesiid, Synanthedon tipuliformis, on currants. More than 99% of the borers in currant cuttings were killed, thus allowing the planting and establishment of borer-free plantations in isolated areas (Bedding & Miller 1981a, Miller & Bedding 1982). In infested areas, 90% population reduction of the borers has been achieved (Miller & Bedding 1982).

However, not all borer species can be controlled with the nematodes. The grape root borer, Vitacea polistiformis, a sesiid pest boring into grapes near or below the soil level, has not been successfully controlled with S. feltiae (All et al. 1981). The limiting factors are the soil acting as a barrier to prevent the nematode access to the gallery and a degree of host tolerance to the nematode. Another borer, Cossus cossus, has not been controlled by S. bibionis, S. feltiae, or H. bacteriophora (Deseo 1982).

Other agricultural insects in cryptic habitats are also susceptible to nematodes. The navel orangeworm, Amyelois transitella, a serious pest of almonds, spends part of its life cycle in the almond hull and nut. Application of S. feltiae at the time of hull split in one test resulted in up to 100% larval mortality; and in another test 55% reduction of the pest and 34% reduction in almond damage occurred (Lindegren et al. 1978). In trials against the codling moth, Cydia pomonella, on apple trees, Dutky (1959) obtained 60% mortality of prepupae in bark crevices. Ustimenko-Bakumovskaya & Ishevskij (1979) obtained 80-100% mortality of codling moth on branches during the fall and 40-75% in summer. Similarly, 90% mortality of overwintering codling moth was obtained in winter application compared to 32% mortality in summer applications of S. feltiae (Kaya et al. 1984).

Application of S. feltiae to corn ears to control corn earworm larvae, Heliothis zea, yielded high mortality but did not prevent damage (Tanada & Reiner 1962). In another test, nematodes applied in early June resulted in 88% mortality (Bong & Sikorowski 1983). Despite high larval mortality, economic damage to ears was not prevented. However, infectives survived inside the silk channels and application of the nematodes during the silking period might serve as a good prophylactic agent against this insect.

Application of S. feltiae to artichoke plume moth larvae infesting artichokes (flower buds) was not successful (Tanada & Reiner 1960) but applications during the vegetative growth phase was effective in controlling 100% of the older larvae and was more effective than the commonly used insecticide, methidathion (Bari & Kaya 1984). Nematodes were less effective against first instars which feed peripherally on the vegetative shoots where the drier microclimate would inhibit survival and mobility of the nematode. Older larvae tunnel and feed on the main stem and leafstalk within galleries, providing an excellent habitat for nematode infection. Insecticides are not effective when the older instars are protected within the plant. Furthermore, artichokes are grown in the cool coastal fogbelt of California, a perfect environment for the use of the nematodes.

These examples show that nematodes can provide good to excellent control of insects that occur in cryptic habitats. Although the environment is important, the behavior of the insect must also be taken into consideration.

Soil

Using nematodes against soil-inhabiting insects has been advocated because of the large number of potential hosts for control and of the excellent chances for nematode survival. Accordingly, a number of field tests have been conducted against these types of insects.

The first attempts to control insects with nematodes used *S. glaseri* against the Japanese beetle, *Popillia japonica*, in the 1930's. Initial results were encouraging with high beetle mortality (Glaser 1932). Subsequent applications were not as encouraging against this insect (Glaser & Farrell 1935, Glaser et al. 1940) or on grass grubs (Hoy 1955). One of the reasons for the ineffectiveness of these tests (1935 to 1955) may have been that the associated bacterium of the nematode was destroyed by the use of antimicrobial agents in the rearing media (Poinar 1979). A trial conducted by Kain et al. (1982) using nematodes cultured with the associated bacterium resulted in 66% reduction of the grass grubs.

Soil application with *S. feltiae* reduced cabbage maggot, *Delia brassicae*, damage but was not as effective as the chemical pesticide (Welch & Briand 1961b). However, Georgis et al. (1983) found that *H. bacteriophora* was more effective in controlling the cabbage maggot than *S. feltiae*. Against the seedcorn maggot, *D. platura*, attacking tobacco, *S. feltiae* was as effective as diazinon (Cheng & Bucher 1972).

The use of *S. feltiae* against lepidopterous pupae in soil has not been overly successful (Jaques et al. 1968, Lewis & Raun 1978). Under laboratory conditions, Kaya & Hara (1980, 1981) showed that soil pupating lepidopterous pupae were less susceptible than prepupae. However, adults emerging in soil infested with *S. feltiae* were infected by the nematode as they worked their way through the soil to the surface (Kaya & Grieve 1982).

S. feltiae has effectively reduced beetle larvae such as wireworms (Kovacs et al. 1980, Toba et al. 1983) and late 4th instars of the Colorado potato beetle (Toba et al. 1983, Veremchuk & Danilov 1976, Welch & Briand 1961a). Generally, soil application of this nematode caused higher mortality of the potato beetle than foliar applications and may suggest that an integrated approach of chemical or biological insecticide applied to foliage may be followed by nematode application to soil to reduce the residual population.

Tests against the pecan weevil, Curculio caryae, (Tedders et al. 1973) and rootworm larvae, Diabrotica sp., in corn (Poinar et al. 1983) using S. feltiae showed considerable promise. Against the pecan weevil, 67% larval mortality was recorded which was higher than recorded for three different fungal species. Against the rootworms, the nematode treatment was significantly more effective than the insecticide treatment.

Bedding & Miller (1981b) demonstrated that H. heliothidis was very effective (up to 100% mortality) against larvae of the black vine weevil, Otiorynchus sulcatus, a serious pest of nursery and agricultural crops. Simons (1981) also reported up to 100% mortality with an undescribed heterorhabditid in potted plants. Application of S. feltiae and S. glaseri against the strawberry root weevil, O. ovatus, infesting raspberry plantings provided 65% mortality (Georgis & Poinar 1984).

These examples demonstrate that nematodes have the potential to be important biological control agents against certain soil insects. The effectiveness of heterorhabditids against the black vine weevil suggests that these nematodes may have greater potential than steinernematids in soil. However, more research is needed with other nematode species and their various strains.

APPLICATION TECHNOLOGY

Nematodes are not harmed by standard insecticide spray equipment utilizing large spray nozzles. Important considerations in using this equipment are flushing the spray tank of harmful pesticide residues, keeping the nematodes in suspension, and maintaining temperatures below 32°C.

Steinernematids can withstand pressures up to 1000 psi (Dutky 1974). Accordingly, nematodes have been applied with small pressurized sprayers, mist blowers and helicopters (Lindegren et al. 1981a). In Australia, L.A. Miller (personal communication) has developed a sprayer for use with currants where the excess spray is collected and recycled into the spray tank. Delivery of nematodes with overhead sprinkler systems has been attempted with artichokes, but the loss of nematodes was too great to give effective control (M.A. Bari & H.K. Kaya, unpublished data).

Against plant-boring insects in commercial orchards, steinernematids have been delivered into galleries with a syringe (Deseo 1982, Lindegren et al. 1981b), cotton swab plug (Foschi & Deseo 1983), oil can, and back pack sprayer (Lindegren & Barnett 1982).

Soil treatment can be accomplished by using a drench, injecting nematodes into the soil (L.A. Miller, personal communication), or applying the nematodes with liquid fertilizer (Poinar et al. 1983). Drip irrigation systems offer a possible avenue for distribution of nematodes. Although not very practical, placement of nematode-containing cadavers has been tried (Welch & Briand 1961b).

RECOMMENDATIONS FOR RESEARCH

1. **Systematics** The systematics of steinernematids needs to be clarified. Separation of species according to morphological characters is difficult. Current species identification is primarily based on lengths of the infectives which overlap in some cases (Wouts et al. 1982). Employing electrophoretic techniques or DNA probes may reveal specific characters.
2. **Standardization** Differences in species and strain effectiveness have been apparent from several reports (Bedding & Miller 1981b, Bedding et al. 1983, Molyneux et al. 1983, Silverman et al. 1982). Moreover, past field and laboratory trials have used different rates, strains, test insects, and plants. Environmental conditions have also varied. Currently, one reason researchers cannot compare results among trials is that nematode use has not been standardized. To do so, the nematodes must be standardized so that their effectiveness can be compared.

   Standardization of nematodes can be accomplished using a laboratory bioassay. Essentially, the technique has already been developed by Molyneux et al. (1983). The only difference would be that a standard insect such as the last stage housefly (*Musca domestica*) larva would be incorporated into the bioassay. Accordingly, the mortality rates of the housefly due to various nematode species or strains can then be incorporated into International Housefly Units or some other convenient international unit of comparison. Standardization will provide a means of checking product quality over time, before shipping, after shipping, between productions, between methods of production, between laboratories, etc.

   The reasons for selecting the housefly are as follows: It is available worldwide, is easily reared in the laboratory, and is not overly susceptible to steinernematids and heterorhabditids. A highly susceptible insect species is not desired because it would not give the true measure of effectiveness.
3. **Safety** The safety of the bacterial complex to non-target organisms is a continual concern. Although the associated bacteria do not appear hazardous, they should be monitored. During mass production in vivo or in vitro, or in storage, potentially hazardous microorganisms might contaminate the culture. Adequate quality control for producing and storing the nematode-bacterial complex will be a necessary safeguard to avoid such events from occurring. Presently, there are no guidelines for checking commercial products for contamination with hazardous microorganisms.
4. **Environmental Studies** Although much is known concerning the ability of *S. feltiae* to cope in its physical environment, a void exists in our knowledge of its relationship to the biotic environment and to the interactions that occur between the physical and biotic environments. In essence,

much of the research has focussed on the effects of one or two environmental factors on the one nematode. This type of research has provided an important database from which more complex studies need to be conducted with *S. feltiae* as well as the other species.

5. **Efficacy** The commercialization of steinernematids has developed rapidly since EPA ruled that the nematode-bacterial complex is exempt from registration. However, efficacy data are lacking for many of the projected target insects. Efficacy data are essential to support advertising claims. The conditions under which the nematodes are or are not effective need to be delineated. This type of data will require extensive field testing because laboratory data do not necessarily predict field effectiveness. Continued failures or inconsistent results for undetermined reasons will be detrimental to the long-range use of the nematodes for insect control. It is hoped that scientists and commercial companies will take a responsible stance on this issue and not be overzealous to profit at the user's expense.
6. **Genetic Improvement of Nematodes** The selection or development of nematode strains that can undergo anhydrobiosis will be a great challenge. Apparently, steinernematids and heterorhabditids do not have this capability; however, selection for this trait has not been seriously attempted. The possibility of inserting the genes that confer this trait from one nematode species to another would be a long term project, but with the advances that are being made it is not an unrealistic goal.

   The search for more virulent and pathogenic strains of nematodes continues. Artificial selection of nematodes to enable them to persist in the environment for longer periods is also worth more effort. These traits probably exist naturally in the field and a method for screening them in the laboratory is needed. Selection of heterorhabditids that can be stored and shipped well also needs to be undertaken.
7. **Storage and Shipping** Storage and shipping procedures for steinernematids have been developed (Bedding 1984). However, storage and shipping of heterorhabditids require more research.

CONCLUSION

The future of entomogenous nematodes in agriculture is bright, especially for the use of steinernematids and heterorhabditids in cryptic and soil habitats. Progress has been substantial in recent years, but much research lies ahead before the full potential of nematodes in agriculture is realized.

ACKNOWLEDGEMENTS

I thank Phyllis G. Hotchkin and Tom Burlando for the discussions and critical review of this manuscript. Special thanks are extended to Jennifer L. Swan for her editorial comments.

REFERENCES

J.N., M.C. Saunders, J.D. Dutcher and A.M. Javid. 1981. Susceptibility of grape root borer larvae, Vitacea polistiformis (Lepidoptera: Sesiidae) to Neoaplectana carpocapsae (Nematoda: Rhabditida): Potential of host kairomones for enhancement of nematode activity in grape vineyards. Misc. Publ. Entomol. Soc. Amer. 12:9-14.

Bari, M.A., and H.K. Kaya. 1984. Evaluation of the entomogenous nematode Neoaplectana carpocapsae (= Steinernema feltiae) Weiser (Rhabditida: Steinernematidae) and the bacterium Bacillus thuringiensis Berliner var. kurstaki for suppression of the artichoke plume moth (Lepidoptera: Pterophoridae). J. Econ. Entomol. 77:225-229.

Bedding, R.A. 1979. Manipulating the entomogenous-mycetophagus nematode, Deladenus siricidicola, for biological control of the woodwasp Sirex noctilio in Australia, pp. 144-147. In: W. E. Waters (Ed.), Current topics in forest entomology. USDA Gen. Tech. Rept. WO-8.

Bedding, R.A. 1981. Low cost in vitro mass production of Neoaplectana and Heterorhabditis species (Nematoda) for field control of insect pests. Nematologica 27:109-114.

Bedding, R.A. 1984. Large scale production, storage and transport of the insect-parasitic nematodes Neoaplectana spp. and Heterorhabditis. Ann. Appl. Biol. 101:117-120.

Bedding, R.A., and R.J. Akhurst. 1975. A simple technique for the detection of insect parasitic rhabditid nematodes in soil. Nematologica 21:109-110.

Bedding, R.A., and L.A. Miller. 1981a. Disinfecting blackcurrant cuttings of Synanthedon tipuliformis, using the insect parasitic nematode, Neoaplectana bibionis. Environ. Entomol. 10:449-453.

Bedding, R.A., and L.A. Miller. 1981b. Use of a nematode, Heterorhabditis heliothidis, to control black vine weevil, Otiorynchus sulcatus, in potted plants. Ann. Appl. Biol. 99:211-216.

Bedding, R.A., and A.S. Molyneux. 1982. Penetration of insect cuticle by infective juveniles of Heterorhabditis spp. (Heterorhabditidae: Nematoda). Nematologica 28:354-359.

Bedding, R.A., A.S. Molyneux and R.J. Akhurst. 1983. Heterorhabditis spp., Neoaplectana spp., and Steinernema kraussei: Interspecific and intraspecific differences in the infectivity for insects. Exp. Parasitol. 55:249-257.

Bong, C.F.J., and P.P. Sikorowski. 1983. Use of the DD 136 strain of Neoaplectana carpocapsae Weiser (Rhabditida: Steinernematidae) for control of corn earworm (Lepidoptera: Noctuidae). J. Econ. Entomol. 76:590-593.

Buchanan, R.E., and N.E. Gibbons (Eds.). 1974. Bergey's manual of determinative bacteriology. 8th ed. Williams & Wilkins, Baltimore.

Burman, M., and A.E. Pye. 1980a. Neoaplectana carpocapsae: Movements of nematode populations on a thermal gradient. Exp. Parasitol. 49:258-265.

Burman, M., and A.E. Pye. 1980b. Neoaplectana carpocapsae: Respiration of infective juveniles. Nematologica 26:214-219.

Chamberlin, H.C., and S.R. Dutky. 1958. Tests of pathogens for the control of tobacco insects. J. Econ. Entomol. 51:560.

Cheng, H.H., and G.E. Bucher. 1972. Field comparison of the neoaplectanid nematode DD 136 with diazinon for control of Hylemya spp. on tobacco. J. Econ. Entomol. 65:1761-1763.

Crowe, J.H., and K.A.C. Madin. 1975. Anhydrobiosis in nematodes: Evaporative water loss and survival. J. Exp. Zool. 193:323-334.

Deseo, K.V. 1982. Prove di lotta col nematoda entomopatogeno (Neoaplectana carpocapsae Weiser) contro i rodilegno Cossus cossus L. e Zeuzera pyrina L. (Lepidoptera: Cossidae). Atti Gior. Fitopathol. 3:3-10.

Dutky, S.R. 1959. Insect microbiology. Adv. Appl. Microbiol. 1:175-200.

Dutky, S.R. 1974. Nematode parasites, pp. 576-590. In: F.G. Maxwell and F.A. Harris (Eds.), Proc. summer institute on biological control of plant insects and diseases. University Press Mississippi, Jackson.

Dutky, S.R., J.V. Thompson and G.E. Cantwell. 1964. A technique for the mass propagation of the DD 136 nematode. J. Insect Pathol. 6:417-422.

Fedorko, A., M. Kamionek, J. Kozlowska and E. Mianowska. 1977a. The effects of some carbamide herbicides on nematodes from different ecological groups. Pol. Ecol. Stud. 3:23-28.

Fedorko, A., M. Kamionek, J. Kozlowska and E. Mianowska. 1977b. The effects of some insecticides on nematodes of different ecological groups. Pol. Ecol. Stud. 3:79-88.

Fedorko, A., M. Kamionek, J. Kozlowska and E. Mianowska. 1977c. The effect of vydate-oxamyl on nematodes of different ecological groups. Pol. Ecol. Stud. 3:89-93.

Finney, J.R. 1981. Potential of nematodes for pest control, pp. 603-620. In: H.D. Burges (Ed.), Microbial control of pests and plant diseases 1970-1980. Academic Press, N.Y.

Foschi, S., and K. V. Deseo. 1983. Risulti di lotta con nematodi entomopatogeni su Zeuzera pyrina L. (Lepidopt.; Cossidae) nel 1982. La difesa delle piante 6:153-156.

Fox, C.J.S., and R.P. Jaques. 1966. Preliminary observations on biological insecticides against imported cabbageworm. Can. J. Sci. 46:497-499.

Gaugler, R. 1981. The biological control potential of neoaplectanid nematodes. J. Nematol. 13:241-249.

Gaugler, R., and G.M. Boush. 1978. Effects of ultraviolet radiation and sunlight on the entomogenous nematode, Neoaplectana carpocapsae. J. Invertebr. Pathol. 32:291-296.

Gaugler, R., and G.M. Boush. 1979. Laboratory tests on ultraviolet protectants of an entomogenous nematode. Environ. Entomol. 8:810-813.

Gaugler, R., and H.K. Kaya. 1983. Bibliography on the entomogenous nematode family Steinernematidae. Bibliogr. Entomol. Soc. Amer. 1:43-64.

Gaugler, R., and D. Molloy. 1981. Field evaluation of the entomogenous nematode, Neoaplectana carpocapsae, as a biological control agent of black flies (Diptera: Simuliidae). Mosq. News 41:459-464.

Georgis, R., and G.O. Poinar. 1983a. Effect of soil texture on the distribution and infectivity of Neoaplectana carpocapsae (Nematoda: Steinernematidae). J. Nematol. 15:308-311.

Georgis, R., and G.O. Poinar. 1983b. Effect of soil texture of the distribution and infectivity of Neoaplectana glaseri (Nematoda: Steinernematidae). J. Nematol. 15:329-332.

Georgis, R., and G.O. Poinar. 1983c. Vertical migration of Heterorhabditis bacteriophora and H. heliothidis (Nematoda: Heterorhabditidae) in sandy loam soil. J. Nematol. 15:652-654.

Georgis, R., and G.O. Poinar. 1984. Field control of the strawberry root weevil, Nemocestes incomptus, by neoaplectanid nematodes (Steinernematidae: Nematoda). J. Invertebr. Pathol. 43:130-131.

Georgis, R., G.O. Poinar and A.P. Wilson. 1983. Practical control of the cabbage root maggot, Hylemia brassicae, (Diptera: Anthyomyiidae) by entomogenous nematodes. IRCS Med. Sci. Microbiol. Parasitol. Infect. Dis. 11:322.

Glaser, R.W. 1932. Studies on Neoaplectana glaseri, a parasite of the Japanese beetle (Popillia japonica). N.J. Dept. Agr. Circ. 211, 34 pp.

Glaser, R.W., and C.C. Farrell. 1935. Field experiments with the Japanese beetle and its nematode parasite. J. N.Y. Entomol. Soc. 43:345-371.

Glaser, R.W., E.E. McCoy and H.B. Girth. 1940. The biology and economic importance of a nematode parasite in insects. J. Parasitol. 26:479-495.

Gorsuch, A.M. 1982. Regulations for the enforcement of the Federal Insecticide, Fungicide, and Rodenticide Act exemption from regulation of certain biological control agents. Fed. Register 47:23928-23930.

Hara A.H., and H.K. Kaya. 1982. Effects of selected insecticides and nematicides on the in vitro development of the entomogenous nematode Neoaplectana carpocapsae. J. Nematol. 14:486-491.

Hara, A.H., and H.K. Kaya. 1983a. Toxicity of selected organophosphate and carbamate pesticides to infective juveniles of the entomogenous nematode Neoaplectana carpocapsae (Rhabditida: Steinernematidae). Environ. Entomol. 12: 496-501.

Hara, A.H., and H.K. Kaya. 1983b. Development of the entomogenous nematode, Neoaplectana carpocapsae (Rhabditida: Steinernematidae), in insecticide-killed beet armyworm (Lepidoptera: Noctuidae). J. Econ. Entomol. 76:423-426.

Hoy, J.M. 1955. The use of bacteria and nematodes to control insects. N.Z. Sci. Rev. 13:56-58.

Jaques, R.P. 1967. Mortality of five apple insects induced by the nematode DD 136. J. Econ. Entomol. 60:741-743.

Jaques, R.P., H.T. Stultz and F. Huston. 1968. The mortality of the pale apple leafroller and winter moth by fungi and nematodes applied to soil. Can. Entomol. 100:813-818.

Kain, W.M., R.A. Bedding and C.J. van der Mespel. 1982. Preliminary evaluations of parasitic nematodes for grass grub (Costelytra zealandica White) control in central Hawke's Bay of New Zealand. N.Z. J. Exp. Agric. 10: 447-450.

Kamionek, M., I. Maslana and H. Sandner. 1974a. The survival of invasive larvae of Neoaplectana carpocapsae Weiser in a waterless environment under various conditions of temperature and humidity. Zesz. Probl. Post. Nauk Roln. 154: 409-412.

Kamionek, M., H. Sandner and H. Seryczynska. 1974b. The combined action of Beauveria bassiana (Bals./Vuill) (Fungi Imperfecti: Moniliales) and Neoaplectana carpocapsae Weiser (Nematoda: Steinernematidae). Bull. Acad. Pol. Sci. 22: 253-257.

Kamionek, M., H. Sandner and H. Seryczynska. 1974c. Combined action of Paecilomyces farinosus Dicks (Brown et Smith) (Fungi: Imp.: Moniliales) and Neoaplectana carpocapsae Weiser, 1955 (Nematoda: Steinernematidae) on certain insects. Acta Parisitol. Pol. 22:357-363.

Kaya, H.K. 1977. Development of the DD-136 strain of Neoaplectana carpocapsae at constant temperatures. J. Nematol. 9:346-349.

Kaya, H.K. 1978a. Infectivity of Neoaplectana carpocapsae and Heterorhabditis heliothidis to pupae of the parasite Apanteles militaris. J. Nematol. 10:241-244.

Kaya, H.K. 1978b. Interaction between Neoaplectana carpocapsae (Nematoda: Steinernematidae) and Apanteles militaris (Hymenoptera: Braconidae), a parasitoid of the armyworm, Pseudaletia unipuncta. J. Invertebr. Pathol. 31:358-364.

Kaya, H.K. 1980. Granulosis virus in the intestinal lumen of Neoaplectana carpocapsae: Retention of infectivity after treatment with formaldehyde or high pH. J. Invertebr. Pathol. 35:20-25.

Kaya, H.K. 1984. Effect of the entomogenous nematode Neoaplectana carpocapsae on the tachinid parasite Compsilura concinnata (Diptera: Tachinidae). J. Nematol. 16:9-13.

Kaya, H.K., and N.A. Brayton. 1978. Interaction between Neoaplectana carpocapsae and a granulosis virus of the armyworm Pseudaletia unipuncta. J. Nematol. 10:350-354.

Kaya, H.K., and B.J. Grieve. 1982. The nematode Neoaplectana carpocapsae and the beet armyworm Spodoptera exigua: Infectivity of prepupae and pupae in soil and of adults during emergence from soil. J. Invertebr. Pathol. 39:192-197.

Kaya, H.K., and A.H. Hara. 1980. Differential susceptibility of lepidopterous pupae to infection by the nematode Neoaplectana carpocapsae. J. Invertebr. Pathol. 36:389-393.

Kaya, H.K., and A.H. Hara. 1981. Susceptibility of various species of lepidopterous pupae to the entomogenous nematode Neoaplectana carpocapsae. J. Nematol. 13:291-294.

Kaya, H.K., and P.G. Hotchkin. 1981. The nematode Neoaplectana carpocapsae Weiser and its effect on selected ichneumonid and braconid parasites. Environ. Entomol. 10:474-478.

Kaya, H.K., J.L. Joos, L.A. Falcon and A. Berlowitz. 1984. Supression of the codling moth (Lepidoptera: Olethreutidae) with the entomogenous nematode, Steinernema feltiae. J. Econ. Entomol. 77:(In Press).

Kaya, H.K., and R.C. Reardon. 1982. Evaluation of Neoaplectana carpocapsae for biological control of the western spruce budworm, Choristoneura occidentalis: Ineffectiveness and persistence of tank mixes. J. Nematol. 14:595-597.

Khan, A., W.M. Brooks and H. Hirschmann. 1976. Chromonema heliothidis n.gen., n.sp. (Steinernematidae: Nematoda), a parasite of Heliothis zea (Noctuidae: Lepidoptera) and other insects. J. Nematol. 8:159-168.

Kovacs, A., K.V. Deseo, G. Poinar and A. de Leonardis. 1980. Prove di lotta contro insetti con applicazione di nematoda entomogeni. Atti Gior. Fitopathol. 1:449-456.

Landazabal, J., F. Fernandez and P.A. Figueroa. 1973. Control biologico de Spodoptera frugiperda J.E. Smith, con el nematodo Neoaplectana carpocapsae en maiz (Zea mays). Acta Agron. Palmira 23:41-70.

Lewis, L.C., and E.S. Raun. 1978. Laboratory and field evaluation of the DD-136 strain of Neoaplectana carpocapsae for control of the European corn borer, Ostrinia nubilalis. Iowa State J. Res. 52:391-396.

Lindegren, J.E., and W.W. Barnett. 1982. Applying parasitic nematodes to control carpenterworms. Calif. Agric. 36(11/12):7-8.

Lindegren, J.E., C.E. Curtis and G.O. Poinar. 1978. Parasitic nematode seeks out naval orangeworm in almond orchards. Calif. Agric. 32(6):10-11.

Lindegren, J.E., J.E. Dibble, C.E. Curtis, T.T. Yamashita and E. Romero. 1981a. Compatibility of NOW parasite with commercial sprayers. Calif. Agric. 35(3/4): 16-17.

Lindegren, J.E., T.T. Yamashita and W.W. Barnett. 1981b. Parasitic nematode may control carpenterworm in fig trees. Calif. Agric. 35(1/2):25-26.

MacVean, C.M., J.M. Brewer and J.L. Capinera. 1982. Field tests of antidesiccants to extend the infection period of an entomogenous nematode, Neoaplectana carpocapsae, against the Colorado potato beetle. J. Econ. Entomol. 75:97-101.

Miller, L.A., and R.A. Bedding. 1982. Field testing of the insect parasitic nematode, Neoaplectana bibionis (Nematoda: Steinernematidae) against the currant borer moth, Synanthedon tipuliformis (Lep.: Sesiidae) in black currants. Entomophaga 27:109-114.

Milstead, J.E. 1981. Influence of temperature and dosage on mortality of seventh instar larvae of Galleria mellonella (Insecta: Lepidoptera) caused by Heterorhabditis bacteriophora (Nematoda: Rhabditoidea) and its bacterial associate Xenorhabdus luminescens. Nematologica 27:167-171.

Molyneux, A.S., R.A. Bedding and R.J. Akhurst. 1983. Susceptibility of larvae of the sheep blowfly Lucilia cuprina to various Heterorhabditis spp., Neoaplectana spp., and an undescribed steinernematid (Nematoda). J. Invertebr. Pathol. 42: 1-7.

Moore, G.E. 1965. The bionomics of an insect-parasitic nematode. J. Kans. Entomol. Soc. 38:101-105.

Moyle, P.L., and H.K. Kaya. 1981. Dispersal and infectivity of the entomogenous nematode, Neoaplectana carpocapsae Weiser (Rhabditida: Steinernematidae), in sand. J. Nematol. 13:295-300.

Mracek, Z., and K. Spitzer. 1983. Interaction of the predators and parasitoids of the sawfly, Cephalica abietis (Pamphiliidae: Hymenoptera) with its nematode Steinernema kraussei. J. Invertebr. Pathol. 42:397-399.

Nash, R.F., and R.C. Fox. 1969. Field control of the Nantucket pine tip moth by the nematode DD-136. J. Econ. Entomol. 62:660-663.

Nickle, W.R. 1981. Mermithid parasites of agricultural pest insects. J. Nematol. 13:262-266.

Nickle, W.R. (Ed.). 1984. Plant and insect nematodes. Marcel Dekker, N.Y. 925 pp.

Obendorf, D.L., B. Peel, R.J. Akhurst and L.A. Miller. 1983. Non-susceptibility on mammals to the entomopathogenic bacterium, Xenorhabdus nematophilus. Environ. Entomol. 12:368-370.

Petersen, J.J. 1982. Current status of nematodes for the biological control of insects. Parasitology 84:177-204.

Platzer, E.G. 1981. Biological control of mosquitoes with mermithids. J. Nematol. 13:257-262.

Poinar, G.O. 1975. Entomogenous nematodes. E.J. Brill, Leiden, Netherlands. 317 pp.

Poinar, G.O. 1979. Nematodes for biological control of insects. CRC Press, Boca Raton, FL. 277 pp.

Poinar, G.O., J.S. Evans and E. Schuster. 1983. Field test of the entomogenous nematode, Neoaplectana carpocapsae, for control of corn rootworm larvae (Diabrotica sp., Coleoptera). Protection Ecol. 5:337-342.

Poinar, G.O., R.T. Hess and A. Cole. 1980. Replication of an iridovirus in a nematode (Mermithidae). Intervirology 14:316-320.

Poinar, G.O., and R. Leutenegger. 1968. Anatomy of the infective and normal third-stage juveniles of Neoaplectana carpocapsae Weiser (Steinernematidae: Nematoda). J. Parasitol. 54:340-350.

Prakasa Rao, P.S., P.K. Das and G. Padhi. 1975. Note on compatibility of DD-136 (Neoaplectana dutkyi), an insect parasitic nematode with some insecticides and fertilisers. Indian J. Agric. Sci. 45:275-277.

Pye, A.E., and M. Burman. 1978. Neoaplectana carpocapsae: Infection and reproduction in large pine weevil larvae, Hylobuis abietis. Exp. Parasitol. 46:1-11.

Pye, A.E., and M. Burman. 1981. Neoaplectana carpocapsae: Nematode accumulations on chemical and bacterial gradients. Exp. Parasitol. 51:13-20.

Schmiege, D.C. 1963. The feasibility of using a neoaplectanid nematode for control of some forest insect pests. J. Econ. Entomol. 56:427-431.

Seryczynska, H. 1975. Defense reactions of the Colorado potato beetle (Leptinotarsa decemlineata Say) against the fungi Beauveria bassiana (Bals/ Vuill), Paecilomyces farinosus (Dicks/Brown et Smith) and nematodes Neoaplectana carpocapsae. Bull. Acad. Pol. Sci. 23:351-353.

Shephard, M.R.N. 1974. Arthropods as final hosts of nematodes and nematomorphs. An annotated bibliography 1900-1972. Farnam Royal: Commonwealth Agric. Bur., Tech. Comm. 45, 248 pp.

Silverman, J., E.G. Platzer and M.K. Rust. 1982. Infection of the rat flea Ctenocephalides felis (Bouche) by Neoaplectana carpocapsae Weiser. J. Nematol. 14: 394-397.

Simons, W.R. 1981. Biological control of Otiorhynchus sulcatus with heterorhabditid nematodes in the glasshouse. Neth. J. Pl. Path. 87:149-158.

Simons, W.R., and G.O. Poinar. 1973. The ability of Neoaplectana carpocapsae (Steinernematidae: Nematodea) to survive extended periods of desiccation. J. Invertebr. Pathol. 22:228-230.

Tanada, Y., and C. Reiner. 1960. Microbial control of the artichoke plume moth, Platyptilia carduidactyla (Riley) (Pterophoridae, Lepidoptera). J. Insect Pathol. 2:230-246.

Tanada, Y., and C. Reiner. 1962. The use of pathogens in the control of the corn earworm, Heliothis zea (Boddie). J. Insect Pathol. 4:139-154.

Tedders, W.L., D.J. Weaver and E.J. Wehunt. 1973. Pecan weevil: Suppression of larvae with the fungi Metarrhizium anisopliae and Beauveria bassiana and the nematode Neoaplectana dutkyi. J. Econ. Entomol. 66:723-725.

Toba, H.H., J.E. Lindegren, J.E. Turner and P.V. Vail. 1983. Susceptibility of the Colorado potato beetle and the sugarbeet wireworm to Steinernema feltiae and S. glaseri. J. Nematol. 15:597-601.

Ustimenko-Bakumovskaya, E.R., and S.S. Ishevskij. 1979. Pathogenicity of Neoaplectana carpocapsae Weiser - agriotos Veremchuk for orchard pests, pp. 217-218. In: Proc. Intern. Coll. Invertebr. Pathol. September 11-17, 1978, Prague.

Veremchuk, G.V., and I.V. Issi. 1970. On the development of a microsporidian of insects in the entomopathogenic nematode Neoaplectana agriotos (Nematoda: Steinernematidae). Parazitologiya 4:3-7.

Veremchuk, G.V., and L.G. Danilov. 1976. Entomopathogenic nematodes. Zash. Rast. 8:22.

Webster, J.M. 1980. Biocontrol: The potential of entomophilic nematodes in insect management. J. Nematol. 12:270-278.

Webster, J.M., and J.F. Bronskill. 1968. Use of Gelgard M and an evaporation retardant to facilitate control of larch sawfly by a nematode-bacterium complex. J. Econ. Entomol. 61:1370-1373.

Welch, H.E. 1958. Test of a nematode and its associated bacterium for the control of the Colorado potato beetle *Leptinotarsa decemlineata* (Say). Ann. Rep. Entomol. Soc. Ontario 88:53-54.

Welch, H.E. 1971. Various target species: Attempts with DD-136. Biological control programmes against insects and weeds in Canada, 1959-1968. Commonw. Inst. Biol. Control Tech. Comm. 4:62-66.

Welch, H.E., and L.J. Briand. 1961a. Tests of the nematode DD 136 and an associated bacterium for control of the Colorado potato beetle, *Leptinotarsa decemlineata* (Say). Can. Entomol. 93:759-763.

Welch, H.E., and L.J. Briand. 1961b. Field experiment on the use of a nematode for the control of vegetable crop insects. Proc. Entomol. Soc. Ontario 91: 197-202.

Wouts, W.M., Z. Mracek, S. Gerdin and R.A. Bedding. 1982. *Neoaplectana* Steiner, 1929 a junior synonym of *Steinernema* Travassos, 1927 (Nematoda: Rhabditida). Syst. Parisitol. 4:147-154.

DISCUSSION: Entomogenous nematodes for insect control in IPM systems.

DISCUSSION LEADER: G.C. Smart

Limits to the use of entomogenous nematodes for insect control in IMP systems involve concerns about quality control, compatibility with all crop management practices, and choice of nematode species or strain. At present, quality control is self-regulated by those companies producing and selling nematodes. Improved storage and production methods are desirable if application methods are to become simpler. Nematodes appear to be compatible with chlorinated hydrocarbon and organophosphate insecticides, but additional work should be conducted to determine whether pesticide selectivity can be exploited in IPM systems. It seems likely that additional species and strains can be discovered that are more effective in specific situations; genetic selection or genetic engineering techniques might produce improved strains.

ESTIMATING THE ABUNDANCE AND IMPACT OF ARTHROPOD NATURAL ENEMIES IN IPM SYSTEMS

L.T. Wilson

Few individuals involved with agricultural production question the economic importance of natural enemies in their role of helping to regulate the abundance of herbivorous arthropods. Fewer can provide quantitative estimates of their abundance in and short-term impact on crop systems. Most sampling procedures for estimating the abundance of natural enemies were developed by researchers to address research problems and, as such, are often inappropriate for use in commercial management of IPM systems. Farmers and scouts seldom have the resources, either time or equipment, necessary to use most of the present sampling procedures. Very few of the available procedures were developed with the goal of obtaining the most reliable estimates for a given cost. As our knowledge of cropping systems expands, it becomes increasingly important that appropriate sampling procedures for herbivores and their natural enemies be developed and presented in an easily usable form.

Despite the lack of guidelines for sampling approaches for natural enemies, there are many prospective methods from which to choose. Southwood (1978) presents a thorough review of a wide range of sampling procedures. Kogan & Herzog (1980) likewise present the state of the art in sampling procedures for assessing the abundance and impact of soybean arthropods. Although most of the information presented in these and other sources is based on research conducted on pests of certain crops, many are suitable for other pests and their natural enemies. However, many suffer some of the failings mentioned above.

The objective of this review is to focus not on particular sampling procedures, but instead to address the underlying components which must be understood as a prerequisite for developing sampling procedures. The components to be discussed include: 1) spatial distribution (or dispersion) patterns, 2) sample size estimation, 3) defining the sample unit, and 4) sampling frequency and forecasting. If an individual has a strong grasp of these general areas, then a sampling program that will be appropriate for the pest or natural enemy species concerned can be more easily developed. The section on estimating the impact of

ISBN 0-12-357030-1

natural enemies, although not as quantitative, will hopefully provide some insights. I have attempted to cite relevant references; however, the wealth of information on specific aspects of sampling precludes an inclusive review.

CONCEPT OF SPATIAL DISTRIBUTION PATTERNS

A key component in developing sampling procedures for natural enemies and pest species in crop systems is an understanding of their distribution patterns within a crop system. A species distribution pattern can be expressed in terms of a variance: mean ratio ($s^2/\bar{x}$) and can be categorized as either clumped (aggregated, contagious, or underdispersed), characterized by $s^2/\bar{x} > 1$; random (Poisson distribution), characterized by $s^2/\bar{x} = 1$; or uniform (regular or overdispersed), where $s^2/\bar{x} < 1$. For the wide range of sample units used in agricultural research and commercial management of agricultural systems, most species have clumped distribution patterns, with random and uniform distributions being extremely rare. Taylor et al. (1978) in examining the distribution patterns of 102 species found only two cases of a species having a distribution pattern which could not be distinguished from random at all densities tested, and two with a uniform distribution pattern. It should be kept in mind that a species distribution pattern is biologically governed and in addition affected by sample unit size. As the sample unit size is progressively reduced or expanded in comparison to the "biologically relevant" sample unit, the apparent distribution pattern will in general tend increasingly towards the random (Morisita 1959). As an example, the clumped distribution pattern of predaceous mites on almond leaves could easily be overlooked were the sampler to pool leaves in creating a sample unit.

Variance-Mean Relationships

To estimate a species distribution pattern requires that a number of sample units be examined and the number of individuals per unit be counted. The sample variance and mean are then estimated for each sample using standard statistics. If repeated samples are taken through time or over a range of population densities, a relationship between sample variance and mean which is specific for the particular species or age class will become apparent.

Fracker & Brischle (1944), Hayman & Lowe (1961), and Taylor (1961) each independently developed an equation which describes this relationship.

$$s^2 = a \cdot x^b \quad (1)$$

where $\underline{a}$ and $\underline{b}$ are species- and sample unit-specific coefficients which together describe, and can be used to classify, a species distribution pattern. Taylor's power law, as the above equation is now called, has proven extremely useful in determining appropriate sample sizes for use in estimating the abundance of various species.

Taylor's a and b coefficients are often estimated by a log-log transformation of the variance-mean data, where b is the slope of the transformed data, and a equals the anti-log of the transformed intercept (Taylor 1961). Miller (1971) indicated that this procedure can give biased estimates of a and b. This problem can be overcome by using iterative regression techniques (either weighted or unweighted) which produce a and b techniques directly from the variance-mean data. Some algorithms are, however, subject to iteration on local maximums which can result in a bias which exceeds that observed with the log-log procedure.

Taylor et al. (1978) consider the coefficient b of Taylor's power law to be constant for a species, with a affected by sample unit size. Banerjee (1976), however, disputes this hypothesis, and results from Wilson et al. (1983a, 1984b) and Wilson & Room (1982, 1983) indicate that both a and b, and correspondingly a species distribution pattern, are subject to change due to age-specific dispersal and mortality and sample unit size. Using Equation (1), it is easy to demonstrate that a high b value ($b>1$) would not necessarily imply a clumped distribution pattern, since a low a value ($a \ll 1$) could result in a uniform distribution pattern for a wide range of densities. A species can progress from having a uniform distribution pattern to a random pattern then a clumped pattern as density increases. Likewise a species distribution pattern may be highly clumped for smaller sample units, and random as the size of the unit increases. In general, a values for most natural enemies using most sampling procedures range from 1.0-2.5, while b values range from 1.0-1.2. The coefficients for herbivores have a much higher range reflecting their greater propensity towards clumping.

Proportion Infested--Mean Density Relationship

Although it is commonly accepted that the $s^2/\bar{x}$ relationship is useful for categorizing a species distribution pattern, it is less commonly known that the relationship between the proportion of sample units infested with a species and its density can be used for the same purpose.

When a species is randomly distributed, the proportion of infested sample units, p, is estimated by Equation (2).

$$p = 1 - e^{-\bar{x}} \qquad (2)$$

When a species with a clumped distribution pattern is compared with one which is randomly distributed but at equal densities, p will be smaller for the species having the clumped distribution pattern. The reverse occurs comparing a uniformly distributed species with one which is randomly distributed (Fig. 1). Wilson & Room (1983), with the objective of developing a field usable

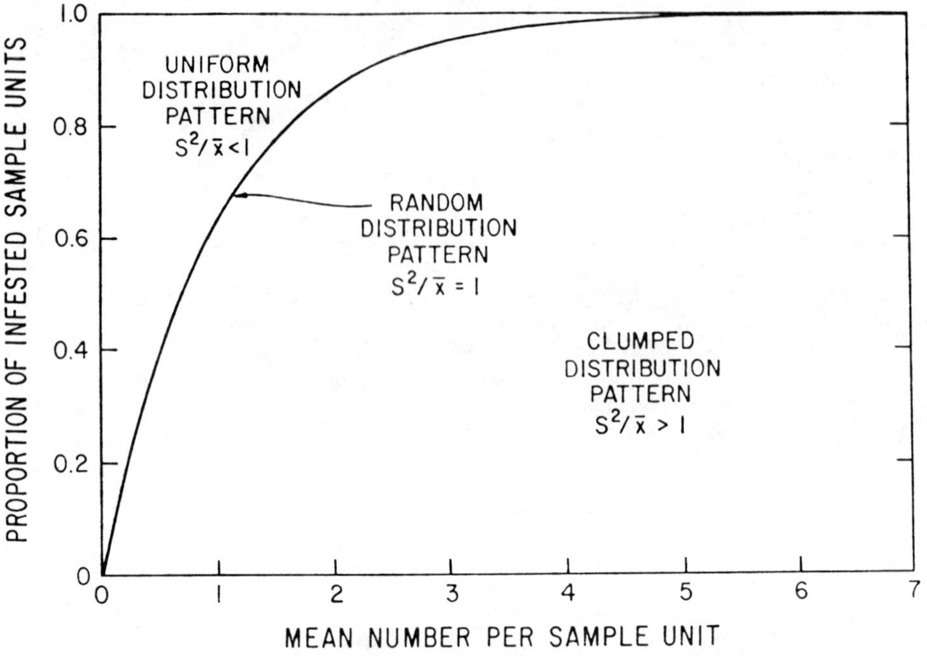

Fig. 1. The impact of species distribution pattern on the mean ($\bar{x}$) - proportion infested (p) relationship.

sampling procedure, derived the following equations describing the relationship between the proportion of infested sample units and the variance and mean per sample unit.

$$p = 1 - e^{-\bar{x} \log_e (s^2/\bar{x})/(s^2/\bar{x}-1)} \quad (3)$$

When Taylor's power law is substituted for the variance,

$$p = 1 - e^{-\bar{x} \log_e (a\bar{x}^{b-1})/(a\bar{x}^{b-1}-1)} \quad (4)$$

For Equations (3) and (4), as $s^2/\bar{x}$ and correspondingly $a\bar{x}^{b-1}$ approach unity, as a limit both equations collapse to Equation (2), the non-zero terms of the random or Poisson distribution function.

If estimates of Taylor's coefficients are available for a species, then a p-$\bar{x}$ curve can be estimated by simply solving Equation (4) for a range of densities. In practical terms, this means that once baseline data relating p to $\bar{x}$ have been obtained, then a quick and reliable estimate of $\bar{x}$ can be obtained in future samples by recording only the proportion of infested sample units. A shortcoming of the presence-absence sampling method is apparent when p = 1. Once p reaches unity, all the sample units contain one or more organisms and it is impossible to estimate density. This limitation is not serious since neither pests nor natural enemies reach such levels in most agricultural systems and generally management recommendations are reached prior to such levels.

SAMPLE SIZE ESTIMATION

Research Sampling (Parameter Estimation)

Ruesink & Kogan (1982) distinguish between data used for research purposes, where emphasis is placed on the reliability of a parameter estimate, and data used for commercial management of a crop system, where the farmer or scout is primarily interested in classifying whether a pest population has reached a level requiring control or whether the pest is likely to reach damaging levels with existing natural enemy levels. When a researcher is deriving a parameter estimate, a considerable investment in sampling time may be required. The greater the reliability necessary for the estimate, the larger must be the number of samples. Too often the number of samples taken is not related to the number required and the reliability of the estimates is often too low or too high. This equates to poor data at one extreme and unnecessary costs at the other.

Karandinos (1976) presents a series of equations for use in estimating sample sizes. Ruesink (1980) and Wilson & Room (1982) independently incorporated Taylor's power law into Karandinos' sample size equation. Equation (5) presents the generalized form of the enumerative sample size equation, which can be used for organisms having distribution patterns ranging from highly clumped to uniform.

$$n = t^2_{\alpha/2} \quad D^{-2} \cdot a \cdot \bar{x}^{b-2} \tag{5}$$

where:

$t^2_{\alpha/2}$ = standard normal variate for a two tailed confidence interval.

D = a fixed proportion of the mean and is used to define 1/2 of the confidence interval (1/2 C.I. = $D\bar{x}$).

a and b are Taylor's coefficients.

Equation (6), from Karandinos (1976), is similar to Equation (5), but is used for presence-absence (binomial) sampling.

$$n = t^2_{\alpha/2} \cdot D^{-2} \cdot p^{-1}q \tag{6}$$

where:

D = a fixed proportion of p and is used to define 1/2 of the confidence interval (1/2 C.I. = Dp).
p = the proportion of infested sample units.
q = 1-p.

A general feature of Equation (5) is that species whose distribution patterns are more clumped require a greater sample size for a given level of reliability. This is due to a relatively greater $s^2/\bar{x}$ ratio (generally larger a and b coefficients). Also, fewer samples are required at higher densities for a given level of reliability. Equation (6) for presence-absence sampling does not make use of Taylor's power law, and unlike enumerative sampling, sample size first increases then decreases as density and p increase. This is due to a small confidence interval about p corresponding to a very large confidence interval about $\bar{x}$, at p values near one.

Commercial Monitoring (Decision Estimation)

When a field is monitored to determine a pest's economic status, or to determine whether a natural enemy species is at a level capable of supressing a pest's population density, the further the mean or the proportion of infested units from the action threshold, the smaller the sample size required to estimate whether the population lies above, or conversely, below the threshold. As with any estimation process, there are errors associated with making management decisions. The two types of errors in decision-making are: 1) the probability of concluding that the population level is above the threshold when the reverse is true (α error) and 2) the probability of concluding that the population level is below the threshold when the reverse is true (β error). An acceptable level of error is, in theory, determined for α by minimizing the total cost involved with monitoring and the costs involved with spraying when not required (labor, machinery, pesticides, resurgence, secondary outbreaks) and for β by minimizing the total cost involved with monitoring and the cost of pest damage due to not spraying when justified (Wilson 1982). Accurate evaluation of the potential for natural enemies to regulate the rate of population increase of pest species is important when determining economically appropriate and error rates. In the case of α errors (that is, spraying when not justified) secondary outbreaks of normally innocuous herbivores and resurgence of primary pests can largely be attributed to disruption of natural enemies and subsequent uncontrolled growth of the pest population. A quantitative understanding of the impact of natural enemies of pest species would enable a more reliable field evaluation of the potential for pests to cause damage.

Equations (7) and (8), which incorporate the and error rates were derived from the central limit theorem (L.T. Wilson, unpublished data; Wilson et al. 1983b).

$$n = t^2_{\alpha \text{ or } \beta} \cdot (\bar{x}-T_i)^{-2} \cdot a \cdot \bar{x}^b \quad (7)$$

$$n = t^2_{\alpha \text{ or } \beta} \cdot (p-T_i)^{-2} \cdot p \cdot q \quad (8)$$

where:

T_i = the threshold (action or economic) for the ith stage of crop growth respectively expressed as a mean density per unit for enumerative sampling [Equation (7)] or as a proportion of infested units for presence-absence sampling [Equation (8)].

In the context of sequential sampling, these equations, or other sequential sampling equations (Wald 1947), can be used to generate upper and lower control decision lines for a range of arthropods in crop systems (Allen et al. 1972, Hearn et al. 1981, Sterling 1975, Waters 1955). However, Equations (7) and (8), as Wilson (1982) indicates, do not suffer from an arbitrary second threshold inherent with the sequential sampling procedures developed by Wald (1947).

Although sequential sampling procedures intrinsically quantify the errors involved with the pest management decision process, and can result in substantial savings in cost (time) when compared with other quantitative procedures (Sterling 1975, Wald 1947), their utility is limited as the number of species or age classes being monitored increases. The greater this number, the more likely are one or more of the species to be close to the action threshold requiring that the scout or farmer sample up to an upper sample number limit. As a general guideline, if five or fewer species or age classes are being recorded with a single monitoring procedure using a common sample unit, then sequential sampling will result in a savings in sampling time. If one is using, say, two different monitoring procedures such as a sweep net for some arthropods and plant terminal counts for others, then it is practical to monitor up to 10 species using sequential sampling.

Environmental heterogeneity may result in the development of biased estimates and should be considered when developing monitoring programs. The more variable a field, the more likely it is that a farmer will spray when not appropriate and not spray when appropriate. Two obvious solutions to this problem are: 1) avoid sampling in areas which are not representative of the field and where possible divide highly variable fields into separate management units, and 2) increase the sample size when in doubt. The latter is accomplished in sequential sampling by using lower error rates to account for the field variability biasing these errors upwards. It is also a good practice to increase the distance between samples in proportion to a field's heterogeneity.

DEFINING THE SAMPLE UNIT

Individuals involved with the use or development of sampling procedures are often concerned with not catching all of the organisms present on the sample unit, believing that the procedure with the greater catch-efficiency is the best. The fallacy here lies in the fact that the procedures that are the most catch-efficient are usually time-consuming and often involve the use of costly equipment. All sampling procedures differ in their catch efficiency (for each species) as well as their ease and cost of operation. The number of sampling procedures which can be used to estimate the abundance of pest species and their natural enemies is overwhelming, and if the problem of determining which procedure is the most appropriate is addressed subjectively, it is unlikely that the best procedure will be chosen.

Objectively defined, the best or most cost-reliable sampling procedure provides an estimate with a defined level of reliability for the least cost. To determine and compare the cost-reliability of two or more sampling procedures requires the following information:

1. Relative or absolute efficiency estimates.
2. Variance-mean relationship (enumerative) or proportion infested-mean relationship (presence-absence).
3. Cost (time) required to collect and examine each sample unit.
4. Cost (time) required to move between sample sites.

Sample Method Efficiency

Few methods provide a truly absolute estimate of density per sample unit, particularly across all of the species or age-classes sampled. When used to sample predators on cotton, the sweep net has an average efficiency of approximately 10% (derived from Smith et al. 1976). A 10% efficiency value means that a sampler is only detecting an average of one in every ten predators in the area sampled. This percentage is, however, both species- and age-specific and crop stage-specific (McGroarty & Croft 1978, Wilson & Gutierrez 1980). For some cotton predators the efficiency of the sweep net has been reported to be as low as 3.5% (Wilson & Gutierrez 1980). Unless the scout or researcher realizes the inefficiency of the sampling procedure, a very distorted view of the importance of natural enemies develops. Factors such as sample unit size and time of day (possibly a temperature or light effect) have also been shown to influence the relative efficiency of several sampling procedures (Dumas et al. 1962, 1964; Gertsch & Riechart 1976; Hutchison & Pitre 1982;

Shepard et al 1974). Depending upon the distribution of the species within the plants and the part of the plant which is sampled, a variable proportion of the individuals will be recovered (Byerly et al. 1978). Bishop (1981), Fye (1972), Lesar & Unzicker (1978), Nyffler (1982), Wilson & Gutierrez (1980) and others have confirmed that the within-plant distribution patterns of natural enemies, although overlapping, are often very distinct and that the catch efficiency of the sampling procedure depends on the part of the plant from which they are sampled. Procedures, such as using the sweep net to sample vegetation, are more efficient at capturing natural enemies and pest species which are located in the outer and upper canopy of the crop.

The first step in comparing the relative efficiency of two sampling procedures is to treat the most consistently efficient procedure over the range of densities and age class sampled as the independent variable and the second as the dependent variable. The two must be scaled to the same unit or unit area (Morris 1955). As an example, if a 50-sweep sample unit procedure is compared with a single plant visually inspected sample unit procedure, then one must be scaled to the other. For cotton a 50-sweep sample unit (0.4-m diameter) covers the area equivalent to approximately 200 plants when at a density of 10 plants/m^2. Therefore for both of these methods to have the same efficiency, the sweep method has to catch 200 times as many per its unit as would the single plant procedure. After conducting this regression where the sweep counts are first multiplied by the correction factor 0.005 (=1/200), if the intercept does not differ significantly from zero, possibly implying a constancy in relative efficiencies for the range of densities compared, then it may be advisable to force the regression through the origin (Zar 1974).

$$\bar{x}_2 C_{2,1} = b_{2,1} \cdot \bar{x}_1 \quad (9)$$

where:

$b_{2,1}$ = the relative efficiency of procedure 2 compared to procedure 1.

$C_{2,1}$ = the scalar for equating the catches to the same unit area.

For a large number of sampling procedures, the above method can be used to derive relative efficiency estimates. The difficulty that some researchers have encountered when attempting to correlate different procedures can largely be attributed to low density estimates often combined with an inadequate number of samples. The lower the density, the greater the relative variability about an estimate; therefore, the difficulty in correlating counts, and thus the need for a larger number of samples.

Relative efficiency estimates are necessary when deriving thresholds for use with different sampling procedures. From a commercial perspective the need is obvious. It does little good to talk about action thresholds except within the context of the sampling procedure being used. It is not uncommon to attend research and grower meetings where two or more individuals are describing the level at which they spray their fields or plots, and to find that they were each talking about different units. This confusion inevitably leads to at least some commercial people using the lowest number, but frequently applying it to the wrong sampling procedure. The confusion is often accentuated because the relative efficiency of each of the methods is either not known, poorly known, or not available beyond a small number of researchers. Until efficiency estimates are available, it is difficult to quantitatively compare and determine which of two or more procedures is the best. When possible, relative sampling procedures should be compared with a procedure which captures most of the organisms in the area sampled. Such comparisons allow subsequent near absolute density estimation, using sampling procedures having low catch efficiencies, but likely the advantages of ease of operation and low cost.

Relative Cost-Reliability

Costs required to obtain an estimate or make a decision with a given level of reliability can be compared to determine which of two sampling procedures is better (Wilson et al. 1982, 1983c).

$$C_1/C_2 = n_1 \cdot (\Theta_1 + \Phi_1)/[n_2 \cdot (\Theta_2 + \Phi_2)] \qquad (10)$$

where:

- C_i = cost for a given level of reliability, for the ith sampling procedure.
- Θi = the time (cost) required to examine an individual sample unit using the ith sampling procedure.
- Φi = the time (cost) required to move from sample unit to sample unit for the ith procedure.
- n_i = the number of samples required for an estimate with a given level of reliability. For purposes of parameter estimation Equations (5) (enumerative) and (6) (presence-absence) are appropriate for estimating the n values; while Equations (7) and (8) are correspondingly used for commercial monitoring.

When the ratio of C_1/C_2 is greater than 1.0, procedure one costs more for a given reliability level than does procedure two. When two enumerative procedures are compared, Equation (10) requires Taylor's coefficients which will differ for each species and for each sampling procedure. It is possible to compare an enumerative and a presence-absence procedure (the latter not using Taylor's power law). When such a comparison is made, it is necessary to use $\underline{D}$ values which give comparable confidence intervals for Equations (5) and (6). The problem does not arise in comparing the sample size equations used in commercial monitoring since the somewhat arbitrary aspect of choosing $\underline{D}$ is replaced by the difference between the action threshold $\underline{T}_i$, and the estimated density, $\bar{x}$, or proportion of infested sample units, $\underline{p}$.

For some species it is possible for the C_1/C_2 ratios to be greater than 1.0 for some densities and less than 1.0 for others. This is partially due to sampling costs changing with density. In enumerative sampling, samples with high densities may be extremely costly to count. This indicates that one method may be better at low densities and another better at higher ones. Density also has the effect of changing variance, which directly affects the number of samples required for a given level of reliability. When the cost-reliability of two or more sampling procedures are compared for more than one species, one can determine which procedure is the best by taking a weighted average of the C_1/C_2 ratios. If this average is greater than one, then method two is best, and vice versa. For pest management purposes it may be better to use the sampling procedure which is the most cost-reliable for the key pest, which may be the best or at least an adequate procedure for a majority of other arthropod species. If information is required for several arthropods, some of which cannot be cost reliably sampled with the same procedure, then more than one sampling procedure should be used. Whether to use one or more sampling procedures, and which ones, should be determined in light of the costs and feasibility for use.

SAMPLING FREQUENCY AND FORECASTING

Sampling frequency is largely subjectively determined and often has little bearing on the phenology of the crop, pests or natural enemies. In its simplest form, phenology models which use pheromone traps to establish a biofix can be used to determine when subsequent samples might be taken. Tummala & Haynes (1977) predict that as our use and dependence upon such models increases, there will be a concomitant decrease in costs associated with monitoring. The degree to which monitoring can be reduced, however, depends on the reliability of these forecasting techniques and upon the current level of monitoring, which may be far less than required based on the relative costs of sampling and making inappropriate management decisions. The number of unknowns or "black boxes" in our understanding of pest management systems promoted calendar-type sampling in the same manner as it promoted a calendar or prophylactic spray strategy.

Sampling frequency can be quantitatively estimated several ways. From an optimization perspective, however, the following should be considered:

1. The phenology of the pest species.

2. The rate at which a pest's population is increasing (affected by abiotic and biotic factors), and the proximity of pest densities to economic or action thresholds.
3. The damage potential and cost of controlling the pest species at different crop stages. This relates to the economics of incremental costs and incremental benefits of control actions (Headley 1982).

Fig. 2 illustrates results from a simple population growth rate model (L.T. Wilson, unpublished data) which meet the three requirements listed above. This procedure is appropriate for univoltine and multivoltine species or for species having short generation times. The first four samples in this example were taken four weeks apart, with the first sample (biofix) taken at approximately $500^{o}D > 12^{o}C$ (from planting). The timing of the fifth and subsequent samples is based on a linear projection from the previous ones. The procedure illustrated suggests that the fifth sample be taken when the upper 95% confidence interval for the estimated proportion of infested leaves reaches the action threshold, or after $200^{o}D > 12^{o}C$, whichever comes first. The $200^{o}D$ (ca. 2 weeks during growing season) maximum time interval for the subsequent sample is a conservative value which could be increased as one's confidence in the prediction procedure increases. The projected time interval between successive samples decreases as the estimated population approaches the action threshold. This phenomenon is very much analogous to sequential sampling where the closer the population is to the action threshold, the greater the number of samples required to estimate with a given level of reliability whether the population is above or below the threshold.

Although fairly simple, this modeling approach, by basing the potential for pest damage on the rate of pest population increase, intrinsically incorporates and integrates mortality caused by natural enemies and other mortality agents. A feature of this type of forecasting approach is that the slower a pest population is increasing, independent of the individual factors responsible for limiting the rate of increase, less often are population estimates required.

Hearn and his co-workers in Australia (Hearn & Room 1979, Hearn et al. 1981, Ives et al. 1984, Room 1979, Wilson et al. 1984a) have carried the use of fairly simple pest phenology models one step further by developing a management tool which incorporates the damage impact of pests on the crop population model (cotton) and predation by natural enemies on the pests. Although they do not attempt to optimize sampling frequency, the resulting information is used to estimate whether control action is needed "today," during the next week, and when the subsequent sample should be taken. The feasibility of using a model which incorporates some detailed biology is demonstrated by this system being commercially purchased and used on 25% of Australia's cotton acreage (Thomson, personal communication).

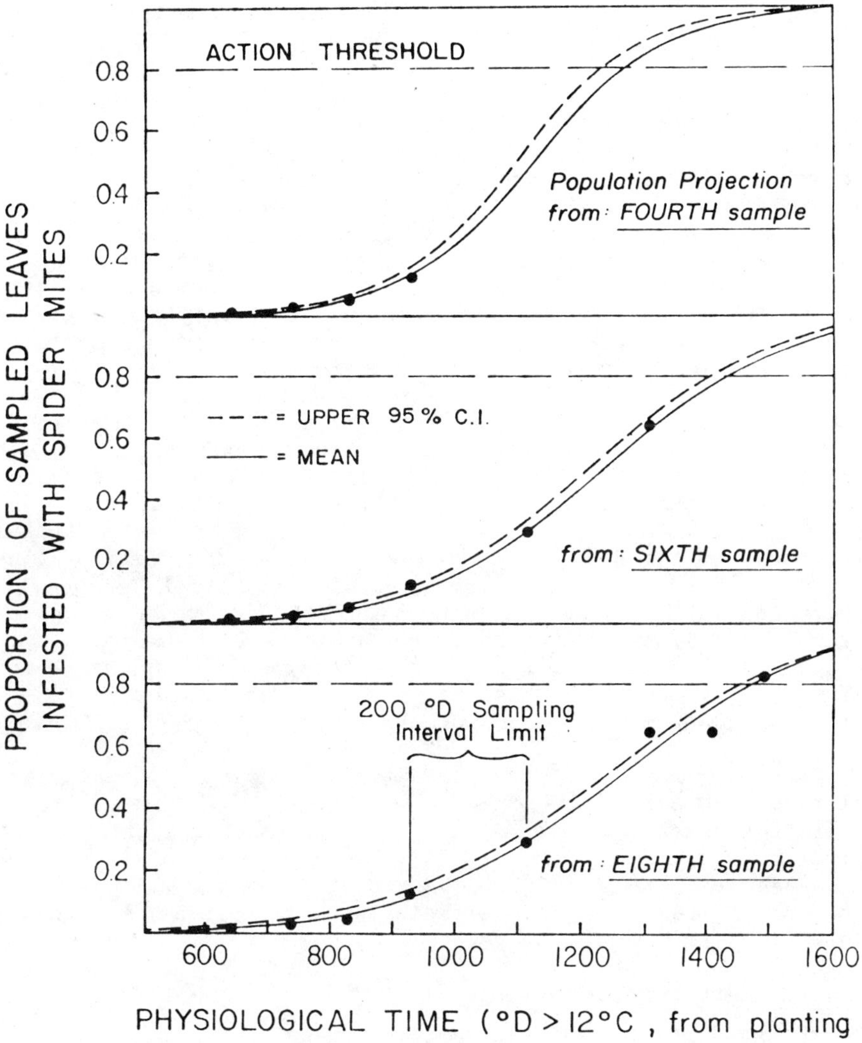

Fig. 2. Sequential updating for use in determining sampling frequency.

At a research level, Gutierrez and his co-workers (Gutierrez et al. 1975, 1976, 1977a, b, 1984; Wang et al. 1977) have examined in considerable detail the crop-pest and pest-natural enemy interactions within the alfalfa ecosystem. Although multitrophic models such as developed by Gutierrez and his co-workers could be converted for use in day-to-day tactical crop management, their structure makes them more suitable for addressing broader strategic management questions.

IMPACT OF NATURAL ENEMIES IN IPM SYSTEMS

Southwood (1978) and others cited in his book give an excellent coverage of life tables, feeding rates, and prey preference, precluding the necessity of covering direct mortality estimation procedures in detail. In passing, however, it is relevant to discuss a deficiency inherent with most direct mortality estimation methods. Estimates of age-specific mortality rates and the quantification of the contribution to mortality by each mortality factor are very sensitive to the frequency with which samples are taken. Commonly, target organisms are gathered from the field, and subsequent emergence of parasites recorded in the laboratory. Unless repeated frequently, this procedure underestimates total mortality since the organisms are protected from any mortality factors while in the laboratory. Likewise, mortality due to say a key parasite may be over- or underestimated: overestimated since many of the hosts might be destroyed in the field by other factors before parasite emergence; underestimated because the laboratory might protect the host during part of a vulnerable stage. Mortality such as that due to ovipositional probing and host-adult parasite feeding is often responsible for more mortality than caused by developing parasites, and is not well estimated except by detailed field observations (DeBach & Bartlett 1974).

It is not uncommon to see life table analyses expressing mortality due to various factors as percentages. Unless related to population size, and the potential for population growth, high mortality levels may be of little value. Mortality in a pest management system is sufficient when it contains the pest(s) at sub-economic levels which, depending upon the system, may require a high percentage or a low percentage.

Indirect Mortality Estimates

Impact on pest population rate of increase As discussed earlier, the rate at which a pest's population increases reflects the combined impact of abiotic and biotic factors. Although there is at least one multitrophic level model that has some utility for yield prediction (Gutierrez et al. 1984), such models by their nature require years of detailed experimentation and are likely to remain largely as strategic tools and limited to only a few crops. Lack of tactical commercial utility, of course, does not preclude the importance of detailed natural enemy-pest models for use in better understanding basic ecological principles.

When addressing the issues of natural enemy impact upon pest species, researchers might best attack this complexity by selecting crops having one or a few number of pests. This selection is of course dependent on funding. Also, when focusing upon the actual process of predation or parasitism, the experimental arena should be the field or as similar to the field as possible. As Kaiser (1983) demonstrates in his predatory mite feeding studies, laboratory results are only a rough indicator of behavior under field conditions, and should therefore be extrapolated to the field with caution. Instead of addressing each natural enemy separately, it might be better to collapse natural enemy impact information to a common unit. Although this approach may not be as ecologically interesting as modeling each individual natural enemy species, it is more likely to pay off in terms of implementability, particularly when developing management models incorporating predator impact. The potential for and degree of control exerted by the complex of natural enemies can be quantified by comparing growth rates under a range of natural enemy levels. Establishing different levels of control can be accomplished using mechanical or chemical exclusion, or for some species by hand removal or trapping of natural enemies (DeBach & Bartlett 1974). DeBach & Bartlett (1974) and DeBach & Huffaker (1971) indicate some of the potential problems of habitat modification which are inherent to but distinct for each of these four techniques.

Impact of pest distribution pattern Wilson et al. (1984b) were able to show that in the presence of the predatory mite, *M. occidentalis*, *Tetranychus* spider mites became less clumped in their distribution pattern on a single leaf sample unit basis when expressed as a proportion infested-mean density curve of the type illustrated in Fig. 1. Furthermore, they found that estimates of net mortality per predator could be obtained by estimating the change in prey density required for the p-$\bar{x}$ curve without predators to be shifted to equal the curve with predators, and by estimating the number of predators responsible for the shift in prey density. These data indicate that net mortality per predator decreases at lower prey densities. This was verified by conducting an additional analysis with the same data but in this case determining which predator/ prey ratio was required at each prey density to result in a reduction in prey the next sampling period. It appears that the lower efficiency of *M. occidentalis* at low densities is likely due to increased searching time (Wilson et al. 1984b).

Several researchers have made use of the predator/prey ratio concept in evaluating the need for chemical intervention. Tanigoshi et al. (1983) determined that a 0.5 to 0.1 predator/prey ratio for the predaceous mite-spider mite complex on apples was sufficient to provide biological control. Flaherty et al. (1981) use a similar ratio in managing spider mites in grapes. Wilson et al. (1984b) likewise obtained similar results on almonds, although at lower prey densities a greater predator/prey ratio is necessary for successful biological control due to lower predator efficiency. Li (1982) reported the use of the predator/prey ratio approach for managing planthoppers on rice in China, where the ratio used was dependent upon the species of spiders in the field and the stage of crop growth.

Experimental design-functional relationship Many of the structured experimental approaches used in entomological field research use conventional agronomic designs. The problem is that replication in the classical sense, that is in space, severely limits the number of treatments that can be examined and correspondingly makes it difficult to develop functional responses with so few data points. This in turn leads to research aimed at determining differences and not functional relationships. One way to use the same number of plots but to look at more factors is to use a non-replicated factorial or split-plot design. Two approaches could be used to analyze this data. The first is a factorial analysis of variance using the highest order interaction as the error term. This particular approach has some limitations in that it will only provide reliable error estimates when the highest order interaction is not significant (Steel & Torrie 1960, Zar 1974), and only if the underlying experimental plots are relatively homogeneous. However, a significant highest order interaction leads at worst to the error term being inflated and some significant differences not being detected when present.

The second approach to analyzing this design is to look not at the particular treatment and its effect on yield, but instead to look at a concomitant variable such as pest density and use regression techniques to relate the parameters being measured. This technique is also applicable for analyzing data generated with the more standard experimental designs. This second approach is also sensitive to plot heterogeneity and, depending upon the particular treatments affected, the regression slope(s) and intercept(s) and significant level(s) may be biased.

CONCLUSION

The continued adoption of IPM practices by the agribusiness sector is predicated upon these ecosystem-based principles being conveyed in a simple, clear and concise manner. A common misconception is that a practice or procedure will be adopted if superior to what is currently practiced. Unless the alternative IPM prodedures are perceived to be better than current methods, then they will not be adopted. Implementation of IPM practices is, of course, facilitated by the failure of current management practices. Fortunately, the logical extension of this line of reasoning is not the promotion of disasters in agriculture. Instead, through education in the form of well thought-out research and extension programs and sound IPM teaching programs at our universities, we can help increase the ecological understanding of IPM systems as well as reduce the differences between perception and reality and give IPM practices a fair chance.

RECOMMENDATIONS

Three general areas that deserve increased attention are:

1. Further development of efficient and cost-reliable estimation procedures;
2. Development and implementation of forecasting models which incorporate natural enemies in the treatment decision process; and
3. Experimentation aimed at estimating functional relationships and not just significant differences. When possible, an attempt should be made to balance the implementation and research components of an IPM system.

ACKNOWLEDGMENTS

Special thanks to R. Bugg, D. Gonzalez, M.P. Hoffman, M.A. Hoy, C.B. Huffaker, P.C. Matteson, and K. Yeargan for critically reviewing the manuscript, R. Bugg for access to his reference collection, to R. Plant for helpful comments on iterative regression and forecasting techniques, and to P. Kaplan and C. Haider for typing the manuscript.

REFERENCES

Allen, J., D. Gonzalez and D.V. Gokhale. 1972. Sequential sampling plans for the bollworm, Heliothis zea. Environ. Entomol. 1:771-780.

Banerjee, B. 1976. Variance to mean ratio and the spatial distribution of animals. Experientia 32:993-994.

Bishop, A.L. 1981. The spatial dispersion of spiders in a cotton ecosystem. Aust. J. Zool. 29:15-24.

Byerly, K.F., A.P. Gutierrez, R.E. Jones and R.F. Luck. 1978. Comparison of sampling methods for some arthropod populations in cotton. Hilgardia 46: 257-282.

DeBach, P., and B.R. Bartlett. 1974. Methods of colonization, recovery, and evaluation, pp. 402-426. In: P. DeBach and E.I. Schlinger (Eds.), Biological control of insect pests and weeds. Chapman & Hall, London.

DeBach, P., and C.B. Huffaker. 1971. Expermental techniques for evaluation of the effectiveness of natural enemies, pp. 113-140. In: C.B. Huffaker (Ed.), Biological control. Plenum Press, N.Y.

Dumas, B.A., W.P. Boyer and W.H. Whitcomb. 1962. Effect of time of day on surveys of predaceous insects in field crops. Fla. Entomol. 45:121-128.

Dumas, B.A., W.P. Boyer and W.H. Whitcomb. 1964. Effect of various factors on surveys of predaceous insects in soybeans. J. Kan. Entomol. Soc. 37: 192-201.

Flaherty, D.L., M.A. Hoy and C.D. Lynn. 1981. Spider mites, pp. 111-125. In: D.L. Flaherty, F.L. Jenson, A.N. Kasimatis, H. Kido and W.J. Moller (Eds.), Grape pest management. Agric. Sci. Pub. 4105, Univ. of Calif. Press, Berkeley.

Fracker, S.B., and H.A. Brischle. 1944. Measuring the local distribution of Ribes. Ecology 25:281-303.

Fye, R.E. 1972. Preliminary investigation of vertical distributions of fruiting forms and insects on cotton plants. J. Econ. Entomol. 65:1410-1414.

Gertsch, W.J., and S.E. Riechart. 1976. The spatial and temporal partitioning of a desert spider community, with description of new species. Amer. Mus. Novitiates 2604:1-25.

Gutierrez, A.P., J.U. Baumgaertner and C.G. Summers. 1984. Multitrophic models of predator-prey energetics: II. A case study in an alfalfa ecosystem. Can. Entomol. 116:(In Press).

Gutierrez, A.P., G.D. Butler, Jr., Y. Wang and D. Westphal. 1977a. A model for pink bollworm in Arizona and California. Can. Entomol. 109:1457-1468.

Gutierrez, A.P., J.B. Cristensen, C.M. Merrit, W.B. Loew, C.G. Summers and W.R. Cothran. 1976. Alfalfa and the Egyptian alfalfa weevil (Coleoptera: Curculionidae). Can. Entomol. 108:635-648.

Gutierrez, A.P., L.A. Falcon, W. Loew, P.A. Leipzig and R. van den Bosch. 1975. An analysis of cotton production in California: A model for Acala cotton and the effects of defoliators on its yields. Environ. Entomol. 4:125-136.

Gutierrez, A.P., T.F. Leigh, Y. Wang and R. Cave. 1977b. An analysis of cotton production in California: Lygus hesperus injury--an evaluation. Can. Entomol. 109:1375-1386.

Hayman, B.I., and A.D. Lowe. 1961. The transformation of counts of the cabbage aphid (Brevicoryne brassicae (L.)). New Zealand J. Sci. 4:271-278.

Headley, J.C. 1982. The economics of pest management, pp. 69-91. In: R.L. Metcalf and W.H. Luckmann (Eds.), Introduction to insect pest management. John Wiley & Sons, N.Y.

Hearn, A.B., and P.M. Room. 1979. Sequential analysis of crop development for pest management in cotton. Prot. Ecol. 1:265-277.

Hearn, A.B., P.M. Ives, P.M. Room, N.J. Thompson and L.T. Wilson. 1981. Computer-based cotton pest management in Australia. Field Crops Res. 4:321-332.

Hutchison, W.D., and H.N. Pitre. 1982. Diurnal variation in sweepnet estimates of Geocoris punctipes (Say) (Hemiptera: Lygaeidae) density in cotton. Fla. Entomol. 65:578-579.

Ives, P.M., L.T. Wilson, P.O. Cull, W.A. Palmer, C. Haywood, N.J. Thomson, A.B. Hearn and A.G.L. Wilson. 1984. Field use of SIRATAC: An Australian computer-based pest management system for cotton. Prot. Ecol. 6:1-21.

Kaiser, H. 1983. Small scale spatial heterogeneity influences predation success in an unexpected way: Model experiments on the functional response of predatory mites (Acarina). Oecologia 56:248-256.

Karandinos, M.G. 1976. Optimum sample size and comments on some published formulae. Bull. Entomol. Soc. Amer. 22:417-421.

Kogan, M., and D.C. Herzog (Eds.). 1980. Sampling methods in soybean entomology. Springer-Verlag, N.Y. 587 pp.

Lesar, C.D., and J.D. Unzicker. 1978. Soybean spiders: Species composition, population densities, and vertical distribution. Ill. Nat. Hist. Surv. Biol. Notes 107, 14 pp.

Li, L.-Y. 1982. Integrated rice insect pest control in the Guangdong Province of China. Entomophaga 27:81-88.

McGroarty, D.L., and B.A. Croft. 1978. Sampling the density and distribution of Amblyseius fallacis in the ground cover of Michigan apple orchards. Can. Entomol. 110:785-794.

Miller, W.E. 1971. Discussion section, pp. 372-377. In: G.P. Patil, E.C. Pielou and W.E. Waters (Eds.), Statistical ecology. Penn. State Univ. Press, University Park Vol. 1.

Morisita, M. 1959. Measuring of the dispersion of individuals and analysis of the distributional patterns. Mem. Fac. Sci., Kyuchu Univ. Ser. E (Biol.) 2: 215-235.

Morris, R.F. 1955. The development of sampling techniques for forest insect defoliators, with particular reference to the spruce budworm. Can. J. Zool. 33:225-294.

Nyffler, M. 1982. Field studies on the ecological role of the spiders as insect predators in agroecosystems (abandoned grassland, meadows, and cereal fields). Ph.D. Thesis, Swiss Fed. Inst. Technol., Zurich.

Room, P.M. 1979. A prototype on-line pest management system for cotton in the Namoi Valley, New South Wales. Prot. Ecol. 4:245-264.

Ruesink, W.G. 1980. Introduction to sampling theory, pp. 61-78. In: M. Kogan and D.C. Herzog (Eds.), Sampling methods in soybean entomology. Springer-Verlag, N.Y.

Ruesink, W.G., and M. Kogan. 1982. The quantitative basis of pest management: Sampling and measuring, pp. 315-352. In: R.L. Metcalf and W.H. Luckmann (Eds.), Introduction to insect pest management. John Wiley & Sons, N.Y.

Shepard, M., V. Waddill and S.G. Turnipseed. 1974. Seasonal abundance of predaceous arthropods in soybeans. J. Ga. Entomol. Soc. 9:120-126.

Smith, J.W., E.A. Stadelbacher and C.W. Gantt. 1976. A comparison of techniques for sampling beneficial arthropod populations associated with cotton. Environ. Entomol. 5:435-444.

Southwood, T.R.E. 1978. Ecological methods. Chapman & Hall, London. 524 pp.

Steel, R.G.D., and J.H. Torrie. 1960. Principles and procedures in statistics. McGraw-Hill, N.Y. 481 pp.

Sterling, W.L. 1975. Sequential sampling of cotton insect populations, pp. 133-135. In: Proc. 1975 Beltwide Cotton Prod. Res. Conf. Natl. Cotton Counc., New Orleans, LA.

Tanigoshi, L.K., S.C. Hoyt and B.A. Croft. 1983. Basic biological and management components for mite pests and their natural enemies, pp. 153-202. In: B.A. Croft and S.C. Hoyt (Eds.), Integrated management of insect pests of pome and stone fruits. Wiley-Interscience, N.Y.

Taylor, L.R. 1961. Aggregation, variance and the mean. Nature 189:732-735.

Taylor, L.R., I.P. Woiwod and J.N. Perry. 1978. The density-dependence of spatial behavior and the rarity of randomness. J. Anim. Ecol. 47:383-406.

Tummala, R.L., and D.L. Haynes. 1977. On-line pest management systems. Environ. Entomol. 6:339-349.

Wald, A. 1947. Sequential analysis. John Wiley & Sons, N.Y. 212 pp.

Wang, Y., A.P. Gutierrez, G. Oster and R. Daxl. 1977. A population model for plant growth and development: Coupling cotton-herbivore interaction. Can. Entomol. 109:1359-1374.

Waters, W.E. 1955. Sequential sampling in forest insect surveys. For. Sci. 1: 68-79.

Wilson, L.T. 1982. Development of an optimal monitoring program in cotton: Emphasis on spider mites and *Heliothis* spp. Entomophaga 27:45-50.

Wilson, L.T., and A.P. Gutierrez. 1980. Within-plant distribution of predators on cotton: Comments on sampling and predator efficiencies. Hilgardia 48:3-11.

Wilson, L.T., A.P. Gutierrez and D.B. Hogg. 1982. Within-plant distribution of cabbage looper (Trichoplusia ni (Hubner)) on cotton: Development of a sampling plan for eggs. Environ. Entomol. 11:251-254.

Wilson, L.T., D. Gonzalez, T.F. Leigh, V. Maggi, C. Foristiere and P. Goodell. 1983a. The within-plant distribution of spider mites (Acari: Tetranychidae) on cotton: A developing implementable monitoring program. Environ. Entomol. 12:128-134.

Wilson, L.T., A.B. Hearn, P.M. Ives and N.J. Thomson. 1984a. Integrated pest control for cotton in Australia, pp. 134-143. In: R.E. Frisbie (Ed.), Guidelines for integrated control of cotton pests. FAO, Rome.

Wilson, L.T., M.A. Hoy, F.G. Zalom and J.M. Smilanick. 1984b. Sampling mites on almonds: I. The within-tree distribution and clumping pattern of mites with comments on predator-prey interactions. Hilgardia 52:1-13.

Wilson, L.T., C. Pickel, R.C. Mount and F.G. Zalom. 1983b. Presence-absence sequential sampling for cabbage aphid and green peach aphid (Homoptera: Aphididae) on Brussels sprouts. J. Econ. Entomol. 76:476-479.

Wilson, L.T., and P.M. Room. 1982. The relative efficiency and reliability of three methods for sampling arthropods in Australian cotton fields. J. Aust. Entomol. Soc. 21:175-181.

Wilson, L.T., and P.M. Room. 1983. Clumping patterns of fruit and arthropods in cotton with implications for binomial sampling. Environ. Entomol. 12: 50-54.

Wilson, L.T., F.G. Zalom, R. Smith and M.P. Hoffmann. 1983c. Monitoring for fruit damage in processing tomatoes: Use of a dynamic sequential sampling plan. Environ. Entomol. 12:835-839.

Zar, J.H. 1974. Biostatistical analysis. Prentice Hall, N.J. 620 pp.

DISCUSSION: Estimating the abundance and impact of arthropod natural enemies in IPM systems.

DISCUSSION LEADER: R.E. Stinner

Using time series techniques and control theory to project population growth and sampling interval does not account for situations where influxes of pests or natural enemies occur. Since these influxes often occur with highly mobile and polyphagous species, dramatic and often unpredictable population increases can occur. With increased knowledge, these influxes and their timing can be predicted and sampling intervals can be adjusted to account for this emigration. This approach allows comparisons in population growth slopes between treated and untreated areas, which can provide a measure of natural enemy impact without fully understanding all the mechanisms or species involved. Because this approach is empirical, however, it must be determined for specific crops and geographic regions. Currently it is being used successfully in cotton, almonds and strawberries.

Because of difficulties in sampling natural enemies and assessing their impacts, it is recommended that, where possible, the densities of other pests should be recorded whenever this impact is being assessed with one pest. Furthermore, all sampling procedures should be described in detail so that meaningful comparisons can be made among regions, crops and sampling techniques.

INTEGRATING INFLUENCES OF NATURAL ENEMIES INTO MODELS OF CROP/PEST SYSTEMS

Jerry L. Stimac and Robert J. O'Neil

A model of a crop/pest system is a mathematical representation of the interactions between a crop and the principal pest species affecting the crop. The pests can include functionally different categories of species such as weeds, pathogens, nematodes and insects. The objective of building a crop/pest model is to describe the dynamics of the crop and pests in the context of the environment in which they coexist. The environment includes many factors influencing the growth of the crop and pest populations: weather inputs, such as temperature, rainfall and solar radiation; biological inputs, such as natural enemies of the pests; and production system inputs, such as irrigation, cultivation and application of pesticides. Therefore, the major submodels in a crop/pest system model are models of the crop subsystem, pest subsystems, production tactics, and environmental inputs into these subsystems. Models of some of the pest populations must include the effects of natural enemies and pesticides.

This paper will discuss techniques for integrating the influences of natural enemies into models of crop/pest systems. Since insect pests have a wide range of natural enemies (pathogens, parasites and predators) and because insect populations have been modeled more extensively than other types of pests, we will emphasize methods for incorporating natural enemy effects into insect population models. However, the approach used to represent natural enemy effects on insect pests should be appropriate for other types of pests as well.

GENERAL STRUCTURE OF CROP/PEST SYSTEM MODELS

Models of crop/pest systems are usually site specific. The system boundaries are often defined to be a single field or grouping of homogenous fields in a given area. The spatial and temporal resolution of the crop/pest model depends on the attributes of the pest and natural enemy populations (Stimac & Barfi
and the desired size of the management unit about whicl
sions will be drawn from model analyses. Often crop/pe

are used to simulate the dynamics of a crop and pests in a single field so that decisions can be made regarding pest management and other production practices for that field. Where farmers have multiple fields of the same crop, the spatial resolution of the model may be a group of fields in the same area because this is the appropriate management unit. Whether the spatial resolution of the crop/pest system is defined as a single field or a group of fields, net influxes of the pest and natural enemies into the managment unit must be measurable. Data on rates of net influx of pest and beneficial organisms links a site-specific crop/pest model to the larger agroecosystem.

In general, crop/pest system models consist of a crop growth model, pest population models, and an environmental subsystem which provides information on weather and production tactics to drive the crop and pest models. The crop growth model is composed of a series of submodels of the physiological processes which generate carbohydrate and allocate it to different plant structures (leaves, stems, roots and fruit). Linkage of stresses to these physiological processes is a key component in the crop/pest model because it is the manner in which the crop model integrates the effects of multiple stress factors imposed on the plants by different categories of pests (insects, weeds, nematodes and plant pathogens), and other stresses such as water shortage, wind damage, or adverse temperature conditions.

MODELS OF CROP/PEST SYSTEMS

Crop models have been developed for a variety of agricultural systems. The most extensive models of crops and pests were developed as a result of the Integrated Pest Management Projects (Adkisson & Frisbie 1979, Huffaker 1980). Five crops and their principal pests have been modeled: alfalfa (Gutierrez et al. 1976, Ruesink et al. 1980), apples (Asquith et al. 1980, Croft 1976), cotton (Brown et al. 1976, Curry et al. 1980, Gutierrez et al. 1980, Wang et al. 1977), oranges (Allen 1976, Luck et al. 1980), and soybean (Rudd et al. 1980, Stimac and Barfield 1979, Wilkerson et al. 1983). Most often population models of pests are for insect and mite species and only some of the pest population models include the effects of natural enemies. Many insect pests in crop systems are attacked by a variety of natural enemy species, but most crop/pest models include descriptions for only one or several "dominant" natural enemies. We will discuss crop/pest models of alfalfa, cotton and soybean systems. These crops have been studied extensively, especially with regard to pests and natural enemies. In addition, for each of these crops, there are detailed crop growth models, which serve as the templates for descriptions of crop/pest/natural enemy interactions.

Crop Models

The major objective of a crop model is to describe the growth and maturation of the crop. Carbohydrate, produced by the process of photosynthesis, is distributed to plant structures depending on the supply of photosynthetic material and physiological demand. By apportioning carbohydrates based on supply/demand, the model mimics physiological responses of the crop to relative availability of nutrients. This physiologically-based carbohydrate allocation scheme allows the crop model to integrate the effects of pests, weather, fertilizer applications, and other production practices.

Realistic description of crop dynamics requires that the growth of structural components of the crop (roots, stems, leaves and fruit) be included in the crop model. Although agricultural pests often only attack a single plant structure, the effect of the pest on the crop is influenced by non-target plant structures because of the supply/demand carbohydrate allocation scheme. Crop models that incorporate pest impact must, therefore, have descriptions of all principal plant structures, regardless of whether or not the pests directly attack all of these structures.

A second critical component of a crop model is the description of the crop growth rate. Plant growth rates are affected by temperature, and temperature patterns often differ from year to year and at different geographic locations. Therefore, describing crop development solely as a function of time is often inadequate because temperature is too variable through time. Although other abiotic environmental factors influence crop developmental rates, at a minimum crop models must incorporate the effect of temperature on development. In addition, if pests primarily attack a particular age class of plants, then some form of age structure must be incorporated into the model.

Two modeling methods have been used to describe crop dynamics. The first approach uses von Foerster (1959) continuous equations of the form:

$$\frac{dN}{dt} + \frac{dN}{da} = - MN(t,a) \qquad (1)$$

Changes in the number of plants or a plant structure of age (a) over time (t) is described as a function of M, a complex death function, N(0,a) and N(t,0). Typically, an equation is written for each plant part (i). The Leslie Matrix model (Leslie 1949) has been used as the discrete version of the von Foerster model (von Foerster 1959). The von Foerster-Leslie matrix approach is particularly useful when specific age classes of plant part (i) are attacked by pests. The method has been used in a variety of crops including alfalfa, cotton and soybean.

A second widely used method of crop modeling utilizes the concept of physiological time. Physiological time incorporates the changes in temperature and time into one concept, and is the appropriate temporal scale for both animals and plants (Gutierrez et al. 1984). Crop dynamics can then be described as a function of time and temperature. An excellent example is seen in alfalfa where crop (standing crop) growth rate (dw/dt) is given by:

$$\frac{dw}{dt} = \frac{dp}{dt} - G_r \frac{dQ}{dt} - R_o W - R_s \frac{dc}{dt}, \tag{2}$$

where dp/dt, dQ/dt and dc/dt are rates of photosynthesis, plant tissue growth and carbohydrate reserve utilization. The terms G_r, R_o, R_s, are growth maintenance and starch respiration; and t is physiological time. The changes in standing crop are governed by the relative rates of photosynthesis and respiration. Changes in standing crop are accounted for over physiological time, and thus incorporate both time and temperature. Two key elements of a crop model are the description of carbohydrate allocation to plant parts and the physiological time scale. Physiologically-based crop models form the template for models of pests and natural enemies.

Pest Population Models

The objectives of pest population models are to describe changes in pest density and to couple pest densities to crop damage. Since both abiotic and biotic factors influence population dynamics, both abiotic and biotic factors must be included in any model description. Natural enemies often contribute significantly to pest population dynamics and, therefore, can be key elements in the accurate prediction of pest density changes.

The units of pest population dynamics models are given in terms of numbers or energy (calories or biomass). Models that use energy units explicitly recognize the importance of energy attainment and utilization in population dynamics (Odum 1983) Such models focus at the physiological level in describing pest density changes. Numerically-based models focus at the behavioral or ecological level. Interaction between crop and pest model components depend on the number of pests present.

Models of pests in cotton and alfalfa have focused on energy-based population dynamics models. In these models, predator-prey behaviors are used as an analogy to describe all intertrophic level interactions. A simple model for multitrophic level interactions incorporating the acquisition and assimilation functions is presented using the pea aphid, *Acyrthosiphon pisum*, as an example (Gutierrez et al. 1984). In alfalfa, the pea aphid consumes crop nutrients (Sap) and is in turn preyed upon by predators. The aphids use crop nutrients to satisfy competing physiological demands, for growth, maintenance and reproduction. The general description of aphid dynamics is given by:

$$\frac{dM_{i+1}}{dt} = M_i[1-f_{i+1}(M_i,M_{i+1})](1-B_{i+1}) - Z_{i+1}M_{i+1} - M_{i+1}[1-f_{i+2}(M_{i+1},M_{i+2})] \quad (3)$$

where: $M_i[1-f_{i+1}(M_i,M_{i+1})]$ = aphid (M_{i+1}) consumption rate of alfalfa (M_i)nutrients,
B_i = food conversion efficiency rate,
Z_{i+1}^{i}, M_{i+1} = aphid respiration rate, and
$M_{i+1}[1-f_{i+2}(M_{i+1},M_{i+2})]$ = rate of predation on aphids by arthropod predators (M_{i+2}).

This model (Gutierrez et al 1984) uses calories as the currency of exchange between trophic levels. The model is general enough to describe any trophic level exchange, either plant-herbivore or herbivore-natural enemy. Note the similarity between the expressions describing aphid consumption of the plant and predator consumption of aphids. The utility of this approach is that all interactions are placed on common physiological footing, and that changes in pest densities can be affected by the quality as well as quantity of crop nutrient.

More common in agricultural systems modeling are pest population dynamics models based on numbers. Pest dynamics in cotton (Curry et al. 1980, Gutierrez et al. 1977) have been described by the von Foerster (1959) equation,

$$\frac{dN}{dt} + \frac{dN}{da} = - M\ N\ (t,a) + I \quad (4)$$

where: M_i = mortality rate on stage (i),
N_i = number of individuals in state (i), and
I^i = complex net immigration function.

Leslie (1945) matrices are used to approximate solutions for the partial differential equations for the pest population growth. The interaction between pest and plant depends on the plant structure attacked and the plant response to damage. Birth and death rates of the pest are influenced by the availability of the plant part or parts on which the pest feeds.

Curry et al. (1980) developed a model to describe boll weevil, *Anthonomus grandis*, dynamics. In this model, boll weevil dynamics are described in a manner similar to Gutierrez et al. (1977). Numbers of boll weevils produced in the population are a function of poikilotherm development rate (Sharpe & DeMichele 1977), adult reproductive and longevity rates, and survivorship rates of immature larvae.

Wilkerson et al. (1983) described the dynamics of the velvetbean caterpillar, *Anticarsia gemmatalis*, in soybean with an extension of the von Foerster (1959) model. Age structure within each developmental stage is maintained. Changes in the numbers of each ith stage of the caterpillar are given by:

$$\frac{dN_i}{dt} + \frac{dN_i}{da} = - M_i N_i - T_i N_i + E_i \qquad (5)$$

where: N_i = number of individuals in the ith stage,
M_i = stage specific mortality rate,
T_i = developmental function, and
E_i = number of individuals which enter a stage from either immigration or development from the previous stage.

The values of M_i, stage specific mortality rates, are a function of mortalities induced by natural enemies (predators). The developmental rates, t_i, are estimated using a cumulative distribution function (Stinner et al. 1975). For all larval stages and the adult stage, E_i is calculated as:

$$E_i(0,t) = - T_{i-1}\, N_{i-1} + I_i, \text{ and}$$

$$E_i(a,t) = 0 \qquad a \neq 0,$$

where I_i, immigration into the field, is non-zero only for adults.

RATIONALE FOR INTEGRATING INFLUENCES OF NATURAL ENEMIES INTO CROP/PEST MODELS

Achieving integrated pest management in agricultural systems requires that all pest control tactics be integrated into a comprehensive management strategy in which pest organisms are maintained at acceptably low population densities (Barfield & Stimac 1980). Since naturally-occurring and introduced biological agents exert some level of control on pest populations, these must be considered as a part of an agricultural IPM system. In pest control programs which preceeded the philosophy of IPM, pest populations were suppressed on a calendar schedule or whenever they reached unacceptably high densities. Control tactics were used in a remedial manner to lower pest densities with little regard for the effects that suppressive tactics would have on regulating agents of pests. In IPM programs, pest control tactics are used to supplement the regulatory influence of naturally-occurring agents so that the combination of tactics maintains pest populations at levels that are not economically damaging to the crop (Stimac & Barfield 1979). Naturally-occurring or introduced biological control agents serve two primary functions in crop/pest systems (Herzog et al. 1984): 1) to buffer the crop system against pest invasion and subsequent exponential growth of pest populations, and 2) to act as suppressive agents of pest populations. Crop/pest system models must incorporate the influence that naturally-occurring or introduced biological control agents can have on pest dynamics if the models are going to realistically simulate crop and pest dynamics as observed in the field.

REPRESENTING NATURAL ENEMIES IN MODELS OF CROP/PEST SYSTEMS

The impact of natural enemies in pest dynamics in alfalfa, cotton and soybean varies, as do the models used to describe pest/natural enemy interactions. In general, the changes in pest numbers (N_t) may be described as:

$$N_{t+1,i} = N_{t,i} \Phi_i (1-M_i) + I_i , \tag{6}$$

where: Φ_i = stage (i) developmental rate,
M_i = stage (i) mortality rate, and
I_i = stage (i) net immigration.

Natural enemies affect M_i, the stage specific mortality rate. For insect populations, the survivorship is a function of the levels of mortality imposed by predators (m_{pred}), parasites (m_{par}), pathogens (m_{path}), and other background mortality factors (m_b). Thus, the value of M_i is a function of these biological control agents:

$$M_i = (m_{pred} + m_{par} + m_{path} + m_b). \tag{7}$$

Equation 7 assumes that the biological control agents act independently (Stimac & O'Neil 1983).

The key aspect of mortality induced by natural enemies is the attack equation, which describes the number (or percentage) of pests attacked. When modeling natural enemy influences in crop/pest systems, it is the understanding and formulation of the attack equation that is the critical step. In the next section we present example models of predators in alfalfa and soybean systems. Attack by parasites could be treated in a similar manner (Holling 1959).

Natural Enemy Models

By definition, predation is an intertrophic level interaction. Models of predation, therefore, require at a minimum, predator and prey components. The central objective in any predation modeling effort is to accurately describe predator attacks on prey. It is the attack equation that couples predator and prey and is the template for incorporating relevant predator and prey behaviors. There has been considerable study of predation and predator attack equations in agricultural systems (Frazier et al. 1981). However, most models of predation in agricultural systems lack the driving mechanism for predation, and thus the attack equation has no identifiable physiological, behavioral or ecological basis. Central to understanding predator attack is how searching behaviors influence the rate at which predators discover prey. Since predators must find prey to consume prey, accurate description of searching behavior is paramount to a realistic description of predation.

As with models of pest population dynamics, there are essentially two basic approaches to modeling predator attacks--physiological or behaviorally-based models. A third approach, and one that includes most predation models in agricultural systems, is descriptive. These models simply describe the level of predation at given predator and prey densities, but have little or no biological mechanism.

Physiologically-based models describe searching driven by the level of hunger (Holling 1965). Explicit mathematical relationships are formulated to couple hunger and searching. Additional mathematical structure describes the effects of other relevant behavior (handling time, Holling 1959), but here we concentrate on simple hunger-search equations presently used in models of predation in agricultural systems.

Predation of aphids in alfalfa has been modeled using the hunger-search physiological approach. Applying the basic model of Frazier & Gilbert (1968), Gutierrez et al. (1984) describe predation as:

$$M_a = (1-e^{-bTM_2/M_1})(1-e^{-SM_1/b}), \quad (8)$$

where: M_a = total prey biomass attacked,
M_1 = prey biomass,
M_2 = predator biomass,
T = time,
b = predator demand rate, and
S = predator discovery rate.

In this model, the discovery rate (S) is affected by the demand rate (b). As b increases, predators search more. The relationship between search and demand is given by:

$$S = S_o + V[1 - (M_a^*/b)], \quad (9)$$

where: S_o = discovery rate per unit satiated predator,
V = maximum rate of increase in discovery as hunger increases, and
M_a^*/b = measure of satiation (supply/demand ratio).

As M_a^*/b approaches 1, Equation 9 reduces to:

$$S = S_o, \quad (10)$$

and the predator discovery rate equals the discovery rate of satiated predators.

Because the model (Equation 8) is physiologically-based, changes in the rate of predation may be influenced by the quantity or quality of prey present. By using biomass and energy units, the model explicitly recognizes that not all prey items are of equal quality. Crop, weather or other environmental conditions may indirectly influence predation by changing the quality or quantity of prey items available to predators in this system.

Behaviorally-based predator-prey models focus on describing predation based primarily on the behavioral characteristics of predator and prey. Since searching is a key predator behavior, these models focus on describing the nature of searching in the system. The level of behavioral detail is dependent on the system and, as with other types of predator models, the attack equation is the key to describe predation.

In soybean, a model of predation on larvae of the velvetbean caterpillar, _A. gemmatalis_, has been formulated (O'Neil 1984). The model describes predation as a function of predator searching in a dynamic searching universe. The visually and tactilely searching predators are constantly faced with an ever-changing substrate, the leaf area of the soybean crop. Prey (pest) density, expressed in terms of number/unit leaf area, changes as a function of numerical changes in the prey population as well as leaf area changes. Predators alter their behavior to accomodate the changes in prey density (per unit leaf area).

The equation describing search in soybean is:

$$S = C_1 e^{-C_2 V_L} + C_3, \tag{11}$$

where: V_L = number of prey (velvetbean caterpillar larvae) per unit soybean leaf area,
C_3 = minimum leaf area searched when V_L is maximal,
C_1 = maximum leaf area searched (above C_3) when V_L equals 0, and
C_2 = rate of change in area searched per V_L.

The relationship between searching and predation is:

$$N_a = PV_L S, \tag{12}$$

where: N_a = number of attacks,
P = number of predators, and
V_L and S are as defined for Equation (11).

Thus, the final attack equation is given as:

$$N_a = PV_L(C_1 e^{-C_2 V_L} + C_3). \tag{13}$$

Note that Equation (13) incorporates three trophic levels: the crop and pest (V_L) and predators (P). Crop dynamics are explicitly incorporated here, and therfore factors that influence crop growth (fertilization and weed management) will directly influence predation.

Models of natural enemy-induced mortality have been described for a variety of agricultural systems. We have discussed two such models that represent a current sample of alternative approaches toward modeling predation. In the physiologically-based model, searching is influenced by predator demand. Because the quality of prey is influenced by the crop nutritional state, the crop is implicitly incorporated into predator dynamics. In the behaviorally-based model, the crop is explicitly incorporated and directly influences predator search and therefore attack rates. The driving physiological mechanisms are not explicitly incorporated. The difference between the two approaches reduces to the nature of the description of predator searching in the system. Searching is driven by either hunger (physiological) or induced through plant growth (behavioral). In both approaches, it is searching behavior that is the key in describing predation.

Synopsis of Models of Natural Enemies

Models of natural enemies in crop/pest systems vary in complexity and realism. Like classical predator-prey models (Lotka 1925, Nicholson 1933, Nicholson & Bailey 1935, Volterra 1931), models of natural enemies used in agricultural systems are dependent on our understanding the processes of predation and parasitism.

When first formulated, the Nicholson-Bailey model had several gross simplifying assumptions. Chief among these assumptions was that predator searching efficiency was a species-specific parameter. Research into the nature of predation (Burnett 1954, DeBach & Smith 1941, Ullyett 1949) indicated that predator efficiency changes as a function of prey density. Holling (1959, 1961, 1965, 1966) showed that it was selected components of predator behavior, driven by hunger, that most heavily influenced predation dynamics. Further, investigations led to the description of the impact of predator density (Hassell & Varley 1969), prey dispersion (Hassell 1978, Huffaker 1958), predator switching (Cock 1978, Murdoch 1969), and multiple predator species (Comins & Hassell 1976, Gutierrez et al. 1981, Hassell 1969) on the overall rate of predation. The pattern for the development of models like the Nicholson-Bailey model has been a feedback between experimentation and theoretical development. The catalyst for this protocol has been the focusing of research on the impact of predator (or parasite) and prey behavior on predation dynamics, and how such behavior is influenced by the physiological and ecological system.

In agricultural systems, there have been few models developed that capture the essence of predation. In most crop/pest models, natural enemy effects are often omitted or are treated parametrically. Few models incorporate relevant predator behavior, or the impact of system level changes (see Frazier & Gilbert, Gutierrez et al. 1984, O'Neil 1984, Tamaki et al. 1974). Where predation has been studied, with the objective of mathematically describing the process, new insights into predation in agricultural systems have followed.

Predation in agricultural systems is influenced by the structure and dynamics of the agroecosystem. Crop/pest models can minimize the complexities of agricultural systems, and provide a template for the study of predation. For example, although alternative habitats (Altieri 1979, Baumgaertner et al. 1981, Naranjo 1983, Risch et al. 1982), and alternative prey items Gutierrez et al. 1984, Hagen et al. 1970), have been demonstrated to be vitally importrant to predation dynamics, few crop/pest models have alternative habitats or prey item components (Gutierrez et al. 1979). Although predator searching behavior has long been identified to be one of the key elements in predation dynamics (DeBach 1964, Frazier & Gilbert 1976, Huffaker 1958, Huffaker et al. 1976, Smith 1939, Ullyett 1943), few experimental studies have focused on searching behavior (Gutierrez et al. 1984, O'Neil 1984, Risch 1981). Our understanding of the process of predation detailed in classical predator/prey models has been enhanced by the close coupling of modeling and experimentation. In agricultural systems there has been little progress made in incorporating our empirical understanding into models of natural enemy dynamics. The net result is that we have few adequate models of predation or parasitism of pests in agricultural crop systems and have yet to fully utilize the power of crop/pest models for the study of predation in agricultural systems.

UTILITY OF CROP/PEST MODELS INCORPORATING NATURAL ENEMY EFFECTS

Incorporating the effects of biological control agents into crop/pest models requires extensive knowledge of how the crop, pests, and biological control agents interact (Stimac 1982, Stimac & Barfield 1979). However, once this task is accomplished, the crop/pest/natural enemy model can be used to explore a variety of IPM and crop production strategies. If the criterion for a "good" production strategy is to maximize or stabilize net profit of the producer, the agricultural system model can be used to explore alternative production strategies in which different combinations of control tactics (biological, cultural and chemical) are tested by computer simulation. The objective is to discover the combinations of pest control tactics and crop production practices which give the largest or most stable net profit over the widest range of environmental variation. The environmental variation includes weather conditions and influxes of pest and beneficial species into the production unit.

The primary utility of crop/pest models is to ask "what if" questions about using combinations of crop production practices, such as cultivation and irrigation scheduling, and pest control tactics, such as introduction or manipulation of biological agents and/or application of pesticides. For example, the model may be used to test a particular production strategy in which the crop will be irrigated to avoid severe water stress, or an application of a pesticide will be made if the density of a particular pest exceeds a specified threshold (Boggess et al. 1983, Wilkerson et al. 1983). The criterion for discovery of a viable production or IPM strategy is to maximize net profit. Where the crop/pest model includes the effects of biological control agents, IPM strategies which may include inundative or inoculative releases or conservation and manipulation of biological agents can be evaluated as part of the crop production strategy analyses. For strategies that are indicated by computer simulation to be viable over a wide range of environmental variation, field testing can then be used to validate the conclusions from the model analyses. Strategy analysis utilizing the crop/pest model reduces the number of possible strategies that must be tested by field experimentation (see also Herzog et al. 1984). Since field testing of a production or IPM strategy is expensive, the dollars invested in construction and analysis of crop/pest models incorporating the effects of biological control agents can be justified.

EVALUATING PREDATORS AND INSECTICIDES IN THE SOYBEAN PRODUCTION SYSTEM

The Soybean Integrated Crop Management (SICM) model (Wilkerson et al. 1983) is a physiologically-based model of soybean growth under variable conditions of weather, insect pests, arthropod predation and pesticide application. This crop/pest system model includes the minimal essential components needed to evaluate the value of natural enemies: a mechanistic model of crop growth in the presence or absence of pests, insect population models including the effects of natural enemies, and models of pesticide effects on both insect pests and their natural enemies. The crop model is responsive to abiotic stresses such as water shortage as well as biotic stresses induced by pest species. Therefore, the system model can be used to evaluate production strategies in which pest management tactics include the application of insecticides and the conservation or manipulation of natural enemies.

The value of natural enemies can be assessed for pest management strategies in which no insecticides are used or strategies in which insecticides of different types are applied when pest densities reach some threshold. First, consider the scenario in which no insecticides are used to control the soybean defoliator, *A. gemmatalis*. The primary source of mortality for the larval population is predation from a complex of arthropod species (Elvin et al. 1983, O'Neil 1984). Assuming an average influx of moths into the soybean field beginning two weeks after planting and irrigation to avoid severe water stress, Fig. 1 shows changes in larval densities, soybean defoliation, and yield and net profit when predators are present (Fig. 1A) or absent (Fig. 1B). In comparing the changes in the number of medium and large size larvae when predators are present or absent in the model, we see the moderate suppressive effect that the predators have on the defoliator. When predators are present, peak larval density is 26/row-m of foliage, and defoliation is less than 50% at 96 days after planting. When predators are absent, larval densities peak at about 38 and defoliation is greater than 70% at day 96. The real value of the predators is seen in the differences in yield and especially net profit. When predators are present, soybean yield is 288.6 kg/ha with a net profit of $10.49/ha, vs. a yield of 243.4 kg/ha and a net profit of only $1.09/ha when predators are absent. The value of the predators in this scenario was $9.40/ha.

Despite the beneficial value of the predators, a net profit of $10.49/ha would not be considered acceptable to most soybean producers. As a second scenario, consider the application of insecticide as a control tactic for the velvetbean caterpillar, whenever the density of the defoliator population is not maintained at a low level by the predators. Assuming that an insecticide will be applied whenever the numbers of medium and large larvae exceed 36/row-m, and that predators are present, we can use the crop/pest model to examine the effectiveness of combined tactics (insecticide and biological control with predators). Fig. 2 shows the simulated results for heavy applications of methyl parathion (Fig. 2A) and light applications of carbaryl (Fig. 2B). For the strategy of methyl parathion and predators, both the yield and the net profit are larger than the strategy with only predators (compare Figs. 1A and 2A) but only minimally larger. However, using carbaryl instead of methyl parathion when predators are present gives a substantially higher yield and nearly douples the net profit. Selection of the insecticide and rate of application (see also Hull & Beers; Mullin & Croft, this volume) can have dramatic effects on net profit because of the differential effects they have on the predators.

With heavy applications of methyl parathion, predator survival is less than 10%, whereas with light applications of carbaryl predator survival is greater than 80% (Wilkerson et al. 1983). The high mortality imposed on predators by methyl parathion nearly eliminates the benefit of their presence and the defoliator population resurges quickly from recruitment. Despite several additional applications of methyl parathion, the soybean crop suffers heavy defoliation between days 95-120, when pods are filling. The increase in soybean yield over the strategy in which no insecticides are used is offset by the cost of insecti-

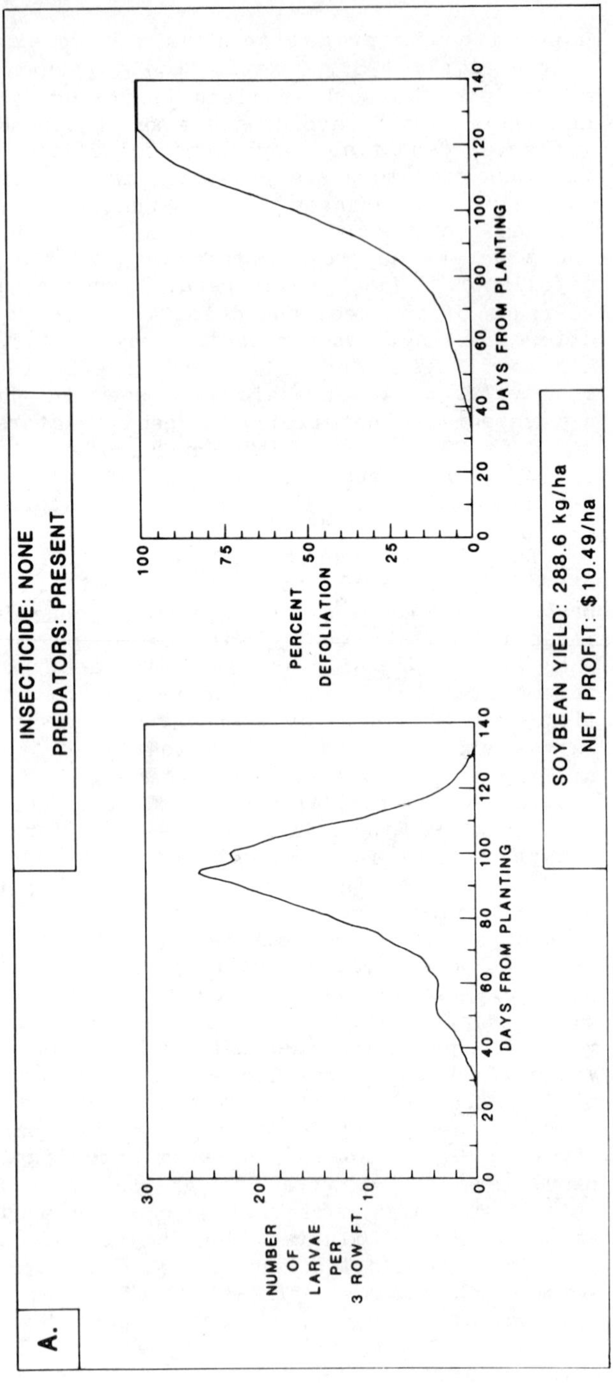
A.
INSECTICIDE: NONE
PREDATORS: PRESENT
NUMBER OF LARVAE PER 3 ROW FT.
0
10
20
30
0
20
40
60
80
100
120
140
DAYS FROM PLANTING
PERCENT DEFOLIATION
0
25
50
75
100
0
20
40
60
80
100
120
140
DAYS FROM PLANTING
SOYBEAN YIELD: 288.6 kg/ha
NET PROFIT: $10.49/ha

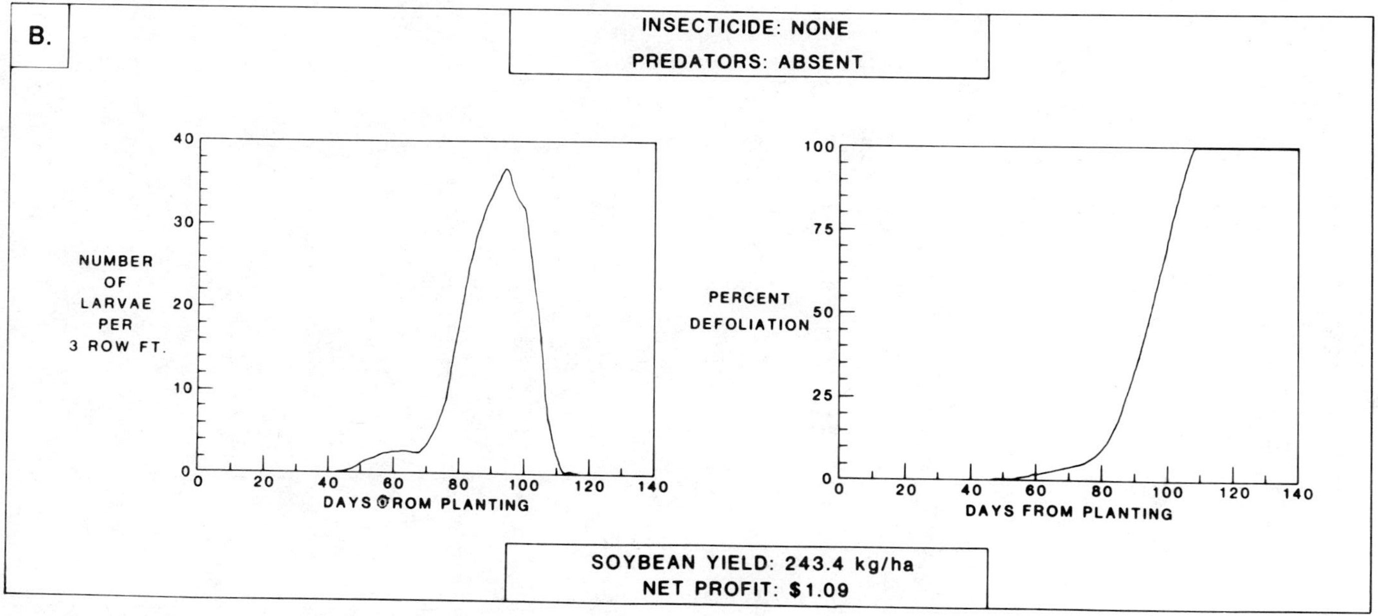

Fig. 1. Simulated changes in the densities of velvetbean caterpillars, defoliation of soybean canopy, yield and net profit when no insecticide is applied and predators are present (A) and absent (B).

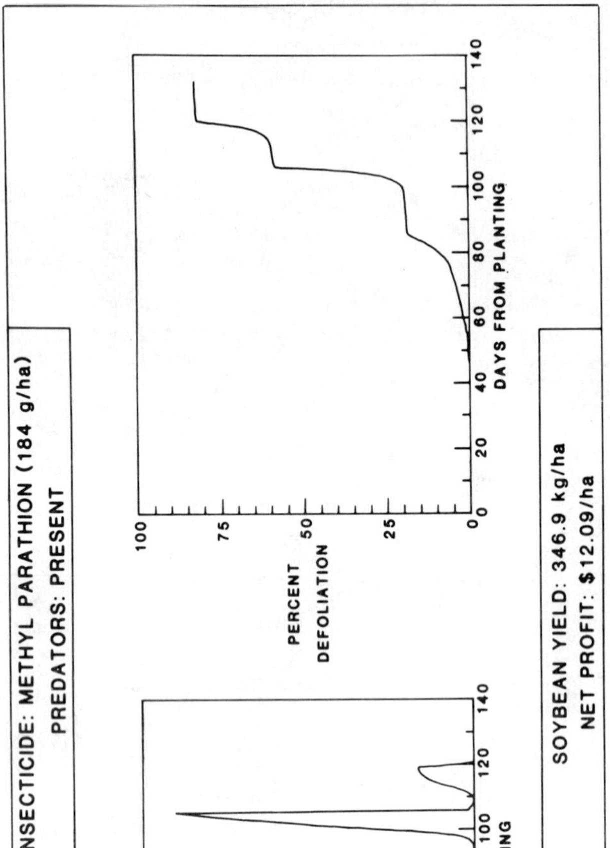
A.
INSECTICIDE: METHYL PARATHION (184 g/ha)
PREDATORS: PRESENT
NUMBER
OF
LARVAE
PER
3 ROW FT.
200
150
100
50
0
0
20
40
60
80
100
120
140
DAYS FROM PLANTING
PERCENT
DEFOLIATION
100
75
50
25
0
0
20
40
60
80
100
120
140
DAYS FROM PLANTING
SOYBEAN YIELD: 346.9 kg/ha
NET PROFIT: $12.09/ha

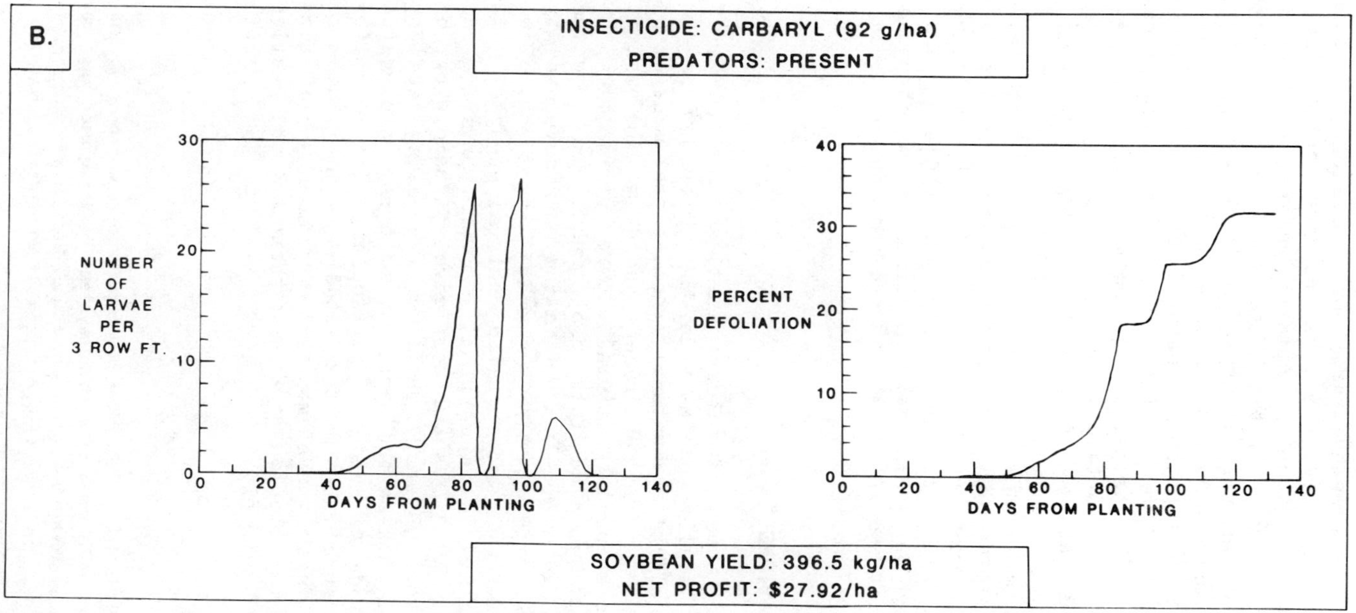

Fig. 2. Simulated changes in the densities of velvetbean caterpillars, defoliation of soybean canopy, yield and net profit when predators are present and insecticide treatments are methyl parathion (A) or carbaryl (B).

cide that is applied in order to compensate for the loss of predator-induced mortality when methyl parathion is used.

The model analysis of predator and insecticide control tactics demonstrates that the value of natural enemies in a crop system can be quantitatively estimated if the influences of natural enemies and pesticides are integrated into pest population models. This integration requires that the impacts of pest control and crop production tactics on natural enemy populations must be described quantitatively. Data such as the mortalities imposed on predators and parasites by application of pesticides at specified rates must be obtained for all of the pesticides that could be used as tactics in a pest management strategy. Selection of pest control tactics can have a dramatic influence on the value of natural enemies and the net profit obtained. Selection of an insecticide which is less detrimental to natural enemies can greatly increase the net profit by allowing indigenous natural enemies to additively contribute to pest mortalities imposed by the pesticide. A similar argument could be made for the interactions of pesticides and introduced natural enemies.

SUMMARY AND RECOMMENDATIONS

Incorporating the influences of natural enemies into models of crop/pest systems is vitally necessary if the system models are going to be used as tools to evaluate alternative pest management and crop production strategies. In recent years, models of mortality of insect pests induced by natural enemies (primarily predators) have been incorporated into pest population dynamics models. Multitrophic level models such as the ones developed by Gutierrez and colleagues (1984) offer a promising approach. However, we must recognize that incorporating natural enemy effects into crop/pest system models is still at a primitive stage and will advance only as quickly as our knowledge of the ecology of natural enemies in crop systems. There is a pressing need for much more experimental work on the behavior and dynamics of natural enemies in agricultural systems, particularly for parasites and pathogens. Natural enemies must be studied within the context of the crop production system so that influences of pest management and crop production on pests and their natural enemies can be described quantitatively.

REFERENCES

Adkisson, P.L., and R.E. Frisbie. 1979. The development of comprehensive, unified, economically and environmentally sound systems of integrated pest management for major crops. Texas A&M Univ. Res. Found. Rept. 3884A, 227 pp.

Allen, J.C. 1976. A model for predicting citrus rust mite damage on Valencia orange fruit. Environ. Entomol. 5:1083-1088.

Altieri, M.A. 1979. The design of pest stable corn agroecosystem based on the management of insect populations through weed management. Ph.D. Thesis, Univ. of Florida, Gainesville.

Asquith, D., B.A. Croft, S.C. Hoyt, E.H. Glass and R.E. Rice. 1980. The systems approach and general accomplishments toward better insect control in pome and stone fruits, pp. 249-317. In: C.B. Huffaker (Ed.), New technology of pest control. John Wiley & Sons, N.Y.

Barfield, C.S., and J.L. Stimac. 1980. IPM: An ecological perspective. BioScience 30:683-689.

Baumgaertner, J.U., B.D. Frazier, N. Gilbert, B. Gill, A.P. Gutierrez, P.M. Ives, V. Nealis, D.A. Raworth and C.G. Summers. 1981. Coccinellids (Coleoptera) and aphids (Homoptera). Can. Entomol. 113:975-1048.

Boggess, W.G., G.D. Lynne, J.W. Jones and D.P. Swaney. 1983. Risk-return assessment of irrigation decisions in humid regions. South. J. Agric. Econ. 15:135-143.

Brown, L.G., J.W. Jones and F.A. Harris. 1976. A simulation study of insect pest management alternatives by integration of Heliothis spp. model and a cotton crop model. Annu. Meeting Amer Soc. Agric. Engin., St. Joseph, MI.

Burnett, T. 1954. Influences of natural temperatures and controlled host densities on oviposition of an insect parasite. Physiol. Zool. 27:239-248.

Cock, M.J.W. 1978. The assessment of preference. J. Anim. Ecol. 47:804-816.

Comins, H.N., and M.P. Hassell. 1976. Predation in multi-prey communities. J. Theor. Biol. 62:93-114.

Croft, B.A. 1976. Pest management systems for phytophagous mites and codling moth, pp. 99-131. In: Proc. 3rd USSR/USA symp. on integrated pest management, Lubbock, TX. Texas A&M Publ., College Station.

Curry, G.L., P.J.H. Sharpe, D.W. DeMichele and J.R. Cate. 1980. Towards a management model of the cotton-boll weevil ecosystem. J. Environ. Mgt. 11:187-223.

DeBach, P. 1964. Biological control of insect pests and weeds. Rienhold, Cambridge, MA. 844 pp.

DeBach, P., and H.S. Smith. 1941. The effect of host density on the rate of reproduction of entomophagous parasites. J. Econ. Entomol. 34:741-745.

Elvin, M.K., J.L. Stimac and W.H. Whitcomb. 1983. Estimating rates of arthropod predation on velvetbean caterpillar larvae in soybeans. Fla. Entomol. 66:319-330.

Frazier, B.D., and N. Gilbert. 1976. Coccinellids and aphids: A quantitative study of the impact of adult ladybirds (Coleoptera: Coccinellidae) preying on field populations of pea aphids (Homoptera: Aphidae). J. Entomol. Soc. Brit. Col. 73:33-56.

Frazier, B.D., N. Gilbert, V. Nealis and R.D. Raworth. 1981. Control of aphid density by a complex of predators. Can. Entomol. 113:1035-1041.

Gutierrez, A.P., J.U. Baumgaertner and K.S. Hagen. 1981. A conceptual model for growth, development and reproduction in the ladybird beetle, Hippodamia convergens (Colepotera: Coccinellidae). Can. Entomol. 109:1457-1468.

Gutierrez, A.P., J.U. Baumgaertner and C.G. Summers. 1984. Multitrophic models of predator-prey energetics. Can. Entomol. 116:(In Press).

Gutierrez, A.P., G.D. Butler, Y. Wang, and D. Westphal. 1977. The interaction of pink bollworm (Lepidoptera: Gelechiidae), cotton and weather: A detailed model. Can. Entomol. 109:1457-1468.

Gutierrez, A.P., J.B. Christensen, C.M. Merritt, W.B. Loew, C.G. Summers and W.R. Cothran. 1976. Alfalfa and the Egyptian alfalfa weevil (Coleoptera: Curculionidae). Can. Entomol. 108:635-648.

Gutierrez, A.P., Y. Wang and R.E. Jones. 1979. Systems analysis applied to crop protection. EPPO Bull. 9:133-148.

Hagen, K.S., E.F. Sawall and R.L. Tassan. 1970. The use of food sprays to increase effectiveness of entomophagous insects. Proc. Tall Timbers Conf. Ecol. Anim. Control. Habitat Mgt. 2:59-81.

Hassell, M.P. 1969. A population model for the interaction between Cyzenis albicans (Fall.) (Tachinidae) and Operophtera brumata (L.) (Geometridae) at Wytham, Berkshire. J. Anim. Ecol. 38:567-576.

Hassell, M.P. 1978. The dynamics of arthropod predator-prey systems. Princeton Univ. Press, Princeton, N.J. 237 pp.

Hassell, M.P., and G.C. Varley. 1969. New inductive population model for insect parasites and its bearing on biological control. Nature 223:1133-1136.

Herzog, D.C., J.L. Stimac, D.C. Boucias and V.H. Waddill. 1984. Compatibility of biological control in soybean insect management. In: P.L. Adkisson and Shijun Ma (Eds.), Proc. Joint U.S. Nat. Acad. Sci./Chinese Acad. Sci. Symp. on Biologicial Control of Insects. Sci. Sinica (In Press).

Holling, C.S. 1959. Some characteristics of simple types of predation and parasitism. Can. Entomol. 91:385-398.

Holling, C.S. 1961. Principles of insect predation. Annu. Rev. Entomol. 6:163-182.

Holling, C.S. 1965. The functional response of predators to prey density and its role in mimicry and population regulation. Mem. Entomol. Soc. Can. 45:3-60.

Holling, C.S. 1966. Functional response of invertebrate predators to prey density. Mem. Entomol. Soc. Can. 48:1-86.

Huffaker, C.B. 1958. Experimental studies on predation: Dispersion factors and predator-prey oscillations. Hilgardia 27:343-383.

Huffaker, C.B. (Ed.). 1980. New technology of pest control. John Wiley & Sons, N.Y. 500 pp.

Huffaker, C.B., F.J. Simmonds and J.E. Laing. 1976. The theory and empirical basis of biological control, pp. 42-80. In: C.B. Huffaker and P.S. Messenger (Eds.), Theory and practice of biological control. Academic Press, N.Y.

Leslie, P.H. 1945. On the use of matrices in certain population mathematics. Biometrika 33:183-212.

Lotka, A.J. 1925. Elements of physical biology. Williams & Wilkins, Baltimore. 460 pp.

Luck, R.F., J.C. Allen and D. Baasch. 1980. The systems approach to research and decision-making in the citrus ecosystem, pp. 366-396. In: C.B. Huffaker (Ed.), New technology of pest control. John Wiley & Sons, N.Y.

Murdoch, W.W. 1969. Switching in general predators: Experiments on predator specificity and stability of prey populations. Ecol. Mono. 39:335-354.

Naranjo, S.E. 1983. Influence of a weed complex in soybean on the predatory, ovipositional and plant feeding responses of Geocoris punctipes Say. M.S. Thesis, Univ. of Florida, Gainesville.

Nicholson, A.J. 1933. The balance of animal populations. J. Anim Ecol. 2:132-178.

Nicholson, A.J., and V.A. Bailey. 1935. The balance of animal populations. Proc. Zool. Soc. Lond. 1935:551-598.

Odum, H.T. 1983. Systems ecology: An introduction. John Wiley & Sons, N.Y. 644 pp.

O'Neil, R.J. 1984. Measurement and analysis of arthropod predation on velvetbean caterpillar, Anticarsia gemmatalis Hubner. Ph.D. Thesis, Univ. of Florida, Gainesville.

Risch, S. 1981. Ants (Hymenoptera: Formicidae) as important predators of rootworm (Coleoptera: Chrysomelidae) eggs in the neotropics. J. Econ. Entomol. 74:88-90.

Risch, S.J., R. Wrubel and D. Andow. 1982. Foraging by a predaceous beetle, Coleomegilla maculata (Coleoptera: Coccinellidae), in a polyculture: Effects of plant density and diversity. Environ. Entomol. 11:949-950.

Rudd, W.G., W.G. Ruesink, L.D. Newsom, D.C. Herzog, R.L. Jensen and N.F. Marsolan. 1980. The systems approach to research and decision making for soybean pest control, pp. 99-122. In: C.B. Huffaker (Ed.), New technology of pest control. John Wiley & Sons, N.Y.

Ruesink, W.G., C.A. Shoemaker, A.P. Gutierrez and G.W. Fick. 1980. The systems approach to research and decision making for alfalfa pest control, pp. 217-247. In: C.B. Huffaker (Ed.), New technology of pest control. John Wiley & Sons, N.Y.

Sharpe, P.J.H., and D.W. DeMichele. 1977. Reaction kinetics of poikilotherm development. J. Theor Biol. 64:649-670.

Smith, H.S. 1939. Insect populations in relation to biological control. Ecol. Mono. 9:311-320.

Stimac, J.L. 1982. History and relevance of behavioral ecology in models of insect population dynamics. Fla. Entomol. 65:9-16.

Stimac, J.L., and C.S. Barfield. 1979. Systems approach to pest management in soybeans, pp. 249-259. In: F.T. Corbin (Ed.), World soybean research conference II: Proceedings. Westview Press, Boulder, CO.

Stimac, J.L., and R.J. O'Neil. 1983. Modeling the impact of natural enemies on insect pests in soybeans, pp. 78-87. In: H.N. Pitre (Ed.), Natural enemies of arthropod pests in soybeans. South. Coop. Ser. Bull. 285.

Stinner, R.E., G.D. Butler, J.S. Bacheler and C. Tuttle. 1975. Simulation of temperature-dependent development in population dynamics models. Can. Entomol. 107:1167-1174.

Tamaki, G., J.U. McGuire and J.E. Turner. 1974. Predator power and efficacy: A model to evaluatee their impact. Environ. Entomol. 3:625-630.

Ulyett, G.C. 1943. Some aspects of parasitism in field populations of Plutella maculipennis. Curt. J. Entomol. Soc. South Afr. 6:65-80.

Ulyett, G.C. 1949. Distribution of progeny by Chelonus texanus Cress (Hymenoptera: Braconidae). Can. Entomol. 81:25-44.

Voltera, V. 1931. Variations and fluctuations of the number of individuals in animal species living together, pp. 409-448. In: R.N. Chapman (Ed.), Animal ecology. McGraw-Hill, N.Y.

von Foerster, H. 1959. Some remarks on changing populations. In: F. Stohlman (Ed.), The kinetics of cellular proliferation. Grune & Stratton, N.Y.

Wang, Y., A.P. Gutierrez, G. Oster and R. Daxl. 1977. A population model for plant growth and development: Coupling cotton-herbivore interactions. Can. Entomol. 109:1359-1374.

Wilkerson, G.G., J.W. Mishoe, J.W. Jones, J.L. Stimac, D.P. Swaney and W.G. Boggess. 1983. SICM Florida soybean integrated crop management model. Version 4.2. Univ. Florida, Dept. Agric. Engin. AGE Rept. 81-1, 216 pp.

DISCUSSION: Integrating influences of natural enemies into models of crop/pest system.

DISCUSSION LEADER: R.E. Stinner

The question was raised as to the funding of modelling research at the expense of implementation. Modeling of agricultural pest systems is, however, relatively new and has required not only extensive validation studies, but also the development of new methodologies. In many research programs, funds allocated to modeling have been used to both determine which biological parameters are most important (from sensitivity analyses) and their values (from biological research).

Although there are few examples of insect pathogen models incorporated in crop models, this is now a major concern in several crop x pest modeling efforts. As has usually proven to be the case, certain key information, however, is lacking. For example, in soybeans with velvetbean caterpillar and the fungus Nomuraea, crop growth effects on the distribution of conidia are not known yet are critical to modeling infection spread.

Since modeling is still a relatively young "discipline," similarities among models are only now becoming apparent. For example, models of natural enemy effects by Gutierrez' group and Stimac's group have used different units (energy vs. density). Only when these models are converted to the same units do they collapse to very similar models. We have observed the initial proliferation of models and are only now recognizing the "sameness" in such models.

Since virtually all of these models are deterministic, the question of variance estimation was raised. Through the use of multiple years of weather data, a given strategy may be simulated and the results used to generate variances for use with techniques such as stochastic dominance for problems such as risk assessment.

Recommendations for research in modeling of natural enemy impacts include: Concentrated effort on field models of general predation and parasitism. This requires efforts to:

1. Quantify mortality curves for the beneficial species for different rates of pesticide applications,
2. Better describe searching and foraging behaviors of predators and parasites and how the crop and production practices influence these behaviors,
3. Develop better means of measuring or predicting influxes of pests and natural enemies,
4. Develop process-oriented crop models for a wider variety of crops, and
5. Recognize difference and importance of both end uses of models: field prediction and strategy analyses.

Biological Control of Weeds

THE USE OF NATURAL AND GENETICALLY ALTERED STRAINS OF PATHOGENS FOR WEED CONTROL

R. Charudattan

The subject of biological control of weeds with plant pathogens has been recently reviewed (Charudattan & Walker 1982). The principles of plant pathology relevant to biological control of weeds (Holcomb 1982), the effect of pathogens on natural plant populations (Quimby 1982), the relationship between genetic variation in weeds and biological control (Barrett 1982), the hazards and benefits of genetic heterogeneity in pathogens to weed control (Leonard 1982), principles of creating epidemics (Shrum 1982), aspects of commercialization of microbial weed control agents (Bowers 1982, Churchill 1982), and the regulation of microbial biological control agents (Charudattan 1982) have been discussed. This review will be confined to the rapidly developing area of biological weed control with pathogens in agricultural integrated pest management (IPM) systems. Examples of biological control of nonagricultural weeds will be discussed only to illustrate specific concepts. So far only natural strains of pathogens have been used as weed control agents; the empirical knowledge gained from these agents will be reviewed. Although opportunities exist for improving naturally-occurring pathogens by conventional strain selection procedures and recombinant genetic manipulative techniques, presently only a limited attempt has been made to produce improved strains for weed control. An overview of some of the techniques available for producing genetically altered pathogens will be discussed to suggest research possibilities.

Although bacterial and viral pathogens have been examined (Templeton 1982a), all of the microbial weed control agents that are in use or under development in North America are fungal pathogens (Table 1). Fungi have a number of advantages as weed control agents. They are the most commonly encountered pathogens of plants, many are destructive, most can be mass cultured and formulated, and they can be integrated into ongoing IPM and crop production systems (Smith 1982). Unlike the bacteria and viruses which usually enter the host through openings or with the help of vectors, fungi are capable of actively penetrating the host.

ISBN 0-12-357030-1

Table 1. Selected list of pathogens under consideration as biological weed control agents in North America.[a]

Principal Weed	Pathogen	Target Crop[b]	Strategy[c]	Location[d]	Investigator(s)	Status[e]
Alternaria cassiae	Cassia obtusifolia, C. occidentalis Crotalaria spectabilis	Soybean	MH	AR; FL; IL; MS; NC; SC	TeBeest, Smith; Charudattan, Sprenkel; Kenney; Walker, Boyette; Quimby; Van Dyke, Haning, Worsham; Ridings	3
A. macrospora	Anoda cristata	Cotton	MH	AR; MS	TeBeest, Smith, Walker, Boyette	2
Alternaria sp.	Silybum marianum	Rangelands	MH	CA	Lindow	1
A. tenuissim f.sp. euphorbiae	Euphorbia esula	Rangelands	MH	ND	Littlefield	1
Araujia mosiac virus	Morrenia odorata	Citrus	CB	FL	Charudattan	1
Aschochyta pteridis	Pteridium aquilinum	Rangelands	MH	CA	Lindow	2

(continued)

Table 1. (Continued).

Principal Weed	Pathogen	Target Crop[b]	Strategy[c]	Location[d]	Investigator(s)	Status[e]
Cercospora rodmanii	Eichornia crassipes	Waterways	MH	FL	Charudattan	3
Cercospora sp.	C. obtusifolia	Soybean	MH	FL	Hofmeister	1
Cercospora sp.	C. occidentalis	---	MH	FL	Hofmeister	1
Colletotrichum cocodes	Abutilon theophrasti	Corn	MH	VT; PQ	Gotlieb; Watson	2
C. dematium	Ipomoea purpurea	Soybean	MH	NC	Van Dyke	1
C. malvarum	Sida spinosa	Cotton	MH	AR	TeBeest, Templeton, Smith	2
Cryptomycina sp.	P. aquilinum	Range-lands	A	CA	Lindow	1
Fusarium lateritium	A. theophrasti, A. cristata, and S. spinosa	Corn; Cotton	MH	IL; MS	Kenney; Walker, Boyette	2

(continued)

Table 1. (Continued).

Principal Weed	Pathogen	Target Crop[b]	Strategy[c]	Location[d]	Investigator(s)	Status[e]
F. solani f. sp. *cucurbitae*	*Cucurbita texana*	Soybean	MH	AR	Templeton, Boyette, Weidemann, Yu	2
Puccinia canaliculata	*Cyperus esculentus*	Corn	A	FL; GA; MD	Freeman; Phatak; Bruckart	2
P. heterospora	*A. cristata*	---	A	SC	Ridings	1
Septoria sylibi	*S. marianum*	Range-lands	MH	CA	Lindow	1
Sphacelotheca holci (*S. cruenta*)	*Sorghum halapense*	Soybean; Corn	A	CA; LA	Lindow; Holcomb	2

[a] Compiled from research reports submitted in 1984 to the Cooperative Regional Research Project, S-136, on Biological Control of Weeds with Plant Pathogens.

[b] The most important crop affected; there are others. --- = The weed is of minor importance in any crop or the location indicated.

[c] The strategy likely to be employed; based on the type of pathosystem involved. CB = Classical biocontrol; A = augmentation; MH = microbial herbicide.

[d] State where the research is in progress. PQ = Province of Quebec.

[e] 1 = laboratory and greenhouse evaluations are in progress; 2 = field efficacy has been demonstrated; 3 = under commercial development.

Hence, the discussion that follows will be confined to fungal pathogens.

BIOLOGICAL CONTROL STRATEGIES: APPLICABILITY TO AGRICULTURAL IPM SYSTEMS

Biological weed control with plant pathogens makes use of three approaches termed classical, augmentative, and microbial herbicide strategies (Charudattan 1984; Templeton 1982a, b). The classical strategy involves deployment of pathogens that are capable of building up to epidemic (epiphytotic) levels following a limited number of inoculum releases. Target weeds are usually exotic and lack a complement of natural enemies in their adventive ranges. Pathogens are obtained from the native range of the weed and are usually rusts or other fungi that have efficient inoculum dissemination capacity through airborne spores. When conditions are favorable, the introduction of the pathogen into the new range of the weed results in a severe disease outbreak, stress on the weed, and weed control. Examples of weeds controlled successfully with this strategy include rush skeletonweed, *Chondrilla juncea*, with the rust *Puccina chondrillina* in Australia (Cullen et al. 1973), blackberries, *Rubus* spp., with the rust *Phragmidium violaceum* in Chile (Oehrens 1977), and hamakua pamakani (river eupatorium), *Ageratina riparia*, in Hawaii with a deuteromycete, *Cercosporella ageratinae* (Trujillo 1976).

The classical strategy is best suited for controlling weeds in undisturbed areas, rangelands and waterways. It is not suitable for agricultural lands where man-made disturbances to the weed's habitat are likely to disrupt epidemics, thereby reducing efficacy of the pathogen. However, it may be possible to integrate classical biological control pathogens with sublethal rates of chemical herbicides in agricultural IPM systems. This needs to be tested.

Augmentation has been attempted with certain pathogens that fail to yield severe epidemics in their native or naturalized homes due to certain biological or ecological constraints on disease development. Insufficiency in disease pressure may be overcome by augmenting the disease build-up. Dispersing inoculum at disease-conducive periods (Phatak et al. 1983), providing suitable humectants to maintain adequate moisture for infection, and the use of automatic, timed-release of spores (Dyer et al. 1982) are methods that have been used for augmenting biological control pathogens. However, generally the inoculum used in augmentation cannot be artificially cultured or applied as an inundative dose over a large portion of the weed population. The rust disease caused by *Puccinia obtegens* was augmented in the croplands of Montana to control Canada thistle, *Cirsium arvense* (Dyer et al. 1982). Yellow nutsedge, *Cyperus esculentus*, has been reported to be controlled with *Puccinia canaliculata* in corn in Georgia by the augmentation strategy (Phatak et al. 1983).

Augmentation may be useful in agricultural IPM systems. It has been suggested that P. canaliculata could be used as a component of integrated weed management systems for yellow nutsedge (Phatak et al. 1983). However, biocontrol based on augmentation, as in the case of the classical strategy, is slow and depends on the gradual build-up of the epidemic. This drawback probably will limit the usefullness of augmentation in agricultural IPM systems, which are characterized by changing schedules of pest and crop management practices.

The microbial herbicide strategy (Templeton 1982a, b) is the best option for weed control in agricultural IPM systems. The applicability of this strategy to commercial agriculture has been successfully demonstrated in citrus, rice, and soybean by the pioneering work of Ridings, Templeton, Smith, TeBeest, and co-workers (Daniel et al. 1973, Ridings 1985, Templeton & Smith 1977, Templeton et al. 1979). In this strategy, the inoculum of a pathogen is mass cultured, standardized, formulated, and applied to weeds in crops when both the crop and the weed are in early growth. For perennial weeds in perennial crops, the pathogen may be applied at any time when conditions are suitable for disease onset and weed growth. Because this method so far has relied on native, endemic diseases that are adapted to local conditions, environmental constraints on disease development are generally not present or can be surmounted with proper timing of inoculum application. Deficiencies in the amount of natural inoculum available in the field can be overcome by the application of mass-produced inoculum. Both intensity and the speed of the epidemic can be increased to achieve the desired level of weed control by applying inundative doses of inoculum at the proper stage of weed growth (Shrum 1982).

USE OF PATHOGENS FOR WEED CONTROL IN AGRICULTURAL IPM SYSTEMS: STATUS AND PROSPECTS

Natural Strains

Presently two pathogens are registered as microbial herbicides for commercial use in the United States. They are DeVine® and Collego®, used respectively for controlling the milkweed vine, Morrenia odorata, in citrus in Florida and northern jointvetch, Aeschynomene virginica, in rice and soybean in Arkansas and other southeastern states.

Milkweed vine, an exotic weed in Florida citrus groves, was introduced as a potential ornamental plant. Its distribution and economic impact on the Florida citrus industry have been detailed by Ridings (1985); about 120,000 ha of citrus are infested with the vine. DeVine, marketed by Abbott Laboratories, Chicago, IL, consists of a pathotype of Phytophthora palmivora that was isolated from diseased milkweed vines in Florida (Ridings 1985). It was registered for use in 1981. The biological control potential of this pathogen was studied cooperatively by scientists of the Division of Plant Industry of the Florida Department of Agriculture and Consumer Services, University of Florida, and the Abbott Laboratories. Results confirmed a high level of efficacy and safety and the desirability of using this pathogen as a microbial herbicide. Details of research and development of DeVine have been reviewed previously (Ridings et al. 1976, 1982; Ridings 1985).

DeVine is a liquid formulation of P. palmivora containing 6.7 x 10^5 live chlamydospores per ml. Ridings (1985) has reported that highly effective weed mortality could be achieved in the field with 8 chlamydospores per cm^2 of soil. In practice, DeVine is applied at the rate of 1.17 l. of formulation in 450 l. of water per treated ha. It is applied postemergence between May and September, after the vine develops from seeds or from rhizomes following winter-kill of the shoots. Application is made to the soil around citrus trees using a boom sprayer (DeVine Label 1981). The soil must be wet when DeVine is applied; either 5 cm of rainfall or irrigation before and after application should be provided. The pathogen in DeVine infects and wilts seedlings or established vines in 2-10 weeks following application. The vine populations gradually decline in treated areas, and complete control may be achieved in 18 months after one treatment (DeVine Product Brochure 1981).

Several unique technical and marketing concepts were applied to the commercialization of DeVine, the first registered microbial herbicide (Kenney 1985). For example, the rapid loss of viability required Abbott to adopt a marketing technique akin to the handling of fresh milk. DeVine was "made-to-order" and stored and distributed under refrigeration. It was necessary to monitor every lot to assure viability; inventories were not carried-over. However, these problems were minimized because the market was localized.

Certain nontarget plants, watermelon, periwinkle, and Rhododendron champanii, were susceptible to DeVine (DeVine Product Brochure 1981, Ridings 1985) when they were exposed to unnaturally high levels of inoculum at disease-conducive moisture levels. Despite this capacity for infecting nontarget hosts, it was possible to register DeVine with proper label restrictions (Kenney 1985). The fact that the pathogen was native to Florida where it was to be used and that it was not to be recommended for use in areas where the nontarget suscepts are grown were considerations in favor of registration (W.H. Ridings, personal communication). There are no published accounts of the total hectarage treated with DeVine or specific field estimates of efficacy, but Kenney (D.S. Kenney, personal communication) has indicated that user acceptance of DeVine has been good and it is a commercial success.

Although the weed northern jointvetch competes with rice and can reduce yield by 4-9%, a more serious problem is reduction in rice quality. Smith (1985) reported that 5.8% of the rice crop in Arkansas was sold at discounted price annually between 1980 and 1982 due to the presence of seeds of northern jointvetch and hemp sesbania, *Sesbania exalta*, in the milled rice. This represented an annual loss of $3.9 million. Northern jointvetch also is a problem in soybean. Although at normal field densities, northern jointvetch does not reduce yield or quality of soybean, control is necessary to prevent seed production and contamination of the land with seeds because of subsequent rotation with rice (R.J. Smith, personal communication).

Collego was the first microbial herbicide registered for use in an annual crop. Scientists of the USDA-ARS, University of Arkansas and the Upjohn Company were involved in the development of Collego. The history of research and development of Collego has been reviewed by Bowers (1982, 1985), Churchill (1982), Smith (1985) and Templeton et al. (1982). The pathogen *Colletotrichum gloeosporioides* f.sp. *aeschynomene*, which causes an anthracnose, was originally found in 31 rice growing counties of Arkansas, where the disease was present at endemic levels (Smith et al. 1973). The pathogen causes lesions, mainly on stems, and girdles the stem. Wilting of the entire stem ensues (Collego Technical Manual 1982). Infection usually occurs seven to ten days after fungus application, and plant death may occur four to five weeks later. The pathogen kills only northern jointvetch, although infection occurs in several species of *Aeschynomene*, *Lathyrus*, *Lupinus*, and certain cultivars of *Pisum sativum* (Daniel et al. 1973., TeBeest 1984a, Templeton et al. 1982).

Collego is a postemergence microbial herbicide. It consists of dried spores of the pathogen obtained from submerged cultures (Churchill 1982, Collego Technical Manual 1982). It has two components, and a packet of activated charcoal is supplied for cleaning pesticide residues from application equipment (Bowers 1985, Collego Technical Manual 1982). The component A is a spore rehydrating agent that is necessary to assure good viability. Component B contains the spores, at least 75.7×10^{10} per package, and inert ingredients. One bag of spores and 946 ml of the rehydrating agent are sufficient to treat 4.04 ha. Collego should be applied to northern jointvetch plants before flowering, when they are 20-61 cm tall and are emerging through the rice canopy. Rice may be at any stage of growth, and rice fields should be flooded before Collego application. Soybean should be treated after the initiation of flowering; fields should be irrigated just prior to application to maintain high levels of humidity (Collego Technical Manual 1982).

The efficacy of Collego (Table 2) from several years of field tests has been summarized by Smith (1985). The efficacy has been extremely good, and the majority of growers who used Collego in 1982 used it again in 1983 (Bowers 1985). Also, if 2,4,5-T and silvex, the two herbicides currently used to control jointvetch, are removed from the market due to environmental concerns (Smith 1985, Templeton 1985), the market for Collego is likely to improve further. Nonetheless, the overall acceptance of Collego has been good, but perhaps less than anticipated (Bowers 1985).

Collego stands as an excellent example of industry-public sector cooperation and one of the landmark developments in microbial herbicides and weed science.

While discussing the prospects of biological control of weeds with plant pathogens, Templeton (1982a) stated that we may expect increasing interest and continued progress toward practical utilization of plant pathogens in future weed control technology. A similar view was expressed by Quimby & Walker (1982). Table 2 lists current projects reported under the Cooperative Regional Research Project, Biological Control of Weeds with Plant Pathogens. The majority of the pathogens in this list are candidates for the microbial herbicide strategy, are native to North America, and are intended to be used against weeds of crops. Evidently, there is active research on this subject and

Table 2. Efficacy of Colletotrichum gloeosporioides f.sp. aeschynomene as a biocontrol agent for northern jointvetch in rice and soybean in Arkansas (from Smith 1985)[a]

Crop	Year	No. of Tests	Ha treated	Average % control
Rice	1972	1	1	95
	1973	4	32	97
	1974	8	191	76
	1975	17	254	94
	1976	20	76	93
	1977	19	218	98
	1978	13	112	94
	1979	17	157	96
	1980	22	214	86
	1981	15	248	98
	1982	2	18	90
Soybean	1976	4	15	100
	1977	6	48	99
	1978	2	20	100
	1979	5	39	100
	1980	7	67	91
	1981	5	58	100

[a] The fungus was applied aerially to rice and soybean fields in nonreplicated trials. Fields were flooded (rice) or were wet from irrigation or rain (soybean) just before applying the mycoherbicide. Reproduced from Weed Sci. Suppl. with permission from the author and the publisher.

the prospects for developing additional microbial herbicides for agricultural systems are very good.

Integration of Microbial Biocontrol Agents With Other Pesticides

Much of what we know on this subject has come from the work of Smith (1982, 1985) on Collego in rice and soybean. In Arkansas, rice and soybean crops are infested with a spectrum of weeds and are attacked by several disease and insect pests. Weed, disease, and insect control practices rely on several chemical pesticides. Generally, it can be expected that microbial herbicides will be sensitive to fungicides and certain other types of pesticides. Therefore, microbial herbicides should be carefully integrated with IPM and crop production systems in a practical and effective manner.

Collego is incompatible with liquid nitrogen fertilizers, insecticides, fungicides, and certain herbicides such as MCPA, 2,4-D, 2,4-DB, 2,4,5-T and molinate (Collego Technical Manual 1982). Germination of spores will be reduced if Collego is combined with these pesticides. Collego should not be applied to northern jointvetch that has been treated with phenoxy herbicides, and fungicides should not be applied within one week following Collego application (Collego Technical Manual 1982, Smith 1985).

Under proper conditions, Collego has been successfully integrated into rice and soybean pest management and production systems. Tables 3 and 4 list the types of integration of Collego and pesticides that have been tested (Smith 1985). In summary, Collego and herbicides can be effectively integrated. Tank mixtures of acifluorfen, bentazon and Collego have been successfully used to control a spectrum of weeds. The insecticides carbofuran and malathion could be integrated without affecting the activity of Collego. The fungicide triphenyltin hydroxide, used for controlling rice sheath blight caused by *Rhizoctonia solani*, can be applied sequentially about one week after Collego treatment. However, benomyl, used for controlling rice blast incited by *Pyricularia oryzae*, will affect Collego activity when applied within three weeks after the fungus application. Benomyl, triphenyltin hydroxide and Collego may be applied sequentially to control rice diseases and northern jointvetch, and benomyl and Collego may be used sequentially against soybean diseases and the weed.

DeVine should not be tank-mixed with wetting agents, fertilizers, or other pesticides (DeVine Label 1981). The germination of chlamydospores of *P. palmivora* is inhibited when the spores are mixed with certain herbicide products containing bromacil, diuron, glyphosate, paraquat and simazine. Sequential applications of spore suspensions and chemical herbicides avoids such inhibition. Both glyphosate and the microbial herbicide are active when the former is applied three weeks prior to the latter (M.E. Kannwisher-Mitchell, personal communication).

Table 3. Control of northern jointvetch and hemp sesbania in rice with tank-mixtures of Colletotrichum gloeosporioides f.sp. aeschynomene (CGA) and acifluorfen (from Smith 1985).[a]

Herbicide	Rate (billion spores or kg/ha)	%Control[b] Northern jointvetch 1981	1982	1983	Avg	Hemp sesbania 1981	1982	1983	Avg
2,4,5-T	1.1	100	100	97a	99	100	100a	100a	100
Propanil + 2,4,5-T	3.4+0.84	100	100	97a	99	100	100a	100a	100
Propanil + acifluorfen	3.4+0.14	83	67	45b	65	100	100a	94ab	98
CGA + acifluorfen	190+0.14	95 ns	97 ns	53b	82	99 ns	81b	91b	90

[a] Herbicide treatments were applied at midseason when rice, northern jointvetch, and hemp sesbania were 48-, 59- and 65-cm tall, respectively.

[b] Values followed by the same letter are not significantly different at the 5% level by Duncan's multiple range test. ns = Values in the columns not significantly different at the 5% level. Reproduced with permission from author and publisher.

The concept of integrating a chemical herbicide and a microbial herbicide to control a target weed can be illustrated by the example of waterhyacinth, Eichornia crassipes. The microbial herbicide candidate, Cercospora rodmanii, and sublethal rates of diquat or 2,4-D (0.3% and 6.4%, respectively, of the current label rates) can be combined to achieve a better control of waterhyacinth than is possible with the microbial herbicide alone (Charudattan 1985, unpublished data). In this case, the integration is intended primarily to improve the efficacy of the microbial herbicide. However, if it becomes practical, the approach could offer a means to reduce the amount of chemical herbicides reaching the aquatic environment (Charudattan 1985).

In the Pacific Northwest, attempts have been made to integrate Puccinia chondrillina, the classical biological control agent for skeletonweed, and chemical herbicides (Lee 1985). The weed poses problems in forage fields and threatens to invade wheat fields (Ehrensing & Isaacson 1977). Integrated control of

Table 4. Interaction of Colletotrichum gloeosporioides f.sp. aeschynomene (CGA) and various rice pesticide treatments on northern jointvetch (from Smith 1985).[a]

Pesticide treatment[b]	Rate (billion spores or kg/ha)	Control[c] %
Untreated check	---	0 d
CGA	190	72 b
Benomyl	0.56	2 d
Fentin hydroxide	0.56	0 d
Propanil	2.2	91 ab
Acifluorfen	0.14	8 d
CGA fb benomyl[d]	190, 0.56	49 c
CGA fb fentin hydroxide[d]	190, 0.56	81 b
CGA fb propanil[e]	190, 2.2	100 a
CGA fb acifluorfen[e]	190, 0.14	84 b

[a] From 1981 and 1982 trials at Colt and Stuttgart, Arkansas.
[b] CGA treatments applied when northern jointvetch plants just emerged through rice canopy. Northern jointvetch and rice plants ranged from 46- to 63- and 46- to 65-cm tall, respectively. fb = followed by.
[c] Average control of four experiments based on dry weights of weeds harvested at rice maturity. Values followed by the same letter are not significantly different at the 5% level by Duncan's multiple range test.
[d] Benomyl and fentin hydroxide applied twice at indicated rates each time, first application 1 to 2 weeks after CGA treatment and second application 2 weeks after the first.
[e] Propanil and acifluorfen applied 1 week after CGA.

the weed using rust infections and low rates of picloram appears feasible (Lee 1985).

Integration of Two Microbial Herbicide Candidates

One of the disincentives to commercialization of microbial herbicides is the specificity of the candidates. Since microbial herbicides are usually specific to one weed, as discussed above, it is necessary to use some chemical herbicides in addition to the microbial herbicides to control the spectrum of weeds in a crop. The economic return to the seller from a highly selective herbicide is also limited. These problems may be overcome by using mixtures of pathogens of different weeds as demonstrated by Boyette et al. (1979), who obtained 94% and 100% control of northern jointvetch and winged waterprimrose, Jussiaea decurrens, in rice with aerial application of a mixture of Colletotrichum gloeosporioides f.sp. aeschynomene and C. gloeosporioides f.sp. jussiaeae.

Walker and colleagues (Walker 1981b, personal communication; Walker & Sciumbato 1981a) have found that both Fusarium lateritium and Alternaria macrospora were capable of controlling spurred anoda, Anoda cristata. When the pathogens were combined in sequential applications, a synergistic disease interaction occurred, and the level of host damage was higher than when the pathogens were applied individually. The interaction was influenced by inoculum concentrations, inoculum ratios, and the sequence of inoculations.

Control of Multiple Weeds With Single Microbial Herbicides

Pathogens that have narrow host ranges but which attack more than one economically important weed are under evaluation. Walker and others (1981b, 1982, personal communication) are examining the herbicidal potential of Fusarium lateritium (a pathogen of velvetleaf, Abutilon theophrasti; prickly sida, Sida spinosa; and spurred anoda) and Alternaria cassiae (a pathogen of sicklepod, Cassia obtusifolia; coffee senna, Cassia occidentalis; and showy crotalaria, Crotalaria spectabilis). Presumably, the use of a single microbial herbicide selective for these weeds, which are often a problem in the same crop in the Southeast, will improve the commercial prospects of that herbicide.

Integration of Arthropods and Microbial Biocontrol Agents

There are no examples of weed control by integrated use of arthropods and pathogens in agricultural systems. The case of waterhyacinth is used to illustrate the potential usefullness of this method. At least five arthropods, including three that were introduced into the U.S. for biological control purposes, have had significant impact on waterhyacinth populations in the Southeast (Charudattan 1985). Of these, two species of weevils, Neochetina eichorniae and N. bruchi, are widespread and may have contributed to the decline of waterhyacinth populations in the Southeast (Charudattan 1985). Generally, the arthropods have a subtle and gradual effect in limiting waterhyacinth populations. When arthropod attacks are combined with diseases, such as Cercospora rodmanii, the resulting level of stress can produce almost complete control of the weed within one growing season (Charudattan 1985, unpublished data). Since the pathogen is most effective when waterhyacinth growth rates are retarded (R. Charudattan et al., unpublished data), the integration of arthropods and the pathogen is a way of improving the overall efficacy of the biological control attempt against this weed (Charudattan 1985).

Biological control of skeletonweed in Australia has been accomplished with the integrated use of Puccinia chondrillina and arthropods (Waphshere 1974). The pathogen is the most effective agent and it is followed by the gall midge Cystiphora schmidti and a gall-forming mite Aceria chondrillinae. These three agents have been effectively used also in the western United States (Lee 1985). A similar approach could be used in crops, for example against naturalized weeds, by integrating classical insect biocontrol agents and pathogens. With respect to the pathogens, the classical and augmentative strategies may be effective in perennial crops, whereas in annual crops, the microbial herbicides may be needed (see section on Biological Control Strategies).

Application Technology

Microbial inocula can be formulated in a number of ways. The existing application methods and equipment are generally adequate for delivering microbial weed control agents, except perhaps in the case of agents used in the augmentation strategy. For example, Dyer et al. (1982) have used a temperature-controlled mechanism for timed-release of spores of Puccinia obtegens in the field. In agricultural IPM systems, any added cost of specialized equipment probably would discourage the use of microbial herbicides. Therefore, the present emphasis is to develop formulations that can be applied with existing application technology.

As previously stated, DeVine and Collego are applied with boom sprayers mounted on tractor and winged-aircraft, respectively. It may be convenient and cost-effective for some farmers to apply Collego to soybean with land-based equipment. Accordingly, Smith (R.J. Smith, personal communication) has evaluated a tractor-mounted boom-nozzle sprayer and a spinning disc applicator for delivering tank-mixtures of Collego and chemical herbicides. The boom-nozzle sprayer appeared to be more effective than the spinning disc applicator. Quimby & Fulgham (1984) have developed a tractor-mounted spray system to deliver simultaneously, but through separate nozzles, pesticides and fertilizers that are incompatible with microbial herbicides. They have designed a logarithmic sprayer which can be used in conjunction with a roller pump sprayer. Any combination of over-the-top and/or directed treatments could be applied through a series of valves. In this way, combinations of incompatible materials could be applied concomitantly by mixing them in intersecting spray patterns with one pass through the field.

Walker & Connick (1983) have developed a method of encapsulating fungal mycelia in alginate gels to enable broadcast application of certain microbial herbicides. Blended mycelia and broth from liquid cultures are mixed with sodium alginate, and the mixture pelletized by dropwise extrusion into a calcium chloride bath. The resulting beaded gels are dried into pellets for prolonged storage. The entrapped mycelium remains viable and sporulates profusely and repeatedly when alternately moistened and dried. Even fungi that are difficult to sporulate in culture could be induced to sporulate on the gel surface (C.D. Boyette, personal communication). The pelletized microbial herbicides can be applied over the plants or the soil, the latter during cultivation. Floating inoculum pellets have also been developed and successfully tested (Boyette et al. 1983), which may be useful against aquatic weeds.

Herbicides are usually applied to waterhyacinth with handheld nozzles connected to pump sprayers aboard air-boats, helicopters, or winged aircraft. Aerial, land-based, and air-boat based application of *Cercospora rodmanii* using boom and hand-held, high pressure nozzle sprayers have been successful (R. Charudattan, unpublished data; Theriot et al. 1981). Two types of sprayers, powered by either a high-pressure piston pump or a low pressure impeller pump, have been found to be suitable for *C. rodmanii* (Theriot et al. 1981).

Resistance Build-up

Resistance of weeds to biological control pathogens has been of some, but not serious, concern. Host resistance was found to be a factor in limiting rust (*Puccinia obtegens*) infections on Canada thistle (Turner et al. 1981). In Australia, three ecotypes of rush skeletonweed varying in their resistance to *Puccinia chondrillina* apparently existed when the pathogen was released (Hull & Groves 1973). Following the successful establishment of *P. chondrillina* as a biological control agent, a decline in the numbers of the predominant host type that was susceptible to the rust and an increase in the numbers of two other resistant host types occurred (Burdon et al. 1981).

Resistance to biological control agents is not likely to be a problem in weeds with small gene pools of resistance genes such as clonally propagated weeds and weeds with narrow genetic diversity (Barrett 1982, Burdon & Marshall 1981, Day 1974). To the contrary, plants with outbreeding systems are more likely to thwart biocontrol agents through recombination of genes and evolution of resistant genotypes (Barrett 1982, Levin 1975). In the case of weeds with prolonged seed dormancy, the build-up of resistant biotypes is not expected to be noticable for many years (Templeton 1985). However, when resistant types do become abundant, it should be possible to control them with virulent strains of pathogens selected to match the resistant host genotypes (Leonard 1982).

Genetically Altered Strains

Status and prospects Altered strains are defined as those created in the laboratory using the conventional mutation and strain-selection processes or by the new recombinant genetic techniques. They may be genetically stable or unstable with respect to the desired feature (for discussions on related topics, see Beckendorf & Hoy; Faulkner & Boucias; Hoy; Napoli & Staskawicz, this volume).

Although no genetically altered strain has so far been put to practical use, an attempt is underway (TeBeest 1984b). With increasing use of pathogens for weed control, the need to improve natural strains into more desirable types possessing weed control and pesticide-resistance qualities is becoming evident. This is especially true in agricultural systems in which any incompatibility between the microbial weed control agent and the ongoing pest and crop management strategies will limit the efficacy and practicability of the microbial agent. Genetic improvement of the microbial agent to suit agricultural needs, therefore, should be explored.

The prospects for genetic improvement of microbial weed control agents are considered good (Day 1974, Leonard 1982, Sands & Rovira 1972, Yoder 1983). As Leonard (1982) has pointed out, one advantage of using pathogens for controlling weeds is that if resistant weed biotypes become dominant or evolve in a population, the pathogen strains can be continually selected to match the newly dominant host types. Likewise, it may be possible to improve other aspects of microbial weed control agents. For example, the resistance or tolerance of strains to chemical pesticides used in crop production systems may be increased, so that the microbial and chemical pesticides can be applied simultaneously without affecting the activity of the microbial agent. Virulence of strains, whether it is controlled by toxins and enzymes or by host-pathogen recognition and compatibility factors, may be increased. Host specificity may be altered to narrow or increase the host range of pathogenic strains. Strains may be adapted to adverse environmental conditions, such as temperature extremes and suboptimal moisture. The sporulation capacity of strains may be increased to facilitate production of primary (fermentation culture) and secondary (on the infected plant) inoculum. The incubation time required for disease development and the lag phase of the disease curve may be reduced to increase the rate of epidemic by using strains with rapid reproductive rates. The persistence of the agent in nature may be controlled by using engineered strains with long or short survival ability.

Techniques A number of old and modern techniques are available to produce genetically altered strains of pathogens. Conventionally, N-methyl-N'-nitro-N-nitrosoguanidine, ethylmethanesulfonate, and ultraviolet radiation have been most commonly used for fungal mutagenesis (Bridges 1976). Strain improvement can also be achieved by sexual and parasexual recombination (Sidhu 1983, Tinline & MacNeill 1969). Virulence may be altered by repeated passage of a pathogenic strain through a host as has been demonstrated with *Septoria nodorum* and *Phytophthora infestans* (Cunfer 1984, Turkensteen 1973). Increased quantities of certain enzymes and toxins involved in pathogenesis can be induced in some fungi by specific cultural techniques and selection processes (Beraha et al. 1983, Keen & Horton 1966, Nelson et al. 1963, Scheffer et al. 1964). The adapted strains may be more pathogenic to the target host than are the parental strains.

Among the new recombinant genetic techniques, protoplast fusion (Alfoldi 1982, Peberdy 1979), micro-injection of genetic information into somatic cells (Celis et al. 1980), microcell mediated gene transfer (Ruddle 1980), chromosome mediated gene transfer (Ruddle 1980), and DNA-mediated gene transfer (Case 1982, Ruddle 1980) are potentially useful tools for producing altered strains of pathogens. A plasmid DNA from Fusarium oxysporum f.sp. lycopersici has been successfully maintained in the bacterium Escherichia coli K-12; this plasmid may be a suitable vector for gene cloning in fungi (Martini et al. 1978).

The protoplast fusion technique in particular is being used increasingly in fungal biotechnology (Peberdy 1979). The significance of the technique is that the removal of the cell wall facilitates the artificial recombination of genetic material into living cells. The protoplasts then regenerate cell walls and are reconverted into growing cells or hyphae (Peberdy 1979). Protoplast fusion has been used or is being explored for a variety of purposes. These include: 1) production of fruiting bodies such as perithecia (Leslie 1983) and mushrooms (Abe et al. 1982, Yamada et al. 1983); 2) production of heterokaryotic mycelium (Dales & Croft 1977, Leslie 1983); 3) fusion of identical/opposite mating types (Becher & Bottcher 1983); 4) interspecific crosses that would not normally occur (Anne & Eyssen 1978, Becher & Bottcher 1983, Minuth & Esser 1983); 5) intergeneric crosses (Minuth & Esser 1983); 6) production of certain primary and secondary metabolites such as cellulases (Toyama et al. 1983), the sex hormone zearalenone (Leslie 1983), and gibberellins (Harris 1982); and 7) transmission of cytoplasmic genetic determinants (Maraz & Ferenczy 1980, Peberdy 1979).

Fungicide Resistance In view of our understanding of the genetics of fungicide resistance, we may be able to make significant progress in the selection of fungicide-resistant strains that could be used simultaneously with crop-protectant fungicides. This is of urgent interest to those who developed Collego. As stated previously (see section on Integration), the natural strain of the pathogen used in Collego is incompatible with benomyl, and the fungicide should not be used within three weeks after mycoherbicide application (Collego Technical Manual 1982). In an attempt to develop a fungicide resistant strain, TeBeest (1984b) has induced benomyl tolerance in the fungus by treating spores with ethylmethanesulfonate and selecting for in vitro benomyl tolerance. Four mutants that have have been selected appear to be genetically stable, pathogenically specific, and potentially useful as control agents for northern jointvetch. The four strains have been repeatedly cultured and passed through the host without loss of benomyl tolerance (TeBeest 1984b).

Although resistance of fungal pathogens to benomyl is well known and probably occurs frequently in situations of heavy benomyl use (Delp 1980), it may not be equally easy to obtain resistance to certain other types of fungicides. To understand how to select for fungicide resistance, we may review what is known about fungicide resistance in fungal pathogens of crops (Delp 1980, Skylakakis 1983). Although there has been an increase in fungicide resistance associated with increasing use of systemic fungicides, the site-specificity rather than systemicity of fungicides appears to be the cause of resistance in fungi. Thus it should be easier to induce resistance to site-specific fungicides than to multisite inhibitors. This is because the latter class of fungicides affect numerous metabolic sites, and therefore multiple changes in the fungal genome will be required to cause resistance. On the other hand, alterations in a single gene may be sufficient to induce change in the site of action of a site-specific fungicide.

The genetic basis of fungicide resistance should be identified in order to facilitate production of fungicide-resistant strains through genetic engineering. Delp (1980) listed nine fungi in which the gene(s) controlling resistance to certain fungicides have been identified. The fungicides involved include aromatic hydrocarbons, benzimidazoles, dodine, carboxamides, cycloheximide, imazalil, and kasugamycin. Although the chromosomal location of the gene(s) is known, plasmid-mediated fungicide resistance also is possible. Single, double, or multiple genetic control of fungicide resistances have been reported. It should be easier to change a single gene than a multiple gene system, especially if the genes are not closely linked or are on different chromosomes.

General and specific biochemical mechanisms of fungicide resistance have been identified (Delp 1980). These mechanisms include increased metabolism, reduced permeability (exclusion), detoxification (breakdown or binding), decreased conversion of the fungicide chemical into a toxicant, and modification of the sensitive sites. These mechanisms, in addition to in vitro assays on fungicide-containing plates, may be used as markers for selection.

Delp (1980) has mentioned the phenomenon of transient resistance to fungicides (training), which results from nongenetic adaptation. This phenomenon is commonly observed in culture and is lost after the fungus is grown in the absence of the fungicide. This type of adaptation among pathogens appears to pose little threat to commercial agriculture, and may be useful for microbial herbicides. If the transient resistance can be retained for one generation of a microbial herbicide agent (i.e., from mass culture to one cycle of host infection in the field), the trained fungus can be applied to weeds with no consequence towards build-up of the fungicide-resistant strain in nature. Moreover, this approach may be attractive from the commercial standpoint of repeat sales of the trained strain.

Although resistance of crop-destructive fungi to fungicides is a difficult problem in agriculture, attempts to select fungicide-resistant strains of pathogens for weed control should not be viewed askance. If weed-specific pathogens are used for induction of resistance to fungicides, the dangers to nontarget plants will be minimized. As further safeguards, attempts to induce fungicide resistance in nonspecific fungi should be avoided and every fungicide-resistant strain intended for use must be subjected to rigorous host range tests, even though the parent strain may have already have been tested and found to be host-specific. The purpose of the latter tests will be to determine whether any genetic change towards altered host specificity also accompanied the genetic change to fungicide resistance.

Toxin Production The genetics of toxin production is fairly well understood in several species of *Alternaria*, *Helminthosporium*, *Periconia*, and *Phyllosticta* that produce host specific toxins (Scheffer & Livingston 1984, Yoder 1980). However, to date there are no reports of successful cloning of genes controlling toxin production in plant pathogenic fungi (Yoder 1983). Yet it may be possible to isolate chromosomal genes that control toxin production and transfer them to toxin-less strains.

Yoder (1983) has considered the use of toxin-producing pathogens for weed control and has stated several interesting concepts that are relevant to the discussion here. First, he has regarded toxins as either pathogenicity (qualitative) factors or virulence (quantitative) factors (Yoder 1980, 1983). The qualitative factors may be required for producing disease in the first place, and the quantitative factors may control the amount or severity of the disease. Presumably, it will be possible to select fungi for either altered pathogenicity or virulence. However, I believe that because of regulatory problems and public concerns, only fungi with proven pathogenic specificity to target weeds are likely to be used for strain improvement. Therefore, alterations in the level of virulence by means of increased toxin production, rather than conferring pathogenic potential to nonpathogens, may be considered for immediate research (see Napoli & Staskawicz, this volume, regarding regulatory constraints on releasing genetically altered microbial biological control agents into the environment).

Yoder (1983) has suggested that genes controlling the production of nonspecific toxins may be cloned and transferred to a host-specific pathogen. The virulence of the transformed strain of the pathogen thus may be improved, provided that the transformed cells also receive one or more genes that confer self-resistance to the toxin. The host-specific pathogen "carrying a foreign gene for production of a nonspecific toxin will be highly virulent towards the pest but will cause no damage to nonhosts or to other organisms" because of its specificity. Yoder also believes that engineering pathogens for host-specific toxins would pose no apparent environmental threat. Since these toxins are extremely specific, only sensitive hosts will be affected.

In summary, because research on recombinant genetics of pathogenic fungi in general and the production of genetically altered pathogens for weed control in particular are at a preliminary stage, these discussions on selection of strains have been unavoidably speculative. Much work needs to be done towards identifying the chromosomal and extrachromosomal determinants of pathogenicity, virulence, sporulation, host specificity, and host range before progress can be made in genetic engineering of pathogens for weed control. Therefore, it is my view that unless specific gene-targeted improvements are desired, our immediate needs for creating improved fungal strains for weed control can be served by random mutation and selection. Protoplast fusions between intraspecific, interspecific, and intergeneric weed-pathogens deserve consideration to produce highly virulent and multi-targeted weed-pathogens. Protoplast fusion following the induction of mutations also is exceptionally useful for certain selection processes, and this approach has been used to obtain desired strains of an antibiotic-producing actinomycete, *Micromonospora rosaria* (Ryu et al. 1983).

RECOMMENDATIONS

Several research needs in biological control of weeds with plant pathogens have been previously identified (Charudattan 1983, Templeton 1985, Worsham 1982). In my view the following should be considered for immediate and expanded research efforts and implementation.

1. More microbial biological control agents for agriculturally important weeds should be developed from natural strains of pathogens. Many major weeds have not been surveyed adequately for pathogens. Each weed-pathogen system will have unique ecological, pathological, and practical problems. As such they can serve as model systems and present us with opportunities to learn more about the principles of microbial weed control.
2. Studies on the compatibility/incompatibility of chemical pesticides and microbial herbicides should be stressed. We need to understand further how best to integrate these agents in field practice and how to achieve complementarity and synergism between different microbial herbicide candidates, between microbial herbicides and insect biological control agents, and between microbial herbicides and chemical herbicides.
3. Research should focus on the development of improved formulations, adjuvants, and protectants for microbial herbicides.
4. Research should be initiated on the molecular and genetic bases of pathogenicity and pesticide resistance in selected weed-pathogen systems. Species of *Alternaria* and *Helmintho-*

sporium may be good candidates for research since considerable information is already available on the genetic bases of parasitism and pathogenicity of these fungi.

5. The feasibility of mutation-selection followed by protoplast fusion, and intraspecific, interspecific, and intergeneric protoplast fusions as means of creating desirable strains of pathogens should be evaluated.
6. Attempts should be made to discover host-specific and nonspecific toxins for weed control.
7. Education of scientists and user-public is necessary for effective utilization of microbial herbicides. For example, persons unfamiliar with plant pathological principles need to be taught how to handle microbial herbicides, the specific environmental requirements for infection, proper methods of integration of microbials and chemical pesticides, and the correct identification of the weed hosts.

ACKNOWLEDGMENTS

The author thanks Drs. D.C. Loschke, S.G. Pueppke, R.J. Smith, D.O. TeBeest, G.E. Templeton, and H.L. Walker for critical review of this manuscript and suggestions for its improvement.

REFERENCES

Abe, M., H. Umetsu, T. Nakai and D. Sasage. 1982. Regeneration and fusion of mycelial protoplasts of Tricholoma matsutake. Agric. Biol. Chem. 46:1955-1957.

Alfoldi, L. 1982. Fusion of microbial protoplasts: Problems and perspectives, pp. 59-86. In: A. Hollaender (Ed.), Genetic engineering of microorganisms for chemicals. Plenum Press, N.Y.

Anne, J., and H. Eyssen. 1978. Isolation of inter-species hybrids of Penicillium citrinum and Penicillium cyaneofulvum following protoplast fusion. FEMS Microbiol. Lett. 4:87-90.

Barrett, S.C.H. 1982. Genetic variation in weeds, pp. 73-98. In: R. Charudattan and L.H. Walker (Eds.), Biological control of weeds with plant pathogens. John Wiley & Sons, N.Y.

Becher, D., and F. Bottcher. 1983. The cell type of Rhodosporidium toruloides after protoplast fusion between strains of identical and opposite mating type. Curr. Microbiol. 9:297-300.

Beraha, L., V. Wisniewski and E.D. Garber. 1983. Avirulence and reduced extracellular enzyme activity in Geotrichum candidum. Bot. Gaz. 144:461-465.

Bowers, R.C. 1982. Commercialization of microbial biological control agents, pp. 157-173. In: R. Charudattan and H.L. Walker (Eds.), Biological control of weeds with plant pathogens. John Wiley & Sons, N.Y.

Bowers, R.C. 1985. Commercialization of Collego--an industrialist's view. Weed Sci. Suppl. 33:(In Press).

Boyette, C.D., G.E. Templeton and R.J. Smith. 1979. Control of winged waterprimrose (Jussiaea decurrens) and northern jointvetch (Aeschynomene virginica) with fungal pathogens. Weed Sci. 27:497-501.

Boyette, C.D., H.L. Walker and G.L. Scimbato. 1983. Production of floating inocula of Rhizoctonia solani and R. oryzae. Phytopathology 73:842 (Abstr.).

Bridges, B.A. 1976. Mutation induction, pp. 7-14. In: K.D. MacDonald (Ed.), Second international symposium on the genetics of industrial microorganisms. Academic Press, London.

Burdon, J.J., R.H. Groves and J.M. Cullen. 1981. The impact of biological control on the distribution and abundance of Chondrilla juncea in southeastern Australia. J. Appl. Ecol. 18:957-966.

Burdon, J.J ,. and D.R. Marshall. 1981. Biological control and the reproductive mode of weeds. J. Appl. Ecol. 18:649-658.

Case, M.E. 1982. Transformation of Neurospora crassa utilizing recombinant plasmid DNA, pp. 87-100. In: A. Hollaender (Ed.), Genetic engineering of microorganisms for chemicals. Plenum Press, N.Y.

Celis, J.E., K. Kaltoft and R. Bravo. 1980. Microinjection of somatic cells with micropipettes and peg-erythrocyte ghost mediated microinjection, pp. 1-28. In: J.E. Celis, A. Graessmann and A. Loyter (Eds.), Transfer of cell constituents into eukaryotic cells. Plenum Press, N.Y.

Charudattan, R. 1982. Regulation of microbial weed control agents, pp. 175-188. In: R. Charudattan and H.L. Walker (Eds.), Biological control of weeds with plant pathogens. John Wiley & Sons, N.Y.

Charudattan, R. 1983. Biological control of weeds with plant pathogens, pp. 23-25. In: S.L. Battenfield (Ed.), Proceedings of the national interdisciplinary biological control conference. Las Vegas, Nevada.

Charudattan, R. 1984. Microbial control of plant pathogens and weeds. J. Ga. Entomol. Soc. 19(Suppl. 2):(In Press).

Charudattan. R. 1985. Integrated control of waterhyacinth with a pathogen, insects, and herbicides. Weed Sci. Suppl. 33:(In Press).

Charudattan, R., and H.L. Walker (Eds.). 1982. Biological control of weeds with plant pathogens. John Wiley & Sons, N.Y. 293 pp.

Churchill, B.W. 1982. Mass production of microorganisms for biological control, pp. 139-156. In: R. Charudattan and H.L. Walker (Eds.), Biological control of weeds with plant pathogens. John Wiley & Sons, N.Y.

Collego Technical Manual. 1982. Tuco, The Upjohn Company, Agricultural Division, Kalamazoo, MI. 24 pp.

Cullen, J.M., P.F. Kable and M. Catt. 1973. Epidemic spread of a rust imported for biological control. Nature 244:462-464.

Cunfer, B.M. 1984. Change of virulence of Septoria nodorum during passage through barley and wheat. Ann. Appl. Biol. 104:61-68.

Dales, R.B.G., and J.H. Croft. 1977. Protoplast fusion and the isolation of heterokaryons and diploids from vegetatively incompatible strains of Aspergillus nidulans. FEMS Microbiol. Lett. 1:201-203.

Daniel, J.T., G.E. Templeton, R.J. Smith and W.T. Fox. 1973. Biological control of northern jointvetch in rice with an endemic fungal disease. Weed Sci. 21: 303-307.

Day, P.R. 1974. Genetics of host-parasite interaction. W.H. Freeman, San Francisco. 238 pp.

Delp, C.J. 1980. Coping with resistance to plant disease control agents. Plant Dis. 64:652-657.

DeVine Label. 1981. Chemical & Agricultural Products Division, Abbott Laboratories, Chicago, IL. 1 p.

DeVine Product Brochure. 1981. Chemical & Agricultural Products Division, Abbott Laboratories, Chicago, IL. 4 pp.

Dyer, W.E., S.K. Turner, P.K. Fay, E.L. Sharp and D.C. Sands. 1982. Control of Canada thistle by a rust, Puccinia obtegens, pp. 243-244. In: R. Charudattan and H.L. Walker (Eds.), Biological control of weeds with plant pathogens. John Wiley & Sons, N.Y.

Ehrensing, D.T., and D.L. Isaacson. 1977. Biological control of skeletonweed: Basic considerations for initiating a program in Oregon with a list of references. Oregon Dept. Agric., Weed Control Bull. 2, 10 pp.

Harris, G.M. 1982. Protoplasts from Giberella fujikuroi. Phytopathology 72: 1403-1407.

Holcomb, G.E. 1982. Constraints on disease development, pp. 61-72. In: R. Charudattan and H.L. Walker (Eds.), Biological control of weeds with plant pathogens. John Wiley & Sons, N.Y.

Hull, V.J., and R.H. Groves. 1973. Variation in Chondrilla juncea L. in southeastern Australia. Aust. J. Bot. 21:113-135.

Keen, N.T., and J.C. Horton. 1966. Induction and repression of endopolygalacturonase synthesis by Pyrenochaeta terrestris. Can. J. Microbiol. 12:443-453.

Kenney, D.S. 1985. DeVine--the way it was developed--an industrialist's view. Weed Sci. Suppl. 33:(In Press).

Lee, G.A. 1985. Progress on classical biological and integrated control of rush skeletonweed (Chondrilla juncea) in the Western U.S. Weed Sci. Suppl. 33:(In Press).

Leonard, K.J. 1982. The benefits and potential hazards of genetic heterogeneity in plant pathogens, pp. 99-112. In: R. Charudattan and H.L. Walker (Eds.), Biological control of weeds with plant pathogens. John Wiley & Sons, N.Y.

Leslie, J.F. 1983. Some genetic techniques for Giberella zeae. Phytopathology 73:1005-1008.

Levin, D.A. 1975. Pest pressure and reconbination systems in plants. Amer. Nat. 109:437-451.

Maraz, A., and L. Ferenczy. 1980. Selective transfer of fungal cytoplasmic genetic elements by protoplast fusion. Curr. Microbiol. 4:343-345.

Martini, G., G. Grimaldi and J. Guardiola. 1978. Extrachromosomal DNA in phytopathogenic fungi, pp. 197-200. In: H.W. Boyer and S. Nicosia (Eds.), Genetic engineering. Elsevier, Amsterdam.

Minuth, W., and K. Esser. 1983. Intraspecific, interspecific, and intergeneric recombination in β-lactam producing fungi via protoplast fusion. Europ. J. Appl. Microbiol. Biotechnol. 18:38-46.

Nelson, R.R., R.P. Scheffer and R.B. Pringle. 1963. Genetic control of toxin production in Helminthosporium victoriae. Phytopathology 53:385-387.

Oehrens, E. 1977. Biological control of the blackberry through the introduction of rust, Phragmidium violaceum, in Chile. FAO Plant Protect. Bull. 25:26-28.

Peberdy, J.F. 1979. Fungal protoplasts: Isolation, reversion, and fusion. Annu. Rev. Microbiol. 33:21-39.

Phatak, S.C., D.R. Sumner, H.D. Wells, D.K. Bell and N.C. Glaze. 1983. Biological control of yellow nutsedge with the indigenous rust fungus Puccinia canaliculata. Science 219:1446-1447.

Quimby, P.C. 1982. Impact of diseases on plant populations, pp. 47-60. In: R. Charudattan and H.L. Walker (Eds.), Biological control of weeds with plant pathogens. John Wiley & Sons, N.Y.

Quimby, P.C., and F.E. Fulgam. 1984. A sprayer for mycoherbicide research. Proc. Annu. Meet. So. Weed Sci. Soc., Hot Springs. 37:393 (Abstract).

Quimby, P.C., and H.L. Walker. 1982. Pathogens as mechanisms for integrated weed management. Weed Sci. Suppl. 30:30-34.

Ridings, W.H. 1985. Biological control of stranglervine in citrus--a researcher's view. Weed Sci. Suppl. 33:(In Press).

Ridings, W.H., D.J. Mitchell, C.L. Schoulties and N.E. El-Gholl. 1976. Biological control of milkweed vine in Florida citrus groves with a pathotype of *Phytophthora citrophthora*, pp. 224-240. In: T.E. Freeman (Ed.), Proc. IV intern. symp. biol. control of weeds. Univ. of Florida, Gainesville.

Ridings, W.H., C.L. Schoulties, M.E. Kannwischer, S.H. Woodhead and N.E. El-Gholl. 1982. Control of stranglervine in citrus groves with *Phytophthora palmivora*: Studies on citrus susceptibility, p. 240. In: R. Charudattan and H.L. Walker (Eds.), Biological control of weeds with plant pathogens. John Wiley & Sons, N.Y.

Ruddle, F.H. 1980. Gene transfer in eukaryotes, pp. 295-309. In: J.E. Celis, A. Graessmann and A. Loyter (Eds.), Transfer of cell constituents into eukaryotic cells. Plenum Press, N.Y.

Ryu, D.D.Y., K.S. Kim, N.Y. Cho and H.S. Pai. 1983. Genetic recombination in *Micromonospora rosaria* by protoplast fusion. Appl. Environ. Microbiol. 45: 1854-1858.

Sands, D.C., and A. D. Rovira. 1972. Modifying the virulence and host range of weed pathogens, p. 305. In: Proc. intern. conf. plant pathogenic bacteria, Cent. Agric. Publ. Document., Wageningen, Netherlands.

Scheffer, R.P., and R.S. Livingston. 1984. Host-selective toxins and their role in plant diseases. Science 223:17-21.

Scheffer, R.P., R.R. Nelson and R.B. Pringle. 1964. Toxin production and pathogenicity in *Helminthosporium victoriae*. Phytopathology 54:602-603.

Shrum, R.D. 1982. Creating epiphytotics, pp. 113-136. In: R. Charudattan and H.L. Walker (Eds.), Biological control of weeds with plant pathogens. John Wiley & Sons, N.Y.

Sidhu, G.S. 1983. Sexual and parasexual variability in soil fungi with special reference to *Fusarium moniliforme*. Phytopathology 73:952-955.

Skylakakis, G. 1983. Theory and strategy of chemical control. Annu. Rev. Phytopathol. 21:117-135.

Smith, R.J. 1982. Integration of microbial herbicides with existing pest management programs, pp. 189-203. In: R. Charudattan and H.L. Walker (Eds.), Biological control of weeds with plant pathogens. John Wiley & Sons, N.Y.

Smith, R.J. 1985. Biological control of northern jointvetch in rice and soybeans--a researcher's view. Weed Sci. Suppl. 33:(In Press).

Smith, R.J., J.T. Daniel, W.T. Fox and G.E. Templeton. 1973. Distribution in Arkansas of a fungus disease used for bio-control of northern jointvetch in rice. Plant Dis. Rep. 57:695-697.

TeBeest, D.O. 1984a. Additions to the host range of *Colletotrichum gloeosporioides* f.sp. *aeschynomene*. Phytopathology 74:864 (Abstract).

TeBeest, D.O. 1984b. Induction of tolerance to benomyl in *Colletotrichum gloeosporioides* f.sp. *aeschynomene* by ethyl methanesulfonate. Phytopathology 74:864 (Abstract).

Templeton, G.E. 1982a. Status of weed control with plant pathogens, pp. 29-44. In: R. Charudattan and H.L. Walker (Eds.), Biological control of weeds with plant pathogens. John Wiley & Sons, N.Y.

Templeton, G.E. 1982b. Biological herbicides: Discovery, development, deployment. Weed Sci. 30:430-433.

Templeton, G.E. 1985. Mycoherbicide research at the University of Arkansas--past, present and future. Weed Sci. Suppl. 33:(In Press).

Templeton, G.E., and R.J. Smith. 1977. Managing weeds with pathogens, pp. 167-176. In: J.G. Horsfall and E.B. Cowling (Eds.), Plant disease: An advanced treatise. Vol. 1. How disease is managed. Academic Press, N.Y.

Templeton, G.E., R.J. Smith and W. Klomparens. 1980. Communication of fungi and bacteria for biological control. Biocontrol News Inf. 1:291-294.

Templeton, G.E., D.O. TeBeest and R.J. Smith. 1979. Biological weed control with mycoherbicides. Annu. Rev. Phytopathol. 17:301-310.

Templeton, G.E., D.O. TeBeest and J.R. Smith. 1982. Biological control of northern jointvetch, Aeschynomene virginica, with Colletotrichum gloeosporioides, f.sp. aeschynomene. pp. 241-242. In: R. Charudattan and H.L. Walker (Eds.), Biological control of weeds with plant pathogens. John Wiley & Sons, N.Y.

Theriot, E.A., R.A. Theriot and D.R. Sanders. 1981. Evaluation of the infectivity of a Cercospora rodmanii formulation using two application systems. Misc. Pap. A-81-2, Environmental Laboratory, U.S. Army Engineer Waterways Experimental Station, Vicksburg, MS. 23 pp.

Tinline, R.D., and B.H. MacNeill. 1969. Parasexuality in plant pathogenic fungi. Annu. Rev. Phytopathol. 7:147-170.

Toyama, H.A., Shinmyo, and H. Okada. 1983. Protoplast formation from conidia of Trichoderma reesei by cell wall-lytic enzymes of a strain of Trichoderma viride. J. Ferment. Technol. 61:409-411.

Trujillo, E.E. 1976. Biological control of hamakua pamakani with plant pathogens. Proc. Amer. Phytopathol. Soc. 3:298 (Abstr.).

Turkensteen, L.J. 1973. Partial resistance of tomatoes against Phytophthora infestans, the late blight fungus. Agricultural Research Report 810. Inst. Phytopathol. Research, Wageningen, Netherlands. 88 pp.

Turner, S.K., P.K. Fay, E.L. Sharp and D.C. Sands. 1981. Resistance of Canada thistle (Cirsium arvense) ecotypes to a rust pathogen (Puccinia obtegens). Weed Sci. 29:623-624.

Walker, H.L., and G.L. Sciumbato. 1981a. Alternaria macrospora, a fungus that kills spurred anoda. USDA-ARS and Miss. Agric. Forest Exp. Stn. Res. Rep. 6(9), 2 pp.

Walker, H.L. 1981b. Fusarium lateritium: A pathogen of spurred anoda (Anoda cristata), prickly sida (Sida spinosa), and velvetleaf (Abutilon theophrasti). Weed Sci. 29:629-631.

Walker, H.L. 1982. Seedling blight of sicklepod caused by Alternaria cassiae. Plant Dis. 66:426-428.

Walker, H.L., and W.J. Connick. 1983. Sodium alginate for production and formulation of mycoherbicides. Weed Sci. 31:333-338.

Wapshere, A.J. 1976. Effectiveness: A comparison of prediction and results during the biological control of Chondrilla, pp. 124-127. In: T.E. Freeman (Ed.), Proc. IV intern. symp. biol. control of weeds. Univ. of Florida, Gainesville.

Worsham, A.D. 1982. Discussion of topics, pp. 219-236. In: R. Charudattan and H.L. Walker (Eds.), Biological control of weeds with plant pathogens. John Wiley & Sons, N.Y.

Yamada, O., Y. Magae, Y. Kashiwagi, Y. Kakimoto and T. Sasaki. 1983. Preparation and regeneration of mycelial protoplasts of Collybia veltipes and Pleurotus ostreatus. Eur. J. Appl. Microbiol. Biotechnol. 17:298-300.

Yoder, O.C. 1980. Toxins in pathogenesis. Annu. Rev. Phytopathol. 18:103-129.

Yoder, O.C. 1983. Use of pathogen-produced toxins in genetic engineering of plants and pathogens, pp. 335-353. In: T. Kosuge, C.P. Meredith and A. Hollaender (Eds.), Genetic engineering of plants. Plenum Press, N.Y.

ARTHROPODS FOR WEED CONTROL IN IPM SYSTEMS

E.A. Bernays

A copious literature attests to the difficulty of achieving biological control of crop weeds with arthropods (Altieri & Doll 1978, Batra 1982, Goeden 1978, Huffaker 1978), and the constraints are certainly formidable. The conventional subdivisions of biological control are relevant, however, in agroecosystems and they probably require separate treatment:

1. Classical biological control involving introduction of host-specific natural enemies of exotic weeds with self-perpetuation and eventual permanent control.
2. Augmentative biological control involving mass propagation and periodic release of exotic or native enemies.
3. Conservation biological control involving management systems which enhance and conserve existing natural enemy complexes.

In any system of integrated pest management such biological control strategies have been discussed as being among the important inputs but as far as weeds are concerned, the question is premature largely because the value of biological control of weeds in crops, using arthropods at least, is itself in doubt. It is complicated by the need to apply insecticides against recalcitrant arthropod pests which may also reduce the numbers of insects on the weeds, although the reality of the problem has not actually been monitored and its intuitive importance could be overrated. However, since many phloem feeders and mites increase more rapidly on plants exposed to insecticides (hormoligosis), some careful manipulations could theoretically aid the weed.

With annual crops the overall problems are obvious but in particular, the relative instability and the disturbance of the habitat too readily destroy the useful organisms, which must find alternative food sources elsewhere or have a life history highly synchronized not only with that of the weed, but also with that of the crop. In annual crops classical biological control of weeds with arthropods appears to be exceptionally difficult and has never been fully achieved. Since annual crops commonly have a complex of weeds, the control of any one may simply influence the remaining species composition whose overall competition with

ISBN 0-12-357030-1

the crop would be unlikely to change much (Fryer 1981). Some minor successes in controlling dominant weeds have been claimed with regular mass releases, but it is considered far too costly and difficult to achieve because inexpensive and effective rearing technology is currently lacking (Schroeder 1981). Finally, conservation biological control has simply not had a reasonable assessment, probably because every different situation requires a very large data base of ecological and general biological facts--not only of the target weeds and potential herbivore control agents, but of the competing species at both levels. Progress is being made in this field and ecological management in its widest sense, including competitive replacement of certain weeds, may open new doors perhaps more fruitful than the use of arthropods alone. In annual crops realistic possibilities to use arthropod natural enemies only exist where weeds are widespread outside the cropping system to maintain a pool of colonizing individuals between harvest periods, and in no till systems where disturbance is minimal. A more stable environment in perennial crops also offers more hope, though success has rarely been achieved.

On the status and limits to the use of arthropods in these systems for weed control I have the disadvantage and advantage of being an outsider, and I shall address some questions which appear relevent to an outsider with more basic experience in plant-insect interactions. Lack of success in using arthropod natural enemies in agricultural IPM systems is not just a reflection of little effort (Pemberton 1981); some new approaches may be needed.

DOES A STUDY OF THE WEEDS GIVE US ANY HELP?

Taxonomy

Some families of plants are more likely to become weeds than others and a comparison of the weed species in different families, given by Holm et al. (1977), and the total numbers of species in the different plant families (from Willis 1960) indicate biases in some areas (Fig. 1). Clearly ecological requirements and growth characteristics in some families make them intrinsically unlikely to ever become weeds, as in Orchidaceae or certain tree families. In other cases the relevent characteristics are less clear. Why are the Labiateae, Leguminoseae and Umbelliferae under-represented as weeds, while others such as the Polygonaceae, Gramineae and Cyperaceae over-represented? It is interesting that the three large plant families under-represented are in the more primitive Diapetalae, and, indeed, almost all the important weed species are in more recently evolved plant families (Takhtajan 1969). By definition they are more aggressive colonists, perhaps a character that might be expected in more recent and rapidly-evolving groups. The expected correlates of variability, versatility and breeding systems are not clearly related.

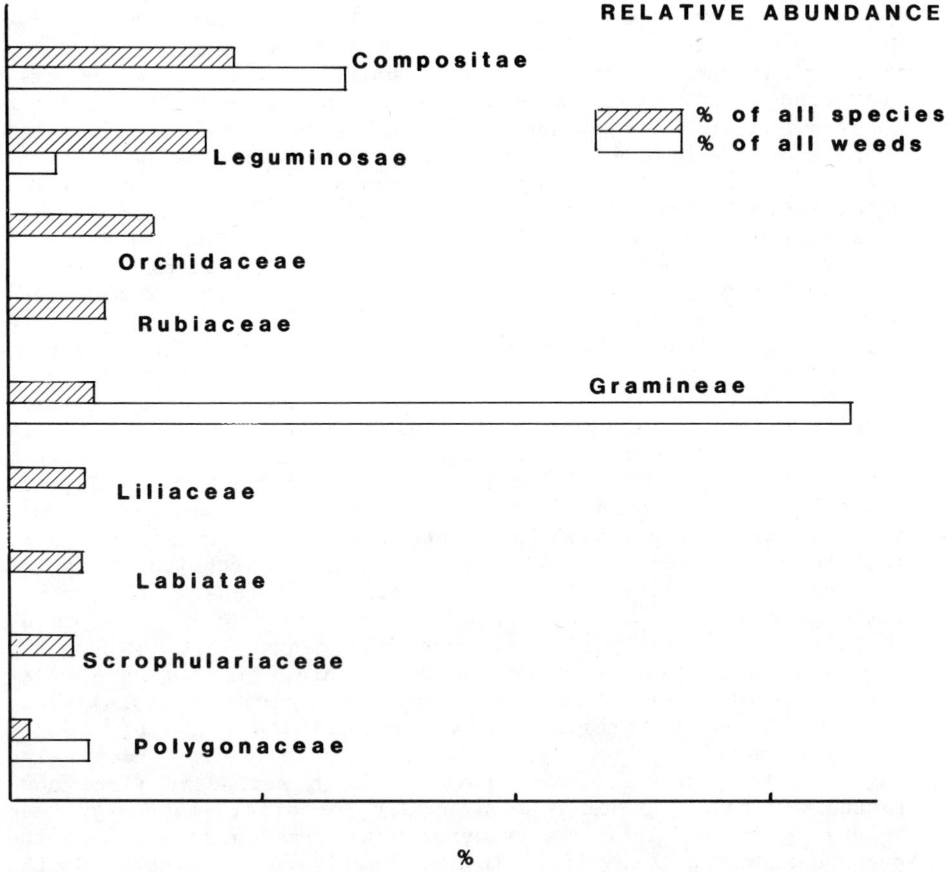

Fig. 1. Number of species in selected families as a percentage of the total number of angiosperm species (open bars); and number of major weed species in the same selected families as a percentage of total major weed species (hatched bars). Data from Holm et al. (1977) and Willis (1960).

For example, the Gramineae are commonly self-fertilizing and many reproduce vegetatively or apomictically (Stebbins 1972).

The use of herbicides and low till systems has significantly increased the relative importance of grass weeds (Fryer 1981), but the importance of the Gramineae probably relates also to its widespread occurrence, far surpassing that of any other family. Besides genetic peculiarities and physiological aggressiveness, grasses have, of course, well known features for resisting the impact of grazing damage: in particular a complex of mechanical features and the presence of unexposed meristematic tissue.

Recently Janzen (1984) has also suggested that many grasses may depend on vertebrate grazers for seed dispersal, so that the whole plant rather than a fruit is being used almost as a lure. The concentration of insect feeding deterrents in seedling grasses and relative absence in mature grasses may be a related phenomenon (Bernays & Chapman 1976). In any case it suggests an overall tolerance of grasses to grazing, after their initial establishment when survival and competitive forces are at their peak (Silcock 1975).

Grasses certainly predominate as weeds in crops including perennial crops, yet real conflicts arise as to the weed status because several major weeds are elsewhere important pasture grasses, erosion control species or lawn grasses. Bermudagrass, Cynodon dactylon, is an example: it has many biotypes; some useful, some weeds; and many biotype crosses produce sterile hybrids indicating a great genetic diversification.

The importance of the Graminaceous weeds in relation to any other family, means that the likelihood of their successful biological control should be objectively assessed. Certain characters are especially notable: along with sedges, which are also important, grasses have a relatively simple architecture which appears to severely limit species richness through niche reduction (Strong & Levin 1979), and thus reducing the choice of potential arthropod biological control agents. It is probably also relevant that among herbaceous plants, grasses have relatively low nitrogen levels, which may limit their potential value as food for insect herbivores and add to the cause of species impoverishment (Mattson 1980) (Fig. 2). Research to date also suggests that grasses are impoverished in secondary plant compounds which may commonly be necessary for stable monophagy (see below). In fact there is evidence that grasses do not have the preponderance of specific insect herbivores typical of the dicotyledons. In the British fauna for example, which has been extensively studied and has a large computer data base of insect food plants, a very clear difference exists between grass feeders and dicot feeders, the former generally being less specific than the dicotyledon feeders (Table 1). The indications are that grasses may be the most difficult group to find arthropod control agents for, though Pemberton (1981) suggests that grasses should be controllable because of the known specificity of some of the cereal pests, and, by analogy, there may be potential control agents of grass weeds. However, ecological and physiological constraints are a factor here which under pressure may break down. The possibility of control agents shifting to cereal crops must also be considered. On balance it appears that Gramineae would be relatively unrewarding as a group on which to undertake control using arthropods.

The Compositae is the largest plant family and, by contrast with grasses, the percentage of important weeds in this family is roughly what would be expected. The family is extremely rich in all manner of secondary plant compounds (Cronquist 1977) which

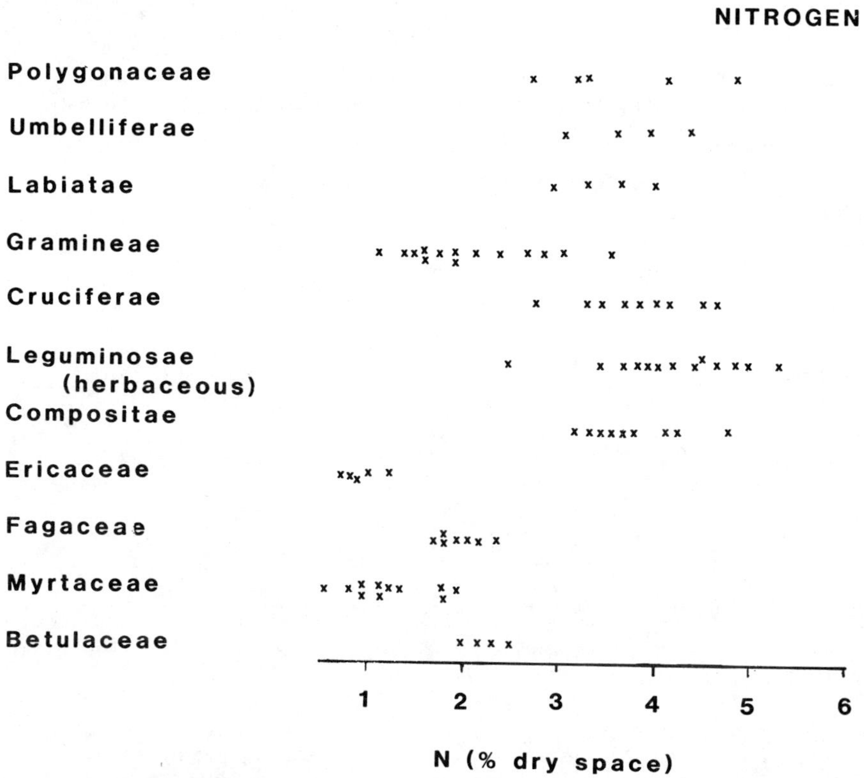

Fig. 2. Midseason values for N content in healthy leaves of plant species in different families (United Kingdom, 1979-82; Bernays, unpublished).

may be key signals for specific herbivores (Schoohoven 1981) and composites such as *Carduus nutans* and *Eupatorium adenophorum* have indeed been the targets of some of the successful biological control programs (Burdon & Marshall 1981). Chemical fingerprinting may give us clues in selecting suitable arthropods or in reducing the biological assay burden as suggested by Lawton & Schroder (1978), but no progress has been made in this field. Composites, some of the most fruitful targets for attempting biological control with arthropods, are currently the subject of several biological control projects.

Reproduction

Another approach to weed biology was made by Burdon et al. (1981). They suggested that sexually reproducing plants were less amenable to biological control than asexually reproducing

Table 1. Percentage of British herbivorous insects restricted to one genus in dicot-feeders vs. grass feeders (data courtesy of L.K. Ward, Institute of Terrestrial Ecology, Furzebrook Experimental Station, Dorset, U.K.)

Taxonomic Group	% Restricted to One Genus (Number of Spp.)			
	Nongrasses		Grasses	
Hemiptera	46	(250)	8	(36)
Homoptera	74	(800)	20	(181)
Thysanoptera	65	(82)	11	(32)
Butterflies	13	(61)	0	(17)
Larger moths	21	(634)[a]	27	(72)
Microlepidoptera	56	(1261)	23	(86)
Coleoptera	40	(935)	9	(62)
Hymenoptera	74	(577)	35	(55)
Diptera	69	(371)	35	(114)

[a] There are a very large number in the Noctuidae and Geometridae which are polyphagous.

plants. However, in using data on successes, Opuntia species tend to swamp the argument, there being many asexually reproducing species of this genus, and if Opuntia is omitted from the tables, differences are barely significant. Asexual reproduction was considered by these authors to be a disadvantage to the plant because of the loss of the segregative and recombinative aspects of variation, and the reduction in a large heterozygous gene pool. It was thought that variation and change would be slower in such cases, giving the herbivore a selective advantage. On the other hand apomictic and selfing plants are genetically variable but generally homozygous. Within a population there may be many different homozygous genotypes each giving an adaptive advantage to deal with the heterogeneity of the habitat. Further, plant populations will purge any mutation without immediate advantages in the homozygous state, and useful characters will be more quickly stabilized and perpetuated (Berg 1972). Facultative apomixis, which is common in grasses, may perhaps provide the best of both worlds for adaptability.

Overall, a study of weeds is, therefore, proving to be of some use, and at present may help us at least not to waste time on groups less likely to be fruitful. The possibilities are little explored because the complexities of plant sciences cannot be undertaken by entomologists or even, perhaps, by plant taxonomists. A case could perhaps be made for improved interaction of biological control scientists with plant physiologists and geneticists; a far-sighted biological control group might fruitfully employ or consult with such personnel.

ARE SOME GROUPS OF ARTHROPODS MORE USEFUL THAN OTHERS?

Host Choice

Although behavioral acceptance of a plant is modifiable by developmental, hormonal, neural and experiential influences, genetic stability is, of course, the predominant aspect of specificity in host plant choice, and specificity is at the heart of any program on biological control of weeds. Can the study of plant recognition mechanisms give us any understanding of stability?

The sensing and encoding activity of the peripheral sensory system determines what is available to the central nervous system and it is in the decoding of this information that host recognition occurs. Because of the technical difficulties, most work has centered on the first stage (sensing) where considerable differences do occur, and there is some recent work on the possible encoding process.

Studies on many isolated caterpillar species indicate the common occurrence of taste cells with high sensitivity to particular compounds, signifying either key host compounds or strong deterrent compounds (Schoonhoven 1973, 1981). Such cells are presumed to be "hard wired" to the central nervous system where motor commands to feed are driven with a minimum of integration or processing. The balance of inputs from such cells, together with phagostimulant information from sugar receptors, for example, appears to drive feeding activity in a surprisingly simple way in *Pieris brassicae* larvae (Blom 1978). The application of this approach to study the variation or differences at or below the species level has only been tackled in any detail with one group, the small ermine caterpillars of the genus *Yponomeuta*. This genus is commonly associated with Celastraceae, especially *Euonymus*. Divergence has apparently involved sympatric host shifts, current subspecies groups being separated largely by host plant preferences (Gerrits-Heybroek et al. 1978). One of the species has shifted to *Salix*, which contains the deterrent phenolic compound phloridzen. This material causes firing in particular receptors in only those species which do not feed on *Salix*. On the other hand, dulcitol, a characteristic sugar alcohol in *Euonymus*, stimulates taste cells in species feeding on *Euonymus*, but will also do so in some of the species which have shifted to Rosaceae where there is no dulcitol. The result is interpreted as indicating the ancestral relationship of Rosaceae-feeding species is with the *Euonymus*-feeding species (van Dron-gelen 1978). It also suggests that processes other than the peripheral sensory ones are important in preference changes.

Recent work indicates that far more complex and varied coding takes place with many species. For example, plant deterrents may act by inhibiting the activity of phagostimulant receptors so that in such cases the mechanism of perception of the deterrent may simply be that the insect does not perceive sugar to be present (Dethier 1982). Moreover, closely related species such as *Spodoptera littoralis* and *S. exempta* appear to perceive deterrent compounds in quite different ways (W.M. Blaney & M. Simmonds, personal communication), while even *Pieris brassicae* and *P. rapae* with very similar preferences have major receptor differences (Schoonhoven 1981).

Real plants are of course very different from the simple mixtures presented in such studies and the sensory coding in nature is possibly very different from the suggestions provided by these models. Some patterns of multireceptor activity in *Manduca sexta* caterpillars have been examined by Dethier & Crnjar (1982) who found that in this insect, decoding of host-related information apparently showed two consecutive phases, with the first having a general message and the second containing the more critical message. Whether changes in one component of such a code have a significant effect in the overall message is not known.

Single mutations can have large effects on insect sensory systems (Ward 1977) and have been demonstrated to influence chemoreception (Falk & Atidia 1975), so that at the level of larval behavior rapid changes in host preference are possible, at least in theory. Moreover genetic strains of some caterpillar species have been identified which have very different sensitivity to secondary compounds (van Drongelen & van Loon 1980, Wieczorek 1976). Such variation is probably common. Hsiao (1978, 1982) found an Arizona population of *Leptinotarsa decemlineata* uniquely capable of utilizing an unusual *Solanum* sp.; breeding studies revealed that host adaptation was polygenic, and that the species generally was enzymatically variable. Further, wide differences occur in host use in different geographical populations of many polyphagous species (Fox & Morrow 1981). From such studies one can conclude that the mechanism of host choice is different in different insects, generally variable and probably labile, so that host-shifts can be expected. The timing, however, is not predictable.

Physiological Suitability

Behavior of the caterpillars is only one aspect of the specificity question. Postingestive phenomena, including differential mortality, are important determinants of the genetic makeup of the next generation, yet the extent to which preferences reflect differential physiological ability to survive on plants has never been widely surveyed. However, in the studies available there are examples where there is a connection and others where there is not. To return to *Yponomeuta*, the species *Y. padellus*, although a Rosaceae feeder, shows positive responses to the supposed ancestral food plant, Euonymus, but larvae fed only with Euonymus will die (Mentink 1974). In such cases, there are at least two components to the adaptive response: behavioral discrimination and the ability to survive on the ingested food. Thus Bush (1975) proposed a minimum of two gene changes necessary for host shifts. Two gene changes may not be required where detrimental postingestive effects do not (or would not) occur. Linkage relationship(s) between host selection and survival genes have been proposed by White (1973), and Pluthero (Pluthero & Singh 1982) found that a single gene change in Drosophila could influence both avoidance of and resistance to certain insecticides.

Discrimination against a plant does not necessarily signify a specific lack of ability to use a plant. For example, Manduca sexta larvae with some of their taste receptors removed will eat and survive well on certain plants not normally accepted (Waldbauer 1964). In Locusta migratoria some highly deterrent compounds were found not to have deleterious effects when artificially placed in the gut (Cottee & Mordue 1982), so that one could imagine that a sensory change alone (measured behaviorally) is all that would be required to change from a host plant without such a chemical to one with it. Similarly, caterpillars of two genotypes of Alsophila pometaria were shown to be different in their preference behavior, yet no different in their ability to grow on certain trees (Futuyma et al. 1984).

Gould (1983), however, considers that overcoming detrimental postingestion effects (such as adaptation to different nutrient profiles or increased capacity to excrete or detoxify secondary plant products) are complex, normally necessary and that this would be the most difficult step in adapting to new host plants. If different host-associated herbivore genotypes result in differences in host preference and not in metabolic fitness factors, the preferences may be a major cause of the evolution of host specificity (Futuyma et al. 1984). Relatively rapid host switching resulting from other ecological pressures are also possible (Fox & Morrow 1981).

Adult Behavior

The behavior and physiology in the adult must be considered in addition to that of the larvae. In many instances the oviposition site of the adult female largely determines what the larvae will feed on and a third factor is brought in. Her choice of host plant may be more sophisticated than that of larvae by the use of more sensory modes. This includes such factors as long distance visual and odor cues, followed by mechanoreception and contact chemoreception by antennae, mouthparts, tarsi and ovipositor sensilla. Her choice of host plant is not always matched by larval preference or larval survival rate. In some cases, preferences are broader than that of the larvae but more often they are very narrow and the insects highly selective. Thus correspondence between female oviposition and larval preference and survival can be extraordinarily high (Rausher 1981). It is not always high, however; in Leptinotarsa decemlineata, females prefer Solanum rostratum over all other Solanaceae, yet other species can be better host plants for larval growth and survival (Hsiao 1978). In such a case it would appear that the possibility of exploring new hosts is limited by rigid oviposition behavior, and adaptive changes to a new host are unlikely in the presence of the preferred host. On the other hand, in the cowpea bruchid, Callosobruchus macalatus, host selection behavior of adults could be altered by selective laboratory breeding while differential development and survival on different plants is not altered by similar artificial selection pressure although the conditions were different (Wasserman & Futuyma 1981).

It may be relevant that in Yponomeuta olfactory studies suggest large differences between closely related species in response to volatile plant constituents (van der Pers 1978), while taste receptor differences in larvae are much less (van Drongelen 1978). According to Bush & Diehl (1982), and contrary to the results of Wasserman & Futuyma (1981), host shifts may occur frequently. It is possible that larval acceptance of foods and at any rate their ability to survive on different plants is quite plastic and that only changes in adult preference behavior is required in many cases to effect a host shift.

The Choice for Introductions

The general value of different arthropod groups for the biological control of weeds lies to some extent in their average degrees of specificity. The Endopterygota and Hemipteroidea are relatively restricted in their host ranges compared with the Orthopteroidea. This restriction is perhaps related to the small numbers of chemoreceptors with greater receptor specificity (Chapman 1982). In any case, screening methods currently in use will quickly separate the more specific, potentially useful species. New questions relate perhaps to whether, among the specific insects, some are more likely to be better candidates than others, which could then be rejected more readily. At present no definite answers can be given but some possible research areas may be suggested. For example, rigidly tuned receptors for host specific compounds may be worth searching for. Such receptors occur in one of the most successful rangeland weed-controlling organisms--Chrysolina spp. Here both adult and larva are "turned on" by hypericin, a quinone restricted to its host plant Hypericum. A rigid behavior pattern becomes more important for biological control in crop systems than in rangeland situations where an introduction is less likely to get a foothold on other plants in a biome already partitioned among the native fauna.

A consideration of host shifts should always be undertaken: it is most likely to occur between taxonomically related groups, enhanced particularly by resource availability. For example,

Buddleia, introduced into the United Kingdom without herbivores, now has its complex of herbivore species. Most are polyphagous, but there are striking taxonomic similarities between the insect species on it in the U.K. and the original fauna on it in its native habitat in South Africa (Southwood & Kennedy 1983). Similar examples to *Buddleia* can be found in the literature. In introducing arthropods to control weeds some advantages are already present. Species choices can utilize such knowledge as relatedness, and the resource utilization which is, by definition, underutilized. The stability record is good so far but in crop situations where biological control has been little attempted, there is the additional problem that an extra underutilized resource, namely the crop, is near at hand. From the discussion above it seems that the possibility of host shifts will depend on the rigidity of the association of adult behavioral, larval behavioral and larval physiological attributes. Investigation of such details could be done but is probably too complex to undertake thoroughly for every candidate arthropod. It is certain that some will have a higher risk attached than others, particularly where variance levels are high in host-related behavior and physiology. Since genetic changes are also possible, no risk rating will be precise.

A particularly interesting case exists in the genus *Diabrotica* (Chrysomelidae) with a possible host shift of a specific insect to a range of crops. These beetles have apparently coevolved with the native Cucurbitaceae and are very strongly stimulated to feed on plants in this family by specific triterpenes, the cucurbitacins (Howe et al. 1976). Some are effective at less than 10 parts per million, and will cause compulsive feeding even on neutral substrates such as silica gel and filter paper (Metcalf et al. 1982). The insects correspondingly aggregate more on those cucurbit species having higher concentrations of the cucurbitacins. It would seem that this coevolved insect-plant association of native North American organisms has certain characteristics looked for in biological control projects. In spite of the existence of "key" chemicals, however, which one would expect to indicate a high degree of specificity at least to plants containing them, several *Diabrotica* species have broad host ranges and are pests particularly of corn, as well as of other crops. Thus, *Diabrotica undecimpunctata howardi* is a serious pest of corn as indicated by its common name, southern corn rootworm, yet is a species responding particularly strongly to cucurbitacins. Moreover, it feeds and develops well on artificial diets without such chemicals (Singh & Howe 1971). The implication is that *Diabrotica* species, originally host specific on plants containing cucurbitacins, have spread onto readily available crop resources which were underutilized. It may be that a widening host range was occurring before the existence of the crops, but the lesson is probably there: host specificity is not fixed. Risk ratings cannot be given but the risk is there. The stability of host specificity is not a foregone conclusion even with well-tested and highly specific arthropods, and the risk may be enhanced in cropping systems over that in rangeland because of the protected resource availability. It perhaps would only be a matter of time, as least in some cases, before introduced, arthropod species invaded crop plants. The risk is also present, however, with native fauna and should not alone be a reason for rejecting introduced biological control agents.

RECOMMENDATIONS

The other side of the stability coin is that controlled modifications can be made which are potentially useful. We do have the chance to artificially select qualities that are required and some simple experiments have been performed with arthropod herbivores with respect to simple receptor change (Stadler 1982) and alterations in host preference (Gould 1983). Selection for such characters as temperature tolerance has been successful for certain parasitoid introductions (White et al. 1970). What may take many generations in a complex natural setting may be achieved in a realistic time frame within artificial selection regimes. The use of mutagens is within reach and other lessons are to be learned from those employing recombinant gene techniques. The eventual cost-effectiveness of such approaches cannot be predicted but preliminary studies by academic scientists may well indicate some fruitful lines of research. Can arthropods with suitable life history patterns be induced to prefer weed species? Can close relations of arthropods attacking the weed be induced to prefer the weed? Such work should be encouraged with the hope of some practical value in terms of weed control with arthropods.

Genetic improvement techniques in relation to predators and parasitoids are discussed elsewhere (Beckendorf & Hoy, this volume), and their use for phytophagous arthropods is equally feasible. Numerous examples of new crop pests and new biotype development on resistant crops attests to the likelihood that useful changes will be possible also with arthropods feeding on weeds. Initially, judicious choice of weeds should be made based on some knowledge of their biological suitability for control. Secondly, oligophagous arthropods should be chosen which are: 1) already attacking the weed or 2) attacking closely-related plants. For the latter case choice should be made particularly from taxa where related arthropods utilize related plants elsewhere. With such a choice, selection pressure may be artificially given to improve the insect's performance and virulence on the weed.

Collaboration with geneticists will be essential to cope with several aspects of the technique, but in particular the maintenance of adequate life history traits and overall variability. To help avoid the loss of genetic variability and useful biological features, field cages as well as laboratory populations should be maintained. Considerable research will be required on the arthropod's behavior, physiology and ecology since a balanced selection program will depend on an extensive knowledge of the species in all three areas.

It would seem at present that the use of pathogens may deserve the greater effort from biological control specialists (Schroeder 1981) and the ecological management of weeds by farmers to favor the less competitive weeds or even useful species, may have the longest lasting benefit. This latter approach may involve an element of conservation of biological control agents to tip the balance, but it is primarily a study for plant ecologists.

Finally, the issue of conflicting interests with introduced biological control agents must be addressed. It is currently of great concern to biological control scientists and, in particular, to conservationists, who see movement of herbivores from weeds to related native flora (Andres 1981). The voice of conservationists is unlikely to diminish. This is not a problem to ignore, and should be faced squarely and acceptable standards set for the possible payoffs. Research on classical biological control of weeds should be reconsidered and augmentative or conservation biological control, perhaps with innovative breeding techniques of arthropods, should be explored in greater depth.

In conclusion, I suggest the following:

1. A reduced input on classical biological control in agricultural land at least and no effort on Gramineae.
2. Encouragement of research on plant competition in crops so that least harmful or potentially useful weeds may be identified.
3. An integrated research effort between weed ecologists and entomologists to establish areas where arthropods can be fruitfully used to alter the weed balance in favor of the crop.
4. Cooperative approaches with plant pathologists should be discussed.
5. Initiation of research projects on genetic improvement of selected species of the native fauna to increase their virulence against weeds.

ACKNOWLEDGMENTS

I wish to thank the following for help and discussion: L.A. Andres, M.A. Hoy, M.L. Higgins and C.B. Huffaker.

REFERENCES

Altieri, M.A., and J. Doll. 1978. Some limitations of weed biocontrol in tropical crop ecosystems in Colombia, pp. 74-81. In: T.E. Freeman (Ed.), Proc. IV intern. symp. on biological control of weeds. Center for Environ. Programs, Inst. of Food and Agric. Sci., Univ. of Florida, Gainesville.

Andres, L.A. 1981. Conflicting interests and the biological control of weeds, pp. 11-20. In: E.S. Del Fosse (Ed.), Proc. V intern. symp. on biol. control of weeds. CSIRO, Melbourne, Australia.

Batra, S.W.T. 1982. Biological control in agroecosystems. Science 215:134-139.

Berg, A.R. 1972. Grass reproduction, pp. 335-350. In: V.B. Younger and C.M. McKell (Eds.), The biology and utilization of grasses. Academic Press, N.Y.

Bernays, E.A., and R.F. Chapman. 1976. Antifeedant properties of seedling grasses. Symp. Biol. Hung. 16:41-46.

Blom, F. 1978. Sensory activity and food intake: A study of input output relationships in two phytophagous insects. Neth. J. Zool. 28:277-340.

Burdon, J.J., and D.R. Marshall. 1981. Biological control and the reproduction mode of weeds. J. Appl. Ecol. 18:649-658.

Burdon, J.J., D.R. Marshall and R.H. Groves. 1981. Aspects of weed biology important to biological control, pp. 21-30. In: E.S. Del Fosse (Ed.), Proc. V intern. symp. on biol. control of weeds. CSIRO, Melbourne, Australia.

Bush, G.L. 1975. Sympatric speciation in phytophagous parasitic insects, pp. 187-206. In: P.W. Price (Ed.), Evolutionary strategies of parasitic insects and mites. Plenum Press, N.Y.

Bush, G.L., and S.R. Diehl. 1982. Host shifts, genetic models of sympatric speciation and the origin of parasitic insects, pp. 297-306. In: J.H. Visser and A.K. Minks (Eds.), Proc. fifth intern. symp. on insect-plant relationships. PUDOC, Wageningen, Netherlands.

Chapman, R.F. 1982. Chemoreception: The significance of receptor numbers. Adv. Insect Physiol. 16:247-355.

Cottee, P., and J.A. Mordue. 1982. An investigation into the physiological actions of feeding deterrents, pp. 397-380. In: J.H. Visser and A.K. Minks (Eds.), Proc. fifth intern. symp. on insect-plant relationships. PUDOC, Wageningen, Netherlands.

Cronquist, A. 1977. On the taxonomic significance of secondary metabolites in angiosperms. Plant Syst. Evol. Suppl. 1:179-189.

Dethier, V.G. 1982. Mechanism of host plant recognition. Entomol. Exp. Appl. 3-1:49-56.

Dethier, V.G., and R.M. Crnjar. 1982. Candidate codes in the gustatory system of caterpillars. J. Gen. Physiol. 79:549-569.

Falk, R., and J. Atidia. 1975. Mutations affecting taste perceptions in *D. melanogaster*. Nature 254:325-328.

Fox, L.R., and P.A. Morrow. 1981. Specialization: Species property or local phenomenon. Science 211:887-893.

Fryer, J.D. 1981. Weed control practices and changing weed problems, pp. 403-414. In: J.M. Thresh (Ed.), Pests, pathogens and vegetation. Pitman, London.

Futuyma, D.J., R.P. Cort and I. Van Noordwijk. 1984. Adaptation to host plants in the fall canker worm (*Alsophila pometaria*) and its bearing on the evolution of host affiliation in phytophagous insects. Amer. Nat. 123:287-295.

Gerrits-Heybroek, E.M., W.M. Herrebout, S.A. Ulenberg and J.T. Wiebes. 1978. Host plant preference of five species of small ermine moths (Lepidoptera: Yponomeutidae). Entomol. Exp. Appl. 24:160-168.

Goeden, R. 1978. Biological control of weeds, pp. 357-414. In: C.P. Clausen (Ed.), Introduced parasites and predators of arthropod pests and weeds: A world review. USDA Agric. Handb. 480.

Gould, F. 1983. Genetics of plant-herbivore systems: Interactions between applied and basic study, pp. 599-653. In: R.F. Denno and M.S. McClure (Eds.), Variable plants and herbivores in natural and managed systems. Academic Press, N.Y.

Holm, L.G., D.L. Plucknett, J.V. Pancho and J.P. Hergerger. 1977. The world's worst weeds, distribution and biology. Univ. Press of Hawaii, Honolulu. 609 pp.

Howe, W.L., J.R. Sanborn and A.M. Rhodes. 1976. Western corn rootworms and spotted cucumber beetle associations with *Cucurbita* and cucurbitacin. Environ. Entomol. 5:1043-1048.

Hsiao, T.H. 1978. Host plant adaptations among geographic populations of the Colorado potato beetle. Entomol. Exp. Appl. 24:237-247.

Hsiao, T.H. 1982. Geographic variation and host plant adaptation of the Colorado potato beetle, pp. 315-324. In: J.H. Visser and A.K. Minks (Eds.), Proc. fifth intern. symp. insect plant relationships. PUDOC, Wageningen, Netherlands.

Huffaker, C.B. 1978. An overview of biological control, with particular commentary on biological weed control, pp. 3-12. In: T.E. Freeman (Ed.), Proc. IV intern. symp. on biological control of weeds. Center for Environ. Programs, Inst. of Food & Agric. Sci, Univ. of Florida.

Janzen, D. 1984. Dispersal of small seeds by big herbivores: Foliage is the fruit. Amer. Nat. 123:338-353.

Lawton, J.H., and D. Schroder. 1978. Some observations on the structure of phytopohgous insect communities: The implications for biological control, pp. 57-73. In: T.E. Freeman (Ed.), Proc. IV intern. symp. on biol. control of weeds. Cent. for Environ. Programs, Inst. of Food & Agric. Sci., Univ of Florida.

Mattson, W.J. 1980. Herbivory in relation to plant nitrogen content. Annu. Rev. Ecol. Syst. 11:119-161.

Mentink, J.M. 1974. Onderzoek naar de voedselvoorkeur bij *Yponomeuta cagnagellus, padellus* en *vigintipunctatus*. Meyendel. Med. 3:111-114.

Metcalf, R.L., A.M. Rhodes, R.A. Metcalf, J. Ferguson, E.R. Metcalf and P.-Y. Lu. 1982. Cucurbitacin contents and diabroticite (Coleoptera: Chrysomelidae) feeding upon *Cucurbita* spp. Environ. Entomol. 11:931-937.

Pemberton, R.W. 1981. International activity in biological control of weeds: Patterns limitations and needs, pp. 57-72. In: E.S. Del Fosse (Ed.), Proc. V intern. symp. on biol. control of weeds. CSIRO, Melbourne, Australia.

Pluthero, F.G., and R.S. Singh. 1982. Insect behavioral responses to toxins: Practical and evolutionary considerations. Can. Entomol. 116:57-68.

Rausher, M.D. 1981. Host plant selection by *Battus philenor* butterflies: The roles of predation, nutrition, and plant chemistry. Ecol. Monogr. 51:1-20.

Schoonhoven, L.M. 1973. Plant recognition by lepidopterous larvae. Symp. Roy. Entomol. Soc. London 6:87-99.

Schoonhoven, L.M. 1981. Chemical mediators between plants and phytophagous insects, pp. 31-50. In: D.A. Nordlund, R.L. Jones and W.J. Lewis (Eds.), Semiochemicals: Their role in pest control. John Wiley & Sons, N.Y.

Schroeder, D. 1981. Biological control of weeds, pp. 41-78. In: W.W. Fletcher (Ed.), Recent advances in weed research. Commonw. Agric. Bur., Slough.

Silcock, R.G. 1975. Factors influencing the establishment of perennial grasses on the lateritic red earths (mulga soils) of south-western Queensland. M.S. Thesis, University of New England, New South Wales, Australia.

Singh, Z., and W.L. Howe. 1971. Feeding, longevity, and fecundity of the adult western corn rootworm fed artificial diets. J. Econ. Entomol. 64:1136-1137.

Southwood, T.R.E., and C.E.J. Kennedy. 1983. Trees as islands. Oikos 41:359-371.

Stadler, E. 1982. Sensory physiology of insect-plant relationships--roundtable discussion, pp. 81-91. In: J.H. Visser and A.K. Minks (Eds.), Proc. Fifth Intern. Symp. Insect-Plant Relationships. PUDOC, Wageningen, Netherlands.

Stebbins, L.G. 1972. The evolution of the grass family, pp. 1-18. In: V.B. Younger and C.M. McKell (Eds.), The biology and utilization of grasses. Academic Press, N.Y.

Strong, D.R., and D.A. Levin. 1979. Species richness of plant parasites and growth form of their hosts. Amer. Nat. 114:1-22.

Takhtajan, A.L. 1969. Flowering plants: Origin and dispersal. Oliver, Edinburgh.

van der Pers, J.C.N. 1978. Responses from olfactory receptors in females of three species of small ermine moths (Lepidoptera: Yponomeutidae) to plant odours. Entomol. Exp. Appl. 24:394-398.

van Drongelen, W. 1978. Contact chemoreception of host plant specific chemicals in larvae of various Yponomeuta species (Lepidoptera). J. Comp. Physiol. 134:265-279.

van Drongelen, W., and J.J.A. van Loon. 1980. Inheritance of gustatory sensitivity in F_1 progeny of crosses between Yponomeuta cagnagellus and Y. malinellus (Lepidoptera). Entomol. Exp. Appl. 28:199-203.

Waldbauer, G.P. 1964. The consumption, digestion, and utilization of solanaceous and non-solanaceous plants by larvae of the tobacco hornworm (Protoparce sexta Johan.) (Lepidoptera: Sphingidae). Entomol. Exp. Appl.7: 253-269.

Ward, S. 1977. Invertebrate neurogenetics. Annu. Rev. Gen. 11:415-450.

Wasserman, S.S., and D.J. Futuyma. 1981. Evolution of host plant utilization in laboratory populations of the southern cowpea weevil, Callosobruchus maculatus. Evolution 35:605-617.

White, E.B., P. DeBach and M.J. Garber. 1970. Artificial selection for genetic adaptation to temperature extremes in Aphytis lingnanensis Compere (Hymenoptera: Aphelinidae). Hilgardia 40:161-192.

White, M.J.D. 1973. Animal cytology and evolution. Cambridge Univ. Press, London. 375 pp.

Wiezcorek, H. 1976. The glycoside receptor of the larvae of Mamestra brassicae L. (Lepidoptera, Noctuidae). J. Comp. Physiol. 106:153-176.

Willis, J.C. 1960. A dictionary of flowering plants and ferns. Cambridge Univ. Press, London. 1245 pp.

DISCUSSION: Biological control of weeds by pathogens and arthropods.

DISCUSSION LEADER: L.A. Andres

In agricultural cropping areas we are usually confronted with a complex of weed species of which the grasses often predominate. Bioherbicides have been developed for postemergence use in the control of the dicotyledonous strangler vine (Morrenia odorata) on citrus in Florida, water hyacinth (Eichornia crassipes) in Florida, and northern jointvetch (Aeschynomene virginica) in rice in Arkansas and Louisiana. A question was raised whether bioherbicides have also been developed for use against monocotyledonous weeds. Although there are efforts underway to identify pathogens on johnsongrass (Sorghum halapense), crabgrass (Digitaria spp.), and nutsedge (Cyperus rotundus), thus far no bioherbicides have been developed for monocotyledons. There is concern that the plant pathogens associated with grasses may not be sufficiently host specific for use.

In some instances, plant pathogens are known to be vectored by plant-feeding arthropods. Several participants suggested that selected or engineered pathogens could be placed into arthropods that are host specific to particular weeds or weed groups and which would then vector the pathogens to these weeds. Presently, there is no effort underway to make use of vectors to enhance weed control with pathogens.

Biological control with naturalized as well as indigenous organisms depends to a large extent on their host specificity and assurance that plants of recognized value will not be attacked. Comments were made on two aspects of the host specificity question: 1) the stability of the specific host relationship, and 2) the concern that the removal of the targeted weed leads only to its replacement by another weed.

The stability and basis of the host specific relationship between the biological control arthropod and its host is always in question. Host shifts to non-related plants have been documented among other arthropods and there is no guarantee that they won't happen with weed-arthropods in the future. Even holometabolous arthropods could shift to other plants though this may take longer to accomplish if the species' developmental stages are linked to particular plant characteristics. Thus far, the safety record of biological control of weeds belies any host switching problems of this nature, although some weed-feeding arthropods have increased their host range to include closely related plant species to which they had not been previously exposed. Other examples of expanded host ranges may be recorded in the future despite the underlying stability of host utilization patterns that characterize many of our weed insects. At this time, there is insufficient evidence of host switching among herbivorous arthropods to justify reducing efforts to control weeds in the classical manner. Studies of the arthropod/host relationship should continue to minimize any future risks.

Host specific control agents may eliminate one weed out of a species complex, but others will soon fill the vacated niche. This is one reason why weed problems are so hard to resolve and is a problem that is faced by anyone who attempts to control weeds with a single approach solution. Ideally we want to arrive at a solution that results in the least loss from weeds (e.g., production losses, control costs) even if it means maintaining low levels of some weeds to preclude more noxious species. Under these circumstances even host specific natural enemies that reduce or slow the growth of a plant or in some way dull its competitive edge, can be of value. In this way, it becomes possible to reduce a weed's ability to cause additional loss without exchanging it for another weed.

Weeds are a major component of all agricultural systems and certainly affect the quality and diversity of the crop environment. Although focus will remain on the pest nature of weeds and their control, they are an ecological phenomenon and the role they play in the agroecosystem and their management must be viewed from that perspective. Weeds and the means for their control will certainly impact on other biological controls present in the fields.

Attempts to use arthropods in the classical biological control approach against agricultural weeds have been limited. The method has worked best against biennial-perennial weeds in areas where varying amounts of plant cover can be tolerated and is encouraged, such as rangelands and pastures, versus those areas that are monocropped and where all plants other than the crops are unwanted. Biological control has been successful where a combination of stresses exceeds the weed's stress tolerance, whether from the natural enemies or environment. Unfortunately, in crop areas, environmental stress on weeds is often minimal, a resulting spin-off of the use of irrigation and fertilization to optimize crop growth. Crop areas are also often characterized by high degrees of disturbance which tends to interrupt the life cycle of weed-control arthropods.

Despite these drawbacks, arthropods might have value against weeds where immediate or complete control is not needed (Aldrich 1984). 1) Arthropods might be of value in preventing weed seed production. There is generally a large guild of arthropod species that feed and develop on the flowers, seeds and fruits of weeds and which could be drawn upon for use. 2) Biological control might also be used to reduce the competitiveness of the subdominant weed species that are always a part of the weed community and hich remain "behind" the dominants, until the latter have been removed or controlled for one reason or another. 3) Natural enemies may also be of use against those reservoirs of weeds along ditchbanks, roadsides, and areas between fields. In all three of these situations the focus is on the reduction of weeds and not complete control. It is often wise to consider biological control as an adjunct to other weed management strategies and that its value should be judged over the long term, rather than for its immediate control capabilities (Aldrich 1984).

Considering the wide array of crops grown throughout the U.S., the wide array of crop habitats, the wide aray of weed species, growth forms and weed communities, plus varying control requirements, there must be a number of situations where biological control would be of value. Shifting agricultural practices will also create new combinations of weeds and problems where biological control could find use. Concern over soil erosion and the advent of reduced tillage has led to an increase in perennial weed species in some areas. In theory, perennial weeds are more amenable to biological control by arthropods because of the more stable food source and habitat they represent. The comment was made that non-tillage has also led to an increased use of herbicides. In response, it was noted that herbicides not only affect weeds and crops, but may also affect the development of plant pathogens residing in the soil which may in turn put increased pressure on the weeds.

In the past, biological control of weeds has been dominated by entomologists. Increasing numbers of plant pathologists and now nematologists are directing attention to the control of weeds. G.C. Smart (personal communication) notes that nematodes have been used or are being considered to control several weeds: 1) Solanum elaeagnifolium, a perennial weed in the south estern U.S. is attacked by the gall-forming Nothanguina phyllobia; 2) another gall-forming nematode, Paranguina picridis, has been released in limited quantities against Acroptilon repens (= Centaurea repens) in Canada; 3) Amsinckia intermedia, an annual weed in the Central Valley of California is parasitized by Anguina amsinckia which causes galling of the fruit and terminal apical meristem; and 4) the buds of several aquatic plants are found to be parasitized by Aphelenchoides fragariae and an unidentified species in Florida, while the roots were attacked by several species of Hirschmanniella.

A question was also raised concerning conflicting interests and the fact that arthropods may extend their range onto related non-target native plants found in the weed habitat. Until a protocol can be developed for resolving conflicts of this nature, it may be prudent to focus attention on the biological control of those naturalized weeds of agricultural and ecological importance that have few, if any, native plant relatives.

REFERENCE

Aldrich, R.J. 1984. Weed-crop ecology, principles in weed management. Breton Publ., North Scituate, MA., 465 pp.

Biological Control of Plant Pathogens

FOLIAR ANTAGONISTS: STATUS AND PROSPECTS

Steven E. Lindow

This review of the status and prospects of foliar antagonism in disease control cannot hope to be comprehensive because of the diversity of foliar plant pathogens and the complexity of the interactions of microorganisms on plants. This review will consider general concepts relating to the unique characteristics of plant diseases affecting aerial plant parts which distinguish them from diseases of roots, or of stresses due to insects and nematodes. Whereas the majority of detailed studies of antagonism, and biological control of plant diseases has concerned soilborne pathogens, biological control of foliar diseases has also been extensively reviewed (Baker & Cook 1974; Blakeman 1981; Blakeman & Brodie 1976; Blakeman & Fokkema 1982; Cook 1977; Dubas & Bulit 1981; Gibbins 1972; Henis & Chet 1975; Reinecke 1981; Spurr 1981a, b). In addition, reviews of the diversity of microbial interactions on leaf surfaces as well as the unique physical and chemical environment of leaf surfaces have been recently made (Blakeman 1971, 1973, 1981; Burrage 1976; Frossard 1981).

Foliar plant diseases are caused by pathogens that are quite different taxonomically. While the majority of foliar plant diseases are caused by eucaryotes (a diversity of sexually and non-sexually reproducing fungi), procaryotes including aerobic and facultatively anaerobic eubacteria, mycoplasmas, viruses and, most recently, viroids also cause foliar plant diseases. The biology, reproductive strategies and potential interactions of these dissimilar plant pathogens will obviously vary greatly. This review will be concerned primarily with bacterial and fungal pathogens causing diseases of the foliar portions of plants because they predominate on leaves.

Unlike many insect pests of plants, the presence of most plant pathogens of plants often cannot be correlated directly with disease. The physical and biological environment of leaf surfaces and of leaves themselves largely dictate the efficiency with which pathogens found associated with plants can actually cause disease symptoms. It is not uncommon to find that whereas many potential plant pathogen propagules, fungal spores or bacterial cells, can be found on plants, disease resulting from the presence of these propagules may occur infrequently or be absent

BIOLOGICAL CONTROL
IN AGRICULTURAL IPM SYSTEMS

ISBN 0-12-357030-1

(Leben 1965; Spurr & Welty 1972, 1975). While the physical environment may be of paramount importance to many plant pathogens in initiation of disease, some plant diseases may be dictated more by the chemical or biological environment of leaf surfaces or may be relatively insensitive to these effects. While the presence of even a single arthropod pest of plants, particularly sucking or chewing insects, can be associated with some (although probably imperceptively small) impact on that plant host, the mere presence of pathogens on plant surfaces may have no impact (Leben 1965) unless disease is initiated. Many physical factors such as duration of moisture on leaf surfaces, intensity of sunlight, temperature, relative humidity, wind velocity (Blakeman 1981, Jarvis & Slingsby 1977), chemical factors (including nutrient depositions), plant inhibitors of microbial growth (Blakeman 1971), self germination inhibitors found in many fungi (Baker & Cook 1974), mineral nutrients involved in microbial growth or regulation of microbial development (Swinburne 1981), and antibiotic-like compounds produced by microorganisms on leaf surfaces (Burchill & Cook 1971) affect disease development and may be modified to control plant disease. These factors control the sequence of events including fungal spore germination, mycelial growth or bacterial cell division, plant invasion, and production of toxins or other biologically active compounds that ultimately initiate symptom development within the plant. Any factor which interrupts this chain of events will reduce the probability with which infective propagules found on plant surfaces will lead to plant disease and, therefore, plant stress.

Many plant pathogens, particularly plant pathogenic bacteria, exist on many plants at least for a portion of their life in a commensal relationship with potential host plants. For example, many bacterial pathogens colonize the surface of healthy plants and develop very large epiphytic populations on plants on which they are potentially pathogenic (Crosse 1971, Leben 1965, Schroth et al. 1974). Spores and other structures of fungal plant pathogens can also be found on potential host plants (Blakeman 1981, Fokkema et al. 1975, Spurr & Welty 1972). The factors which trigger a change from a purely commensal to a parasitic phase of many plant pathogens is not well understood but may involve specific environmental triggers.

Since the presence of many plant pathogens cannot be directly correlated with plant disease, several different avenues for disease control exist. Many disease control strategies, including the use of fungicides and bactericides and cultural techniques, attempt to limit the growth or survival of potential sources of infective inoculum on plant surfaces (Wood & Tveit 1955, Zentmeyer & Bold 1977). However, disease control has also been obtained by interference with spore germination or bacterial cell division on leaf surfaces by plant derived chemicals and by biological competition (Rotem et al. 1976). Nutrients required for spore germination and infection, or bacterial cell division can also be modified by biological control agents either naturally found on leaf surfaces (Andrews et al. 1983; Blakeman & Brodie 1977; Blakeman & Parbery 1977; Farabee & Lockwood 1958; Fokkema & van der Meulen 1976; Fokkema et al. 1975; McBride & Hayes 1979; Warren 1972a, b) or induced via augmentation (to be discussed later).

EVIDENCE FOR NATURALLY OCCURRING BIOLOGICAL CONTROL

Biological control has been defined as "any conditions under which, or practice whereby, survival or activity of a pathogen is reduced through the agency of any other living organism (except man itself) with the result that there is reduction in the incidence of the disease caused by the pathogen". It is obvious from this definition that man through indirect effects on crop growth such as fertilization, irrigation practices, variety selection, plant breeding, weed control, planting date selection, and a myriad of other cultural techniques, can influence the microbiota of plant surfaces and, therefore, indirectly achieve biological control of plant disease (Baker & Cook 1974, Cook 1977, Zentmeyer & Bold 1977). Cultural management of disease, which might be considered a component of integrated pest management, has been practiced by plant pathologists and is a major tool in the management of foliar plant diseases. This review, however, will not discuss the well known growing practices which minimize foliar plant diseases but will concentrate on direct modification of leaf surface microflora by man, and on the complex interactions which occur naturally on plants that minimize plant diseases.

Increases in plant disease after treatment of plants with pesticides, referred to as iatrogenic diseases (Griffiths 1981, Griffiths & Berrie 1978), offers indirect evidence of natural biological control (antagonism) of plant disease-inciting organisms on plants. A thorough review of iatrogenic plant diseases has recently appeared (Griffiths 1981). Many examples of plant diseases which become more severe following crop protection chemical application have been noted (Altman & Campbell 1977, Griffiths & Berrie 1978, Katan & Eshel 1973, Kavanaugh 1974), but many of these observations are anecdotal. The causes of increased epiphytotics following perturbations of plants with only a few crop protection chemicals have been studied in detail. Plant disease can be induced or exacerbated through chemicals which act on 1) the host plant; 2) the pathogen; or 3) the ecosystem, including the microbial interactions in which host and pathogen coexist. Effects of (1) and (3) will be most important in this review. Various mechanisms of increased plant disease due to changes of the host plant have been postulated. These mechanisms include: 1) changes in host composition and structure, such as a decrease in cuticle thickness by pesticide formulations (Natti & Hervey 1956); 2) alteration in sugar content of plants treated with herbicides (Horsfall & Dimond 1957); 3) leakage of metabolites to the host surface, such as can also occur following herbicide treatment (Altman 1972, Altman & Campbell 1977); or 4) changes in natural defense mechanisms such as reduction in phytoalexin production by plants (Romig & Sasser 1972).

Several examples of foliar iatrogenic diseases following fungicide treatment have been reported (Carter & Price 1974, Firman & Waller 1977, Fokkema et al. 1975, Jackson 1970, Jordan 1973, van der Zweep 1970). Increased infection of rye leaves by *Cochliobolus sativus* following treatment with benomyl (Bashi & Fokkema 1977; Fokkema et al. 1975, 1979) is typical of diseases favored following pesticide treatment. Benomyl reduced the naturally occurring saprophytic microflora approximately 10-fold. Thus the antagonistic capacity of leaves normally colonized with saprophytic microflora was reduced. Many other examples of iatrogenic diseases also appear to be due to disruption of competition of the pathogen with other saprophytic microorganisms that have been displaced by pesticide treatment. Thus while the effect of pesticides on non-target microorganisms have been relatively unstudied, and a great need exists for increased study of this phenomenon (Andrews 1981, Andrews & Kenerley 1978), the diseases studied thoroughly to date indicate that natural biological control of foliar plant diseaes may be quite common.

More indirect evidence that biological control might operate (although possibly at low levels) on plants is obtained by examination of the saprophytic microflora on plants. These saprophytes are found in association with pathogens and possess traits thought to be important in antagonism of these parasites. Hyperparasites can be observed to attack hyphae, sporulating structures, or vegetative cells of plant pathogenic fungi and bacteria in the field, therefore reducing inoculum (Kranz 1981, Scherff 1973, Vidaver 1976). Important species of hyperparasitic fungi include *Darluca filum*, *Tuberculina vinosum*, *Verticillium lecanii*, *Ampelomyces quisqualis* and *Trichoderma* spp. (Blakeman & Fokkema 1982, Kranz 1981). While such hyperparasitic fungi can be readily found in the field, their effectiveness in controlling disease development in the field is largely unstudied. Numerous viruses can also be found to infect plant pathogenic fungi (Lemke & Nash 1974). A more detailed recent examination of viruses affecting the important pathogen *Endothia parasitica* in southern Europe and the United States has been made (Day et al. 1977, Dickinson & Skidmore 1976, Grente & Sauret 1969, Jaynes & Elliston 1980). *Bdelovibrio bacteriovorus*, which attacks plant pathogenic bacteria, have been readily isolated from soils. As with hyperparasites of fungi, the relative importance of natural biocontrol of plant pathogenic bacteria with these parasites has been relatively unstudied (Scherff 1973, Vidaver 1976).

Nutrient composition by saprophytic bacteria such as *Erwinia herbicola*, have also been indicated to reduce nutrient levels in water films on leaf surfaces to levels which may reduce mycelial growth, germ-tube formation, or lesion development, particularly with necrotrophic plant pathogens such as *Botrytis cinerea* (Blakeman 1973, Blakeman & Brodie 1977, Blakeman & Fraser 1971,

Clark & Lorbeer 1977, Dickinson & Skidmore 1976, Hsieh & Buddenhagen 1974). Many different species of microorganisms including filamentous fungi, yeast, bacteria, and actinomycetes found on plants have been reported to produce antibiotics in vitro (Baker & Cook 1974, Blakeman & Fokkema 1982, Leben 1964, Leben & Daft 1965, Leben et al. 1965). Such antibiotic production may either kill or reduce the growth of pathogenic fungi and/or bacteria on plant surfaces. Therefore, both direct evidence of increased disease following modification of leaf surface microbial communities, and evidence of microorganisms having phenotypes capable of affecting plant pathogens indicate that natural biological control (antagonism) occurs on plants. It is also clear that minor perturbations in the microbial balance on plants due to cultural conditions could easily achieve modest levels of biological control of disease. Numerous microorganisms have potential for achieving biological control of plant disease when used either singly or in combination with other microorganisms in deliberate microbial modification schemes.

APPLICATION OF ANTAGONISTS FOR BIOLOGICAL CONTROL

Many examples of successful introduction of bacteria, yeast, mycelial fungi, or viruses for the reduction of foliar diseases caused by both bacteria or fungi have been reported in greenhouse and laboratory studies. Dozens of successful experiments have been reported and are summarized in an excellent recent review (Blakeman & Fokkema 1982). Antagonistic interactions between common leaf surface hyphal fungi, yeasts, bacteria, and a variety of necrotrophic leafspotting pathogens have been demonstrated under controlled environmental conditions by applying the antagonist to the leaves before or at the same time as the pathogen. Control of disease following introduction of antagonistic microorganisms has ranged from barely detectable to excellent, but in many cases has depended on unique environmental conditions (Blakeman & Fokkema 1982, Sleesman & Leben 1976).

Although many examples of successful biological control of foliar diseases have been demonstrated under greenhouse and laboratory conditions, few reports of successful field applications of these antagonists exist. Because of the unique and often complicated interactions of different pathogens, other microorganisms, and their host plant, no attempt will be made in this review to document the wealth of information, except to direct readers to excellent reviews of the topic (Baker & Cook 1974; Blakeman 1981; Blakeman & Brodie 1976, 1977; Blakeman & Fokkema 1982; Corke 1978; Crosse 1971; Dubas & Bulit 1981; Gibbins 1972; Henis & Chet 1975; Reinecke 1981; Spurr 1981a, b). Instead, a more detailed examination of efforts to control two fungal and two bacterial incited diseases will be presented. These examples will exemplify the restrictions of, and the strategies often used with, biological control agents and will provide a basis for general conclusions on the prospect of the use of biological control agents in foliar diseases.

Control of Brown Spot of Tobacco

Brown spot of tobacco is caused by the pathogenic fungus, *Alternaria alternata*. This disease is economically important in tobacco and develops during the middle to latter part of the growing season (Spurr 1973, Spurr & Welty 1972). Preliminary work indicated that tobacco leaves could be colonized both internally and externally by several saprophytic fungi, and externally by many different types of bacteria (Spurr & Welty 1972). Fungal and bacterial residents of tobacco leaves appeared to exist in a commensal relationship with tobacco (Spurr & Welty 1972, 1975). Modification of leaf surface saprophytic fungi and bacteria consistently affected the response of tobacco to inoculations with pathogenic strains of *A. alternata* (Spurr & Welty 1975). Disease severity on tobacco leaves treated with ethanol could be as much as twice that of untreated leaves when inoculated with the same number of *A. alternata* conidia (Spurr 1983). This work, therefore, indicated that microflora on leaf surfaces of tobacco interact with the pathogen and often may be the major determining factor in disease development.

Some of the resident microflora on tobacco leaves were isolated, tested, and selected for disease biological control studies (Fravel & Spurr 1977; Spurr 1977, 1978a, b, 1981a, b). Nonpathogenic isolates of *Alternaria* applied to leaves of tobacco before inoculation with the pathogen, *A. alternata*, reduced tobacco brown spot severity 60% in laboratory experiments and 65% in artificially induced field infections (Spurr 1977). As with many such experiments, the level of disease reduction was related directly to the concentration of saprophytic conidia applied. Protection against brown spot infection was maximum in laboratory experiments when the protective conidia were applied to the leaves two to three days before inoculation with the pathogen (Spurr 1977). Non-pathogenic conidia germinated and grew on the leaf surface with no evidence of antagonism towards the pathogen. In the presence of colonization by protective saprophytic *Alternaria* sp., *A. alternata* did not penetrate host plants but instead grew epiphytically on tobacco leaf surfaces.

Brown spot of tobacco and *Cercospora* leaf spot of peanut foliage have been reduced following application of bacterial antagonists. Strains of *Pseudomonas cepacia*, *Bacillus mycoides*, and *Bacillus thuringiensis* have reduced these two foliar leaf spots in both greenhouse and field tests (Fravel & Spurr 1977; Spurr 1978a, b, 1981a, b). Treatment of plants with a concentration of 10^8 antagonist cells/ml provided some control of both of these diseases. Up to 50% field control of tobacco *Alternaria* leaf spot and peanut *Cercospora* leaf spot with bacterial antagonists has been observed (Spurr 1981a, b). It is of some interest that commercial emulsifiable concentrate and wettable powder formulations of B. thuringiensis significantly reduced both foliar leaf spots when applied prior to inoculation with the appropriate pathogen. Field tests over three growing seasons with these three bacterial antagonists indicated that foliar biological control varied with field conditions (Spurr 1981a, b). While significant control of disease was often obtained with these biological control agents, control often was impractical (50% control or less).

Biological Control of Apple Scab

Apple scab caused by Venturia inaequalis is a serious disease causing economic losses in nearly all areas in which apples are grown. While the yields of apple may not be significantly reduced by V. inaequalis infection, quality of apples infected with this pathogen is markedly reduced. An important means of controlling this pathogen involves breaking the overwintering cycle by eliminating ascospore production (Heye & Andrews 1983). Ascospores are the primary inoculum for infection of young leaves and fruit in many growing areas during the spring. Eradicative chemicals have long been used to reduce or eliminate ascospore production in pseudothecia which form in senescent leaves on the orchard floor during winter months.

Several reports of antagonism of microorganisms to V. inaequalis have prompted renewed studies for biological control of this disease (Firman & Waller 1977, Heye & Andrews 1983, Simard et al. 1957, Wassermann 1969). In all studies to date, the goal has been to discover microorganisms and elucidate the mechanisms by which pseudothecia development or ascospore maturation might be reduced or eliminated. In early studies mixtures of apple leaf surface microorganisms reduced formation of pseudothecia by V. inaequalis on incubated leaf discs and in leaves placed in the orchard after inoculation with the pathogen (Simard et al. 1957). In these early studies no attempts were made to identify the most effective antagonist within mixtures of microorganisms found associated with pseudothecia on leaves. In more recent studies Andrews and coworkers (Andrews et al. 1983, Heye & Andrews 1983) have studied the quantitative antagonism of numerous microorganisms isolated from apple leaves toward V. inaequalis. Several different assays were used to evaluate antagonism in vitro and in vivo: growth on nutrient agar, germination and germ tube growth on agarose-coated slides, and lesion size, overall symptom development, and conidial production on infected leaves (Andrews et al. 1983). Flavobacterium sp., Cryptococcus sp., Aureobasidium pullulans, Trichoderma viride, Chaetomium globosum, Microsphaeropsis olivacea, and two unidentified actinomycetes were found to be most antagonistic in these studies. The best and most consistent antagonist was C. globosum. The presumptive modes of antagonism for the eight most antagonistic strains appeared to be antibiosis and/or nutrient competition. More extensive field evaluations of these antagonists are currently scheduled (J.H. Andrews, personal communication).

Integrated control of apple scab by modifying the nutrient status of overwintering leaves is also well established. Treatment of senescent apple leaves on trees shortly before leaf fall or of fallen leaves on the orchard floor with a solution of 2% urea greatly reduced ascospore production in the spring (Burchill & Cook 1971). Urea treatment of fallen apple leaves greatly increased the bacterial populations found on decaying leaves (Burchill & Cook 1971). Fluorescent Pseudomonads increased greatly in number on urea-treated leaves. Pseudothecial development may be reduced directly by antibiotics produced by fluorescent Pseudomonads on decaying leaves, or indirectly by degeneration of leaf structure due to enhanced degradation of leaf material in the presence of higher bacterial populations induced by treatment with a good nitrogen source (Burchill & Cook 1971). Current evidence suggests that the latter possibility is the most likely. Treatment of orchards with urea is rapidly becoming an accepted management practice. It is clear this practice is a form of biological control that simply augments control due to naturally occurring bacteria. Under natural conditions these bacteria would lead to an insufficient degradation of leaf structure to significantly reduce pseudothecial development and, therefore, ascospore production in spring.

Biological Control of Fireblight of Pear

Fireblight of pear, caused by *Erwinia amylovora*, is probably the most serious disease of these trees and can lead to death when untreated. Infection of pear occurs primarily through flowers that become colonized by epiphytic populations of *E. amylovora* (Schroth et al. 1974). Control of fireblight continues to be by repeated applications of one or more bactericides, (primarily streptomycin) following the onset of weather favorable for the development of the disease (Schroth et al. 1974). However, repeated bactericide applications are costly and resistance to streptomycin has occurred widely in some areas, limiting the effectiveness of this material (Moller et al. 1981; Schroth et al. 1974, 1979). Concern also exists that the use of antibiotics in agricultural crops could diminish the effectiveness of these materials in medical applications. Therefore, biological control of this disease has been sought (Beer et al. 1980, Schroth et al. 1974). Several investigators have shown that many bacteria are antagonistic to *E. amylovora* in vitro and that co-inoculations of various epiphytic bacteria with *E. amylovora* on pear or apple could largely prevent disease development (Beer et al. 1980; Chatterjee et al. 1969; Erskine & Lopatecki 1975; Farabee & Lockwood 1958; Goodman 1965, 1967; Lindow 1982d; McIntyre et al. 1973; Parker 1936; Riggle & Klos 1972; Thomson et al. 1976; Wrather et al. 1973). Parker (1936) attempted to control fireblight with saprophytic bacteria but obtained inconsistent results. Other more recent workers have obtained partial control of fireblight by preinoculation of blossoms with E. herbicola or other saprophytic bacteria. Most of these studies, however, have been hindered because the population dynamics of the pathogen or the applied antagonistic microorganism were not followed. In studies in which population dynamics were followed closely, an inhibition of E. amylovora populations was strongly dependent on populations of antagonistic bacteria (Thomson et al. 1976). Apparently a high threshold antagonist density on pear leaf and flower surfaces were required for significant effects on pathogen multiplication. Density dependent mechanisms such as colonization of sites and/ or nutrient acquisition may be likely explanations for the reduction of E. amylovora populations following application of antagonistic bacteria.

Biological control of fireblight with applied antagonistic bacteria can approach that of frequent applications of bactericides (Lindow 1982c, Thomson et al. 1976). Fireblight of trees treated with antagonistic bacteria have been reduced from 20 to 85% in various studies and approach the results obtained with treatments of streptomycin on a weekly basis.

Biological Control of Plant Frost Injury

Frost injury is a serious abiotic disease of plants. Losses in plant production in the United States due to frost injury are estimated at over one billion dollars yearly (White & Haas 1975). Frost injury until recently was considered an unavoidable result of physical stress (low temperatures) to plants (Arny et al. 1976, Burke et al. 1976, Cary & Mayland 1970). Some plant frost injury recently, however, has been shown to involve an interaction of Pseudomonas syringae and Erwinia herbicola on leaf surfaces as well as low temperature stress. These bacteria cause the frost sensitive plants on which they reside to become more susceptible to freezing damage by initiating the formation of ice that is required for frost injury (Arny et al. 1976; Burke et al. 1976; Cary & Mayland 1970; Chandler 1958; Lindow 1983a, b; Lindow et al. 1978b, c, 1982b). Many frost sensitive agricultural plants are important agronomic crops which are distinguished from frost hardy plants by their relative inability to tolerate ice formation within their tissues (Arny et al. 1976, Burke et al. 1976, Chandler 1958). Ice formed in and/or on frost sensitive plants spreads rapidly both intercellularly and intracellularly causing mechanical disruption of cell membranes. Thus, most frost sensitive agricultural plants have no significant mechanisms of frost tolerance and must, therefore, avoid ice formation to avoid frost injury (Arny et al. 1976). The discovery of bacterial ice nuclei on plants has indicated several new methods of frost protection based on enhancement of the natural supercooling ability of plants by controlling ice nuclei contributed by ice nucleation active bacteria (Lindow 1983a). Any treatment that reduces the numbers and/or the ice nucleation activity of bacteria is promising as an alternative method of frost control to plants. The role of bacterial ice nuclei and methods of control have been reviewed recently (Lindow 1982b, 1983a, b).

Preliminary work indicates that natural antagonism among leaf surface bacteria may be augmented to favor increased populations of non-ice nucleation active bacterial competitors and, therefore, reduce frost sensitivity of plants (Lindow et al. 1983a, b). Only about 0.05-10% of the total bacteria found on plant surfaces are active as ice nuclei and, therefore, are involved directly in frost injury (Lindow et al. 1978c, 1982a). Competition between these and other epiphytic bacteria on leaf surfaces appears likely based on studies of other ecological niches. However, the degree of natural competition among epiphytic microorganisms appears insufficient to prohibit significant build-up of populations of epiphytic ice nucleation active bacteria on most plants in nature. However, these bacterial competitors may be augmented by direct application of antagonists or by other alterations of the leaf surface microbial ecology to occupy a niche on the plant that might otherwise be colonized by ice nucleation active bacteria.

Antagonistic bacteria have been established on plants by application to seeds or by foliar application to plants prior to colonization of these plants by ice nucleation active bacteria (Lindow 1979, 1983a; Lindow et al. 1983a, b). Bacterial competitors have been selected on the basis of their effective colonization of leaf surfaces. Populations of ice nucleation active bacteria are generally lowest on young vegetative tissues (Lindow 1982c, Lindow et al. 1978c). Treatment of young tissues with non-ice nucleation active bacteria reduces potential increases in populations of ice nucleation active bacteria on plants during periods of low temperatures and, therefore, reduces probability of frost injury. Many bacterial antagonists have effectively colonized emerging and mature plant tissues for a relatively long period of time (1-5 months) following single foliar applications (Lindow 1982c, 1983a). Populations of ice nucleation active bacteria have been reduced under field conditions from 10-100 fold on plants treated with antagonistic bacteria when compared with untreated plants. Frost damage to treated plants have been reduced from 30-95% compared to untreated plants and is directly related to reductions in the logarithm of populations of ice nucleation active bacteria on plants following treatment (Lindow 1983a, Lindow et al. 1978a).

Integrated management of fireblight and frost injury of pear has recently been reported (Lindow 1982d). Antagonistic bacteria applied to pear trees colonized pear flowers and leaves for over three months and reduced significantly the epiphytic populations of *P. syringae* and *E. amylovora*. The control of frost injury of fireblight from single applications of antagonistic bacteria was nearly as good as that achieved by from weekly applications of a mixture of streptomycin and oxytetracycline (Lindow 1982d).

Mechanisms determining effect of biological control of frost injury are not fully known. Some, but not all, bacterial antagonists of ice nucleation active bacteria on plants are antagonistic to these bacteria in vitro. Nearly all mutants of antibiotic producing antagonistic bacteria deficient in production of antibiotics in vitro were still effective in reducing colonization of plants by P. syrinage (Lindow 1982a). Antibiosis, therefore, may play a minor role in antagonism on leaves by other bacteria. Antagonism on leaf surfaces may be a general phenomenon such that non-specific control of more than one target microorganism may be possible.

COMMERCIAL APPLICATION OF BIOLOGICAL CONTROL AGENTS

Although numerous examples of successful biological control of foliar plant diseases have been demonstrated in greenhouse, laboratory, and even in some cases field experiments, commercial application and widespread grower use of these agents has not occurred. This author is aware of no commercialy produced biological agents effective against foliar diseases or abiotic stress. Several biological and non-biological factors can explain why commercial development of biological control agents have not progressed further. Because the interactions of plant pathogens and other microorganisms with their host plant are inherently complex and dependent on unpredictable environmental conditions, field evaluations of biological control agents have often given inconsistent results. While some biological control agents have consistently yielded some degree of disease protection, the degree of disease protection has been variable and often has not compared favorably with that achieved with the use of chemical pesticides. Where chemical pesticides are available for disease control, growers would be unwilling to accept risks associated with incomplete disease control afforded by biological control agents even though their cost of application and other indirect costs may be lower.

Large agrochemical companies historically have not viewed biological control agents as being sufficiently profitable to be considered for development. Even though the cost of developing biological control agents might be less than comparable chemical fungicides or bactericides due to existing EPA regulations, the potential markets for many biological control agents might also be correspondingly small. Many biological control agents evaluated to date appear to have rather unique targets, i.e., might protect against a single plant pathogen causing disease on a single crop plant. Unless that biocontrol agent affected disease of an important crop plant the potential market for the use of such an agent would be small.

Many agrichemical corporations require some degree of patent protection before they are willing to proceed with advanced testing and development of biological control agents for disease control. Most biological agents of foliar plant pathogens have been discovered by non-industrial scientists. Unless these scientists or their institutions take some steps to obtain patent protection for the process of disease control using these microorganisms, such patent protection would, therefore, not be available to prospective corporations with the resources to develop further such discoveries through licensing agreements with the inventor or their institution. Therefore, to enable commercial development of discoveries of biological control agents, researchers must be aware of the importance of patenting such discoveries in addition to carrying out definitive biological research.

Many major agrichemical corporations have historically had little experience with biological agents. Only recently has renewed interest in biotechnology shaken historical views of biological agents among many large corporations. While smaller companies with an interest in biological agents may have a difficulty in financially supporting the continued research and development required for commercialization of such agents, larger corporations with financial expertise and marketing capabilities often have not had personnel or corporate policies which were conducive for development of biological agents. This latter trend, however, appears to be changing.

RECOMMENDATIONS

A major limitation of the current use of biological control agents of foliar diseases has been inconsistent results when applied under field conditions. Many of the inconsistent results obtained in the use of these microbial agents may be due to a lack of appreciation of the complex ecology into which they are being introduced and the mechanisms by which they are exerting biological control against pathogens.

1. Although considerable study of ecological interactions among microorganisms on the aboveground parts of plants has already been done, much more will be needed to understand the limitations with which biological control might be practiced. Without understanding the complex interactions microorganisms exhibit on plants we cannot currently predict or understand the successes or the failure of many current biological control efforts.
2. Major mechanisms of antagonism of saprophytic microorganisms and pathogens must also be clearly understood so that more rational approaches for the selection of efficient biological control microorganisms can be undertaken. Selecting potential biological control organisms because of phenotypes which may be unrelated to their ability to antagonize pathogens on leaves may be misleading at best and is likely to be counterproductive.

3. Much work needs to be completed in various aspects of industrial microbiology relating to the development of biological control agents after further research has been done in their discovery and testing. Because of the current absence of many large scale commercial productions of microorganisms for biological disease control in agriculture, it is difficult for scientists and corporate executives to assess the ease and, therefore, cost with which new biological control agents will be commercialized. Many questions such as genetic stability, large scale culturing (such as in fermentation), storage techniques, length of survival following culturing, must all be determined and must all be favorable to obtain a commercially viable product.

This review would be misleading if it left the reader with the opinion that biological control of foliar diseases was an unlikely prospect in future integrated pest management schemes. It is clear that many indirect growing practices already exert considerable biological control over foliar diseases. Commercial application of biological control also appears feasible. The complex microbial ecology of plant surfaces need not be a hinderance to exploitation of biological control. Indeed, the extensive biological diversity found on plants offers a wealth of potentially exploitable microorganisms. With advances in understanding the ecology of microorganisms on leaf surfaces this reviewer feels confident that suitable microorganisms can be identified to improve crop productivity.

REFERENCES

Altman, J. 1972. Increased glucose exudate and damping off in sugar beets in soils treated with herbicides. Phytopathology 62:743.

Altman, J., and C.L. Campbell. 1977. Effect of herbicides on plant diseases. Annu. Rev. Phytopathol. 15.:361-385.

Andrews, J.H. 1981. Effects of pesticides on non-target microorganisms on leaves, pp. 283-304. In: J.P. Blakeman (Ed.), Microbial ecology of the phylloplane. Academic Press, N.Y.

Andrews, J.H., F.M. Berbee and E.V. Nordheim. 1983. Microbial antagonism to the imperfect stage of the apple scab pathogen, Venturia inaequalis. Phytopathology 73:228-234.

Andrews, J.H., and C.M. Kenerley. 1978. The effects of a pesticide program on non-target epiphytic microbial populations of apple leaves. Can. J. Microbiol. 24:1058-1072.

Arny, D.C., S.E. Lindow and C.D. Upper. 1976. Frost sensitivity of Zea mays increased by application of Pseudomonas syringae. Nature 262:282-284.

Baker, K.F., and R.J. Cook. 1974. Biological control of plant pathogens. W.H. Freeman, San Francisco. 433 pp.

Bashi, E., and N.J. Fokkema. 1977. Environmental factors limiting growth of Sporobolomyces roseus, an antagonist of Cochliobolus sativus, on wheat leaves. Trans. Brit. Mycol. Soc. 68:17-25.

Beer, S.V., J.L. Norelli, J.R. Rundle, S.S. Hodges, J.R. Palmer, J.I. Stein and H.S. Aldwinckle. 1980. Control of fireblight by non-pathogenic bacteria. Phytopathology 70:459.

Blakeman, J.P. 1971. The chemical environment of the leaf surface in relation to growth of pathogenic fungi, pp. 255-268. In: T.F. Preece and C.H. Dickinson (Eds.), Ecology of leaf surface microorganisms. Academic Press, N.Y.

Blakeman, J.P. 1973. The chemical environment of leaf surfaces with special reference to spore germination of pathogenic fungi. Pestic. Sci. 4:575-588.

Blakeman, J.P. (Ed.). 1981. Microbial ecology of the phylloplane. Academic Press, N.Y.

Blakeman, J.P., and I.D.S. Brodie. 1976. Inhibition of pathogens by epiphytic bacteria, pp. 529-557. In: C.H. Dickinson and T.F. Preece (Eds.), Microbiology of aerial plant surfaces. Academic Press, London.

Blakeman, J.P., and I.D.S. Brodie. 1977. Competition for nutrients between epiphytic micro-organisms and germination of spores of plant pathogens on beetroot leaves. Physiol. Plant Pathol. 10:29-42.

Blakeman, J.P., and N.J. Fokkema. 1982. Potential for biological control of plant diseases on the phylloplane. Annu. Rev. Phytopathol. 20:167-192.

Blakeman, J.P., and A.K. Fraser. 1971. Inhibition of *Botrytis cinerea* spores by bacteria on the surface of chrysanthemum leaves. Physiol. Plant Pathol. 1:45-54.

Blakeman, J.P., and D.G. Parbery. 1977. Stimulation of appressorium formation in *Colletotrichum acutatum* by phylloplane bacteria. Physiol. Plant Pathol. 11:313-325.

Burchill, R.T., and R.T.A. Cook. 1971. The interaction of urea and microorganisms in suppressing development of perithecia of *Venturia inaequalis* (Cke.) Wint, pp. 471-483. In: T.F. Preece & C.H. Dickenson (Eds.), Ecology of leaf surface microorganisms. Academic Press, N.Y.

Burke, M.J., L.A. Gusta, H.A. Quamme, C.J. Weiser and P.H. Li. 1976. Freezing and injury to plants. Annu. Rev. Plant Physiol. 27:507-528.

Burrage, S.W. 1976. Aerial microclimate around plant surfaces, pp. 173-184. In: C.H. Dickinson and T.F. Preece (Eds.), Microbial ecology of aerial plant surfaces. Academic Press, N.Y.

Carter, M.V., and T.V. Price. 1974. Biological control of *Eutypa armeniacae*. II. Studies of the interaction between *E. armeniacae* and *Fusarium lateritium*, and their relative sensitivities to benzimidazole chemicals. Aust. J. Agric. Res. 25:105-119.

Cary, J.W., and H.F. Mayland. 1970. Factors influencing freezing of supercooled water in tender plants. Agron. J. 62:715-719.

Chandler, W.H. 1958. Cold resistance in horticultural plants: A review. Proc. Amer. Soc. Hortic. Sci. 64:552-572.

Chatterjee, A.K., L.N. Gibbins and J.A. Carpenter. 1969. Some observations on the physiology of *Erwinia herbicola* and its possible implication as a foliar antagonistic to *Erwinia amylovora* in the "Fireblight syndrome." Can. J. Microbiol. 15:640-642.

Clark, C.A., and J.W. Lorbeer. 1977. Comparative nutrient dependency of *Botrytis squamosa* and *B. cinerea* for germination of conidia and pathogenicity on onion leaves. Phytopathology 67:212-218.

Corke, A.T.K. 1978. Microbial antagonisms affecting tree diseases. Ann. Appl. Biol. 89:89-114.

Cook, R.J. 1977. Management of the associated microbiota, pp. 145-166. In: J.G. Horsefall and E.B. Cowling (Eds.), Plant Disease. Vol. 1. How disease is managed. Academic Press, N.Y.

Crosse, J.E. 1971. Interactions between saprophytic and pathogenic bacteria in plant disease, pp. 283-290. In: T.F. Preece and C.H. Dickinson (Eds.), Ecology of leaf surface microorganisms. Academic Press, London.

Day, P.R., J.A. Dodds, J.E. Elliston, R.A. Jaynes and S.L. Agnostakis. 1977. Double strand RNA in *Endothia parasitica*. Phytopathology 67:1393-1396.

Dickinson, C.H., and A.M. Skidmore. 1976. Interactions between germinating spores of *Septoria nodorum* and phylloplane fungi. Trans. Brit. Mycol. Soc. 66:45-46.

Dubas, B., and J. Bulit. 1981. Filamentous fungi as biological control agents on aerial plant surfaces, pp. 353-367. In: J.P. Blakeman (Ed.), Microbial ecology of the phylloplane. Academic Press, N.Y.

Erskine, J.M., and L.E. Lopatecki. 1975. *In vitro* and *in vivo* interactions between *Erwinia amylovora* and related saprophytic bacteria. Can. J. Microbiol. 21:35-41.

Farabee, G.J., and J.L. Lockwood. 1958. Inhibition of *Erwinia amylovora* by bacterium species isolated from fireblight cankers. Phytopathology 48: 208-211.

Firman, I.D., and J.M. Waller. 1977. Coffee berry disease and other *Colletotrichum* diseases of coffee. Phytopathol. Pap., C.M.I. 20, 53 pp.

Fokkema, N.J., J.G. den Houter, Y.J.C. Kosterman and A.L. Nelis. 1979. Manipulation of yeasts on field-grown wheat leaves and their antagonistic effect on *Cochliobolus sativus* and *Septoria nodorum*. Trans. Brit. Mycol. Soc. 72:19-29.

Fokkema, N.J., J.A.J. van der Laar, A.L. Nelis-Blomberg and B. Schippers. 1975. The buffering capacity of the natural mycoflora of rye leaves to infection by *Cochliobolus sativus* and its susceptibility to benomyl. Neth. J. Plant Pathol. 81:176-186.

Fokkema, N.J., and F. van der Meulen. 1976. Antagonism of yeast-like phyllosphere fungi against *Septoria nodorum* on wheat leaves. Neth. J. Plant Pathol. 82:13-16.

Fravel, D.R., and H.W. Spurr. 1977. Biocontrol of tobacco brown-spot disease by *Bacillus cereus* subsp. *mycoides* in a controlled environment. Phytopathology 67:930-932.

Frossard, R. 1981. Effect of guttation fluids on growth of microorganisms on leaves, pp. 213-226. In: J.P. Blakeman (Ed.), Microbial ecology of the phylloplane. Academic Press, N.Y.

Gibbins, L.N. 1972. Relationship between pathogenic and non-pathogenic bacterial inhabitants of aerial plant surfaces, pp. 15-24. In: H.P. Maas Geesteranus (Ed.), Proc. 3rd int. conf. on plant pathogenic bacteria. Wageningen, 14-21 April, 1971. Cent. Agric. Publ. Doc., Wageningen, Netherlands.

Goodman, R.N. 1965. *In vitro* and *in vivo* interaction between components of mixed bacterial cultures isolated from apple buds. Phytopathology 55:217-221.

Goodman, R.N. 1967. Protection of apple stem tissue against *Erwinia amylovora* infection by avirulent strains and three other bacterial species. Phytopathology 57:22-24.

Grente, J., and S. Sauret. 1969. L'hypovirulence exclusive phenomene original en pathologie vegetale. C.R. Acad. Sci. Paris 268:3173-3176.

Griffiths, E. 1981. Iatrogenic plant diseases. Annu. Rev. Phytopathol. 19:69-82.

Griffiths, E., and A.M. Berrie. 2978. Iatrogenic diseases of plants. Ann. Appl. Biol. 89:122-125.

Henis, Y., and I. Chet. 1975. Microbiological control of plant pathogens. Adv. Appl. Microbiol. 19:85-107.

Heye, C.C., and J.H. Andrews. 1983. Antagonism of Athelia bombacina and Chartomium globosum to the apple scab pathogen, Venturia inaqualis. Phytopathology 73:650-654.

Horsfall, J.G., and A.G. Dimond. 1957. Interaction of tissue sugar, growth substances and disease susceptibility. Z. Pflanzenkr. Pflanzenpathol. Pflanzenschutz 64:415-421.

Hsieh, S.P.Y., and I.W. Buddenhagen. 1974. Suppressing effects of Erwinia herbicola on infection by Xanthomonas oryzae and on symptom development in rice. Phytopathology 64:1182-1185.

Jackson, N. 1970. Evaluation of some chemicals for control of stripe smut of Kentucky bluegrass turf. Plant Dis. Reptr. 54:168-170.

Jarvis, W.R., and K. Slingsby. 1977. The control of powdery mildew of greenhouse cucumber by water sprays and Ampelomyces quisqualis. Plant Dis. Reptr. 61: 728-730.

Jaynes, R.A., and J.E. Elliston. 1980. Pathogenicity and canker control by mixtures of hypovirulent strains of Endothia parasitica in American chestnut. Phytopathology 70:453-456.

Jordan, V.W.L. 1973. The effects of prophylactic spray programmes on the control of pre- and post-harvest diseases of strawberry. Plant Pathol. 22:67-70.

Katan, J., and Y. Eshel. 1973. Interactions between herbicides and plant pathogens. Residue Rev. 45:145-177.

Kavanagh, T. 1974. The influence of herbicides on plant disease. II. Vegetables, root crops and potatoes. Sci. Proc. R. Dublin Soc. Ser. B 3:251-265.

Kranz, J. 1981. Hyperparasitism of biotrophic fungi, pp. 327-352. In: J.P. Blakeman (Ed.), Microbial ecology of the phylloplane. Academic Press, N.Y.

Leben, C. 1964. Influence of bacteria isolated from healthy cucumber leaves on two leaf diseases of cucumber. Phytopathology 54:405-408.

Leben, C. 1965. Epiphytic microorganisms in relation to plant disease. Annu. Rev. Phytopathol. 3:209-230.

Leben, C., and G.C. Daft. 1965. Influence of an epiphytic bacterium on cucumber anthracnose, early blight of tomato, and northern leaf blight of corn. Phytopathology 55:760-762.

Leben, C., G.C. Daft, J.D. Wilson and H.F. Winter. 1965. Field tests for disease control by an epiphytic bacterium. Phytopathology 55:1375-1376.

Lemke, P.A., and C.H. Nash. 1974. Fungal viruses. Bacteriol. Rev. 38:29-56.

Lindow, S.E. 1979. Frost damage to potato reduced by bacteria antagonistic to ice nucleation active bacteria. Phytopathology 69:1036 (Abstract).

Lindow, S.E. 1982a. Role of antibiosis in antagonism against ice nucleation active bacteria by epiphytic bacteria. Phytopathology 72:986 (Abstract).

Lindow, S.E. 1982b. Epiphytic ice nucleation active bacteria as incitants of frost injury and plant disease, pp. 344-362. In: G. Lacy and M. Mount (Eds.), Phytopathogenic prokaryotes. Academic Press, N.Y.

Lindow, S.E. 1982c. Population dynamics of epiphytic ice nucleation active bacteria on frost sensitive plants and frost control by means of antagonistic bacteria, pp. 395-416. In: P.H. Li and A. Sakai (Eds.), Plant cold hardiness. Academic Press, N.Y.

Lindow, S.E. 1982d. Integrated control of frost injury and fire blight of pear with antagonistic epiphytic bacteria. Phytopathology 72:946 (Abstract).

Lindow, S.E. 1983a. Methods of preventing frost injury through control of epiphytic ice nucleation active bacteria. Plant Dis. 67:327-333.

Lindow, S.E. 1983b. The role of bacterial ice nucleation in frost injury to plants. Annu. Rev. Phytopathol. 21:363-384.

Lindow, S.E., D.C. Arny, W.R. Barchet and C.D. Upper. 1978a. The role of bacterial ice nuclei in frost injury to sensitive plants, pp. 249-261. In: P.H. Li and A. Sakai (Eds.), Plant cold hardiness and freezing stress. Academic Press, N.Y.

Lindow, S.E., D.C. Arny and C.D. Upper. 1978b. Erwinia herbicola: An active ice nucleus incites frost damage to maize. Phytopathology 68:523-527.

Lindow, S.E., D.C. Arny and C.D. Upper. 1978c. Distribution of ice nucleation active bacteria on plants in nature. Appl. Environ. Microbiol. 36: 831-838.

Lindow, S.E., D.C. Arny and C.D. Upper. 1982b. Bacterial ice nucleation: A factor in frost injury to plants. Plant Physiol. 70:1084-1089.

Lindow, S.E., D.C. Arny and C.D. Upper. 1983a. Biological control of frost injury I: An isolate of Erwinia herbicola antagonistic to ice nucleation active bacteria. Phytopathology 73:1097-1102.

Lindow, S.E., D.C. Arny and C.D. Upper. 1983b. Biological control of frost injury II: Establishment and effects of an antagonistic Erwinia herbicola isolate on corn in the field. Phytopathology 73:1102-1106.

Lindow, S.E., S.S. Hirano, D.C. Arny and C.D. Upper. 1982a. The relationship between ice nucleation frequency of bacteria and frost injury. Plant Physiol. 70:1090-1093.

McBride, R.P., and A.J. Hayes. 1979. Interactions of the leaf pathogen Meria laricis with nutrients and microorganisms of the larch phylloplane. Nova Hedwigia 31:507-517.

McIntyre, J.L., J. Kuc and E.B. Williams. 1973. Protection of pear against fire blight by bacteria and bacterial sonicates. Phytopathology 63:872-877.

Moller, W.J., M.N. Schroth and S.V. Thomson. 1981. The scenario of fire blight and streptomycin resistance. Plant Dis. 65:563-568.

Natti, J.J., and G.E.R. Hervey. 1956. Influence of insecticide and fungicidal sprays on downy mildew of broccoli. Phytopathology 46:242.

Parker, K.G. 1936. Fireblight: Overwintering, dissemination, and control of the pathogen. N.Y. Agric. Exp. Sta. Mem. 193:1-42.

Reinecke, P. 1981. Antagonism and biological control on aerial surfaces of the Gramineae, pp. 383-395. In: J.P. Blakeman (Ed.), Microbial ecology of the phylloplane. Academic Press, N.Y.

Riggle, J.H., and E.J. Klos. 1972. Relationship of Erwinia herbicola to Erwinia amylovora. Can. J. Bot. 50:1077-1083.

Romig, W.R., and M. Sasser. 1972. Herbicide predisposition of snap beans to Rhizoctonia solani. Phytopathology 62:785-786 (Abstract).

Rotem, J., B.G. Clare and M.V. Carter. 1976. Effects of temperature, leaf wetness, leaf bacteria and leaf bacterial diffusates on production and lysis of Rhynchosporium secalis spores. Physiol. Plant Pathol. 8:297-305.

Scherff, R.H. 1973. Control of bacterial blight of soybean by Bdellovibrio bacteriovorus. Phytopathology 63:400-402.

Schroth, M.N., S.V. Thomson, D.C. Hildebrand and W.J. Moller. 1974. Epidemiology and control of fireblight. Annu. Rev. Phytopathol. 12:389-412.

Schroth, M.N., S.V. Thomson and W.J. Moller. 1979. Streptomycin resistance in Erwinia amylovora. Phytopathology 69:565-568.

Simard, J., R.L. Pelletier and J.G. Coulson. 1957. Screening of microorganisms inhabiting apple leaf for their antibiotic properties against Venturia inaequalis (Cke.) Wint. Quebec Soc. Prot. Plants Rep. 39:59-67.

Sleesman, J.P., and C. Leben. 1976. Microbial antagonists of Bipolaris maydis. Phytopathology 66:1214-1218.

Spurr, H.W. 1973. An efficient method for producing and studying tobacco brown-spot disease in the laboratory. Tobacco Sci. 17:145-148.

Spurr, H.W. 1977. Protective applications of conidia of nonpathogenic Alternaria sp. isolates for control of tobacco brown-spot disease. Phytopathology 67: 138-132.

Spurr, H.W. 1978a. Field biocontrol of two foliar diseases by a bacterial antagonist. Phytopathol. News 12:74 (Abstract).

Spurr, H.W. 1978b. Control of foliar diseases with the insect bioagent Bacillus thuringiensis. Phytopathol. News 12:129 (Abstract).

Spurr, H.W. 1981a. Introduction of microbial antagonists for the control of foliar plant pathogens, pp. 323-332. In: Symposia in agricultural research. Vol. 5. Biological control in crop production. Allanheld, Osmun Publishers, Totowa, N.J.

Spurr, H.W. 1981b. Experiments on foliar disease control using bacterial antagonists, pp. 369-381. In: J.P. Blakeman (Ed.), Microbial ecology of the phylloplane. Academic Press, N.Y.

Spurr, H.W. 1983. Ethanol treatment--a valuable technique for foliar biocontrol studies of plant disease. Phytopathology 69:773-776.

Spurr, H.W., and R.E. Welty. 1972. Incidence of tobacco leaf microflora in relation to brown spot disease and fungicidal treatment. Phytopathology 62:916-920.

Spurr, H.W., and R.E. Welty. 1975. Characterization of endophytic fungi in healthy appearing leaves of Nicotiana spp. Phytopathology 65:417-422.

Swinburne, T.R. 1981. Iron and iron chelating agents as factors in germination, infection and aggression of fungal pathogens, pp. 227-243. In: J.P. Blakeman (Ed.), Microbial ecology of the phylloplane. Academic Press, N.Y.

Thomson, S.V., M.N. Schroth, W. J. Moller and W. O. Reil. 1976. Efficacy of bactericides and saprophytic bacteria in reducing colonization and infection of pea flowers by Erwinia amylovora. Phytopathology 66:1457-1459.

Van der Zweep, W. 1970. Effects of herbicides on susceptibility of plants to pests and diseases, pp. 917-919. In: Proc. 10th Brit. weed control conf.

Vidaver, A.K. 1976. Prospects for control of phytopathogenic bacteria by bacteriophages and bacteriocins. Annu. Rev. Phytopathol. 14:451-465.

Warren, R.C. 1972a. Interference by common leaf saprophytic fungi with the development of Phoma betae lesions on sugarbeet leaves. Ann. Appl. Biol. 72: 137-144.

Warren, R.C. 1972b. The effect of pollen on the fungal leaf microflora of Beta vulgaris L. and on infection of leaves by Phoma betae. Neth. J. Plant Pathol. 78:89-98.

Wassermann, O. 1969. Untersuchungen uber den antibiotischen Einfluss von Bakterien und Pilzen auf das Wachstum von V. inaequalis (Cke.) Wint. in vitro. Zentralbl. Bakteriol. Parasitenkd. Infektionskr. Hyg., Abt. 2; 123:30-55.

White, G.F., and J. E. Haas. 1975. Assessment of research on natural hazards. MIT Press, Cambridge, MA. 495 pp.

Wood, R.K.S., and M. Tveit. 1955. Control of plant diseases by use of antagonistic organisms. Bot. Rev. 21:441-491.

Wrather, J.A., J. Kuc and E.B. Williams. 1973. Protection of apple and pear fruit against fireblight with nonpathogenic bacteria. Phytopathology 63:1075-1076.

Zentmeyer, G.A., and J.G. Bold. 1977. Management of the environment, pp. 121-144. In: J.G. Horsfall and E.B. Cowling (Eds.), Plant disease--an advanced treatise. Vol. I. How disease is managed. Academic Press, N.Y.

SOIL ANTAGONISTS IN IPM SYSTEMS

M.N. Schroth and J.G. Hancock

This review of biological control of soil plant pathogens in IPM systems is relatively philosophical and conceptual, since it is intended to encourage a better understanding of the problems and complexities of research unique to plant pathology. The volume of subject matter dealing with biological control of plant disease is too enormous to allow a general review of this subject. There also are ample reviews (Baker 1980, Baker & Cook 1974, Burr & Caesar 1984, Cook 1977, Cook & Baker 1983, Papavisas & Lumsden 1980, Schroth & Hancock 1981, Schroth et al. 1984, Suslow 1982), which focus on the many components of the field.

Historically, biological control represents a major research area in plant pathology which includes extensive, complementary research contributions from such related fields as soil microbiology, bacteriology, mycology, plant breeding and the basic plant sciences. The placement of "biological control" into an IPM context does not add a new dimension for discussion because the approach today in plant pathology, as it has been historically, is interdisciplinary and integrative, and focuses on the interactions among the plant (host), causal agent (pathogen), and the biotic and abiotic environment. This comprehensive approach has led to the economical control of hundreds of major plant diseases, mostly by nonchemical means, such as rusts and mildew diseases of grains, and bacterial diseases of legumes and cotton (Baker 1980, Cook & Baker 1983, Fry 1982). While the phrase IPM is new to plant pathology, the concept and practice is as old as the discipline. Nevertheless, the concept of integrated pest management of insects, and its extension to include all pests has been useful to the agriculturally related sciences; it has provided an awareness of the importance of considering all aspects of pest control, including environmental and socio-economic factors as well as the production-related components.

Entomology and plant pathology are sometimes referred to as sister disciplines because they have the common goal of controlling pests. On the other hand, because of the nature of the pests, the two disciplines have developed both similar and dissimilar strategies and methods to control them with the resultant evolution of distinct disciplinary strengths and philosophies;

ISBN 0-12-357030-1

entomologists have close ties with zoology and some with botany, whereas plant pathologists identify closely with microbiology and botany. Plant pathology is by its very nature an interdisciplinary science. With this in mind, the purpose of this presentation is to compare and contrast some of the major differences in biological control IPM strategies used by plant pathologists and entomologists and to provide state-of-the-art examples of the many different tactics used in the biological control of plant pathogens. Because of our personal experiences, we intend to examine the research needs necessary to better understand the highly complex interactions between plant roots and soil microorganisms. Lastly, recommendations for future research are made.

WHAT CONSTITUTES BIOLOGICAL CONTROL?

Entomologists generally use a more restricted definition of biological control than plant pathologists, limiting it to "the action of parasites, predators, or pathogens in maintaining another organism's population density at a lower average than would occur in their absence" (Baker & Cook 1974, Cook & Baker 1983). While plant pathologists do not necessarily agree as to what should be included in biological control, the preponderance favor a relatively broad definition that includes such techniques as crop rotation, alteration of soil pH, use of organic soil amendments as well as the use of one organism against another (Baker, this volume). Cultural practices are considered forms of biological control when they modify the biotic environment to disfavor disease.

The broad definition of biological control in plant pathology seems appropriate when considering mechanisms involved in suppression of populations of pathogens. In contrast to insect pests, biological control of a typical soil-borne plant pathogen is the result of the interaction of individual inoculum units, such as an oospore, chlamydospore, or microsclerotium of a fungus, with large populations of diverse organisms, all within a small volume of soil, e.g., a cubic centimeter. Thus, any cultural method (such as crop rotation, use of organic amendments, tillage, flooding) which causes population shifts in the soil with concomitant increases in antagonists is just as valid a biological control tactic as directly using an antagonistic organism against a pest. In fact, the benefits from large-scale population shifts favoring many broad spectrum antagonists are more likely to be sustained longer than that obtained from a single specialized antagonist.

The ubiquitous pathogenic soil-borne fungus, *Verticillium dahliae*, presents a good example to illustrate the complexity of biological control in plant pathology and to explain why our definition tends to be more comprehensive. A typical soil contains approximately 10^7 to 10^9 microorganisms per gram. One microsclerotium of *V. dahliae* per gram of soil is sufficient to cause severe disease of cotton (Huisman 1982). The determining factor influencing pathogenesis by Verticillium is not a specific predator, parasite, or competitor but the result of its interactions with a population of diverse organisms as affected by the abiotic environment and the susceptibility of the host. The environment has a mediating effect on all interactions, as with the biological control of insects, since it affects the susceptibility of the host as well as the population dynamics of the pathogen and associative microflora. Despite the complexity of this root-microbe environment, plant pathologists and farmers of antiquity learned by trial and error that cultural practices could be used to improve plant health, thus indirectly managing populations to disfavor disease causal agents.

BIOLOGICAL CONTROL STRATEGIES

There are other substantial differences which affect concepts and strategies in the biological control of insects and plant pathogens. Antagonists of plant pathogens, for example, are largely passive and relatively nonmobile and make accidental contact with pathogens. Therefore, biological control agents of plant pathogens must be applied directly to the locality where infection will occur. Although this is the same with pathogens of insects, the ingestion of one propagule by an insect may cause death whereas great populations of antagonists are usually required to suppress plant pathogens. Antagonists must be in great concentrations at the infection court, such as a leaf surface to control a leaf-infecting pathogen, or seed, roots or subterranean stems to control soil pathogens. Antagonists, however, are sufficiently mobile to locally colonize plant parts such as roots if inoculated onto seed at high densities. The broadcasting of specific microflora directly to field soils to control diseases is unrealistic and economically unfeasible considering the populations that would have to be used.

Plant pathologists have employed classical biological control techniques to control weeds by importing parasites from the natural range of the host (Schroth & Hancock 1981, Templeton et al. 1979), but this has not been a tactic in disease control. In fact there is little evidence to date that indicates there are specific microorganisms found only in certain geographical areas in the world that are responsible for controlling a particular pathogen. Many microorganisms in the soil are relatively ubiquitous in nature, whereas others are of restricted occurrence. Some are present in certain soil types, and others occur only under specific environmental and/or cultural conditions. Thus, whereas one might easily identify an organism or group of organisms of interest as biological control agents, such as in suppressive soils because of greater populations, they are usually present elsewhere but large populations may be disfavored because of the biotic or abiotic environment. Our knowledge of microorganisms of economic importance indicate that most, especially those which can be termed soil inhabitants, are widely distributed. Extensive dissemination of microorganisms has occurred throughout the world through the actions of wind, water and animals and the distribution of objects such as plant stock, farming equipment, seed, and other objects that may harbor soil and microflora. Thus, the absence of a pathogenic microorganism from any particular region may be the result of an environment that disfavors its establishment rather than its not being introduced.

Another difference between controlling insects and plant pathogens concerns the importance of the mechanisms that affect their populations. Biological control of plant pathogens has been affected principally by competition and antibiosis, and sometimes hyperparasitism, whereas biological control of insects is mainly by predation, parasitism and disease. Although predation of inocula occurs by nematodes, insects, amoebae and other animals, its importance as an implementable biological control tactic is highly conjectural.

Most efforts at biological control of plant pathogens are directed at the use of antagonists that are endemic to an area. The intent is to identify the best strains that will be antagonistic to a specific pathogen and to apply them in a way, such as in an altered environment, that would favor their activities at the expense of the pathogen. This approach, called the bioherbicide tactic in the case of weed control, may eventually be relatively effective but requires an intensification of research efforts. While the approach of foreign exploration for more effective biological control agents has not been ignored by plant pathologists, there does not appear to be any particular advantage over searching for antagonists in the variety of habitats that exist in the U.S.

There are many well documented examples of natural biological control of plant diseases operating in nature, such as in the case of disease suppressive soils (Cook & Baker 1983; Shipton 1972, 1979). This term is used to designate agricultural or wildland soils where certain specific soil-borne diseases are absent, or occur only to a low degree even when the pathogen is present or naturally introduced. Disease suppression has been found to depend on elements of the soil microbiota (Cook & Baker 1983). It is expected that if the principal antagonistic microorganism can be identified in these soils as well as the important associated biotic and abiotic factors in the environment that favor their populations, then it should be possible to use these microorganisms to control diseases in so-called "disease conducive soils." Most likely this would occur for only relatively short periods of time since there would be strong natural pressures to return to the inherent biological equilibrium which would again favor disease unless the abiotic environment could be altered to retard regression.

INDIRECT MANGEMENT OF MICROBIAL POPULATIONS

When examining Baker and Cook's definition of biological control (Baker & Cook 1974, Cook & Baker 1983) and the many examples presented, it is clear why plant pathologists generally consider IPM as a new hat for an old concept. Thousands of papers, including books, have been written on the control and management of plant diseases by biological means involving the manipulation of the host and environment. Knowledge from these studies has led to the development of unique and effective methods to manage disease. For example, Horsfall & Cowling's (1977) Plant Disease: An Advanced Treatise, Vol. 1, features at least six chapters dealing with disease management. A recent book by Cook & Baker (1984) on biological control of plant disease also emphasizes the relationship between management and biological control. Effective control tactics have involved such approaches as crop rotation, altering soil pH, use of certain fertilizers and organic amendments, tillage methods that modify soil structure, irrigation practices, planting dates, use of trap or inhibitory plants, and manipulation of the environment to promote maximum resistance of the host. Table 1 lists various methods used to control diseases with a few examples selected from hundreds where these methods have been used. Plant pathologists and farmers for centuries have used these approaches because of economic reasons and the general lack of effective chemical controls. Whether or not all of these practices are forms of biological control depend on ones definition of biological control (Baker, this volume). We prefer the more restricted definition that attributes disease reduction directly or indirectly to the means of living microorganisms or macroorganisms. Therefore, we would consider as biological control tactics those practices in Table 1 which result in the manipulation of antagonistic microflora to disfavor disease. This would exclude crop spacing, trap and decoy crops, and time of planting unless they involved antagonists. Time of planting, for example, may markedly affect the activity of antagonists and hence disease suppression since the activity of soil microflora is greatly influenced by temperature and other environmental parameters.

DIRECT MANIPULATION OF SPECIFIC MICROORGANISMS TO CONTROL OTHERS

There are numerous examples of the direct use of a specific microorganism to inhibit or reduce populations of a pathogen and to suppress disease. However, there are only a few cases, principally the control of Agrobacterium tumefaciens (Kerr & Tate 1984) and Fomes annosus (Rishbeth 1979), where specific microorganisms are directly used commercially to control a plant disease. This is not surprising since biological control in nature is the result of the multiple, additive, diverse activities of populations of organisms. Obviously, some organisms are more

Table 1. Examples of agricultural practices that relate to the biological control of diseases.

Method	Plant Pathogen Controlled	Crop	Reference(s)
Crop rotation	*Cephalosporium gramineum*	wheat	(Bruehl 1968)
	Gaeumannomyces graminis var. *tritici*	wheat	(Cook & Baker 1983)
Crop spacing	*Sclerotinia sclerotiorum*	dry beans	(Steadman et al. 1973)
Decoy crop	*Spongospora subterranea*	potato	(White 1954)
Fallow	*Pseudomonas solanacearum*	banana	(Sequeira 1958)
Fertilizer practices	*Fusarium solani* f. sp. *phaseoli*	dry bean	(Schroth & Hildebrand 1964)
Flooding	*Fusarium oxysporum* f. sp. *cubense*	banana	(Stover 1962)
Monoculture	*Streptomyces scabies*	potato	(Goss & Afanasiev 1938)
	Gaeumannomyces graminis var. *tritici*	wheat	(Gerlagh 1968, Shipton 1972)
Organic amendments	*Phytophthora cinnamomi*	avocado	(Baker & Cook 1974, Cook & Baker 1983)
pH adjustment	*Plasmodiophora brassicae*	crucifer	(Walker 1969)
	Streptomyces scabies	potato	(Walker 1969)
Tillage	*Sclerotium rolfsii*	peanut	(Garren & Duke 1957)
Time of planting	*Fusarium roseum* 'culmorum'	wheat	(Cook & Baker 1983)
	Tilletia controversa	wheat	(Hoffman 1982)
Trap crop	*Meloidogyne* spp.	various	(Huang et al. 1981)
Water management	*Phytophthora* spp.	walnut, cherry	(Schreader 1972, 1977)
	Macrophomina phaseoli	sorghum	(Edmunds 1964)

important than others as antagonists. The challenge has been to identify them and gain a thorough understanding of the mechanisms and factors that affect their antagonistic behavior. This includes a study of the ecology, physiology, and genetics of the interacting organism and a circumscription of the environmental factors that favor their antagonistic behavior. With this knowledge, it should be possible to commercially employ microorganisms as biological control agents for certain pathogens. At present, the problems of application, stability, cost, effective duration of the biological control and other aspects have not been perfected sufficiently to allow commercialization. Table 2 lists some of the many examples of specific organisms used to control pathogens in laboratory, greenhouse, and field studies. Schroth & Hancock (1981) critically reviewed this area and discussed some of the problems associated with the development and commercial use of biological control agents.

DEVELOPMENT OF BIOLOGICAL CONTROLS IN PLANT PATHOLOGY

Because of the nature of the pests, status of knowledge in the plant and microbiological sciences, and because the information applicable for advancing biological control development is dependent upon data from a number of disciplines, a plant pathologist's discussion of research needs may appear to be quite basic and interdisciplinary in outlook. However, this seems to be the progressive course since, in many respects, most of the implementable data and usable strategies have been effectively applied and practiced in farming for years as evidenced by the great number of successful programs (Baker & Cook 1974, Cook & Baker 1983, Fry 1982). The problems remaining today are exceedingly difficult and the development of biological control from the perspective of using specific antagonistic microorganisms against pathogens requires a substantial understanding of the biology of both the pathogens and associated microflora in relation to the plant. Progress is dependent upon the combined contributions from a number of basic disciplines. There are no simple "data gaps" that will result in information that can be quickly implemented. However, there are many recent encouraging findings and leads indicating that specific microorganisms can be used to alter the composition of root microflora (Kloepper & Schroth 1981, Suslow & Schroth 1985, Yuen 1984) and to hold some diseases in check (Cook & Baker 1983). The interdisciplinary approach to biological control in plant pathology is mandated by the nature of the problem and historically this has proven to be an effective research plan.

The following is a brief overview and discussion of research needs, approaches, and problems associated with the development of bacterial biological control agents for soil pathogens using our research program to illustrate points: The first step is to isolate and identify potential key antagonists. Considering that

Table 2. Examples of antagonists used or studied in the biological control of plant pathogens.

Mechanism	Plant	Plant Pathogen	Antagonist (Reference)
Antibiotic	Many	*Agrobacterium tumefaciens*	Avirulent *Agrobacterium* spp. (Ellis et al. 1979, Kerr & Tate 1984)
Competition/ antibiosis	Corn	*Fusarium roseum* 'Graminearum'	*Chaetomium globosum* (Meiler & Taylor 1971, Kommedahl & Mew 1975)
	Pine	*Heterobasidion annosum*	*Peniophora gigantea* (Rishbeth 1979)
	Various	various fungi	*Trichoderma* spp. (Papavizas & Lumsden 1980)
	Various	various fungi	*Bacillus subtilis* (Kommedahl & Mew 1975, Utkhede & Rahe 1980)
	Carnation	*Fusarium oxysporum* f. sp. *dianthi*	*Alcaligenes* spp. (Yuen & Schroth 1984)
	Cotton, Wheat	*Pythium*, *Gaeumannomyces graminis* var. *tritici*, *Pseudomonas tolaasii*, *Fusarium oxysporum* f.sp. *lini*	*Pseudomonas* spp. (Meiler & Taylor 1971; Nair & Fahy 1976; Howell & Stipanovic 1979, 1980; Kloepper et al. 1980b; Scher & Baker 1982; Weller & Cook 1983)
	Apple	*Erwinia amylovora*	*Erwinia herbicola* (Riggle & Klos 1972)
	Tobacco	*Pseudomonas solanacearum*	Avirulent strain of *P. solanacearum* (Chen et al. 1981)
	Many	various fungi	*Gliocladium* spp. (Cook & Baker 1983)
Competition for attachment sites	Many	*Agrobacterium tumefaciens*	Avirulent *Agrobacterium* spp. (Ellis et al. 1979)

(continued)

Table 2. (Continued).

Mechanism	Plant	Plant Pathogen	Antagonist (Reference)
Cross protection	Sweet potato	*Fusarium oxysporum* f. sp. *batatas*	non-pathogenic *F. oxysporum* (Komada 1983)
	Cucurbits	*Fusarium solani* f. sp. *cucurbitae*	Squash mosaic virus (Magyarosy & Hancock 1974)
Hyperparasitism	Many	Various fungi	*Trichoderma* spp. (Papavizas & Lumsden 1980)
	Sunflower, Beans	*Sclerotinia* spp.	*Coniothyrium minitans* (Huang 1980, Trutmann et al. 1982)
	Lettuce	*Sclerotinia* spp.	*Sporodesmium sclerotivorum* (Adams & Ayers 1982)
	Sugar beets	*Pythium* spp.	*Pythium oligandrum* (Cook & Baker 1983)
	Cucumber, Beans	*Rhizoctonia solani*	*Laetisaria arvalis* (Lewis & Papavizas 1980, Odvody et al. 1980)
	Cucumber	mildews	*Ampelomyces grisqualis* (Jarvis & Slingsby 1977)
	Rye, Other Cereals	Ergot	*Fusarium roseum* 'heterosporium' (Mower et al. 1975, Gay & Shattock 1980)
Hypovirulence	Chestnut	*Endothia parasitica*	Mycovirus (Day et al. 1977)
Parasitism	Soybean	*Pseudomonas syringae* pv. *glycinea*	*Bdellovibrio bacteriovorus* (Scherff 1973)
Predation		Various fungi	*Arachnula impatiens* (Old & Patrick 1979)

millions of organisms are in a gram of soil and assuming that many different species and strains play a role in influencing the population size and behavior of a pathogen, there is no one sure approach enabling the identification of the key organisms. Problems concerned with the isolation and differentiation of the various organisms in soil are not within the scope of this review. For example, many of the soil organisms, some possibly of importance as biological control agents, do not readily grow on standard isolation media. Even when a bacterium of interest is isolated from soil, there is only about a 1-10% chance that it can be placed taxonomically into a defined specific or infrasubspecific taxon. Thus, the field of bacterial taxonomy is very important because, how can there be communication among scientists interested in antagonists if there is no specific definition upon which to base such communication?

After an organism is discovered that exhibits some potential to inhibit a pathogen in vivo, its importance as a possible biological control agent depends on understanding some of the following questions: 1) How genetically stable is it and what environmental and cultural factors affect its stability in storage (Schroth et al. 1984)? 2) How can it be stored, transported, and used in commercial agriculture without loss of viability? 3) Is its use compatible with standard agricultural practices? 4) What environmental factors (such as root exudates, soil type, pH, temperature, aeration, moisture, and carbon-nitrogen balance) affect its ability to colonize plant roots and to inhibit pathogens? 5) What characteristics of the organism [such as elaboration of antibiotics, auxins (Loper et al. 1982), osmotolerance, nutritional requirements, moisture tolerance, resistance to metabolites from other microflora, and mobility] relate to its effectiveness as an antagonist? In addition, the population dynamics of the potential antagonist must be studied under various environmental conditions, including the field. It is particularly important to know its distribution and duration on root systems (Bahme & Schroth 1984, Loper et al. 1985) and how this relates to the sites of colonization by the target pathogen.

The environment is a major factor mediating the antagonistic behavior of biological control agents. For example, metabolites that inhibit target organisms such as siderophores (Kloepper et al. 1980a, b; Scher & Baker 1980, 1982) may be elaborated only under certain temperatures and nutrient regimes. Thus, it is very possible that an antagonist may not be effective when the environmental conditions are most conducive for growth of the pathogen.

Technology developed by molecular biologists is having a revolutionary effect on plant pathology as in many other biological fields. This technology provides relatively simple methods to gain an understanding of the genetic basis for microbial behavior, including their activities as biological control agents (Napoli & Staskawicz, this volume). By techniques of mutagenesis, gene deletion, gene amplification, and gene addition it is now possible to identify some of the key factors affecting their antagonistic properties. For example, with pseudomonads, siderophore production appears to be important in plant growth promotion and in the inhibition of several plant pathogens (Kloepper et al. 1980a, b; Scher & Baker 1980, 1982; Yuen & Schroth 1984). Studies of the genetics of pseudomonads indicate that a number of genes control siderophore production thus complicating the engineering of improved strains (Loper et al. 1984b, Moores et al. 1984).

The identification of the key characters important in antagonism and the genetic components that regulate the antagonistic behavior of biological control agents is necessary for making rapid advancement in biological control. Knowledge of these antagonistic factors should enable the development of effective screening programs to identify major antagonists. It should also lead to the understanding of the environmental factors affecting the antagonistic behavior of populations of microorganisms. Lastly, strain improvement of antagonists is possible and probable by using molecular biology techniques.

RELEVANCE OF IPM CONCEPTS TO PLANT PATHOLOGY

The concept and practice of IPM is as important to plant pathology today as it was at its inception as a science. Modern programs are more comprehensive with further integration of companion pest sciences. However, from a plant pathologist's viewpoint, many models that have been proposed for IPM are neither truly integrative or interdisciplinary in nature nor relevant to disease control. They frequently ignore large components of the agroecosystem, either because there are no available data or the time frame for collecting such data is considered too long. When developing plant models it is not realistic to ignore areas where little is known (for example, the complex microbial world of the soil). Root-infecting microorganisms greatly affect the health of plants and, consequently, affect many other physiological and ecological relationships. The relevancy of some of the simulation models to actual growing conditions also is open to question. For example, the excising of leaves or plant parts to simulate insect injury and the disruption of photosynthesis is not an appropriate method when considering the impact of foliar disease on the physiology of the plant.

The concept of economic threshold and its importance as defined by Edwards & Heath (1964) or Headley (1972) which focuses on critical population levels of a pest represents another area of differences between entomology and plant pathology. With many plant pathogens, the concept of economic threshold is not useful or is too simplistic to have much relevance. For example, with most bacterial pathogens and certain fungal pathogens, such as Phytophthora, population density is of little importance since they have the potential to multiply rapidly under favorable environmental conditions. Inversely, there may be a high population and disease does not occur because of the physiological state of the host or because the environmental conditions do not favor disease. With soil pathogens, such as *Fusarium solani* f. sp. *phaseoli* of common bean, the determining factor affecting disease incidence may be nutritional status of the soil (Baker & Cook 1974). Low soil nitrogen results in reduced chlamydospore germination and subsequent infection (Schroth & Hildebrand 1964). Disease progress is so dependent on environmental conditions and the nutritional and genetic status of the host plant that the mere accounting of the presence of the pathogen is often of little value in pest management. In fact, several disease forecasting systems ignore inoculum altogether and stress environmental conditions.

Biases toward quick solutions can be detrimental to the establishment of the knowledge necessary for the development of biological control programs in plant pathology as well as in the other pest disciplines. The use of the technology developed in molecular biology offers outstanding possibilities for making quantum advances in pest control in all disciplines. IPM is both a philosophy and an instrument. Its successful application depends on having comprehensive knowledge of pests, the environment, and the host. Without data developed from fundamental research, the promise of IPM may go unrealized.

RECOMMENDATIONS

General

The development of biological control in plant pathology cannot be advanced by the activities of plant pathologists alone. Its advancement is dependent upon integration of knowledge generated by a variety of disciplines. The principal factor governing the rate of advancement is the same as it is in other biological fields--availability of support monies enabling the continuity of research programs. By the nature of the discipline, most plant pathologists are concerned with problem-solving, and the sources of available funds generally are directed towards activities which produce relatively immediate results. The public appreciates research that focuses on the control of disease problems that are limiting crop production even if the controls represent "stop gap" measures. Understanding the rationale for highly basic research requires considerable sophistication. Thus, few laboratories have the support sufficient to develop and sustain long-term programs to study the highly complex, fundamental aspects of microbial interactions needed for developing successful biological control programs.

Specific

The following are some specific subjects we consider to have particular importance:

1. **Disease impact** In the area of IPM perhaps the greatest need is to have a better definition of the effects of endemic soil-borne diseases on plant health. Constant stress from root-infecting organisms is seriously reducing crop quality and yields with apparent staggering economic losses.
2. **Strain selection** The continued search for superior strains of antagonists and hyperparasites of disease organisms is critical. Emphasis should be placed on methodologies for speeding up screening processes and improving identification of control agents.
3. **Ecology of antagonists** Much emphasis should be placed on learning more of the ecology of **biological control** agents, (e.g., environmental conditions that favor their antagonistic effects as well as their growth and survival).
4. **Physiology and molecular biology of biological control agent** Understanding genetics and physiology of biological control agents requires considerable study. Much of the future success of biological control of plant disease will in part depend on the manipulations of existing agents. Enhancement of antagonistic behavior, whether it be by hyperparasitism, antibiosis, competition or other mechanisms is likely to be the most significant area of research in biological control over the next quarter of a century. However, a prerequisite to strain improvement is the determination of the factors that are important in enabling a microorganism to be an effective antagonist. This undoubtedly involves many different characteristics such as those relating to general rhizosphere competence and ability to survive during unfavorable environmental conditions.

REFERENCES

Adams, P.B., and W.A. Ayers. 1982. Biological control of Sclerotinia lettuce drop in the field by Sporidesmium sclerotivorum. Phytopathology 72:485-488.

Bahme, J.B., and M.N. Schroth. 1984. Colonization dynamics of a rhizobacterium on potato. Phytopathology 74:806.

Baker, K.F. 1980. Microbial antagonism--the potential for biological control, pp. 327-347. In: D.C. Ellwood, M.J. Latham, J.N. Hedger, J.M. Lynch and J.H. Slater (Eds.), Contemporary microbial ecology. Academic Press, London.

Baker, K.F., and R.J. Cook. 1974. Biological control of plant pathogens. W.H. Freeman, San Francisco. 433 pp.

Bruehl, G.W. 1968. Ecology of Cephalosporum striped disease of winter wheat in Washington. Plant Dis. Reptr. 52:590-594.

Burr, T.J., and A.J. Caesar. 1984. Beneficial plant bacteria. CRC Crit. Rev. Microbiol. (In Press).

Chen, W., E. Echandi and H.W. Spurr. 1981. Protection of tobacco plants from bacterial wilt with avirulent bacteriocin-producing strains of Pseudomonas solanacearum, II, pp. 482-492. In: J.C. Lozano (Ed.), Proc. 5th intern. conf. on plant pathogenic bacteria. August 16-21, 1981, Cali, Colombia, Univ. of Missouri.

Cook, R.J. 1977. Management of the associated microbiota, pp. 145-166. In: J.G. Horsfall and E.B. Cowling (Eds.), Plant disease--An advanced treatise. Academic Press, N.Y. Vol. 1.

Cook, R.J., and K.F. Baker. 1983. The nature and practice of biological control of plant pathogens. The American Phytopathological Society, St. Paul, MN. 539 pp.

Day, P.R., J.A. Dodds, J.E. Elliston, R.A. Jaynes and S.L. Anagnostakis. 1977. Double-stranded RNA in Endothia parasitica. Phytopathology 67: 1393-1396.

Edmunds, L.K. 1964. Combined relation of plant maturity, temperature, and soil moisture to charcoal stock rot development in grain sorghum. Phytopathology 54:514-517.

Edwards, C.A., and G.W. Heath. 1964. The principles of agricultural entomology. Chapman and Hall Ltd., Springfield, IL. 418 pp.

Ellis, J.G., A. Kerr, M. van Montagu and J. Schell. 1979. Agrobacterium: Genetic studies on agrocin 84 production and the biological control of crown gall. Physiol. Plant Pathol. 15:311-319.

Fry, W.E. 1982. Principles of plant disease management. Academic Press, N.Y. 378 pp.

Garren, K.H., and G.B. Duke. 1957. The peanut stem rot problem and a preliminary report on interrelations of "non-dirting" weed control and other practices to stem rot and yield of peanuts. Plant Dis. Reptr. 41:424-431.

Gay, C.N., and R.C. Shattock. 1980. Fusarium heterosporum and Claviceps purpurea on Lolium perenne. Trans. Brit. Mycol. Soc. 74:537-542.

Gerlagh, M. 1968. Introduction of Ophiobolus gramminis into new polders and its decline. Neth. J. Plant Pathol. 74(Suppl.2), 97 pp.

Goss, R.W., and M.W. Afanasiev. 1938. Influence of rotations under irrigation of potato scab, rhizoctonia, and fusarium. Nebr. Exp. Sta. Bull. 317:118.

Headley, J.C. 1972. Defining the economic threshold, II, pp. 100-108. In: Pest control strategies for the future. National Research Council, National Acad. Sci., Washington, D.C.

Hoffmann, J.A. 1982. Bunt of wheat. Plant Dis. 66:979-986.

Howell, C.R., and R.D. Stipanovic. 1979. Control of Rhizoctonia solani on cotton seedlings with Pseudomonas fluorescens and with an antibiotic produced by the bacterium. Phytopathology 69:480-482.

Howell, C.R., and R.D. Stipanovic. 1980. Suppression of Pythium ultimum-induced damping-off of cotton seedlings by Pseudomonas fluorescens and its antibiotic, pyoluteorin. Phytopathology 70:712-715.

Huang, H.C. 1980. Control of sclerotinia wilt of sunflower by hyperparasites. Can. J. Plant Pathol. 2:26-32.

Huang, C.S., R.C.V. Tenente, F.C.C. DeSilva and J.A.R Lara. 1981. Effect of Crotalaria spectabilis and two nematicides on numbers of Meloidogyne incognita and Helicotylenchus dihystera. Nematologia 27:1-5.

Huisman, O.C. 1982. Interrelations of root growth dynamics to epidemiology of root-invading fungi. Annu. Rev. Phytopathol. 20:303-327.

Jarvis, W.R., and K. Slingsby. 1977. The control of powdery mildew of greenhouse cucumber by water sprays and Ampelomyces quisqualis. Plant Dis. Reptr. 61: 728-730.

Kerr, A., and M.E. Tate. 1984. Agrocins and the biological control of crown gall. Microbiol. Sci. 1:1-4.

Kloepper, J.W., J. Leong, M. Teintze and M.N. Schroth. 1980a. Enhanced plant growth by siderophores produced by plant growth-promoting rhizobacteria. Nature 286:885-886.

Kloepper, J.W., J. Leong, M. Teintze and M.N. Schroth. 1980b. Pseudomonas siderophores: A mechanism explaining disease-suppressive soils. Curr. Microbiol. 4: 317-320.

Kloepper, J.W., and M.N. Schroth. 1981. Relationship of in vitro antibiosis of plant growth-promoting rhizobacteria to plant growth and the displacement of root microflora. Phytopathology 71:1020-1024.

Komada, H. 1983. Aspects of research on biological control of soil-borne pathogens in Japan. Phytopathology 73:776.

Kommedahl, T., and I.C. Mew. 1975. Biocontrol of corn root infection in the field by seed treatment with antagonists. Phytopathology 65:296-300.

Lewis, J.A., and G.C. Papavizas. 1980. Integrated control of Rhizoctonia fruit rot of cucumber. Phytopathology 70:85-89.

Loper, J.E., C.S. Orser, N.J. Panopoulos and M.N. Schroth. 1984a. Genetic analysis of fluorescent pigment production in Pseudomonas syringae pv. syringae. J. Gen. Microbiol. 130:1507-1515.

Loper, J.E., M.N. Schroth and N.J. Panopoulos. 1982. Influence of bacterial sources of indole-3-acetic acid (IAA) on root elongation of sugar beet. Phytopathology 72:997.

Loper, J.E., T.V. Suslow and M.N. Schroth. 1985. Log normal distribution of bacterial populations in the rhizosphere. Phytopathology 75:(In Press).

Magyarosy, A.C., and J.G. Hancock. 1974. Association of virus-induced changes in laimosphere microflora and hypocotyl exudation with protection to Fusarium stem rot. Phytopathology 64:994-1000.

Meiler, D., and A. Taylor. 1971. The effect of cochliodinol, a metabolite of Chaetomium cochliodes, in the respiration of microspores of Fusarium oxysporum. Can. J. Microbiol. 17:83-86.

Moores, J.C., M. Magazin, G.S. Ditta and J. Leong. 1984. Cloning of genes involved in the biosysthesis of pseudobactin, a high-affinity iron transport agent of a plant growth-promoting Pseudomonas strain. J. Bacteriol. 157:53-58.

Mower, R.L., W.C. Snyder and J.G. Hancock. 1975. Biological control of ergot by Fusarium. Phytopathology 65:5-10.

Nair, N.G., and P.C. Fahy. 1976. Commercial application of biological control of mushroom bacterial blotch. Aust. J. Agric. Res. 27:415-422.

Odvody, G.N., N.G. Boosalis and E.D. Kerr. 1980. Biological control of Rhizoctonia solani with a soil-inhabiting Basidiomycete. Phytopathology 70:655-658.

Old, K.M., and Z.A. Patrick. 1979. Giant soil amoebae, potential biocontrol agents, pp. 617-628. In: B. Schippers and W. Gams (Eds.), Soil-borne plant pathogens. Academic Press, London.

Papavizas, G.C., and R.D. Lumsden. 1980. Biological control of soil-borne fungal propagules. Annu. Rev. Phytopathol. 18:389-413.

Riggle, J.H., and E.J. Klos. 1972. Relationship of Erwinia herbicola to Erwinia amylovora. Can. J. Bot. 50:1077-1083.

Rishbeth, J. 1979. Modern aspects of biological control of Fomes and Armillaria. Eur. Jour. For. Pathol. 9:331-340.

Scher, F.M., and R. Baker. 1980. Mechanism of biological control in a Fusarium-suppressive soil. Phytopathology 70:412-417.

Scher, F.M., and R. Baker. 1982. Effect of Pseudomonas putida and a synthetic iron chelator on induction of soil suppressiveness to Fusarium wilt pathogens. Phytopathology 72:1567-1573.

Scherff, R.H. 1973. Control of bacterial blight of soybean by Bdellovibrio bacteriovorus. Phytopathology 63:400-402.

Schreader, W.R. 1972. Walnut decline project--1972. Univ. Calif., Coop. Ext. San Joaquin Co., Stockton.

Schreader, W.R. 1977. Cherry dieback project. Univ. Calif., Coop. Ext. San Joaquin Co., Stockton.

Schroth, M.N., and J.G. Hancock. 1981. Selected topics in biological control. Annu. Rev. Microbiol. 35:453-476.

Schroth, M.N., and D.C. Hildebrand. 1964. Influence of plant exudates on root-infecting fungi. Annu. Rev. Phytopathol. 2:101-132.

Schroth, M.N., J.E. Loper and D.C. Hildebrand. 1984. Bacteria as biological control agents of plant disease, pp. 362-369. In: M.J. Klug and C.A. Reddy (Eds.), Current perspectives in microbial ecology. ASM Publ. Off., Washington, D.C.

Sequeira, L. 1958. Bacterial wilt of bananas: Dissemination of the pathogen and control of the disease. Phytopathology 48:64-69.

Shipton, P.J. 1972. Take-all in spring-sown cereals under continuous cultivation: Disease progress and decline in relation to crop succession and nitrogen. Ann. Appl. Biol. 71:33-46.

Shipton, P.J. 1979. Natural and induced suppression of soil-borne plant pathogens: Concluding remarks, pp. 201-206. In: B. Schippers and W. Gams (Eds.), Soil-borne plant pathogens. Academic Press, London.

Steadman, J.R., D.P. Coyne and G.E. Cook. 1973. Reduction of severity of white mold disease on Great Northern beans by wider row spacing and determinate plant growth habit. Plant Dis. Reptr. 57:1070-1071.

Stover, R.H. 1962. Fusarium wilt (Panama disease) of bananas and other Musa species. Commonwealth Mycol. Inst. Phytopathol. Paper 4:1-117.

Suslow, T.V. 1982. Role of root-colonizing bacteria in plant growth, pp. 187-223. In: M.S. Mount and G.H. Lacy (Eds.), Phytopathogenic prokaryotes. Academic Press, N.Y., Vol. 1.

Suslow, T.V., and M.N. Schroth. 1985. Inhibition of root colonizing fungi by Pseudomonas rhizobacteria. Phytopathology 75:(In Press).

Templeton, G.E., D.O. TeBeest and R.J. Smith. 1979. Biological weed control with mycoherbicides. Annu. Rev. Phytopathol. 17:301-310.

Trutmann, P., P.J. Keane and P.R. Merriman. 1982. Biological control of Sclerotinia sclerotiorum on aerial parts of plants by the hyperparasite Coniothyrium minitans. Trans. Brit. Mycol. Soc. 78:521-529.

Utkhede, R.S., and J.E. Rahe. 1980. Biological control of onion white rot. Soil Biol. Biochem. 12:101-104.

Walker, J.C. 1969. Plant pathology. McGraw-Hill, N.Y. 3rd Ed. 819 pp.

Weller, D.M., and R.J. Cook. 1983. Suppression of take-all of wheat by seed treatments with fluorescent pseudomonads. Phytopathology 73:463-469.

White, N.H. 1954. Decoy Crops. The use of decoy crops in the eradication of certain soil-borne plant diseases. Austral. Jour. Sci. 17:18-19.

Yuen, G.Y. 1984. The role of rhizosphere bacteria in reducing *Fusarium* wilt of carnation and in promoting plant growth by altering the root microfloral composition. Ph.D. thesis, Univ. California, Berkeley.

Yuen, G.Y., and M.N. Schroth. 1984. Iron competition by *Alcaligenes* sp. reduces *Fusarium* chlamydospore germination. Phytopathology 74:825.

BIOLOGICAL CONTROL OF SOILBORNE PATHOGENS WITH ANTAGONISTS

S.B. Martin, G.S. Abawi and H.C. Hoch

Biological antagonistic interactions involved in the destruction or alteration of inoculum density or inoculum potential of plant pathogens have long been recognized for foliar as well as for soil-borne plant pathogens. However, the preponderance of attention for purposes of biological control has been concentrated on soilborne plant pathogens, primarily soilborne fungi. This is probably because environmental changes in soil systems are much less extreme over short periods than on aboveground surfaces and because of the general availability of effective and inexpensive foliar pesticides (Blakeman & Fokkema 1982). Nevertheless, there are strong indications that biological control of foliar disease is important under natural field conditions (see Lindow, this volume).

An enormous amount of information is available concerning parasites of plant pathogens in both foliar as well as soil pathogen-plant systems. For the purposes of this chapter, however, emphasis will be placed on the use of antagonists for control of diseases induced by soilborne plant pathogens.

Crop losses inflicted by soilborne pests continue to increase and are becoming a limiting factor in stabilizing or maximizing crop yields on a worldwide basis. The latter situation is partly due to intensive land use and also to changing crop production practices as a result of economic circumstances, mechanization, and limited availability of land. It is becoming obvious that root diseases rarely occur as the result of a single pathogen, but more frequently result from a complex of interacting pathogens as influenced by many biotic and abiotic soil factors (Powell 1979, Wallace 1978).

However, such interactions have not been studied sufficiently so that economic threshold densities required for predicting disease development and potential crop loss can accurately be determined. Most control measures available today, including pesticides, cannot realistically lead to the elimination of soilborne plant pathogens. Accordingly, the population density of plant pathogens at which yield loss can no longer be tolerated under the prevailing production conditions should be known.

ISBN 0-12-357030-1

Deliberate control measures should be initiated before or when economic threshold densities are reached. In the past, control of soilborne diseases with single control measures (the "silver bullet" approach) often has been attempted, but with limited success. However, seed and soil treatment with certain pesticides, crop rotation, soil amendments, or other treatments have, at times, improved crop yield or reduced root rot severity. Identifying cultivars has been difficult and very slow. Thus, there is a great need to employ new approaches for the management of soilborne disease complexes. Control of soilborne pathogens and other pests may be most effectively and economically attained by utilizing integrated control strategies. Biological control of soilborne pathogens can and should be considered as a component of any IPM system to be developed; however, biological control measures have not been sufficiently studied and deliberately implemented as principal components of integrated control programs for root disease management (Klassen 1981, Papavizas & Lewis 1981, Papavizas & Lumsden 1980).

There has long been an interest in biological control of plant pathogens as an alternative to or as a part of an IPM system for disease control (Baker & Snyder 1965); however, interest has increased greatly in recent years. This increased interest has resulted primarily from a concern about our environment as it is affected by the escalated use of potentially hazardous pesticides. Increased pesticide costs and pathogen resistance to pesticides also has aroused interest in alternative control measures. Over the past few years, numerous situations have been reported where soilborne plant pathogens were controlled directly with specific biological agents. However, most tests were conducted in the laboratory or greenhouse and only a few have been examined in large field trials. Furthermore, only three systems are reported to be economically feasible. These systems consist of the control of crown gall on a variety of plants with *Agrobacterium radiobacter*-strain 84 (Moore 1979), control of *Sclerotium rolfsii* on peanut with *Trichoderma harzianum* (Backman & Rodriguez-Kabana 1975) and control of *Fomes annosus* with conidia of *Peniophora gigantea* (Rishbeth 1963). Only the first and last examples are currently being used as routine control measures.

A multitude of organisms have been implicated, and in some cases with excellent documentation, as agents responsible for biological control of plant pathogens (Cook & Baker 1983, Lumsden 1981, Papavizas 1981, Papavizas & Lumsden 1980, Schippers & Gams 1979). The mechanism(s) by which these organisms suppress plant pathogens also vary, but generally involve either antibiosis, competition, and/or direct parasitism. Examples of successful biological control of soilborne pathogens have also been attributed to "suppressive soils." The latter involve complex and often not clearly understood mechanisms (Cook & Baker 1983). Numerous fungi are known and have been documented as effective antagonists of several important soilborne plant pathogens. The majority of the fungal antagonists which have been studied the most in recent years belong in the genera Coniothyrium, Gliocladium, Laetisaria, Penicillum, Sporidesmium and Trichoderma. All of these fungi operate as hyperparasites of some target fungi, although this mode of action frequently is not sufficient to explain their total antagonistic activities in soils. The concepts in the use of these antagonists are not new, but what gives them more promise are new methods of delivery, as well as the possibility of their deliberate and knowledgeable integration with other control measures. Research in biological control of soilborne plant pathogens has indicated that biological control often is not highly effective when deployed as a single control strategy. Biological control data available for use of these antagonists against fungal soilborne pathogens will be reviewed briefly. Prospects for the use of soilborne viruses, bacteriocins and hypovirulent strains of fungi as biological control agents effective against soilborne pathogens appear limited at this time and will not be discussed in this chapter. Bacteria have been extensively studied and utilized (mainly experimentally) as biological control agents for suppression of diseases induced by bacterial as well as fungal pathogens. Recent research with bacterial antagonists will be briefly discussed. The objectives of this brief treatment are to highlight potential uses of these antagonists, limiting factors in their commercial applications and further research needs. Examples utilized in this review are not intended to be either comprehensive nor inclusive of published reports.

DESCRIPTIONS OF SELECTED FUNGAL ANTAGONISTS

Trichoderma spp.

Trichoderma species represent the most thoroughly and widely studied fungi that show antagonistic activities toward soilborne (as well as some foliar) plant pathogens. In spite of all the research with these fungi, their use as biological control agents, for the most part, has been limited to laboratory, greenhouse and experimental field plots. Through direct application, Trichoderma spp. have not yet been recognized as commercially acceptable or preferred disease control agents of soilborne pathogens. T. viride, however, has been purposely favored (indirectly) in commercial orchard soils for the control of Armillaria mellea (Baker & Cook 1974). Through soil fumigation, CS_2 at moderate rates T. viride was believed to be responsible for increase in the population of T. viride, thus providing control of A. mellea (Bliss 1951).

Control of other soilborne plant pathogens has been achieved under experimental conditions with Trichoderma spp. T. harzianum, for example, suppressed Sclerotium rolfsii, a causal agent for root and stem diseases of several agricultural crops (Backman & Rodriguez-Kabana 1975, Chet et al. 1979, Elad et al. 1980a,

Wells et al. 1972), but only when applied at relatively high inoculum rates. *T. harzianum* also has been shown to be an effective agent in suppressing *Rhizoctonia solani* (Chet & Baker 1980, Chet et al. 1979, Elad et al. 1980a, Hadar et al. 1979, Liu & Baker 1980). *T. hamatum*, a more recently recognized species with biological control potential (Chet & Baker 1981), is similarly effective against *R. solani* (Chet & Baker 1981; Chet et al. 1981; Harman et al. 1980, 1981) and also against *Pythium* spp. (Chet & Baker 1981; Chet et al. 1981; Harman et al. 1980, 1981). Other soilborne plant pathogenic fungi have been controlled with *T. viride* (= *harzianum*) (Wright 1956, and references therein).

Trichoderma spp. are most frequently associated with soil debris (Danielson & Davey 1973, Gindrat 1979). *T. harzianum* was isolated from decaying sclerotia of *S. rolfsii* (Wells et al. 1972), while *T. hamatum* was obtained from the debris of a highly organic nursery soil in Columbia (Chet & Baker 1981). The fungus can frequently be seen sporulating on decaying seed coats. *Trichoderma* spp. proliferate upon addition of organic amendments to natural soils or where incomplete soil fumigation has been achieved (Bliss 1951, Lewis & Papavizas 1984). *Trichoderma* spp. are most frequently observed existing as hyphae and conidia, the latter of which frequently occur profusely, especially in moist soils (Danielson & Davey 1973, Griffin 1972, Taylor & Parkinson 1964). Chlamydospores also occur, although they have been observed much less frequently. Recently, Lewis & Papavizas (1984) have explored the use of chlamydospores as a more effective means of disseminating *Trichoderma*. They noted that chlamydospores would form on natural substrates (oat kernels and plant parts) buried in natural soil. Previously, such reproductive structures were noted primarily in *Trichoderma* cultures.

The mechanism(s) by which *Trichoderma* spp. control or suppress the activities of soilborne plant pathogens are not fully known. Early on, it was established that antibiotics (viridin) were produced by various species that were lytic to a number of fungi (Weindling 1934, Weindling & Emerson 1936, Weindling & Fawcett 1936). Subsequently, other nonvolatile and volatile antibiotics have been discovered (Dennis & Webster 1971a, b). More recently, *Trichoderma* spp. have been shown to be mycoparasitic toward other fungal hosts (*R. solani*, *Pythium* spp.) (Chet et al. 1981; Harman et al. 1980, 1981; Hubbard et al. 1983; Liu & Baker 1980) and to produce various cell wall degrading enzymes [B-(1-3) glucanase, chitinase] (Chet & Baker 1981, Jones & Watson 1969). Probable events in mycoparasitism probably include trophic growth of *Trichoderma* hyphae to target fungi, lectin-mediated attachment to hyphae and subsequent degradation of hyphae through the activity of cell wall-degrading enzymes. Thus, *Trichoderma* spp. have the ability to suppress pathogenic fungi through antibiotics, enzymes, and/or mycoparasitism. In addition, *Trichoderma* spp. are early colonizers of substrates and can reduce activities of other fungi simply by substrate occupation and depletion. However, none of these mechanisms of biological control have been conclusively established to be effective in *Trichoderma*-plant pathogen interactions. Nearly all of the mechanisms were observed in laboratory culture--not under natural field conditions, much less in soil. An exception is the observation (Hubbard et al. 1983) that parasitism occurs on pea seed surfaces planted in field soil.

Trichoderma spp., especially *T. hamatum* and *T. harzianum*, frequently have been associated with the biotic nature of soils that demonstrate suppressiveness to soilborne plant pathogens. There seems to be a consensus that to be effective the soil must be acidic and that conidia propagule levels be high (10^6-10^8/gram soil). Chet & Baker (1981) reported that the suppressive, highly organic soil from Columbia contained 10^8 *Trichoderma* propagules/g of soil. Significant reductions in disease indices caused by *R. solani* and *Pythium* spp. were noted when a Colorado clay loam soil was adjusted to pH 6 (from pH 8) and amended to contain 10^6 *Trichoderma* propagules/gram soil. Virgin soil contained 10^2 propagules/g and was conducive for diseases caused by *R. solani* and *Pythium* spp. Recently, chlamydospores have been evaluated for biological control. These propagules may be more effective than conidia because they have the ability to proliferate in field soil, while conidia remain quiescent (Lewis & Papavizas 1984). High propagule level and low pH do not necessarily mean the soils will be suppressive. The same *T. hamatum* isolate used to create suppressive soils in Colorado failed when added to New York soils (Hubbard et al. 1983). Hubbard et al. (1983) reported that in New York soils, already low in available iron, *T. hamatum* was inhibited by pseudomonads which competed with *T. hamatum* for iron. This inhibition was overcome by selecting strains resistant to pseudomonad competition (Hubbard et al. 1983).

Trichoderma spp. are delivered to the soil in a variety of ways. Most popular has been the use of *Trichoderma*-colonized wheat bran (Chet et al. 1979), or inactivated seeds (rye, sunflower). Molasses-enriched meals and clay granules were used by Backman & Rodriguez-Kabana (1975). Seeds coated with conidia through the use of such stickers as carboxymethylcellulose are an effective means of delivering the biological control agent near the seed (Harman et al. 1981). Recently, Lewis & Papavizas (1984) have experimented with liquid and solid fermentations for the production of conidia and chlamydospores.

For the most part, isolates of *Trichoderma* obtained from nature have been studied as biological control agents. Papavizas et al. (1982), however, have recently induced laboratory mutants that are tolerant to the fungicide benomyl. More importantly, these mutants showed enhanced biological control capabilities against *P. ultimum*, *R. solani* and *S. cepivorum*. Such mutants may represent a viable means of introducing a new approach into the biological control of soil fungi with *Trichoderma* spp.

Laetisaria arvalis

Laetisaria arvalis, a basidiomycete fungus that resembles *Rhizoctonia* spp., was first isolated from soil planted to sugar beets by Boosalis in the early 1960's (Burdsall et al. 1980). It has been reported to suppress diseases incited by *Rhizoctonia solani* (Lewis & Papavizas 1980, Odvody et al. 1980), *Pythium ultimum* (Hoch & Abawi 1979, Martin et al. 1983) and *Phoma betae* (Martin et al. 1983, 1984). Most studies concerning the biological control capabilities of *L. arvalis* have been conducted in greenhouse or small field plot trials. Commercial application of the fungus to field soils has not yet occurred.

In greenhouse and field trials, *L. arvalis* controlled pre- and postemergence damping-off of table beet as well as, in some instances better than, currently recommended fungicides (Hoch & Abawi 1979; Hoch et al., unpublished data). *L. arvalis* also decreased the incidence of *R. solani* fruit rot of cucumber (Lewis & Papavizas 1980). Odvody et al. (1980) reported that *L. arvalis* increased plant stand of sugar beet in soils infested with *R. solani*.

L. arvalis produces an abundance of small sclerotia which serve as long-lived inocula used to coat seeds or to incorporate into the seed furrow. Also, *L. arvalis* can be grown on a variety of media (spent sugar beet pulp, wheat bran, senescent corn leaves) and incorporated into field soils. Basidiospores of the fungus are also readily produced in culture on appropriate media, but may have little use as field inocula.

L. arvalis, in culture, is a mycoparasite of *R. solani* (Hoch & Abawi 1979, Odvody et al. 1980), but not of *Pythium ultimum* (Hoch & Abawi 1979). The mechanism of action in soil is, however, only speculative, but likely involves mycoparasitism and antibiosis depending on the target fungus. The cytoplasm of *P. ultimum* is lysed, due to the antibiotic nature of 8-hydroxylinoleic acid produced by *L. arvalis* (H.C. Hoch & G.S. Abawi, unpublished data). The fatty acid has little effect on *R. solani* at levels that inhibit hyphal growth of *P. ultimum* (ED_{50} = 20 ppm). Substrate occupation (competition) was also speculated to be of importance (Odvody et al. 1980).

L. arvalis, originally described from Nebraska, has since been isolated from Ohio, New York, Maryland and Georgia. It is obviously widespread among agricultural soils. As a biological control agent it is effective over a broad moisture and temperature (15-20°C) range (Hoch & Abawi 1979).

Selective media and isolation protocols for *L. arvalis* have been developed by Martin et al. (1983, unpublished data) and Papavizas et al. (1983). These enumeration techniques, as well as the sugar beet seed colonization method (Odvody et al. 1980), have allowed researchers to monitor the population levels of *L. arvalis* in greenhouse and field test plots. Martin et al. (1983) determined that in the greenhouse, *L. arvalis* propagule levels, after a brief increase, leveled off at 200 days near the original level of incorporation (ca. 60 propagules/gram soil). Very similar trends were also noted in field plots after two years (Martin et al., unpublished data). Population levels of Pythium spp. remained relatively constant and suppressed in soils augmented with L. arvalis (Martin et al. 1983). It thus appears possible that L. arvalis may remain stable and competitive in field soil for several seasons.

L. arvalis has been studied very little in comparison to some other fungal antagonists and its eventual potential as a biological control agent may not be ascertained for some time. Nevertheless, several studies indicate that it has potential which should be further evaluated. L. arvalis, like other biological control fungi, apparently does not readily colonize root surfaces. This represents a limitation to its use as a root protectant when applied as a seed dressing. Therefore, to protect plant roots from pathogenic fungi, the surrounding soil must be colonized with L. arvalis. Unfortunately, this means massive and economically impractical amounts of material must be incorporated. Hence, other avenues of enhancement should be sought. These could include use of more effective mutants, more prudent selection of the crop to be protected (container-grown plants, greenhouse crops) and other more practical means of building up a L. arvalis population in field soils, essentially to render the soil suppressive to a particular pathogen.

Coniothyrium minitans

Parasitism of sclerotia of Sclerotinia sclerotiorum by Coniothyrium minitans was first reported in 1947 from California (Campbell 1947). Numerous reports have since appeared in the literature detailing the parasitic nature of C. minitans to sclerotia and hyphae of S. sclerotiorum, S. minor, S. trifoliorum, Sclerotium cepivorum, Botrytis cinerea, B. fabae and other soilborne pathogens from several localities in North America, Europe, Asia, New Zealand and Australia (Ahmed & Tribe 1977, Huang & Hoes 1976, Trutmann et al. 1980, Turner & Tribe 1976). Results available have documented for the most part the efficiency of C. minitans in parasitizing and increasing the mortality of sclerotia of the host fungi under controlled (Huang & Hoes 1976, Turner & Tribe 1976) as well as natural field conditions (Huang 1980, Trutmann et al. 1982, Turner & Tribe 1975). However, limited information is available on the effectiveness of C. minitans in reducing disease incidence and increasing yield under greenhouse and field conditions. In greenhouse tests, Ahmed & Tribe (1977) showed that seed or soil treatment with pycnidial dust of C. minitans increased emergence of onion seedlings and decreased the incidence of white rot incited by S. cepivorum. Huang (1980) demonstrated that incorporation of C. minitans inoculum as a band treatment at planting time resulted in reduced incidence of sunflower wilt caused by S. sclerotiorum and increased yield. He also reported that the number of primary infection loci were reduced by C. minitans in fields both artificially and naturally infested with S. sclerotiorum. Recently, Trutmann et al. (1982) found that although spraying bean foliage with spore suspension of C. minitans did not reduce incidence of white mold induced by S. sclerotiorum or affect pod yield, it significantly reduced the number and weight of sclerotia produced per plant. In addition, they found that sclerotia produced on the C. minitans-treated plants had significantly lower carpogenic germination ability, thus further reducing inoculum potential for succeeding crops. In spite of the former encouraging reports, data on the performance of C. minitans in a commercial production mode are essentially lacking at the present time.

Tribe (1957) reported that C. minitans survives in soil for one to two years. Survival structures appear to be spores inside pycnidia and probably stromatal tissues embedded in sclerotia of host fungi. C. minitans is not considered to be a soil inhabitant as it has no documented saprophytic activity in soil (Tribe 1957). The fungus was shown to infect sclerotia of S. sclerotiorum produced in sunflower stems (Huang 1977) and on bean foliage (Trutmann et al. 1982), suggesting its possible adaptation for activity and survival on foliar parts.

The mode of parasitism of sclerotia of host fungi by C. minitans can be best characterized as necrotrophic or destructive mycoparasitism (Lumsden 1981). Hyphae of C. minitans were shown to directly penetrate hyphae and sclerotial cells of host fungi. Growth of C. minitans within the host cells becomes extensive over time resulting in disorganization of the cytoplasm, collapse of cell walls and the eventual production of pycnidia of C. minitans. It was reported that glucanases and chitinases produced by C. minitans may be initially involved in degradation of cell walls and death of cells of host fungi (Jones & Watson 1969, Jones et al. 1974). Parasitized sclerotia become soft, exhibit numerous pycnidia of C. minitans and eventually decay.

Concentrated pycniospore and hyphal fragment suspensions can be easily prepared as C. minitans grows on simple laboratory media such as potato-dextrose agar. The optimum temperature for growth on agar and parasitism of sclerotia is about 20°C (Turner & Tribe 1976). Good growth and parasitism also occurred at 10 but not at 5 or 30°C. Pycniospore and hyphal suspensions have been found to be effective inocula for host fungi in soil or sand media (Tribe 1957, Turner & Tribe 1976) and on foliar plant parts (Trutmann et al. 1982). A pulverized dry cornmeal-sand preparation of C. minitans mixed with soil was highly effective as it resulted in 85-95% of sclerotia being parasitized after 11 weeks (Tribe 1957). Another effective preparation was produced by growing C. minitans on a barley, rye and sunflower seed preparation which had been sterilized in autoclavable plastic bags (Huang 1980). The inoculum preparation was used moist or dried first at room temperature for 2-4 weeks before incorporation. The most practical preparation reported appears to be the pycnidial dust preparation from milled rice (Ahmed & Tribe 1977,

Turner & Tribe 1976). Water-soaked rice was sterilized in heat-resistant plastic bags and spore suspensions of C. minitans added and incubated for 6 weeks. The contents of the bags were air-dried for one week, coarsely ground and stored at 2°C prior to use. Pycnidial dust used as seed treatment in methyl cellulose or added directly to soil was effective for control of white rot of onions (Ahmed & Tribe 1977).

Although C. minitans has been reported from wide geographical areas of the temperate regions of the world, it does not appear to be abundant or present at high population densities over a large area in any one region. For example, Tribe (1957) found that in England the distribution of C. minitans varied greatly from location to location and ranged from 0-42% of sclerotia parasitized. Huang (1977) reported that percent of parasitism of sclerotia with C. minitans was highest toward the later part of the growing season, which reflected the abundance of the sclerotia of the host fungus. He also reported that C. minitans parasitized sclerotia of S. sclerotiorum, even those produced inside root and stem tissues. However, those produced on or inside the roots were parasitized and killed at higher rates (59 and 78%, respectively) than those produced inside the stem (29%). The number of sclerotia of S. sclerotiorum produced on bean plants treated with C. minitans was reduced by 55% as compared to the control. These data do suggest that C. minitans can reduce the sclerotial population of the host fungi effectively and thus should reduce incidence of disease in later years even when such reductions are not attainable during the first season of application. The use of C. minitans in combination with other control measures such as following fungicide applications late in the growing season has been suggested (Trutmann et al. 1982).

Sporidesmium sclerotivorum

This mycoparasite was first identified and documented as a parasite of sclerotia of Sclerotinia minor in 1978 by Uecker et al. (1978). To date, the only other known hosts of Sporidesmium sclerotivorum are S. minor, S. trifoliorum, Sclerotium ceptivorum and Botrytis cinerea (Ayers & Adams 1979a, 1981). S. sclerotivorum has been shown to be a destructive parasite of sclerotia of the host fungi resulting in their decline in artificially and naturally infested soils (Adams & Ayers 1980, 1981; Ayers & Adams 1979a). In soil, conidia of S. sclerotivorum germinate when present in close proximity to sclerotia of a host fungus and can completely destroy it in approximately 10 weeks. The mycoparasite sporulates abundantly on infected sclerotia in soil and results in infection of nearby noninfected sclerotia. S. sclerotivorum was demonstrated to be effective in controlling lettuce drop (reductions of 40 and 83%) caused by S. minor in field plots and was effective for consecutive lettuce plantings over a two-year period after incorporation (Adams & Ayers 1982). In addition, application of S. sclerotivorum at densities of 100 and

1000 conidia/g soil resulted in a 75 or 95% reduction in numbers of sclerotia, respectively.

S. sclerotivorum has been isolated or trapped from field soils in ten states in the U.S. that have known populations of Sclerotinia or Sclerotium species (Adams & Ayers 1982). S. sclerotivorum has a brown-to-black mycelium and produces multiseptate dark macroconidia on simple or branched conidiophores (Ayers & Adams 1979b, Uecker et al. 1978). A low number of chlamydospores are produced in old cultures and the fungus also produces hyaline microconidia which have not been observed to germinate. The mycoparasite was shown to survive in moist or dry soils at room temperature for 15 months (Ayers & Adams 1979b) and to remain effective in field soil for at least two years after introduction (Adams & Ayers 1982).

It has been well documented that the mode of parasitism and destruction of sclerotia of susceptible fungi by S. sclerotivorum is through necrotrophic mycoparasitism (Adams & Ayers 1980; Ayers & Adams 1979a, b). Macroconidia within a few millimeters from viable sclerotia germinate and directly invade sclerotia of host fungi. The mycoparasite grows intracellularly in the sclerotia utilizing primarily the digested glucose products (Adams & Ayers 1983). The hyphae of the mycoparasite that protrude from the sclerotial rind produce conidia and are also capable of growth through the soil to infect sclerotia that are 1 cm or farther away. Parasitized sclerotia become soft and eventually decay. The optimal conditions for mycoparasitic activities of S. sclerotivorum in soil are 20-25°C, pH 5.5-7.5 and soil water potential of -8 bars and higher (Adams & Ayers 1980, Ayers & Adams 1979b). Germination of conidia and mycelial growth on sclerotial agar was maximum at pH 5.0-5.5.

Although S. sclerotivorum can be grown on a sclerotial agar medium (Ayers & Adams 1979a) or a defined medium (Barnett & Ayers 1981), most data collected on mycoparasitic activities of this fungus have utilized an inoculum preparation consisting of sand and viable sclerotia of host fungi (Adams & Ayers 1980; Ayers & Adams 1979b, 1981). This method utilized fine moist sand amended with 1% (w/w) of sclerotia of S. minor and inoculated with the mycoparasite. After appropriate incubation periods, such inoculum preparations have yielded in excess of 10^5 macroconidia of S. sclerotivorum/g which can be directly incorporated into soils of experimental plots. Recently, Adams et al. (1984) reported the use of an inoculum source of S. sclerotivorum that was commercially grown on a vermiculite-based nutrient medium prepared by the Sylvan Spawn Laboratory of Worthington, Pennsylvania. Parasitism and decay of sclerotia of S. minor by S. sclerotivorum was closely related to the initial population density of both organisms and time of incubation. Successful parasitism of each sclerotium by S. sclerotivorum resulted in the average production of 15,000 new macroconidia.

Gliocladium spp.

Several species of Gliocladium (G. roseum, G. catanulatum, G. virens, G. deliquescens) have been shown to be parasitic on a wide variety of plant pathogens including Sclerotinia and Sclerotium species (Huang 1978, Tu 1980), Rhizoctonia solani (Howell 1982, Howell & Stipanovic 1983), Phomopsis sclerotioides (Moody & Gindrat 1977), Botrytis allii (Walker & Mande 1975), Fusarium spp. (Huang 1978) and others (Barnett & Lilly 1962, Hoshioka & Fukita 1969, Ricard et al. 1974).

Howell (1982) reported that G. virens added to soil as an in-furrow treatment significantly reduced damping-off of cotton seedlings incited by R. solani or P. ultimum. He also showed that soil amendment with G. virens resulted in a 63% reduction of the number of viable sclerotia of R. solani, but it had no effect on the population of oospores of P. ultimum. Tu & Vaartaja (1981) demonstrated that G. virens added to artificially infested soil at planting or two months prior to planting reduced Rhizoctonia root rot severity in white beans. Rhizoctonia damping-off of radishes grown in composted hardwood bark was also reduced by amendment with G. virens (Nelson et al. 1983). Gliocladium roseum was reported to be effective in reducing cucumber black root rot caused by P. sclerotioides (Moody & Gindrat 1977) and onion neck rot caused by B. allii (Walker & Mande 1975).

Species of Gliocladium are widespread in occurrence as soil-inhabiting fungi with considerable saprophytic abilities, in addition to being parasites of many fungi. Information concerning the survival of Gliocladium spp. in soil is limited. Chlamydospores are produced by certain species, but the role of conidia and hyphae may also be significant for survival as a result of the saprophytic and mycoparasitic activities exhibited by these species in soil.

Mode of parasitism of plant pathogens by Gliocladium species has been shown to be through necrotrophic mycoparasitism or antibiosis. Tu & Vaartaja (1981) found that hyphal tips of G. virens grew towards and alongside the hyphae and sclerotia of R. solani. G. virens hyphae produced specialized structures that directly penetrated and produced intracellular hyphae in sclerotia and hyphae of R. solani. Parasitized hyphae and sclerotia of R. solani eventually collapsed and became nonviable. Similar mycoparasitism was observed with G. roseum on Botrytis allii (Walker & Mande 1975) and on Eutypa armeniacea (Ricard et al. 1974). Howell & Stipanovic (1983) showed that G. virens hyphae coil around and penetrate hyphae of R. solani. They also showed in the same study that G. virens did not penetrate hyphae of P. ultimum. However, they concluded that the strong inhibition and adverse disintegration of protoplasm of P. ultimum when in contact with G. virens was due to antibiosis.

Gliovirin, an antibiotic with high activity against P. ultimum was later identified by Howell & Stipanovic (1983) from culture filtrate of G. virens. A deficient mutant of G. virens lacking this antibiotic had no effect of *P. ultimum*, whereas a mutant with higher production ability of the antibiotic was more inhibitory. Parasitism of *G. catenulatum* and *G. deliquescens* appears to be through antibiosis as penetrations of host fungi by these parasites were not observed (Hoshioka & Fukita 1969, Huang 1978).

Gliocladium species grow readily on common culture media on which they sporulate abundantly. Tu & Vaartaja (1981) grew *G. virens* on chopped raw oat kernels which were sterilized, inoculated and incubated for two months at room temperature. This inoculum of the mycoparasite was effective when added to soil at a ratio as low as 1% (v/v).

An effective inoculum preparation of *G. virens* was also obtained by growing the parasite on a sterilized medium consisting of peat moss wetted with Czapek's broth and incubated for 10 days at 25°C (Howell 1982). Six grams of this preparation added as an in-furrow application was effective in reducing damping-off on cotton incited by *P. ultimum* of *R. solani*. Application of this preparation of *G. virens* into soil reduced the number of viable sclerotia of *R. solani* from 9.2 to 3.4/g soil. Moody & Gindrat (1977) used a mixture of peat, soil, leaf compost and sand amended with Czapek's broth to prepare an inoculum of *G. roseum*. Conidia of *G. roseum* was added to the autoclaved medium and incubated for 4-7 days at 25°C. This inoculum preparation significantly reduced the incidence of cucumber black root rot. Barley kernels colonized by *G. roseum* were ground, mixed with other conifer bark plus fresh barley flour and formulated into pellets also reduced disease severity (Moody & Gindrat 1977).

Penicillium spp.

Several *Penicillium* spp. (*P. oxalicum*, *P. frequentans*, *P. vermiculatum*, *P. godlewskii*, *P. nigricans*, *P. chrysogenum*) have been shown to be effective as biological control agents (Boosalis 1956, Cook & Baker 1983, Dutta 1981, Ghaffar 1969, Kommedahl & Windels 1978, Liu & Vaughan 1965, Windels & Kommedahl 1978). They have been, for the most part, tested as seed coatings, soil incorporations and as root treatments for protection against diseases incited by *Pythium* spp., *Rhizoctonia solani*, *Sclerotium cepivorum*, *Aphanomyces* sp., *Fusarium* sp. and *Verticillium albo-atrum*. Kommedahl & Windels (1978, 1981) and Windels & Kommedahl (1978) indicated that *P. oxalicum* was very effective, when applied as a seed coating, for protection of pea seeds against various rot fungi. In fact, the protection provided by *P. oxalicum* was as good as that provided by a common protectant fungicide. Seed treatments have reportedly been effective for a number of other crops, e.g., corn, soybeans, table beets and garbanzo beans (Cook & Baker 1983). *Talaromyces flavus*, the perfect state of *P. vermiculatum*, was recently shown to be very effective in preventing *Verticillium* wilt of eggplant under field conditions (Marois et al. 1982). Incidence of wilt was reported to be reduced 76% and 67% in two different fields. Equally important, the crop yield was significantly increased (in weight and number of fruit) by application of *T. flavus*.

Penicillium spp. are very efficient and common soil colonizers. They produce an abundance of conidia and, in part, for this reason, they are frequently isolated from soil debris. The perfect state, when found, produces ascospores, such as that of *T. flavus*.

The mechanism(s) by which *Penicillium* spp. serve as biological control agents are not known with certainty. Mycoparasitism of *R. solani* hyphae was reported for *P. vermiculatum* (Boosalis 1956), although it was difficult to establish whether or not the *R. solani* hyphae were invaded before or after cytoplasmic necrosis. However, because they are efficient colonizers, substrate occupation is most likely a key mode of these biological control agents. The fungi are also well known for their antibiotic production, e.g., griseofulvin from *P. nigricans*, although good evidence that antibiotics from these fungi are operative in soil are lacking. Populations of *Penicillium* sp. can be increased substantially in soil with the addition of organic matter. *Penicillium* spp. as well as *T. flavus* are usually increased in vitro (on potato dextrose agar) for seed coating or root dipping (Kommedahl & Windels 1978, Marois et al. 1982, Wright 1956).

Bacterial Antagonists

Antagonistic bacteria have been extensively studied as biological control agents effective against soilborne pathogens. The foremost example of biological control of a soilborne pathogen is the control of crown gall caused by *Agrobacterium tumefaciens*. The biological control agent is a strain of the related bacterium *A. radiobacter* (strain K-84) identified by Kerr (1980) and associates at the Waite Agricultural Research Institute in Australia. The organism is marketed in New South Wales, New Zealand, California and Oregon. Treatment is effective when seeds, cuttings, or roots of young plants are dipped in a suspension of K-84 immediately before planting. The mechanism of action presumably involves an antibiotic called agrocin 84 that selectively inhibits most pathogenic agrobacteria. For greater detail concerning this system, see Kerr (1980).

Other effective bacterial antagonists have mainly been identified as strains of *Bacillus subtilis*, *Streptomyces* spp., or *Pseudomonas* spp. As with most fungal antagonists, results in field control of disease has been variable. However, recent results with plant growth promoting rhizobacteria (primarily *Pseudomonas* spp. in the *fluorescens*-*putida* group) are encouraging (Burr et al. 1978; Kloepper & Schroth 1981; Kloepper et al. 1980a, b). These bacteria have been identified as biological agents with greater rhizosphere-rhizoplane colonizing capabilities than antagonistic fungi have shown in the past. Equally encouraging is their ability to grow with the advancing root when applied to seeds or seed pieces. Apparently their attributes as early season plant growth promoters relates to the preemption of rhizoplane microflora (including pathogens and "quasipathogens"). Researchers have shown that antibiotic production in vitro as well as aggressive root-colonizing ability are essential attributes of effective growth-promoting strains (Kloepper & Schroth 1981).

Research on root-colonizing pseudomonads including the so-called plant growth promoting rhizobacteria has been vigorous in the past few years. Several crops have been investigated and include potato (Burr et al. 1978, Kloepper & Schroth 1981, Kloepper et al. 1980b), onion (Kawamoto & Lorbeer 1976), sugar beets (Suslow & Schroth 1982), wheat (Weller & Cook 1983) and radishes (Kloepper et al. 1980a). In all of these crops, the antagonists were effective when added on the seed or seed piece.

Mechanisms of action presumably could involve antibiosis and/or competition. The fluorescent pseudomonads associated with plant growth promotion are thought to compete with other rhizoplane and rhizosphere microorganisms by production of siderophores which have greater iron-binding affinities than siderophores produced by the displaced competitors (Kloepper et al. 1980a).

COMPATIBILITY OF ANTAGONISTS WITH OTHER IPM STRATEGIES AGAINST SOILBORNE PATHOGENS

The integration of biological control agents with other control measures and crop production practices has been inadequately studied. However, in plant pathology, the situation is beginning to change.

For example, combined treatments of the fungicide PCNB and *Trichoderma harzianum* decreased the inoculum potential of *Rhizoctonia solani* and increased disease control in comparison to separated treatments for control of damping-off of radish (Henis et al. 1978) or eggplant (Hadar et al. 1979). *Rhizoctonia* fruit rot of cucumber was decreased with combinations of deep plowing to bury infested soil layers and the application of a *Trichoderma* sp. (Lewis & Papavizas 1980). Physical, biological and chemical control was integrated to control soilborne diseases of potato induced by *R. solani* or *Verticillium dahliae* (Elad et al. 1980b). These researchers found that, although solar heating or methyl bromide fumigation significantly reduced diseases of potato caused by *R. solani* and *V. dahliae*, the combination of either solar heating or methyl bromide fumigation followed by *T. harzianum* application improved control over solarization or fumigation alone and additionally controlled *Sclerotium rolfsii*.

As indicated by Schroth & Hancock (this volume), plant pathologists have traditionally and necessarily employed multiple tactics for the control of plant diseases, including biological control in an indirect manner. There have long been an awareness and considerable research effort aimed at integrating biological control measures with other control strategies available against soilborne plant pathogens (Baker 1983, Cook & Baker 1983, Kommedahl & Windels 1981, Papavizas & Lewis 1981). However, studies aimed at deliberately implementing the use of fungal antagonists to manage soilborne plant pathogens below the economic threshold densities in defined IPM systems are few and generally of limited nature (Andrews 1983, Klassen 1981). Nevertheless, many examples are available in the literature relating to the compatibility and proper management of antagonists with other effective control strategies such as cultural practices (deep plowing, seedbed preparations and other tillage systems, organic amendments, crop rotation, intercropping, cover crops, irrigation management, soil nutrient management, soil pH); pesticide use (compatible pesticides, low rate applications of pesticides, nontarget effects of chemicals on pathogens and/or antagonists, introduction of antagonists to chemically- or heat-treated soils for colonization, pesticide resistant antagonists); and resistant cultivars [use of partially resistant (tolerant) cultivars, effect of cultivars on the microbiology of the rhizosphere with regard to pest and antagonist populations]. Biological control measures are compatible in many cases and should be demonstrated to be manageable as a components of management systems for a variety of soilborne fungal pathogens. Eventually, IPM systems developed for the control of soilborne pests will have to be compatible and a component of a total program for soil management. Maintaining high soil productivity will undoubtedly become, if not already, the major limiting factor for efficient crop and animal production. The latter is the challenge for all agricultural disciplines and will be possible only through coordinated multidisciplinary efforts.

RECOMMENDATIONS

It is not difficult to document examples where biological control agents are effective, under limited conditions, in controlling soilborne plant pathogens. Too frequently, however, biological control is not achieved when environmental conditions may vary, in differing locales (soils), or in closely related pathosystems. Several areas of research should be given priority that are currently limiting our abilities to achieve biological control routinely and economically. They are a lack of knowledge:

1. concerning the mechanism(s) of action by which the biological agent controls the activities of the plant pathogen;
2. of how to mass produce and deliver the agents effectively and economically;
3. of how to readily integrate biological control agents with other control measures and crop production practices; and
4. of how to obtain genetically superior isolates.

Foremost among these, we believe, is the need to understand the mechanism(s) of action. Such information has been elucidated with certainty for very few biological control agents. To know how best to grow the agents or when and where to deliver them relies greatly on their mode of antagonism. Closely coupled with knowledge of primary mechanisms of action against specific target pathogens is the necessity of determining optimal nutritional requirements, effective inoculum densities, survivability in response to environmental stresses likely to be encountered and activities of the antagonists under the influence of such stresses in comparison to the target organism(s).

Mass production and application technology are among the most neglected research areas. In many cases, antagonists can be grown on simple substrates such as sugar beet pulp or wheat bran or inactivated seeds, but the optimization of delivery systems using these and other preparations has not been developed. Research in this area in plant pathology has recently been increasing, e.g., the incorporation of antagonists in gels used in liquid drilling of germinated seeds of small seeded crops such as lettuce.

There has been limited research in the exploration for genetically superior isolates of antagonists. As indicated previously, most potential biological control agents in use at this time are naturally occurring selections. Nature will probably remain our most potent source of good antagonists; however, the genetic improvement of antagonists holds much promise (see Napoli & Staskawicz, this volume). Within our technological grasp is the possibility of inserting genes associated with biological control (e.g., antibiotic production) into more efficient root-colonizing organisms. The challenge is to determine primary modes of action and the genes governing those mechanisms.

Other areas where research is needed include:

5. timing and frequency of application of the biological agents;
6. influence of host genotype on antagonist populations and effectiveness; and
7. ways to modify the soil environment to substantially favor the antagonists.

ACKNOWLEDGMENTS

The authors wish to acknowledge partial support from USDA-SEA (59-2361-1-1-750-0), Western Regional Project W-147 and wish to thank G. Harman for suggested additions to the manuscript.

REFERENCES

Adams, P.B., and W.A. Ayers. 1980. Factors affecting parasitic activity of Sporidesmium sclerotivorum on sclerotia of Sclerotinia minor in soil. Phytopathology 70:366-368.

Adams, P.B., and W.A. Ayers. 1981. Sporidesmium sclerotivorum: Distribution and function in natural biological control of sclerotial fungi. Phytopathology 71:90-93.

Adams, P.B., and W.A. Ayers. 1982. Biological control of lettuce drop in the field by Sporidesmium sclerotivorum. Phytopathology 72:485-488.

Adams, P.B., and W.A. Ayers. 1983. Biological and physiological aspects of infection of sclerotia of Sclerotinia species by two mycoparasites. Phytopathology 73:1072-1076.

Adams, P.B., J.J. Marois and W.A. Ayers. 1984. Population dynamics of a mycoparasite, Sporidesmium sclerotivorum, and its host, Sclerotinia minor, in soil. Soil Biol. Biochem. (In Press).

Ahmed, A.H., and H.T. Tribe. 1977. Biological control of white rot of onion (Sclerotium cepivorum) by Coniothyrium minitans. Plant Pathol. 26:75-78.

Andrews, J.H. 1983. Future strategies for integrated control, pp. 471-480. In: T. Kommedahl and P.H. Williams (Eds.), Challenging problems in plant health. Amer. Phytopathol. Soc., St. Paul, MN.

Ayers, W.A., and P.B. Adams. 1979a. Mycoparasitism of sclerotia of Sclerotinia and Sclerotium species by Sporidesmium sclerotivorum. Can. J. Microbiol. 25:17-23.

Ayers, W.A., and P.B. Adams. 1979b. Factors affecting germination, mycoparasitism, and survival of Sporidesmium sclerotivorum. Can. J. Microbiol. 25:1021-1026.

Ayers, W.A., and P.B. Adams. 1981. Mycoparasitism and its application to biological control of plant diseases, pp. 91-103. In: G.C. Papavizas (Ed.), Biological control in crop production. Vol. 5. Allanheld, Osmun Publ., Totowa, N.J.

Ayers, W.A., E.A. Barnett and P.B. Adams. 1981. Germination of macroconidia and growth of Sporidesmium sclerotivorum in vitro. Can. J. Microbiol. 27:664-669.

Backman, P.A., and R. Rodriguez-Kabana. 1975. A system for the growth and delivery of biological control agents in soil. Phytopathology 65:819-821.

Baker, K.F. 1983. The future of biological and cultural control of plant diseases, pp. 422-430. In: T. Kommedahl and P.H. Williams (Eds.), Challenging problems in plant health. Amer. Phytopathol. Soc., St. Paul, MN.

Baker, K.F., and R.J. Cook. 1974. Biological control of plant pathogens. W.H. Freeman, San Francisco. 433 pp.

Baker, K.F., and W.C. Snyder. 1965. Ecology of soil-borne plant pathogens: Prelude to biological control. Univ. of Calif. Press, Berkeley. 571 pp.

Barnett, E.A., and W.A. Ayers. 1981. Nutritional and environmental factors affecting growth and sporulation of Sporidesmium sclerotivorum. Can. J. Microbiol. 27:685-691.

Barnett, H.L., and V.G. Lilly. 1962. A destructive mycoparasite, Gliocladium roseum. Mycologia 54:72-77.

Blakeman, J.P., and N.J. Fokkema. 1982. Potential for biological control of plant diseases on the phylloplane. Annu. Rev. Phytopathol. 20:167-192.

Bliss, D.E. 1951. The destruction of Armillaria mellea in citrus soils. Phytopathology 41:655-683.

Boosalis, M.G. 1956. Effect of soil temperature and green-manure amendment of unsterilized soil on parasitism of Rhizoctonia solani by Penicillium vermiculatum and Trichoderma sp. Phytopathology 46:473-478.

Burdsall, H.H., H.C. Hoch, M.G. Boosalis and E.C. Setliff. 1980. Laetisaris arvalis (Aphyllophorales, Corticiaceae): A possible biological control agent for Rhizoctonia solani and Pythium species. Mycologia 72:728-736.

Burr, T.J., M.N. Schroth and T. Suslow. 1978. Increased potato yields by treatment of seedpieces with specific strains of Pseudomonas fluorescens and P. putida. Phytopathology 68:1377-1383.

Campbell, W.A. 1947. A new species of Coniothyrium parasitic on sclerotia. Mycologia 39:90-95.

Chet, I., and R. Baker. 1980. Induction of suppressiveness to Rhizoctonia solani in soil. Phytopathology 70:994-998.

Chet, I., and R. Baker. 1981. Isolation and biocontrol potential of Trichoderma hamatum from soil naturally suppressive to Rhizoctonia solani. Phytopathology 71:286-290.

Chet, I., G.E. Harman and R. Baker. 1981. Trichoderma hamatum: Its hyphal interactions with Rhizoctonia solani and Pythium spp. Microb. Ecol. 7:29-38.

Chet, I., Y. Hadar, Y. Elad, J. Katan and Y. Henis. 1979. Biological control of soil-borne plant pathogens by Trichoderma harzianum, pp. 585-592. In: B. Schippers and W. Gams (Eds.), Soil-borne plant pathogens. Academic Press, N.Y.

Cook, R.J., and K.F. Baker. 1983. The nature and practice of biological control of plant pathogens. Amer. Phytopathol. Soc., St. Paul, MN. 539 pp.

Danielson, R.M., and C.G. Davey. 1973. The abundance of Trichoderma propagules and the distribution of species in forest soils. Soil Biol. Biochem. 5:485-494.

Dennis, C., and J. Webster. 1971a. Antagonistic properties of species-groups of Trichoderma. I. Production of nonvolatile antibiotics. Trans. Brit. Mycol. Soc. 57:25-39.

Dennis, C., and J. Webster. 1971b. Antagonistic properties of species-groups of Trichoderma. II. Production of volatile antibiotics. Trans. Brit. Mycol. Soc. 57:41-48.

Dutta, B.K. 1981. Studies of some fungi isolated from the rhizosphere of tomato plants and the consequent prospect for the control of Verticillium wilt. Plant Soil 63:209-216.

Elad, Y., I. Chet and J. Katan. 1980a. Trichoderma harzianum: A biological agent effective against Sclerotium rolfsii and Rhizoctonia solani. Phytopathology 70:119-121.

Elad, Y., J. Katan and I. Chet. 1980b. Physical, biological, and chemical control integrated for soilborne diseases in potatoes. Phytopathology 70:418-422.

Ghaffar, A. 1969. Biological control of white rot of onion. II. Effectiveness of Penicillium nigricans (Bain.) Thom. Mycopathol. Mycol. Appl. 38: 113-127.

Gindrat, D. 1979. Biological soil disinfestation, pp. 253-287. In: A. Mulder (Ed.), Soil disinfestation. Developments in agricultural and managed forest ecology. Vol. 6. Elsevier, Amsterdam, Netherlands.

Griffin, D.M. 1972. Ecology of soil fungi. Syracuse Univ. Press, N.Y. 193 pp.

Hadar, Y., I. Chet and Y. Henis. 1979. Biological control of Rhizoctonia solani damping-off with wheat bran culture of Trichoderma harzianum. Phytopathology 69:64-68.

Harman, G.E., I. Chet and R. Baker. 1980. Trichoderma hamatum effects on seed and seedling diseases induced in radish and pea by Pythium spp. or Rhizoctonia solani. Phytopathology 71:569-572.

Harman, G.E., I. Chet and R. Baker. 1981. Factors affecting Trichoderma hamatum applied to seeds as a biocontrol agent. Phytopathology 71:569-572.

Henis, Y., A. Ghaffar and R. Baker. 1978. Integrated control of Rhizoctonia solani of radish: Effect of successive plantings, PCNB, and Trichoderma harzianum on pathogen and disease. Phytopathology 68:900-907.

Hoch, H.C., and G.S. Abawi. 1979. Biological control of Pythium root rot of table beet with Corticium sp. Phytopathology 69:417-419.

Hoshioka, Y., and T. Fukita. 1969. Ultrastructural observations on mycoparasitism of Trichoderma, Gliocladium, and Acremonium to phytopathogenic fungi. Rep. Tottori Mycol. Instit. 7:8-18.

Howell, C.R. 1982. Effect of Gliocladium virens on Pythium ultimum, Rhizoctonia solani, and damping-off of cotton seedlings. Phytopathology 72:496-498.

Howell, C.R., and R.D. Stipanovic. 1983. Gliovirin, a new biological control of Pythium ultimum. Can. J. Microbiol. 29:321-324.

Huang, H.C. 1977. Importance of Coniothyrium minitans in survival of sclerotia of Sclerotinia sclerotiorum in wilted sunflower. Can. J. Bot. 55:289-295.

Huang, H.C. 1978. Gliocladium catenulatum: Hyperparasite of Sclerotinia sclerotiorum and Fusarium species. Can. J. Bot. 56:2243-2246.

Huang, H.C. 1980. Control of Sclerotinia wilt of sunflower by hyperparasitism. Can. J. Plant Pathol. 2:26-32.

Huang, H.C., and J.A. Hoes. 1976. Penetration and infection of Sclerotinia sclerotiorum by Coniothyrium minitans. Can. J. Bot. 54:406-410.

Hubbard, J.P., G.E. Harman and Y. Hadar. 1983. Effect of soilborne Pseudomonas spp. on the biological control agent Trichoderma hamatum, on pea seeds. Phytopathology 73:655-659.

Jones, D., A.H. Gordon and J.S.D. Bacon. 1974. Cooperative action by endo- and exo-B(1-3)-glucanases from parasitic fungi in the degradation of cell wall glucans of Sclerotinia sclerotiorum (Lib.) de Bary. Biochem. J. 140: 47-55.

Jones, D., and D. Watson. 1969. Parasitism and lysis by soil fungi of Sclerotinia sclerotiorum (Lib.) de Bary, a phytopathogenic fungus. Nature 224:287-288.

Kawamoto, S.O., and J.W. Lorbeer. 1976. Protection of onion seedlings from Fusarium oxysporum f.sp. cepae by seed and soil infestation with Pseudomonas cepacia. Plant Dis. Reptr. 60:189-191.

Kerr, A. 1980. Biological control of crown gall through production of agrocin 84. Plant Dis. 64:24-30.

Klassen, W. 1981. The role of biological control in integrated pest management systems, pp. 433-445. In: G.C. Papavizas (Ed.), Biological control in crop production. Vol. 5. Allanheld, Osmun Publ., Totowa, N.J.

Kloepper, J.W., J. Leong, M. Teintze and M.N. Schroth. 1980. Enhanced plant growth by siderophores produced by plant growth-promoting rhizobacteria. Nature 286:885-886.

Kloepper, J.W., and M.N. Schroth. 1981. Development of a powder formulation of rhizobacteria for inoculation of potato seedpieces. Phytopathology 71:590-592.

Kloepper, J.W., M.N. Schroth and T.D. Miller. 1980. Effects of rhizosphere colonization by plant growth-promoting rhizobacteria on potato plant development and yield. Phytopathology 70:1078-1082.

Kommedahl, T., and C.E. Windels. 1978. Evaluation of biological seed treatment for controlling root diseases of pea. Phytopathology 68:1087-1095.

Kommedahl, T., and C.E. Windels. 1981. Introduction of microbial antagonists to specific courts of infection: Seeds, seedlings and wounds, pp. 227-248. In: G.C. Papavizas (Ed.), Biological control in crop production. Vol. 5. Allanheld, Osmun Publ., Totowa, N.J.

Lewis, J.A., and G.C. Papavizas. 1980. Integrated control of Rhizoctonia fruit rot of cucumber. Phytopathology 70:85-89.

Lewis, J.A., and G.C. Papavizas. 1984. Chlamydospore formation by Trichoderma spp. in natural substrates. Can. J. Microbiol. 30:1-7.

Liu, S., and R. Baker. 1980. Mechanisms of biological control in soil suppressive to Rhizoctonia solani. Phytopathology 70:404-412.

Liu, S.Y., and E.K. Vaughan. 1965. Control of Pythium infection in table beet seedlings by antagonistic microorganisms. Phytopathology 68:1087-1095.

Lumsden, R.D. 1981. Ecology of mycoparasitism, pp. 295-318. In: D.T. Wicklow and G.C. Carroll (Eds.), The fungal community, its organization and role in the ecosystem. Marcel Dekker, N.Y.

Marois, J.J., S.A. Johnston, M.T. Dunn and G.C. Papavizas. 1982. Biological control of Verticillium wilt of eggplant in the field. Plant Dis. 66:1166-1168.

Martin, S.B., G.S. Abawi and H.C. Hoch. 1984. Influence of the antagonist Laetisaria arvalis on infection of table beets by Phoma betae. Phytopathology 74: (In Press).

Martin, S.B., H.C. Hoch and G.S. Abawi. 1983. Population dynamics of Laetisaria arvalis and low-temperature Pythium spp. in untreated and pasteurized beet field soils. Phytopathology 73:1445-1449.

Moody, A.R., and D. Gindrat. 1977. Biological control of cucumber black rot by Gliocladium roseum. Phytopathology 67:1159-1162.

Moore, L.W. 1979. Practical use and success of Agrobacterium radiobacter strain 84 for crown gall control, pp. 553-578. In: B. Schippers and W. Gams (Eds.), Soil-borne plant pathogens. Academic Press, N.Y.

Nelson, E.B., G.A. Kuter and H.A.J. Hoitink. 1983. Effects of fungal antagonists and compost age on suppression of Rhizoctonia damping-off in container media amended with composted hardwood bark. Phytopathology 73: 1457-1462.

Odvody, G.N., M.G. Boosalis and E.D. Kerr. 1980. Biological control of Rhizoctonia solani with a soil-inhabiting basidiomycete. Phytopathology 70:655-658.

Papavizas, G.C. (Ed.). 1981. Biological control in crop production. Granada Publ., London. 461 pp.

Papavizas, G.C., and J.A. Lewis. 1981. Introduction and augmentation of microbial antagonists for the control of soilborne plant pathogens, pp. 305-322. In: G.C. Papavizas (Ed.), Biological control in crop production. Allenheld, Osman Publ., Totowa, N.J.

Papavizas, G.C., J.A. Lewis and T.H. Abd-El Moity. 1982. Evaluation of new biotypes of Trichoderma harzianum for tolerance to benomyl and enhanced biocontrol capabilities. Phytopathology 72:126-132.

Papavizas, G.C., and R.D. Lumsden. 1980. Biological control of soil-borne fungal propagules. Annu. Rev. Phytopathol. 18:389-413.

Papavizas, G.C., B.B. Morris and J.J. Marois. 1983. Selective isolation and enumeration of Laetisaria arvalis from soil. Phytopathology 73:220-223.

Powell, N.T. 1979. Internal synergism among organisms inducing disease, pp. 113-133. In: J.G. Horsfall and E.B. Cowling (Eds.), Plant disease. Vol. 4. Academic Press, N.Y.

Ricard, J.L., C. Groslaude and N. Ale-Agha. 1974. Antagonism between Eutypa armeniacea and Gliocladium roseum. Plant Dis. Reptr. 58:983-984.

Rishbeth, J. 1963. Stump protection against Fomes annosus. III. Inoculation with Peniophora gigantea. Ann. Appl. Biol. 52:63-77.

Schippers, B., and W. Gams (Eds.). 1979. Soil-borne plant pathogens. Academic Press, N.Y. 686 pp.

Suslow, T.V., and M.N. Schroth. 1982. Rhizobacteria of sugar beets: Effects of seed application and root colonization on yield. Phytopathology 72: 199-206.

Taylor, G.S., and D. Parkinson. 1964. Studies of fungi in the root region. II. The effect of certain environmental conditions on the development of root surface microfloras of dwarf bean seedlings. Plant and Soil 20:34-42.

Tribe, H.T. 1957. On the parasitism of Sclerotinia trifoliorum by Coniothyrium minitans. Trans. Brit. Mycol. Soc. 40:489-499.

Trutmann, P., P.J. Keane and P.R. Merriman. 1980. Reduction of sclerotial inoculum of Sclerotinia sclerotiorum with Coniothyrium minitans. Soil Biol. Biochem. 12:461-465.

Trutmann, P., P.J. Keane and P.R. Merriman. 1982. Biological control of Sclerotinia sclerotiorum on aerial parts by the hyperparasite Coniothyrium minitans. Trans. Brit. Mycol. Soc. 78:521-529.

Tu, J.C. 1980. Gliocladium virens, a destructive mycoparasite of Sclerotinia sclerotiorum. Phytopathology 70:670-674.

Tu, J.C., and O. Vaartaja. 1981. The effect of the hypoparasite (Gliocladium virens) on Rhizoctonia solani and on Rhizoctonia root rot of white beans. Can. J. Bot. 59:22-27.

Turner, G.J., and H.T. Tribe. 1975. Preliminary field plot trials on biological control of Sclerotinia trifoliorum by Coniothyrium minitans. Plant Pathol. 24:109-113.

Turner, G.J., and H.T. Tribe. 1976. On Coniothyrium minitans and its parasitism of Sclerotinia species. Trans. Brit. Mycol. Soc. 66:97-105.

Uecker, F.A., W.A. Ayers and P.B. Adams. 1978. A new hyphomycete on sclerotia of Sclerotinia sclerotiorum. Mycotaxon 7:275-282.

Walker, J.A., and R.B. Mande. 1975. Natural occurrence and growth of Gliocladium roseum on mycelium and sclerotia of Botrytis allii. Trans. Brit. Mycol. Soc. 65:335-338.

Wallace, H.R. 1978. The diagnosis of plant diseases of complex etiology. Annu. Rev. Phytopathol. 16:379-402.

Weindling, R. 1934. Studies on a lethal principle effective in the parasitic action of Trichoderma lignorum on Rhizoctonia solani and other soil fungi. Phytopathology 24:1153-1179.

Weindling, R., and O.H. Emerson. 1936. The isolation of a toxic substance from the culture filtrate of Trichoderma. Phytopathology 26:1068-1070.

Weindling, R., and H.S. Fawcett. 1936. Experiments in the control of Rhizoctonia damping-off of citrus seedlings. Hilgardia 10:1-16.

Weller, D.M., and R.J. Cook. 1983. Suppression of take-all of wheat by seed treatment with fluorescent pseudomonads. Phytopathology 73:463-469.

Wells, H.D., D.K. Bell and C.A. Jaworski. 1972. Efficacy of Trichoderma harzianum as a biocontrol for Sclerotium rolfsii. Phytopathology 62:442-447.

Windels, C.E., and T. Kommedahl. 1978. Factors affecting Penicillium oxalicum as a seed protectant against seedling blight of pea. Phytopathology 68:1656-1661.

Wright, J.M. 1956. Biological control of a soil-borne *Pythium* infection by seed inoculation. Plant Soil 8:132-140.

MOLECULAR GENETICS OF BIOLOGICAL CONTROL AGENTS OF PLANT PATHOGENS: STATUS AND PROSPECTS

Carolyn Napoli and Brian Staskawicz

Molecular genetics and recombinant DNA technology are powerful research tools which are currently being applied to many biological systems. This technology application can potentially lead to both an understanding of basic biological phenomena and the engineering of organisms with novel, desirable genotypic and phenotypic characteristics. There is considerable interest in applying molecular genetics to the field of biological control, specifically with intent to either improve existing biocontrol agents or to engineer novel biocontrol agents.

Biological control of plant pathogens is a very complex subject upon which a good deal of documentation exists (Baker & Cook 1974, Cook & Baker 1983, Hornby 1983Kerr 1972, Papavizas & Lumsden 1980, Schroth & Hancock 1981, Suslow 1982). Mechanisms have been proposed for biocontrol systems (Baker & Cook 1974, Kerr 1980, Moore & Warren 1979, Schroth & Hancock 1981, Suslow 1982, Vivader 1976); however, knowledge is lacking about the genetic basis underlying these mechanisms. It is, then, no small task to begin to apply a molecular genetic approach to improve systems for which no genetic basis exists. Biological control of plant pathogens is the outcome of a number of undefined multigenic effects and environmental factors whereby a pathogen does not (is unable to) exert a pathogenic effect on a susceptible host. Molecular genetics can be used as a research approach to begin to define the genes which are important in biological control phenomena. An understanding of the genetic basis underlying the mechanisms of biocontrol will provide a basis from which biological systems can be manipulated to enhance biological control phenomena.

This paper will discuss how molecular genetics can be used in establishing a genetic basis for biocontrol mechanisms. This knowledge is an essential prerequisite to speculation on ways to improve biological control agents of plant pathogens. Our emphasis will be on the use of bacteria as biological control agents. Currently genetic engineering of bacteria has been refined to the point that the technology is, in most cases, non-limiting; and gene cloning and expression are both routine and straightforward.

ISBN 0-12-357030-1

The subject of determining host gene effects in biological control will not be addressed in this paper. Plant molecular biology, especially in the area of gene cloning and plant transformation, is making notable advances in technology. However the techniques for identifying and manipulating plant genes are extremely technology-limited at this time.

A major obstacle to the use of engineered microbial biological control agents will be the ability to have these strains approved for release into the environment. This topic will be briefly discussed in this paper.

DEFINING GENES INVOLVED IN AN EFFECTIVE BIOCONTROL AGENT

Antagonism appears to be a principle concept underlying biological control (Baker & Cook 1974, Schroth & Hancock 1981). As a broad category, antagonism can be broken up into parasitism, antibiosis and competition. While aspects of these interactions can be attributed to microbial genes, the host plant and the environment are powerful influences in defining the ecological balance between antagonists and pathogens. The complexity of such a system is nothing short of a nightmare to molecular geneticists who understand problems by focusing in on narrowly defined specifics, with limited variables, and proceeding one gene at a time. The important first step in the application of molecular genetics to such complexity is the choice of the right system to work on. The logical choice would be a system that works in the field and can be manipulated by standard molecular genetic techniques.

The successful control of crown gall disease, caused by *Agrobacterium tumefaciens*, has been effected by *A. radiobacter* strain 84 (Kerr 1972, 1980; Moore & Warren 1979). The most dramatic control has been reported in Australia (Kerr 1980), while varying degrees of success have been reported in other geographical locations (Moore & Warren 1979). Strain 84, while effective in many cases, is not a successful biocontrol agent against *A. tumefaciens* biotype 3 strains which cause crown gall on grapevines (Kerr & Panopoulos 1977).

Initially, control by *A. radiobacter* strain 84 was ascribed to the production of a bacteriocin, agrocin 84, encoded by a plasmid resident in strain 84 (Kerr 1980, Moore & Warren 1979). A number of correlations supported this hypothesis. Pathogenic strains of *A. tumefaciens* which had been shown to be sensitive to agrocin 84 could be controlled in the field, and only *A. radiobacter* strains which produced agrocin 84 were effective biological control agents (Kerr 1980). However, since the initial observations on this subject, it has been demonstrated that agrocin 84 is not the only requirement for biocontrol of *A. tumefaciens* (Ellis et al. 1979). Some strains of *A. radiobacter* that produced agrocin in vitro were not effective in the field as biological control agents (Kerr & Panopoulos 1977). Additionally, an Agrobacterium strain that was the recipient of the agrocin 84 plasmid failed to provide control in the field even though the bacteriocin was produced in vitro (Ellis et al. 1979). Other factors such as multiplication in the rhizosphere in the vicinity of the wound and competition for wound attachment sites may be important (Ellis et al. 1979).

This system lends itself to genetic analysis to begin to determine what single gene or combination of genes is important to a biocontrol mechanism. The genus Agrobacterium is amenable to standard genetic analyses and manipulations. By mutational analysis it would be possible to remove genes, one at a time, to determine what gene products are important. If these genes could be retrieved from a gene library, then it would be possible to restore gene function through complementation analysis and restore effective biocontrol. These results can be assayed in the field to correlate in vitro analysis with in vivo results. In this case, the host, pathogen and hopefully the environmental influences remain constant while the antagonist is altered.

A precise way to isolate mutants in bacteria is to employ transposon mutagenesis (Staskawicz 1983). Prokaryotic transposons are defined genetic entities which are capable of inserting as discrete, nonpermuted DNA segments at random sites in bacterial genomes (Kleckner 1981). Transposons encode one or more functions, one of which is antibiotic resistance. Transposons are very effective mutagenic agents because they cause mutations by insertional inactivation and usually transpose only once in a cell such that the problem of multiple mutagenesis does not occur as is often the case with chemical mutagens. As the transposon encodes for antibiotic resistance, mutagenized cells (which contain a transposon) can be directly selected by antibiotic resistance.

A common way of delivering a transposon to a bacterium, other than Escherichia coli, is to employ a method called suicide plasmid mutagenesis (Simon et al. 1983). The transposon, for example the transposon Tn5, is inserted in a plasmid resident in E. coli. The plasmid is conjugated from the E. coli into a bacterium, for example A. radiobacter. The E. coli plasmid contains a replicon which cannot replicate in Agrobacterium. The Tn5 will transpose at a low frequency from the suicide plasmid into the Agrobacterium genome. This transposition can be either into the chromosome or plasmid DNA of Agrobacterium. Agrobacterium exconjugants which contain a Tn5 can be selected for by Kanamycin (Km) resistance, and the E. coli cells can be selected against by using a rifampicin (rif) resistant derivative of A. radiobacter and incorporating rif, Km into the selection medium.

Transposon mutagenesis not only causes polar mutations but also serves as a molecular tag. A gene library of transposon mutagenized A. radiobacter DNA can be prepared in E. coli by standard recombinant DNA techniques. The particular E. coli cell which contains the mutated gene can be directly selected for by Km resistance (Dahlbeck & Staskawicz, unpublished data). The mutated gene can be nick-translated, labeled with ^{32}P, and used as a probe to identify the wild type gene from a library prepared from unmutagenized DNA. Wild type genes can also be isolated by use of complementation analysis. In this case, the wild type library stored in E. coli is mobilized en masse into a mutant A. radiobacter. Single colony isolates can then be screened for restoration of the gene function.

Transposon mediated mutagenesis, which is widely applicable in molecular genetic studies (de Bruijn & Lupski 1984), could be used to generate mutants of A. radiobacter which do not function as biocontrol agents. This could potentially be a laborious screen as there is no direct selection for an attenuated biological control agent. However, this brute force approach has worked in mutational analysis in host/symbiont systems (Meade et al. 1982). This strategy may prove unsuccessful if a single gene effect is not enough to reduce or eliminate control. However, if such a mutant could be obtained, it could be analyzed for agrocin production, multiplication in the rhizosphere, attachment to wound tissue, etc. This probe can be used to retrieve the wild type gene from a gene library made from strain 84. This gene could also be cloned by complementation. Mutations in agrocin production could be obtained directly, as a suitable and convenient Petri dish inhibition assay is available. Mutants which are unable to attach to wound surfaces could be found, if the appropriate selection, perhaps an affinity column, could be devised.

Competition among antagonists and pathogens appears to be an important mechanism involved in biocontrol. Competition is a difficult problem to investigate on a molecular genetics level. Once again the system is complex (involving a number of interacting microbes), most likely multigenic, and subject to environmental conditions and host effects. However, molecular genetics is possibly the only approach that can provide an understanding of the genes that are necessary for a "highly" competitive bacterium.

Again, transposon mediated mutagenesis could be applied to select rhizosphere or phylloplane bacteria with altered competition phenotypes. These mutants could be tested for multiplication and survival under field conditions. Based on the molecular weight of the E. coli chromosome, a bacterium would have enough genetic material for about $3X10^3$-$4.5X10^3$ average sized proteins. This means that a large number of mutagenized bacteria would have to be screened before a mutant could be identified. The power of molecular genetics lies in applying an appropriate positive selection as a means to identify a single mutant from a large number of mutagenized cells. An alternative approach that would employ a positive selection would be to take a non-competitive bacterium, transfer a gene library from a highly competitive strain into this bacterium, and then select for bacteria which can successfully compete.

DELETING PATHOGENIC TRAITS FROM PLANT PATHOGENIC BACTERIA TO ENGINEER BIOCONTROL AGENTS

Most leaf-spotting pathogenic bacteria are facultative parasites, i.e. they live as epiphytes on the leaf surfaces of both host and non-host plants for a certain portion of their life cycle. Thus we can assume that these types of bacteria have evolved very efficient epiphytic and pathogenic growth properties of which the genetic control is unknown. One approach to define the numbers and type of genes involved is to isolate mutants of a particular organism and screen them for loss of a specific trait, e.g., pathogenicity, virulence, epiphytic colonization of leaf surfaces. The method of transposon mediated mutagenesis can be applied to phytopathogenic bacteria to isolate these mutants. By the methods previously outlined, it would be possible to isolate both mutagenized and wild type genes.

Once a fragment of DNA containing a wild type gene is in hand, in vitro mutagenesis or deletions can be performed (Orser et al. 1984, Ruvkin & Ausubel 1981). This altered gene is then exchanged for the wild type gene and a strain is constructed that is identical to the wild type except for the lack of a specific function. In this manner the engineered organism can be tested for its ability to control the wild type organism. A specific scenario may be the following: If we assume that the construction of non-virulent phytopathogenic bacteria does not alter their ability to compete and colonize during epiphytic growth and that there are limited and specific sites of colonization of the leaf surface, we might be able to spray plants with non-virulent engineered biocontrol strains to prevent the establishment of wild type virulent strains. This seems to be the case with non-ice nucleating mutants of Pseudomonas syringae (Orser et al. 1984). If proven successful a series of nonvirulent strains of other phytopathogenic bacteria may be able to act in a similar manner.

CONSTRAINTS ON THE USE OF GENETICALLY ALTERED MICROORGANISMS AS BIOCONTROL AGENTS

One of the major constraints on the use of genetically altered or engineered biocontrol agents will be the ability to have these organisms released into the environment. Molecular genetic techniques have been used to engineer deletion mutants of *Pseudomonas syringae* and *Erwinia herbicola* that are deficient in ice nucleating activity (INA^-) (Orser et al. 1984). These mutants were approved by the Recombinant DNA Advisory Committee (RAC) in June 1983 for field testing to determine if INA^- mutants can reduce the susceptibility of crops to frost damage. The field tests have been postponed due to a number of legal issues. At the present time, it is difficult to predict what constraints will hinder the release of recombinant organisms into the environment for controlled field testing.

SUMMARY AND RECOMMENDATIONS

The use of microorganisms in the biological control of plant pathogens is currently a very active area of research in both academic and industrial institutions. The application of molecular genetics to selected topics in biocontrol can potentially provide an understanding of the genes involved in biological control phenomena. An establishment of a genetic basis for biological control is a prerequisite to further molecular genetic studies that center on the enhancement of biological control phenomena. It is not necessary to completely understand a biological control system in order to use it successfully in the field. However, this basic understanding is necessary in order to apply molecular genetics as a means to improve a biological control system. Initially, a model system that consistently works under field conditions should be exploited and genetically dissected, as we have made the case for the biological control of crown gall. In the future, the area of biological control can be best approached by collaborations between molecular biologists and plant pathologists. Thus, the plant pathologist can provide the biological material and the practical experience, while the molecular biologist can manipulate the specific genes of interest.

Finally, the release of these engineered strains into the environment has come under much debate. The construction of effective and safe biological control agents should be carefully analyzed and the risks and benefits assessed before a release is allowed. Unfortunately, the ability to predict which biological control agents fall in this category may be very subjective.

REFERENCES

Baker, K.F., and R.J. Cook. 1974. Biological control of plant pathogens. W.H. Freeman, San Francisco. 433 pp.

Cook, R.J., and K.F. Baker. 1983. The nature and practice of biological control of plant pathogens. Amer. Phytopathol. Soc., St. Paul, MN. 539 pp.

de Bruijn, F.J., and J.R. Lupski. 1984. The use of transposon Tn5 mutagenesis in the rapid generation of correlated physical and genetic maps of DNA segments cloned into multicopy plasmids--a review. Gene 27:131-149.

Ellis, J.G., A. Kerr, M. van Montagu and G. Schell. 1979. Agrobacterium: Genetic studies on agrocin 84 production and the biological control of crown gall. Physiol. Plant Pathol. 15:311-319.

Hornby, D. 1983. Suppressive soils. Annu. Rev. Phytopathol. 21:65-85.

Kerr, A. 1972. Biological control of crown gall: Seed inoculation. J. Appl. Bacteriol. 35:493-497.

Kerr, A. 1980. Biological control of crown gall through production of agrocin 84. Plant Dis. 64:24-30.

Kerr, A., and C. Panopoulos. 1977. Biotypes of Agrobacterium radiobacter var. tumefaciens and their biological control. Phytopathol. Z. 90:172-179.

Kleckner, N. 1981. Transposable elements in prokaryotes. Annu. Rev. Genet. 15: 31-404.

Meade, H., S. Long, G. Ruvkin, S. Brown and F. Ausubel. 1982. Physical and genetic characteristics of symbiotic and auxotrophic mutants of Rhizobium meliloti induced by transposon Tn5 mutagenesis. J. Bacteriol. 149:114-122.

Moore, K.W., and G. Warren. 1979. Agrobacterium radiobacter strain 84 and biological control of crown gall. Annu. Rev. Phytopathol. 17:163-179.

Orser, C., R. Lotstein, E. Lahue, D.K. Willis, N. Panopoulos and S. Lindow. 1984. Structural and functional analysis of the Pseudomonas syringae pv. syringae ice region and construction of ice-deletion mutants. Phytopathology 74:(In Press).

Papavizas, G.C., and R.D. Lumsden. 1980. Biological control of soil-borne fungal propagules. Annu. Rev. Phytopathol. 18:389-413.

Ruvkin, G., and F. Ausubel. 1981. A general method for site-directed mutagenesis in prokaryotes. Nature (London) 289:85-88.

Schroth, M.N., and J.W. Hancock. 1981. Selected topics in biological control. Annu. Rev. Microbiol. 35:453-476.

Simon, R., U. Priefer and A. Puhler. 1983. Vector plasmids for in vivo and in vitro manipulations of gram-negative bacteria, pp. 98-106. In: A. Puhler (Ed.), Molecular genetics of the bacteria-plant interaction. Springer-Verlag, N.Y.

Staskawicz, B. 1983. Molecular genetics of plant disease, pp. 199-213. In: J.A. Callow (Ed.), Biochemical plant pathology. John Wiley & Sons, N.Y.

Suslow, T.V. 1982. Role of root colonizing bacteria in plant growth, pp. 187-223. In: M.S. Mount and G.H. Lacy (Eds.), Phytopathogenic prokaryotes. Vol. I. Academic Press, N.Y.

Vivader, A. 1976. Prospects for control of phytopathogenic bacteria by bacteriophages and bacteriocins. Annu. Rev. Phytopathol. 14:451-465.

DISCUSSION: Biological control of plant pathogens.

DISCUSSION LEADER: J.J. Marois

During this session it became obvious that there were many differences within the conceptual framework of the plant pathologists and entomologists; however, the obstacles to successful implementation of biological control into integrated pest management programs were similar. These differences begin with the actual concept of biological control. Whereas entomologists generally limit their concept of biological control to the use of parasites, predators, or pathogens; plant pathologists usually favor a broader concept which includes cultural practices and resistance (Schroth & Hancock, this volume). These concepts are influenced greatly by the biology of the pests considered. Plant pathologists deal with control agents which are practically immobile when compared to insects. The procedure of releasing an individual into a new area is common in the control of insects and sometimes weeds, but almost absent in plant pathology. The fact that plant pathogens may occur in high numbers but not cause disease (Lindow, this volume) makes economic threshold studies, used successfully by entomologists, extremely difficult for plant pathologists. The mechanisms of interaction between the control agent and pest differs also. Whereas plant pathologists concentrate on competition, antibiosis, and hyperparasitism; entomologists rely on predation and parasitism.

The differences in the biology of the pests and control agents have resulted in considerable differences in the approach to biological control. In the selection for potential control agents, plant pathologists search for organisms within the environment it will be used, or in suppressive soils where the disease does not occur even though the host and pathogen are present. Entomologists, however, have relied heavily on the importation and release of control agents from distant areas. Entomologists then depend on alternate prey or target prey to build up and maintain the control agent populations. In contrast, plant pathologists usually rely on the saprophytic abilities of the control agent to not only reduce disease by increasing the level of competition toward the pathogen but also to increase the control agent's populations.

Areas of consideration that plant pathologists seem to be more involved in than entomologists are exploitation of natural biological control in suppressive soils, use of control agents to alter the response of plants to pathogens (Schroth & Hancock, this volume), enhancement of biological control agents by the combination of chemical and biological controls (Baker, this volume), alteration of saprophytic populations with chemicals (Lindow, this volume), and the use of techniques to enhance the genetic components of control agents (Napoli & Staskawicz, this volume). Entomologists emphasize programs to find and identify biological control agents, develop computer simulations of the predator-prey interactions, and utilize economic threshold models.

The development and implementation of biological control of insects, nematodes, weeds, and plant pathogens seem to share several major obstacles. All three areas need more and better data that can be integrated into interdisciplinary efforts to understand the ecology of the systems investigated. Too often organisms are applied to the field and no follow-up studies performed. This approach has resulted in a void of biological data and the principles that could have been derived for application of biological programs. The major parameters of the pest, control agent, and crop system have not been defined. This has led to a stagnant level of scientific development where the procedures to choose a pest problem and select for a control agent are basically the same as those used thirty years ago, even though successful biological control procedures are still rare.

An obstacle more unique to plant pathologists is the complex microscopic ecosystem in which biological control agents for plant disease must survive, multiply, and decrease disease. Disease can be reduced by a reduction in inoculum density of the pest as with insects, by protection of infection sites, or by affecting the host response to the pathogen (induced resistance and cross protection). The latter two cases appear to be more common in plant pathological problems. Whether on leaves (Lindow, this volume) or roots (Schroth & Hancock, this volume), the control agent must be highly competitive. In contrast, entomologists rely on control agents that are parasites or predators, which are more easily sampled and monitored.

The delivery of control agents is critical in plant disease control. The beneficial microorganisms must be placed in an environment where they can survive and interact with the pathogen or plant in a manner as to reduce disease. Favorable environments are usually found where colonization by other microorganisms is reduced, such as young leaves (Lindow, this volume), fumigated soils (Martin et al., this volume), roots of transplants, or fresh wounds (Schroth & Hancock, this volume).

Due to the difficulty of establishing organisms in complex communities it is important for plant pathologists to consider the agroecosystem as a continuum, from relatively simple greenhouse systems where large amounts of energy and chemical inputs are used to maintain a constant agroecosystem, to complex natural forest systems where little or no inputs are available. The simpler, more intensive greenhouse systems are more amenable to biological control in plant pathology. For example, control agents can be applied to steamed or fumigated soils when competition from other microorganisms is low. And because of the uniformity of different greenhouse operations, a control agent that works in one greenhouse should work in another. The agroecosystem continuum can be extended to include annual crops transplanted to fumigated soils, annual crops seeded into fumigated soils, annual crops seeded into nonfumigated soils, perennial fruit and nut crops intensively managed, perennial crops which receive little or no management, and finally unmanaged natural systems. Due to the increase in complexity and decrease in areas available for introduction of control agents, as one moves along the continuum the implementation of biological control is more difficult. This may be a direct contrast to the situation in entomology, where long term crops such as deciduous orchards are considered more amenable to biological control than are the more temporary annual crops.

The use of genetically engineered control agents is more reasonable with present technology for procaryotic microorganisms than eukaryotic insects (Napoli & Staskawicz, this volume). However, for these new technologies to be used as a means to improve control agents the necessary phenotypic and genotypic attributes must be identified, a rapid screening procedure must be available, and protocol for release of genetically engineered organisms into the environment must be developed.

The major obstacles for the implementation of biological control programs for plant diseases into integrated pest management are lack of understanding of the ecology of the pathosystem, mass production and delivery of control agents, selection and enhancement of control agents, and the specificity of control agents to only one disease on one crop in a small geographic area. Those working on the biological control of insects seem to share many of these same problems. It is logical that if we combine our efforts through communication and cooperation we will be better able to achieve our goals.

Biological Control of Nematodes

BIOLOGICAL CONTROL OF NEMATODES: STATUS AND PROSPECTS IN AGRICULTURAL IPM SYSTEMS

S.D. Van Gundy

The current state of development in biological control of nematodes has been reviewed extensively (Kerry 1981a, b, 1984; Mankau 1980, 1981; Sayre 1971, 1980). It is clear from these reviews that the emphasis over the last two decades concentrated on parasitic fungi and was largely descriptive in nature. Plant-parasitic nematodes co-exist in the rhizosphere with many other organisms and it has been relatively easy to isolate and identify many new parasites and predators of nematodes. Even so, there have been no systematic searches for exotic parasites and predators in the countries of origin of specific nematode pests. Little has been done to quantify the role of these biotic agents in the population dynamics of the plant parasitic nematodes and even less has been done to assess the possible use of these biotic agents as management tools. Viruses, bacteria, fungi, protozoans, nematodes, tardigrades, turbellarians, enchytraeids, collembola, and mites have been identified as parasites or predators of nematodes. One or more of these antagonists occur in most agroecosystems and there is evidence that some biological control occurs naturally in these situations (Kerry et al. 1982, Stirling et al. 1979).

Currently there are no methods to quantify the level of natural control except that it is usually increased by the addition of various organic amendments (Goday et al. 1983, Mankau & Das 1974, Mian et al. 1982, Rodriguez-Kabana et al. 1984, Singh & Sitaramaiah 1973). Therefore, until recently the principal applied approach to biocontrol of nematodes has been to study how the addition of organic soil amendments reduce damage caused by nematodes. The two most prevalent hypotheses are 1) that decomposition products released from amendments into the soil are directly toxic or, 2) that increase of the soil microbial and faunal populations by addition of amendments stimulates parasite and predator activity. Both are important and functional hypotheses and continue to support optimism about the prospects for biological control of plant parasitic nematodes.

Of this long list of over 200 pathogens, parasites and predators, 75% of which are fungi, only two products are marketed com-

ISBN 0-12-357030-1

mercially. Royal 300 and Royal 350 which contain nematode-trapping fungi of the genus Arthrobortys, and are reported to control nematodes in mushroom houses and greenhouses in France (Mankau 1981).

Historically, since the discovery of fumigant halogenated hydrocarbons in 1945 (1,3-D, EDB, MB, DBCP) as relatively cheap, quick-acting and effective preplant nematicides, there has been little incentive either in manpower or financial support to search for biological control alternatives. Even in 1978, after the cancellation of DBCP a survey of Agricultural Experiment Stations suggested that only about four staff-years were directed toward biological and 15 staff-years toward chemical nematicides (Van Gundy 1980). There has been an even greater disparity in funding from national and private sources. Recent information supplied by the Cooperative State Research Service on biological control research conducted by State Agricultural Experiment Stations, Forestry Schools, 1890 Institutions, Veterinary Colleges and other cooperating institutions suggest that approximately 76% of the effort in 1982 was in entomology, 16% in plant pathology and nematology, and 6% in weed science (Table 1). It is estimated that the effort in nematology is less than 2%. Biological control effort in entomology appears to be evenly distributed on research on pathogens, predators, and parasitic agents while the major effort in plant pathology-nematology has been on pathogens.

Another deterrent to biological control of nematodes has its foundation in nematode control philosophy based on soil fumigation designed to reduce nematode populations by 90% or more. It has been perceived that biological control could never achieve

Table 1. Research effort on biological agents for control of diseases, parasites and nematodes on agricultural commodities.

SYS - 1982[a]	Insects, Mites & Snails	Diseases & Nematodes	Weeds
Pathogens	31	17	2
Predators	25	1	1
Parasites	34	1	1
Other	7	2	8
TOTAL	97	21	8
% of Total	76	16	6

[a] Supplied by CSRS, Science and Education, USDA.

these expectations ("as good as" and "as cheap as"). Biological control of nematodes was considered only a curiosity instead of a serious approach to nematode control. Fortunately, the expansion of the IPM philosophy in the 1970's has encouraged management approaches with integrated multiple control practices, one of which may be biological control. The losses of some important fumigant nematicides in 1983-84, groundwater contamination by some nonfumigant nematicides, increased cost of chemical control and the lack of new nematicides undergoing the registration process suggest that it is time to exploit the full potential of biological control of nematodes in a management program.

Beyond political and economic problems are a number of scientific barriers to the commercial development of biological control agents of soil nematodes. These scientific barriers include the stability of the natural biological balance of the ecosystem, antagonism of soil organisms to new introductions, distribution of the introductions through the soil profile, establishment of the introductions in agroecosystems, the zone of activity of the biological control agents, the assessment of biological activity and synchronization with nematode biology. They are the important factors that must be considered when attempting any of these classical approaches to biological control: introduction, manipulation/augmentation, and conservation. There is also some mystique about the soil environment that few biologists want to confront. It is difficult to visualize, manipulate, sample and monitor with precision; thus, few scientists are willing to conduct research in one of the most difficult research areas in agricultural sciences.

Balance of the Ecosystem

Although conventional agriculture often contributes to a biological imbalance, the agroecosystem still is strongly buffered against physical, chemical and biological change. The addition of organic matter to soils appears to be the only factor that can rapidly alter all three of these soil properties and may be the reason why the additional organic matter succeeds in altering nematode populations. Unfortunately, most organic additives must be used in such large quantities to make a significant change in the ecosystem that they are neither practical nor economical unless there is an abundant supply availible locally.

Antagonism to Introductions

Fungistasis, antibiosis, competition and parasitism are readily demonstrated major barriers to the introduction of any alien fungus or other organism into soil (Baker, this volume). This is particularly true for nematode-trapping fungi which are inherently non-persistent and are poor competitive saprophytes. So far attempts to add these fungi to field soils have been unsuccessful even in the presence of energy sources. Obviously there are complex interactions between these nematode-trapping fungi, the decomposing organic substances and the remainder of the soil microbial organisms. Similar problems occur with the introduction of bacteria, nematodes or mites into the soil ecosystem.

Zone of Activity and Distribution of Introductions

Ecologically, the majority of microbial activity occurs in the soil A^1 horizon (3-8 cm). Plant parasitic nematode activity is active throughout the entire root zone but concentrated in the top 60 cm of soil. There is a discrepancy between the zone of activity of the biological control agents and their target. Obviously, many of the nematodes in the subsurface layers of soil will escape being captured or devoured either by introduction or by resident predators and parasites. In addition, the soil has a complex physical structure that makes uniform distribution of introductions both horizontally and vertically difficult or impossible. Soil structure is a major barrier even to simple gaseous fumigants and water-soluble non-fumigant nematicides. It is even more difficult to distribute organisms such as nonmotile spores, cells, and mycelium. Motile nematodes and mites might be somewhat more successful in distributing themselves in a given zone although this has never been tested in agroecosystems.

Establishment of Introductions

Before parasitism or predation can occur nematode-trapping fungi must grow and colonize the soil or rhizosphere. These processes require an energy input in the form of some readily available carbon and nitrogen source. The addition or availability of excess energy may intensify the activity of soil microbes which compete with predaceous fungi and reduce the rate and zone activity. So even if these fungi overcome fungistasis their establishment is tenuous and energy dependent. Kerry (1981b) found that in England it took five to seven years for the fungi attacking the cereal cyst nematode, *Heterodera avenae*, to reach effective and economic predacious activity, which occurred at fungal populations of 200-400 propagules per gram of soil. Thus it appears from this example that biological control of nematodes with parasitic fungi will be slow, erratic and not economical for the first years after introduction. That will make grower acceptance extremely difficult.

Assessment of Activity

One of the major obstacles to the acceptance of biological control of plant parasitic nematodes is the lack of satisfactory methods or techniques for the assessment of parasitism or predation. Laboratory studies usually over-estimate the biological activity of a predator because there is no means of escape for the nematode. The importance of nematodes to crop production was never fully appreciated or even accepted in some circles until the discovery of the soil nematicides which selectively controlled nematodes but not other soil microorganisms. A similar technique is urgently needed to document efficiency of biological control agents of nematodes. Kerry et al. (1980) demonstrated that formalin, a weak nematicide but good fungicide, suppressed nematode-parasitic fungi activity and increased nematode numbers. This technique had other side effects and did not appear to be a satisfactory technique for assessing biological control activity.

Lack of Selectivity

Many of the parasites and predators of nematodes are not very selective in their feeding activity. For example, nematode-trapping fungi parasitize any nematode that enters a trap, not just plant parasites. Specificity may only occur in some viral and bacterial parasites. Very little is known at this time about the strain differences, virulence levels or predatory specificity by different isolates of the antagonist. It may be possible to select highly virulent strains for commercial production.

Synchronization with Nematode Biology

Many of the important plant parasitic nematodes spend a major portion of their life cycle within the root, which affords protection from parasitism and predation. Predation takes place over a period of weeks or months and on more than one life stage of the nematode (eggs, larvae and adults) so that it takes more than one crop season to quantify the biological control activity. It would appear that in order to biologically control such nematodes that the control would have to be established within the plant rhizosphere and protect the host roots from nematode entry.

COMPATIBILITY WITH OTHER CONTROL STRATEGIES

Past philosophy on nematode control was to reduce population levels to approach eradication. Such a philosophy is incompatible with biological control since eradication of the host means eradication of the control agent unless it has another food source. Broad spectrum nematicides such as methyl bromide reduce all nematodes and their antagonists to near extinction levels. The 1,3-D fumigant (1,3D) and the nonfumigant organophosphate and carbamate nematicides, although not as toxic to the bacteria and fungi, are destructive to all the invertebrate predators. Conservation of natural parasites and predators usually is not very compatible with soil fumigation. As in most biological control systems, chemical pesticides must be used judiciously. Crop rotation is the standard cultural practice most often used to manage nematode populations. Kerry et al. (1982) found that at least five years of continuous monoculture cropping was necessary

Table 2. Considerations for cyst nematode control in different climatic regions.

Considerations	Temperate	Semi-Tropical
Generations/year	1-2	3-5
Multiplication rates	6-12 X	18-30 X
Damage threshold (eggs/g soil)	5-10	1-2
Control under crop rotation	2-3 yrs	3-5 yrs
Natural control under monoculture	5+ yrs	?

to bring the level of biological control of the oat cyst nematode to an economic level. This monocropping is inconsistent with what we consider to be good agronomic practices. The use of resistant varieties is also a form of crop rotation designed to rapidly reduce nematode populations as well as antagonists. So our current nematode control strategies are not conducive for establishing a long term balance between the pest and its natural enemies. Economically, farmers usually cannot withstand the long term losses that might be required to establish a natural balance in five years with pests in temperate climates and even longer in subtropical climates (Table 2). The future of biological control for plant parasitic nematodes thus lies in the development and use of short term (1-2 years) practices that can be incorporated into the standard control strategies, including chemicals, resistant varieties and cultural practices.

PROSPECTS FOR BIOLOGICAL CONTROL OF SOIL NEMATODES

More than 200 soil antagonists have been identified as potential biocontrol agents. At least 75% are soil fungi that parasitize nematode juveniles, eggs and adults or cysts. The mode of action includes infection by motile zoospores, mycelial trapping of nematodes, endoparasitic mycelial growth in nematodes and eggs, and production of metabolic toxins by the fungus. The fungi are cosmopolitan in distribution and belong to a divergent group of families in all the major taxonomic divisions. A detailed account of these fungi is given by Barron (1975). Most of these nematode-trapping fungi have had a long co-evolution with nematodes and appear to have developed a balanced interrelationship of co-existance, i.e., they seldom destroy the entire nematode biomass. Most of these fungi are also saprophytes and are strongly influenced by the amount and type of organic matter and its decomposition in soils. In addition, organic matter also accelerates many other microbial decomposition processes and may yield metabolites which are toxic to soil nematodes. The mechanisms of these complex interactions are not fully understood. Following the microbial decomposers are a series of protozoa and other invertebrates, all capable of devouring all nematodes in their specific zone of activity. New genetic engineering techniques are on the horizon that might be used for the improvement of effectiveness of biological control agents once their modes of action and the genes which control their action are identified (Napoli & Staskawicz, this volume). In the future it may be possible to introduce genes responsible for the production of specific enzymes or toxins which control nematodes into known antagonists or even in non-antagonists such as soil-rhizosphere bacteria. There are numerous prospects but their realization in the near future will be dependent on a major increase in research manpower and financial support over our current efforts. Concurrent with an increase in effort there needs to be a shift in philosophy from "near eradication" to "management." It has taken nearly 15 years to accomplish such a philosophical change in the use of non-fumigant nematicides in IPM programs. There also has to be a recognition that total reliance on biological control agents is unlikely to be effective but that biological control as one of several tactics integrated into a total management system can be effective. Integration may include the use of nematicides, resistant varieties and cultural practices along with biological control.

Rather than review all the potential antagonists of nematodes which are already thoroughly reviewed in the literature (Kerry 1981a, b, 1984; Mankau 1980, 1981; Sayer 1971, 1980; Stirling & Mankau 1977), I would prefer to critically analyze two specific important nematode-antagonist situations. Such a hypothetical critique may stimulate a new approach to biological control of nematodes.

Root Knot and Cyst Nematodes

The root knot nematodes (Meloidogyne spp.) are important plant parasitic nematodes in the subtropical and tropical regions of the world. The cyst nematodes (Heterodera spp.) are important plant-parasites in the temperate regions. Members of all of the major antagonists have been described affecting these groups. Numerous other IPM tactics are already in use for managing these pests.

Meloidogyne on an annual crop is illustrated in Fig. 1. The adult female of this obligate parasite is usually embedded in the plant host tissue and partially, or totally, protected from predation. The eggs and larvae are found in the soil and are available and susceptible to antagonists. The highest population of larvae and eggs is usually found at harvest or at the end of the crop season. During the fallow or dormant period (usually the

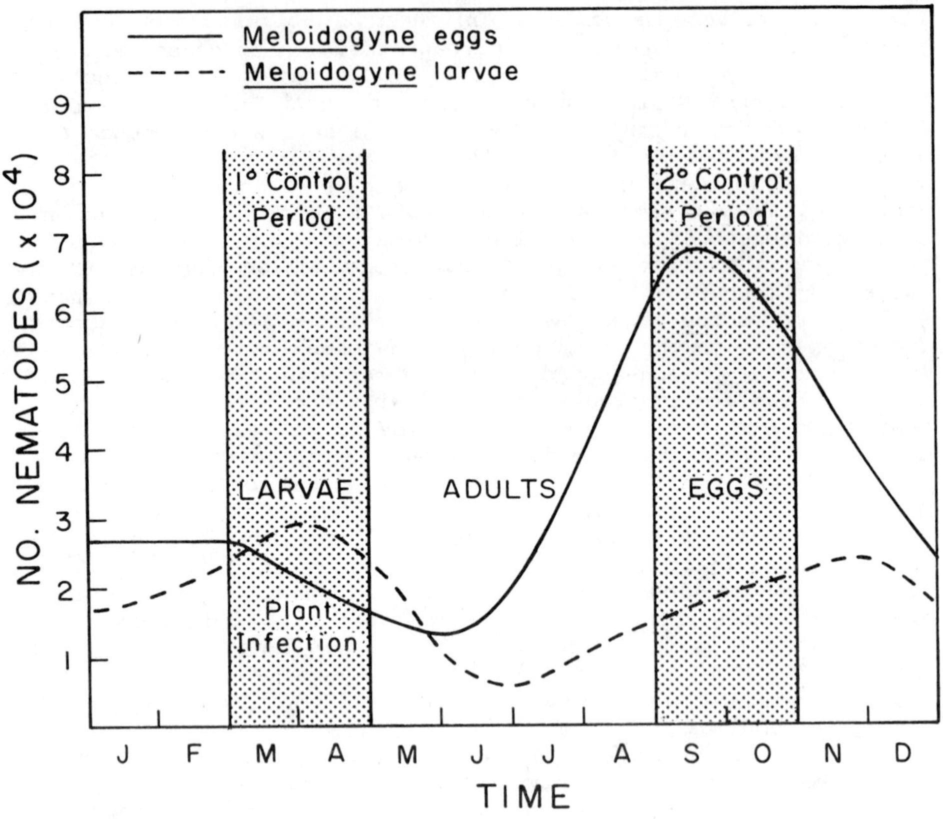

Fig. 1. Diagrammatic illustration of the population dynamics of *Meloidogyne* spp. and critical periods in the disease cycle when primary (1°) and secondary (2°) control practices are used to suppress the population.

coldest portion of the year) the population proceeds through a natural decline from starvation, natural predation and parasitism. Then when the new growing season or planting starts in the spring there is usually a slight rise in the population due to hatch of dormant eggs. This population represents the inoculum potential and the infective stage of the pest. The level of initial infection strongly influences crop yield during the growing season. If populations are low (soil fumigation), or infection is delayed (non-fumigant nematicides), crop yield is maximized. If no control action is taken at the primary control point, the crop may suffer extensive yield loss.

The secondary control action period is near the end or immediately following the growing season. The major benefit of this control action does not occur until the following growing season. Few long-term biological studies have been conducted to measure this carryover effect, although a few isolated observations have noted where root knot was not a serious problem even though it was expected based on previous crop history. Unfortunately, not enough of these observations have been followed and researched to determine what is happening ecologically. There is a serious need for more keen observers and researchers in the area of suppressive soils.

It seems appropriate to categorize the addition of antagonists on their mode of action and activity in the root zone against Meloidogyne. There are those organisms that need to be added or stimulated prior to planting. These include most nematode-trapping fungi, Bacillus penetrans, and microbial decomposers. Their mode of action is to attack or to deter nematode infection. The microbial decomposers and nematode trapping fungi also are enhanced by the addition of organic matter and/or nutrient base so that there is "biomass flush" and a general succession of organisms, many of which have the potential to suppress soil nematode populations. Timing is critical so that maximum predation and parasitism of soil nematode larvae takes place before planting. Additions at planting time, as is often done, are too late to be effective in reducing infection. Although not tested, it would seem that a stimulation of a "biomass flush" late in the growing season might be important in reducing the population during the next growing season. Again timing seems to be critical to this type of action. The use of B. penetrans needs to begin early so that infective larvae are parasitized and either fail to enter the root or so that those that enter become first generation adults that provide additional spores for inoculating reinfecting larvae (Stirling 1984). The lack of motile spores of Bacillus means that placement and concentration of the organism are extremely important.

Ideally, the addition of fungal parasites and invertebrate predators of nematode eggs and adults should take place during the second generation cycle of the nematode pest when eggs and adults are numerous in the rhizosphere (Fig. 1). This will have little effect on crop yield during this growing season but may be important in the subsequent season. Additions of these antagonists at planting time, although convenient, do not appear to be the optimum time for introduction into the agroecosystem.

Of all the antagonists, the rhizobacteria appear the best suited for addition at planting time (Schroth & Hancock, this volume; Zavaleta-Mejia & Van Gundy 1982). They provide some protection from infection by larvae. However, their activity becomes diluted with naturally occurring rhizobacteria and nematode infection is high at the end of the crop season. Crop yields, however, are increased in a similar fashion to non-fumigant nematicides. Recent findings by Merck & Co. (Campbell et al. 1983) suggest that the soil bacteria also may be a source for our new third generation nematicides. A fermentative product, avermectin, from Streptomyces avermitilis, is highly nematicidal.

In field tests it is 10 to 20 times more effective than our non-fumigant nematicides. It has a new mode of action by attacking the gaba system rather than the acetyl cholinesterase system. Certainly, if one soil organism can produce such an effective nematicide, there must be others unidentified in soil that can do the same or even better. The major drawback to avermectin is its insolubility in water. Organisms producing natural nematicides have a high potential for genetic engineering. Various genes for producing enzymes, toxins and even nematicides could be introduced into *Streptomyces*. In the future, with new biotechnology and the right organisms, it may be possible to engineer a rhizosphere-inhabiting organism that lives on the root surface as it grows through the soil and at the same time produces a toxin that in minute quantities kills, repels, or deters nematodes from attacking roots. Venture capital biotechnology firms are already investigating this avenue of biocontrol of nematodes and some product may already be in the developmental stage.

RECOMMENDATIONS

1. There needs to be a major infusion and redirection of research personnel and resources into the area of biological control of nematodes. At the same time, researchers need to focus on the applied aspects rather than the desciptive aspects.
2. Biological control of soil nematodes needs to be developed as a short-term practice (1-2 years) so that it will be more compatible with current agronomic and control strategies. Its results must also be more visible to the agricultural producers.
3. Research needs to focus on strategies that mesh better with the life cycle of the pest so that high concentrations of the antagonists are introduced during periods when ample populations of the pest are present for parasitism and predation.
4. Research needs to be directed to improving the efficiency of parasitism or predation of the biological control agents already identified. This includes the use of new biotechnological methods.
5. Research needs to focus on the activities and the organisms of microbial decomposition. Most of the fungi and bacteria that are so active against nematodes require supplemental energy to increase their establishment and activities. A combination of nutrient supplement and organisms timed to mesh with life cycles of the nematode pest and host may be possible.
6. New "third generation" nematicides should be sought from among the soil bacteria and fungi.
7. Systematic exploration and search for exotic parasites and predators of nematodes, or identification of more effective biotypes, should be conducted.

8. Mass culture techniques for production of important obligate parasites of nematodes should be developed.

REFERENCES

Barron, G.L. 1975. The nematode-destroying fungi. Can. Biol. Publ., Guelph, Ontario, Canada. 140 pp.

Campbell, W.C., M.H. Fisher, E.O. Stapley, G. Abers-Schonberg and T.A. Jacob. 1983. Ivermectin: A potent new antiparasitic agent. Science 221:823-828.

Goday, G., R. Rodriguez-Kabana, R.A. Shelby and G. Morgan-Jones. 1983. Chitin amendments for control of Meloidogyne arenaria in infested soil. II. Effects on microbial population. Nematropica 13:63-74.

Kerry, B.R. 1981a. Fungal parasites: A weapon against cyst nematodes. Plant Dis. 65:390-393.

Kerry, B.R. 1981b. Progress in the use of biological agents for control of nematodes, pp. 79-90. In: G.C. Papavizas (Ed.), Biological control in crop production. BARC Symposium No. 5. Allanheld, Osmun Publ., Totowa, N.J.

Kerry, B.R. 1984. Nematophagus fungi and the regulation of nematode populations in soil. Helminth. Abst. Ser. B 53:1-14.

Kerry, B.R., D.H. Crump and L.A. Mullen. 1980. Parasitic fungi, soil moisture and multiplication of the cereal cyst-nematode. Nematologia 26:57-68.

Kerry, B.R., D.H. Crump and L.A. Mullen. 1982. Natural control of the cereal cyst-nematode, Heterodera avenae Woll., by soil fungi at three sites. Crop Prot. 1:99-109.

Mankau, R. 1980. Biological control of nematode pests by natural enemies. Annu. Rev. Phytopathol. 18:415-440.

Mankau, R. 1981. Microbial control of nematodes, pp. 475-494. In: B.M. Zuckerman, W.F. Mai and R.A. Rohde (Eds.), Plant parasitic nematodes. Vol. III. Academic Press, N.Y.

Mankau, R., and S. Das. 1974. Effect of organic materials on nematode bionomics in citrus and root-knot nematode infested soil. Indian J. Nematol. 4:138-151.

Mian, I., G. Godoy, R.A. Shelby, R. Rodriguez-Kabana and G. Morgan-Jones. 1982. Chitin amendments for control of Meloidogyne arenaria in infested soil. Nematropica 12:71-84.

Rodriguez-Kabana, R., G. Morgan-Jones and B. Ownley Gintis. 1984. Effects of chitin amendments to soil on Heterodera glycines, microbial populations, and colonization of cysts by fungi. Nematropica 14:10-25.

Sayer, R.M. 1971. Biotic influences in soil environment, pp. 235-256. In: B.M. Zuckerman, W.F. Mai and R.A. Rohde (Eds.), Plant parasitic nematodes. Vol. I. Academic Press, N.Y.

Sayre, R.M. 1980. Promising organisms for biocontrol of nematodes. Plant Dis. 64:527-532.

Singh, R.S., and K. Sitaramaiah. 1973. Control of plant parasitic nematodes with organic amendments of soil. Exp. Sta. Res. Bull. Govind Ballabh Pant, Univ. Agric. & Tech., Pantnagar, India 6, 289 pp.

Stirling, G.R. 1984. Biological control of Meloidogyne javanica with Bacillus penetrans. Phytopathology 74:55-60.

Stirling, G.R., M.V. McKendry and R. Mankau. 1979. Biological control of root-knot nematodes (Meloidogyne spp. on peach). Phytopathology 69:806-809.

Stirling, G.R., and R. Mankau. 1977. Biological control of natural parasites of citrus by natural enemies. Proc. Intern. Soc. Citricult. 3:843-847.

Van Gundy, S.D. 1980. Nematology situation in the United States and priority areas of research. EPPO Bull. 10(3):393-402.

Zavaleta-Mejia, E., and S.D. Van Gundy. 1982. Effects of rhizobacteria on Meloidogyne infection. J. Nematol. 14:475A-475B

DISCUSSION: Biological control of nematodes: Status and prospects in agricultural IPM systems.

DISCUSSION LEADER: G.C. Smart

One limitation to the biological control of nematodes in agricultural crops could be due to the fact that nematodes occur as deep as four feet or more below the soil surface. The question was asked whether the reservoir of nematodes remaining deep in the soil could harm the current crop and the concensus was that, in most cases, nematodes are not important at these depths, but in specific cases they may be very important. Thus, the example was given for the stubby root nematode, Paratrichodorus christiei, which transmits tobacco rattle virus. This nematode is distributed down to the water table (which may be five feet) and when nematicides are applied to control nematodes in the upper 18 inches of soil, the reservoir of nematodes remaining causes severe disease in potatoes unless another nematicide is applied just before tubers are formed.

Status and Current Limits to Biological Control in Selected Commodity IPM Systems

CITRUS: CURRENT STATUS OF BIOLOGICAL CONTROL IN FLORIDA

Clayton W. McCoy

Successful biological control of cottony-cushion scale, Icerya purchasi, in Florida and California via the introduction of Rodolia cardinalis and Cryptochaetum iceryae around 1900 stimulated interest and continued research emphasis on classical biological control on citrus throughout this century. According to DeBach (1964), about 40% of the insects controlled biologically up to 1960 were soft scales, armored scales and mealybugs and these occurred mainly on citrus. However, other insects such as citrus blackfly and some species of whiteflies also have been controlled biologically, which implies a broader explanation for successful biological control on citrus. Certainly, the fact that citrus is a perennial host confers a degree of ecological stability that is advantageous to parasites, predators and pathogens. Additionally, subtropical and tropical climatic regions conducive to citrus production also favor survival of natural enemies. Whatever the reason(s), the bottom line appears to be, there is a high probability for success of classical biological control of pests on citrus.

If this hypothesis is incorrect however, future control of citrus pests promises to be costly. In both Florida and California, international commerce has flourished during the past two decades. This commerce involves areas of high population density, thereby increasing the likelihood of exotic pest introductions. According to Sailer (1983), at approximately three-year intervals a foreign pest of major importance becomes established in the U.S. Of these, Florida has more than their share as evidenced by the invasion of the citrus blackfly in 1976 and the Mediterranean fruit fly in 1980 and 1984. After an aborted eradication effort costing $15,000,000, the citrus blackfly was controlled through introduction of parasites, a program that cost less than $300,000. This example points to the very substantial benefits that result from successful application of classical biological control programs as well as the high cost of alternative methods having eradication as their objective. Fortunately, most of the potentially dangerous exotic pests of citrus not present in the United States are species against which classical biological control offers a high probability of success.

ISBN 0-12-357030-1

In the past 40 to 50 years, the present 850,000 acre citrus industry of Florida has evolved into a highly complex monocultural system. The pest complex common to this system is rather typical of humid citrus-growing regions of the world. Numerous phytophagous mites, armored and unarmored scales, mealy bugs, whiteflies, aphids and some opportunistic insects represent the arthropod fauna common to Florida citrus (Griffiths & Thompson 1957, Knapp 1981, McCoy et al. 1976a). A number of chronic virus diseases and other microorganisms occur, amd three phytopathogenic fungi of the fruit and/or foliage are widespread in distribution and play a major role in IPM. The fungus, Mycosphaerella citri, is generally recognized as the most important pest requiring chemical control regardless of marketing objectives.

Classical biological control has realized a number of spectacular successes in this century (Table 1). In addition, augmentation of natural enemies within a defined IPM strategy has been successful in research and extension programs (Knapp 1981, McCoy et al. 1976a). Although integrated pest management is being used in some areas of citrus production today in Florida, the overall systems approach is generally ignored. Citrus growers are by economic necessity strongly oriented to the philosophy of the agrochemical world. This does not mean that they are oblivious to biological control and what it can do for them. Rather, they prefer that it function independently within the programmed three to four annual pesticide applications targeted primarily to control greasy spot and citrus rust mite. Fortunately, this situation does not present a crisis to the industry as a result of pesticide resistance or serious secondary pest outbreaks. Some environmental problems have developed in the past, however, and may occur from time to time in the future.

CURRENT STATUS OF BIOLOGICAL CONTROL

Phytophagous Mites

Citrus rust mite Phyllocoptruta oleivora has been an important pest to citrus since the inception of commercial production in Florida (Hubbard 1885). The injury caused by the rust mite is the result of intra-epidermal feeding on fruit and foliage (McCoy & Albrigo 1975). Leaf and stem injury will, on occasion, cause premature defoliation (McCoy 1976). Fruit injury appears as different forms of surface discoloration and can affect external fruit quality (McCoy & Albrigo 1975), internal fruit quality (McCoy et al. 1976c), fruit drop (Allen 1978) and fruit size (Allen 1979). Pesticidal spray costs per year for citrus rust mite are estimated at 50 to 75 million dollars.

Yothers & Mason (1930) observed a number of predatory insects feeding on the citrus rust mite. Muma (1967) observed Semidalis vicina and Coniopteryx westwoodi feeding on adult mites and found them abundant only at high population densities. Muma & Selhime

Table 1. Cases of biological control of citrus pests by natural enemies in Florida in citrus this century.

Scientific Name	Principal Natural Enemies: Common Name	Name	Type	Origin & Date	Control Results (C = complete, S = substantial, P = partial)	Reference
Chrysomphalus aonidum	Florida red scale	*Aphytis holoxanthus*	Parasite	Hong Kong via California 1960	C	Clancy et al. 1963 Selhime et al. 1969
Lepidosaphes beckii	purple scale	*Aphytis lepidosaphes*	Parasite	Unknown 1958	C	Clancy & Muma 1959
Unaspis citri	snow scale	*Aphytis lingnanesis* (HK-1)	Parasite	Hong Kong via California 1971	S	Selhime & Brooks 1977 Riehl et al. 1980
Icerya purchasi	cottony-cushion scale	*Rodolia cardinalis*	Predator	Australia, via California 1899	C	Muma 1969a
Dialeurodes citri	citrus whitefly	*Ascheronia aleyrodis*	Pathogen	Florida 1977	S	Berger 1921
		Encarsia lahorensis	Parasite		P	McCoy et al. 1980
Dialeurodes citri	citrus whitefly	*Delphastus catalinae*	Predator	California 1917	P	Denmark 1977
Aleurocanthus woglumi	citrus blackfly	*Amitus hesperidum* *Encarsia opulenta* *Encarsia smithi*	Parasite	Mexico 1976	C	Hart et al. 1978 Selhime 1979 Cherry & Pastor 1980
Planococcus citri	citrus mealybug	*Cryptolaemus montrouzeri*	Predator	Australia via California 1930	P	Muma et al. 1975
Aphis spiraecola	spirea aphid	*Leis dimidata quiquedecimmaculata*	Predator	California 1925	P	Denmark 1964
Anastrepha suspensa	Caribbean fruitfly	*Biosteres longicaudatus*	Parasite		P	Baranowski 1977

(1971) concluded that despite the ability of the predatory mite Agistemus floridanus to feed and reproduce on citus rust mite, it too was not capable of regulating this eriophyid. Recently, a new phytoseiid, Euseius mesembrinus, was observed feeding on citrus rust mite; however, its potential is unknown in Florida (Rakha & McCoy, unpublished data).

Although biological control of citrus rust mite with predators appears limited due to their high reproductive potential during the spring and summer, further research on importation of exotic predators should be encouraged in view of the economic importance of the citrus rust mite and the fact that it has virtually never been undertaken.

The fungal pathogen, Hirsutella thompsonii, appears to be the most important natural enemy of the citrus rust mite. It was first reported in association with this eriophyid by Spears & Yothers (1924). Fisher (1950b) described the fungus as H. thompsonii, while both Fisher et al. (1949) and Muma (1955a, 1958) recorded dead mites containing internal hyphae suspected to be the fungus from the field. McCoy & Kanavel (1969) isolated the fungus and confirmed pathogenicity.

In Florida, natural epizootics caused by the fungus occur regularly during the summer. Diseased mites can be found on fruit and foliage throughout the year. Mites move from the old to the new foliage, to small immature fruit in April and increase during May (McCoy 1979). By mid-June, a few diseased mites appear on the new growth. As mite populations reach peaks after mid-June, epizootics develop regularly, lasting two to three weeks. Two to six diseased mites per square centimeter of fruit surface are required to stimulate a decline in mite population density. However, 30 to 50 live mites per square centimeter of fruit surface must be present to obtain the above level of disease. Obviously, high population densities of citrus rust mites are required to initiate field epizootics by H. thompsonii. The epizootics, however, result in a high fungal residue that usually prevents further mite buildup during the fall and winter (McCoy 1981, McCoy et al. 1976b, Riehl et al. 1980).

In the last decade, H. thompsonii has been scrutinized extensively for its biological control potential against the citrus rust mite and other Acarina on citrus, turf, coconut, and greenhouse crops (Gardner et al. 1982, Gerson et al. 1979, Hall et al. 1980, Litvinova et al. 1979, McCoy & Couch 1982, McCoy & Selhime 1977, McCoy et al. 1971, Santos et al. 1981, Urueta 1980, van Brussel 1975, Yen 1974). In Florida, two avenues of biological control research with the fungus have been studied simultaneously: microbial control and habitat manipulation. Both large scale laboratory mass production and industrial methods were developed using semi-solid and submerged fermentation to produce mycelial and conidial formulations of the fungus (McCoy 1981; McCoy et al. 1971, 1975). In 1975-76, Abbott Laboratories developed the first diphasic fermentation method for the fungus. Concurrently, safety testing against non-target organisms provided a data base necessary for registration (McCoy & Heimpel 1980). This research led to the registration of *H. thompsonii* under the trade name Mycar; it received full registration status in 1981.

Since 1969, both fragmented mycelial and conidial formulations of *H. thompsonii* have been applied experimentally in the field (McCoy 1981, McCoy & Couch 1982) with encouraging results. Since *H. thompsonii* is absent from the newly-formed foliage and fruit early in the growing season and becomes regulatory only at high mite population densities, the fungus was applied as a prophylatic control agent early in the season against lower rust mite populations. The key to this strategy was to get the fungus to persist in an infective state on the foliage while the mites built up. Conidiation of the fungal inoculum on foliage generally began to appear after 48 hours as small patches of grey growth under optimum weather conditions.

Results of field trials with commercial preparations of *H. thompsonii* applied at two to four lbs per acre have been effective when weather conditions were optimum for the establishment of the growing fungal propagules on citrus fruit and foliage. In all field trials, applications rarely resulted in lowering the mite population below that of the standard chemical acaricide. However, there was no difference in the percent fruit injury at the end of the season. The addition of medium oil at 0.5-1.0% appears to improve overall knockdown of the mite population and has no apparent detrimental effect on the fungus.

Since *H. thompsonii* sporulates only moderately, it would appear that growth and conidiation from the residue is the key to improving the field performance of Mycar in the future. In addition, adjuvants that stimulate growth and conidiation should be included in the spray when feasible. Considerable research in the area of microbial formulation with highly virulent pathotypes is needed to improve persistence and performance of the fungus in the field.

In addition to Mycar, Avermectin B_1, a natural product macrocylic lactone derived from *Streptomyces avermitilis*, has demonstrated high toxicity to the citrus rust mite at dosages of 0.02-0.04 oz ai/100 gal (McCoy et al. 1982). This biological pesticide appears to be somewhat selective to natural enemies having no detrimental effect to *H. thompsonii* and other entomogenous fungi.

Coincident to studies on the development and evaluation of Mycar as a mycoacaricide, three pest control programs (conventional, integrated and no spray) were tested on Valencia oranges in central Florida (McCoy et al. 1976b, Riehl et al. 1980). Augmentation of *H. thompsonii* against the citrus rust mite was designed as part of the integrated control program to reduce spraying on processed fruit. Numerous assumptions were made in the experiment due to data gaps relating to the pests and their influence on the crop. Fungicides such as copper, which are known to cause increases in citrus rust mite populations in the field (Griffiths & Fisher 1949, McCoy et al. 1976a) and benomyl, which is toxic to entomopathogenic fungi in the laboratory (McCoy et al. 1976a), were not used for greasy spot control. Medium oil, which is generally more innocuous to parasites, predators, and pathogens, was used exclusively. Acaricides known to be least disruptive were generally used to control citrus rust mite. Methidathion, a scalicide known to cause increases in citrus rust mite populations (McCoy 1977) and similar organophosphates, were not used in the study. The oxide form of manganese, which is known to affect the viability of conidia of H. thompsonii, was not used as a nutritional foliar spray (McCoy et al. 1976a).

During the four-year study, the use of selective pesticides and a simplified monitoring program reduced the need for chemical control of citrus rust mite and other arthropod pests. By reducing the number of sprays for citrus rust mite via monitoring, the activity of H. thompsonii was significantly increased in the summer and fall. As expected, the incidence of disease by H. thompsonii was highest in the no spray plots. Although rust mite populations were always higher in the integrated control program, no loss of yield was detected between the conventional and integrated programs after four years (McCoy et al. 1976a, b; Riehl et al. 1980).

Although many growers have applied the above strategy in their processed fruit program, data gaps relating to pest monitoring and long-term effects on the tree limit the biological control opportunities offered via environmental manipulation.

A citrus plant model is under experimental development (Allen 1977, Luck et al. 1980, Stamper & Allen 1979) and a predictive damage model for citrus rust mite damage has been developed for Valencia orange (Allen 1976, 1981). Intensified research is needed to test these models and add the natural enemy components to the total system. The use of H. thompsonii as a microbial insecticide and the augmentation of predators in the citrus system, whether introduced or otherwise, can be used most effectively only through the development of a complete model.

Texas citrus mite Of the species of spider mites found on Florida citrus, the Texas citrus mite, Eutetranychus banksi, is considered the most important today. Injury includes a scratching or etching of the leaves and, when tree moisture is critical, defoliation may occur. During heavy infestations, both fruit and leaves may experience russeting or bronzing. Injury thresholds are not known for any spider mite species.

Predators of the Texas citrus mite were reported by Kanavel & Selhime (1967) and Muma (1955a, 1965b). These included Galendromus floridanus, Paracheyletia bakeri, and Agistemus sp., generally considered to be ineffective as regulating natural control agents. Parasitic fungi were reported by Fisher (1954), Muma (1958) and Selhime & Muma (1966). These publications also dealt with a species described as Entomophthora floridana by Weiser & Muma (1966). According to Muma (1969), this fungus is the critical natural control factor affecting E. banksi populations on Florida citrus. Although its mode of infection is influenced by weather, it produces catastrophic and long lasting epizootics. Because of its localized importance, there are no current investigations on the biological control of Texas citrus mite.

Citrus red mite Panonychus citri occurs in most citrus growing areas of the world. Although its injury, which is similar to the Texas citrus mite, may cause defoliation, it has a narrower distribution than the Texas citrus mite in Florida.

The predators of the citrus red mite were studied by Kanavel & Selhime (1967), Muma (1955a, b, 1958), Muma et al. (1975) and Watson (1926). Although nine predators have been recognized, none, with the possible exception of some phytoseiids, were considered to be important in the natural control of the mite. No definitive studies have been made in Florida to establish more effective predators.

Parasitic fungi were recorded by Fisher (1951), Muma (1955a, 1958), Muma et al. (1975) and Selhime & Muma (1966). An undescribed species of Entomophthora sp. was reported to cause 70% mortality and produce natural control of citrus red mite (Muma 1955a). Selhime & Muma (1966) were successful in infecting citrus red mite with E. floridana. Attempts have been made to innoculate citrus red mite from Florida with a non-inclusion virus found in California with no success (A.G. Selhime, personal communication). Since citrus red mite populations are generally localized or developed because of pesticide misuse, no research is currently being conducted on the biological control of this mite.

Six-spotted mite Eotetranychus sexmaculatus is a sporadic problem on Florida citrus (Pratt & Thompson 1953). Mite populations generally reach a peak by May or June, then rapidly decline to noninjurious levels late in the early summer. They feed in restricted areas of the underside of the leaf nearly always along the midrib. The infested area becomes concave and yellow and the colony is covered with webbing. Leaf drop occurs with high populations.

Predators of the six-spotted mite were reported by Kanavel & Selhime (1967), Muma (1955a, 1958, 1965a), Muma et al. (1975), Selhime et al. (1963) and Watson (1926). Predators most commonly cited were Aleurodothrips fasciapennis, Chrysopa bicarnea, Paracheyletia bakeri, Scolothrips sexmaculatus, Stethorus utilis, and G. floridanus with the latter considered most important. Although destruction of six-spotted mite populations by predation was frequently observed, ecological studies indicate that temperature and humidity are also limiting.

Parasitic fungi were recorded by Muma et al. (1975) and Selhime & Muma (1966). Under grove conditions, an undescribed species of Hirsutella occasionally causes mycosis of six-spotted mites. In the laboratory, Entomophthora floridana was found to infect the mite. Fungi are generally not considered important in the natural control of the six-spotted mite. There are no investigations on the biological control of this mite at this time due to its minor pest status.

Armored Scales A number of armored scale insects have been important pests of citrus in Florida. As previously mentioned, successful classical biological control of Florida red scale and purple scale have reduced their importance significantly today. Generally speaking, citrus snow scale, Unaspis citri, is considered the only armored scale on citrus not adequately controlled by beneficial insects, requiring additional research.

Florida red scale Prior to 1963, Florida red scale was of major importance in Florida. Heavy infestations caused defoliation and fruit injury.

A number of predators have been reported attacking Florida red scale, the more important being Chilocorus stigma, Microweisea coccidivora, C. bicarnea, and A. fascipennis (Muma 1955c, Muma et al. 1975 and Selhime et al. 1963). Although early researchers felt that entomogenous fungi contributed to control of Florida red scale, that is no longer the case (Fisher 1950a, Selhime & Brooks 1977). Prior to 1961, the parasitic wasps, Pseudhomalopoda prima and Aphytis chrysomphali were the most common parasites of Florida red scale, producing satisfactory natural control. In 1960, following reports of success in Israel, A. holoxanthus was introduced into Florida from Riverside, California where the parasite had been reared in large numbers. Five shipments totaling 15,500 adult parasites were received by the USDA and released in seven locations throughout Florida. By the summer of 1964, A. holoxanthus had become dispersed over the entire citrus area and had reduced Florida red scale populations to extremely low levels (Clancy et al. 1963). A. holoxanthus has essentially replaced P. prima in Florida as the predominant parasite of Florida red scale (Selhime et al. 1969). This example of classical biological control of an armored scale is recognized in Florida as a complete success (Table 1).

Purple scale This scale occurs in all citrus growing areas of Florida and at one time was considered a serious pest of both fruit and foliage.

Predators common to Florida red scale also have been found attacking purple scale (Muma et al. 1975, Selhime et al. 1963). Parasitic fungi were also reported attacking purple scale (Muma & Clancy 1961). Hirsutella besseyi and Myiophagus ucrainicus are considered to be the most important fungi affecting purple scale. The hymenopterous parasite, Aphytis lepidosaphes, was found fortuitously in Florida in 1957 and 1958 for the first time (Clancy & Muma 1959). Within a short time, it was established in all citrus areas and reduced purple scale populations to low densities (Selhime & Brooks 1977).

Citrus snow scale For many years, *Unaspis citri* was localized in Florida in a small area in Orange and Seminole Counties. The distribution of the scale increased rapidly during the late 1950's and throughout the 1960's. Heavy infestations cause twig dieback and loss of tree vigor.

Predators and fungi common to other armored scales have also been found attacking citrus snow scale (Muma 1969). Indigenous parasites of the snow scale are *Aphytis* sp. near *lingnanensis* and *Aspidiotiphagus citrinus* (Muma et al. 1975).

In 1971, through collaboration with Paul DeBach, Univ. of California, Riverside, exotic parasites from China were supplied to the USDA, Orlando in low numbers. Subsequently, a culture of *A. lingnanensis* designated HK-1 strain was established in the insectary at Lake Alfred (Selhime & Brooks 1977). From a modest beginning of nine adult parasites, the culture was expanded and releases were initiated throughout the state. Over 6.5 million parasites have been released in Florida. Although parasite establishment appeared widespread, biological control of the scale has been partial. In addition to the HK-1 strain of *A. lingnanensis*, two other strains have been released in Florida: *A. lingnanensis* found on California red scale and a hybrid (R^2-65-23) (a cross between *A. lingnanensis* native to Florida and R-2002 from Puerto Rico).

Since the *A. lingnanensis* strain native to Florida cannot be distinguished taxonomically from the HK-1 strain and other sibling species of the complex, a major limitation to this biological control campaign has involved the recognition of sibling species. Until better diagnostic methods such as biochemical techniques are developed for the separation of sibling species within the *A. lingnanensis* complex, a true measure of biological control of *U. citri* and other armored scales by the respective parasites will be difficult.

Other armored scale insects Glover scale, chaff scale, yellow scale, and a few less important armored scale insects have a few predators and entomogenous fungi common to all armored scales and these exhibit some degree of natural control. However, *Aphytis citrinus*, *A. lounsburyi*, *A. lingnanensis*, *A. hispanicus*, and *A. chrysomphali* appear to be the primary natural enemies controlling these scales (Selhime & Brooks 1977).

Unarmored Scales and Other Coccids Brown soft scale, green scale, Florida wax scale, and black scale have a wide distribution on the foliage of all citrus varieties. These species frequently generate enough honeydew to cause serious sooty mold buildup on fruit. Black scale appears to be the most important, particularly on grapefruit on the east coast where secretions of honeydew produce a seasonal buildup of sooty molds (McCoy & Selhime 1971).

Numerous predators, parasites and entomogenous fungi, many common to all soft scales, have been reported (Muma et al. 1975).

In 1970-71, the USDA Horticultural Laboratory, Orlando, FL, introduced Metaphycus helvolus, a primary parasite of Saissetia oleae, from Fillmore, California. Following the release of less than 2,000 parasites in five locations of Florida, the parasite failed to become established according to subsequent survey. On numerous occasions, however, encapsulated M. helvolus eggs were detected in second stage S. oleae. However, concurrent taxonomic studies of the Saissetia complex by DeLotto (1971) revealed that S. oleae was not the predominant black scale in Florida and therefore M. helvolus is the wrong parasite on the wrong host. Today, survey shows that S. negelecta is the predominant scale species on Florida citrus. In the future, parasites of S. neglecta should be introduced into Florida.

Citrus Mealybug Populations of citrus mealybug differ greatly in intensity from year to year. Scymnus flavifrons and Cryptolaemus montrouzeri, an exotic predator introduced into Florida in 1930, are among the more important predators (Muma et al. 1975). Two hymenopterous parasites are considered to be the most important natural enemies of the citrus mealybug. Because these parasites, Leptomastidea abnormis and Leptomastix dactylopii, are so effective little biological control research has been considered on mealybug.

Aphids Although aphids were considered to be major insect pests during the 1920's, they are of minor importance now even though tristeza is transmitted by the green citrus aphid, Aphis spiraecola, in Florida (Griffiths & Thompson 1957). Aphid injury is usually severe only during the period of spring growth on new leaves. Varieties such as Temple and tangerine are most susceptible.

Numerous natural enemies reduce aphid populations on citrus; ladybeetle predators being the most conspicuous. Cycloneda sanguinea is the most common followed by Hippodamia convergens and Scymnus spp. The chinese ladybeetle, Leis dimidiata quinquedecimmaculata, was introduced into Florida from California in 1925 (Denmark 1964). Lysiphlebus testaceipes is parasitic on adult aphids and the fungus, Entomophthora fresenii, is infectious under favorable weather conditions.

Whiteflies The citrus whitefly, cloudy-winged whitefly, and wooly whitefly are the major injurious species on Florida citrus. Other species are found but are seldom injurious (Muma 1969). A number of predators have been identified attacking the various whitefly species. The species mentioned most frequently are Delphastus catalinae, D. pusillus, C. bicarnea and A. fascipennis (Selhime et al. 1963). None of these predators appear to regulate either citrus or cloudy-wing whitefly. Parasitic fungi have been recorded by Berger (1921) as effective natural control factors of citrus and cloudy-wing whitefly. Species most frequently cited were Aegerita webberi, Aschersonia aleyrodis and A. goldiana. Some have felt that chemical control is generally not needed for whitefly because of the biological control exerted by Aschersonia sp.

Since Aschersonia generally become regulatory after citrus whitefly reaches high densities each year, the hymenopterus parasite Encarsia (= Prospaltella) lahorensis was introduced into the citrus-growing regions in the fall of 1977 and the winter of 1978 (McCoy et al. 1980). By the fall of 1978, the parasite was well-established in the Lakeland-Auburndale area on citrus and ornamental plants. Since that time, the parasite has spread throughout Florida. Although biological control with E. lahorensis has been successful on ornamental plants, it is premature to assess the degree of biological control on commercial citrus.

Interestingly, with the decline of citrus whitefly on ornamentals and citrus, cloudy-winged whitefly, Dialeurodes citrifolii, appears to be increasing in its more preferred southern area of distribution. An excellent opportunity for classical biological control exists for this species through parasite introductions from southeast Asia. Two species of parasitoids, Encarsia perstrenua and E. strenua, are known to attack cloudy-winged whitefly in this region.

Citrus Blackfly In January 1976, the citrus blackfly was detected for the second time in south Florida. The infestation was confined to urban areas of the Fort Lauderdale and Miami region. An eradication effort was conducted using acephate and malathion foliar sprays at the outer limits of the infestation while biological control was initiated in the center of the infestation (Reinert 1976). Between April and November 1976, several species of hymenopterus parasites including Amitus hesperidum, Prospaltella clypealis, E. opulenta, and E. smithi were introduced into the area. Approximately 22,000 A. hesperidum and 2,500 E. opulenta were released in the central area of the blackfly infestation. Although both A. hesperidum and E. opulenta were established initially, A. hesperidum was dominant and most widespread in its distribution during the first part of the program (Hart et al. 1978). It is interesting to note, however, that E. opulenta became predominant once blackfly populations reached low levels (Cherry & Pastor 1980, Dowell et al. 1981, Selhime 1979). Today, E. opulenta performs very well at low host densities even though A. hesperidum and E. smithi remain established.

Fruit Flies The Mediterranean fruit fly, Ceratitis capitata, has entered Florida and has been succesfully eradicated four different times since it was reported for the first time in Florida in 1929. Anastrepha suspensa entered Florida in the 1960's and have spread throughout the southern part of the citrus growing region. Although citrus is not a preferred host, A. suspensa will attack mature fruit.

Sterile insect releases, male annihilation and the use of

protein hydrolysate bait-insecticide sprays are the primary methods for control of tephritids infesting citrus (Baranowski 1977). Numerous attempts to reduce *A. suspensa* populations by release of the hymenopterous parasite, *Biosteres longicaudatus*, has been partially successful (Baranowski 1977).

Citrus Root Weevils In Florida, many species of root weevils occur on commercial citrus; however, six are considered predominant (Woodruff 1979). Although no accurate information is available on the annual economic loss caused by these weevils, larval feeding by the Fullers rose weevil, *Pantomorus cervinus*, the little leaf notcher, *Artipus floridanus*, the citrus root weevils, *Pachnaeus* spp., and the so-called West Indian sugarcane rootstock borer weevil, *Diaprepes abbreviatus*, will reduce tree vigor. The latter species, which also attacks sugarcane, is a more recent introduction into Florida and has received considerable attention from the standpoint of eradication and subsequent chemical control (Schroder & Beavers 1977).

Numerous natural enemies have been reported attacking all developmental stages of root weevils. Many species of ants (Buren & Whitcomb 1977), a predatory mite *Blattisocius keegani* (Beavers et al. 1972), and three species of parasites (Baranowski 1960, Beavers & Selhime 1975) attack the eggs of mainly *D. abbreviatus*. Other predatory arthropods and birds also prey on larvae and adult weevils (Schroder & Beavers 1977, Whitcomb et al. 1982). The entomopathogenic fungi *Metarhizium anisopliae* and *Beauveria bassiana* have been reported from the larvae and adult stages of *D. abbreviatus* in Puerto Rico and Florida (Beavers et al. 1972, Wolcott 1952). In a recent survey, *M. anisopliae* was frequently found infecting the larvae and adults of *Pachnaeus* spp., *P. cervinus*, *Artipus floridanus*, and *D. abbreviatus*. In addition, the fungi *B. bassiana*, *Paecilomyces lilacinus*, and *Aspergillus ochraceous*, in descending order of occurrence, were isolated from *D. abbreviatus* larvae and found to be pathogenic to the larvae of *A. floridanus*, *P. litus*, and *D. abbreviatus*. Nematodes of the genera *Heterorhadditis* and *Steinernema* (= *Neoplectana*) were also found parasitizing the larvae of *Diaprepes*. The fungi and nematodes were most abundant from June through August when weevil larvae are most abundant in the soil and seasonal rainfall highest (Beavers et al. 1982, 1983). Present biological control research is addressing the fungi attacking the larval and adult stages of the citrus root weevil complex. Particular attention should be given to the development of these fungi and nematodes as biological insecticides and devising an IPM strategy for root weevil larval control since no soil insecticides are registered for larvae.

Plant Nematodes Plant-parasitic nematodes co-exist with many other organisms in the rhizosphere, and together they form a complex ecological community in citrus groves (see Van Gundy, this volume). The soil organisms known to parasitize and prey on nematodes include bacteria, viruses, fungi, protozoa, turbellarians, tardigrades, nematodes, collembola and mites (Stirling & Mankau 1977). Most of the interest in biological control of nematodes on citrus has centered on the nematode-trapping fungi. DeWolfe et al. (1954) observed Arthrobotrys oligospora and A. dactyloides preying on Tylenchulus semipenetrans, particularly in mulched citrus groves. In Florida, Tarjan (1961) applied nematophagus fungi and organic matter to citrus in pre- and post-plant treatments but failed to reduce the numbers of Radopholus similis in greenhouse and field tests. Predatory nematodes, mainly Mylonchulus spp., are capable of feeding on large numbers of citrus nematodes (Stirling & Mankau 1977); however their potential as biological control agents remain unknown.

In the future, it would appear that management of parasites of citrus will require the use of biocontrol agents as part of an integrated pest management system designed to minimize losses to Florida citrus. The loss of fumigants and ability of nematicides to enter groundwater make their long-term use somewhat tenuous.

Weeds Biological control of weeds of citrus in Florida has received very little attention. A classical program to introduce natural enemies of Lantana has been recommended by the citrus industry; however, the nursery industry recognizes Lantana as an ornamental and has discouraged the plan.

The milkweed vine, Morrenia odorata, is a major pest in many Florida citrus groves. A native pathotype of the parasitic fungus Phytophthora citrophthora has been developed as a biological insecticide by Abbott Laboratories under the trade name Devine (Burnett et al. 1973). This slow-acting mycoherbicide causes stem rot and decline of the plant and has been used in the citrus industry for nearly 10 years (see also Charudattan, this volume).

Recent biological studies have shown that allelochemics isolated from Lantana are toxic to Morrenia (Achhireddy & Singh 1984). The area of allelopathy appears to have excellent potential for biological control of weeds in the future.

Plant Pathogens Greasy spot, caused by Mycosphaerella citri, is the most important defoliating pest of citrus. This fungus causes lesions on the under surface of leaves of citrus and is particularly severe on grapefruit. No natural enemies are known to attack M. citri. As previously mentioned, some fungicides used to control greasy spot are also detrimental to the parasitic fungi that attack insects and mites. Obviously, one of the limitations to biological control of arthropod pests in Florida citrus is the disruption caused by fungicides used for greasy spot and other plant parasitic fungi. Development of a fungicide which controls phytopathogens without affecting entomopathogens may be difficult since most of the important ones are either ascomycetes or deuteromycetes. Melanose and scab, both fungal diseases that attack and cause lesions on the fruit, have no known natural enemies and no biological control studies have been undertaken. Other more chronic diseases such as tristeza, blight, foot rot, and exocortis, that result in tree decline, are being researched biologically through use of resistant varieties, modification of cultural practice, and budwood certification programs.

RECOMMENDATIONS FOR RESEARCH IN BIOLOGICAL CONTROL OF FLORIDA CITRUS PESTS

1. Introduce exotic insect and mite predators for the biological control of the citrus rust mite.
2. Develop improved formulations and spray technology to maximize the field efficacy of the mycoacaricide, Mycar, for control of the citrus rust mite.
3. Develop novel selective acaricides, insecticides, and fungicides to maximize the conservation of indigenous natural enemies.
4. Determine the long-term effect of greasy spot disease and citrus rust mite independently on citrus varieties harvested for the processing market.
5. Develop models of the greasy spot-citrus rust mite-natural enemy interaction to better define key variables that can be manipulated to maximize alternative pest control by biological control agents.
6. Conduct biosystematic and fundamental genetic studies on specific natural enemies of key economic pests, particularly parasites of the *Aphytis lingnanensis* complex.
7. Introduce exotic natural enemies for the control of citrus snow scale, black scale, cloudy-winged whitefly and citrus root weevils.
8. Develop entomogenous fungi and entomophagous nematodes as biological control agents of citrus root weevils and determine the ecological factors that influence their survival in the soil.
9. Determine the long-term effect of key parasitic nematodes on citrus and expand biological control research particularly in the area of augmentation of existing natural enemies.
10. Investigate the role of allelopathy as a potential measure for the biological control of *Lantana* and other weeds.

REFERENCES

Achhireddy, N.R., and M. Singh. 1984. Allelopathic effects of lantana on milkweed vine. Weed Sci. 33:(In Press).

Allen, J.C. 1976. A model for predicting citrus rust mite damage on Valencia orange fruit. Environ. Entomol. 5:1083-1088.

Allen, J.C. 1977. A tree model for stimulating pest management strategies in citrus. Proc. Intern. Soc. Citricult. 2:507-511.

Allen, J.C. 1978. The effect of citrus rust mite damage on citrus fruit drop. J. Econ. Entomol. 71:746-750.

Allen, J.C. 1979. Effect of citrus rust mite damage on citrus fruit growth. J. Econ. Entomol. 72:195-201.

Allen, J.C. 1981. The citrus rust mite game: A simulation model of pest losses. Environ. Entomol. 10:171-176.

Baranowski, R.M. 1960. Notes on a parasite of the citrus root weevil Pachnaeus litus. Fla. Entomol. 43:197.

Baranowski, R.M. 1977. Current control and management programs for Tephritidae. Proc. Intern. Soc. Citricult. 2:474-481.

Beavers, J.B., H.A. Denmark and A.G. Selhime. 1972. Predation by Blattisocius keegani on egg masses of Diaprepes abbreviatus in the laboratory. J. Econ. Entomol. 65:1483-1484.

Beavers, J.B., D.T. Kaplan and C.W. McCoy. 1982. Natural enemies of Diaprepes abbreviatus larvae in Florida. Proc. Fla. State Hort. Soc. 95:150-158.

Beavers, J.B., C.W. McCoy and D.T. Kaplan. 1983. Natural enemies of subterranean Diaprepes abbreviatus (Coleoptera: Curculionidae) larvae in Florida. Environ. Entomol. 12:840-843.

Beavers, J.B., and A.G. Selhime. 1975. Further attempts to establish the weevil egg parasite, Tetrastichus haitiensis in Florida. Fla. Entomol. 58:29-31.

Berger, E.W. 1921. Natural enemies of scale insects and whiteflies in Florida. Quart. Bull. Fla. State Plant Bd. 5:141-154.

Buren, W.F., and W.H. Whitcomb. 1977. Ants of citrus: Some considerations. Proc. Intern. Soc. Citricult. 2:496-498.

Burnett, H.C., D.P.H. Tucker, M.E. Patterson and W.H. Ridings. 1973. Biological control of milkweed vine with a race of Phytophthora citrophthora. Proc. Florida State Hort. Soc. 86:111-115.

Cherry, R., and S. Pastor. 1980. Variations in population levels of citrus blackfly, Aleurocanthus woglumi (Hom: Aleyrodidae) and parasites during an eradication program in Florida. Entomophaga 25:365-368.

Clancy, D.W., and M.H. Muma. 1959. Purple scale parasite found in Florida. J. Econ. Entomol. 52:1025-1026.

Clancy, D.W., A.G. Selhime and M.H. Muma. 1963. Establishment of Aphytis holoxanthus as a parasite of Florida red scale in Florida. J. Econ. Entomol. 56: 603-605.

DeBach, P. 1964. Successes, trends, and future possibilities, pp. 673-713. In: P. DeBach (Ed.), Biological control of insect pests and weeds. Reinhold Publ., N.Y.

DeLotto, G. 1971. A preliminary note on the black scales of North and Central America. Bull. Entomol. Res. 61:375-376.

Denmark, H.A. 1964. Florida Department of Agriculture's position on parasites and predators introduced into Florida for biological control. Hyacinth Control J. 3:16-19.

DeWolfe, T.A., L.J. Klotz, R.C. Baines and P.W. Moore. 1954. Citrus nematode activity reduced by fungus plants. Citrus Leaves 34:11.

Dowell, R.V., D. Puckett and M. Johnson. 1981. Searching and ovipositional behavior of Amitus hesperidum and Encarsia opulenta parasitoids of the citrus blackfly. Entomophaga 26:233-239.

Fisher, F.E. 1950a. Entomogenous fungi attacking scale insects and rust mites on citrus in Florida. J. Econ. Entomol. 43:305-309.

Fisher, F.E. 1950b. Two new species of Hirsutella Pat. Mycologia 42:290-297.

Fisher, F.E. 1951. An entomophthora attacking citrus red mite. Fla. Entomol. 34:83-88.

Fisher, F.E. 1954. Diseases of citrus insects. Fla. Agric. Exp. Sta. Ann. Rep. 1953, 191 pp.

Fisher, F.E., J.T. Griffiths and W.L. Thompson. 1949. An epizootic of Phyllocoptruta oleivora Ashm. on citrus in Florida. Phytopathology 39:510-512.

Gardner, W.A., R.D. Oetting and G.K. Storey. 1982. Susceptibility of the two-spotted spider mite, Tetranychus urticae Koch, to the fungal pathogen Hirsutella thompsonii Fisher. Fla. Entomol. 65:458-465.

Gerson, U., R. Kenneth and T.I. Muttath. 1979. Hirsutella thompsonii, a fungal pathogen of mites. II. Host-pathogen interactions. Ann. Appl. Biol. 91:29-40

Griffiths, J.T., and F.E. Fisher. 1949. Residues on citrus trees in Florida. J. Econ. Entomol. 42:829-838.

Griffiths, J.T., and W.L. Thompson. 1957. Insects and mites on Florida citrus. Fla. Agr. Exp. Sta. Bull. 591:1-96.

Hall, R.A., N.W. Hussey and D. Marian. 1980. Results of a survey of biological control agents of the coconut mite Eriophyes guerreronis. Oleagineux 35:395-400.

Hart, W.G., A.G. Selhime, D.P. Harlan, S.J. Engle, R.M. Sanchez, R.H. Rhode, C.A. Garcia, J. Caballero and R.L. Garcia. 1978. The introduction and establishment of parasites of citrus blackfly, Aleurocanthus woglumi, in Florida. Entomophaga 23:361-366.

Hubbard, H.G. 1885. Rust of the orange. USDA Rep., pp. 361-373.

Kanavel, R.F., and A.G. Selhime. 1967. Biological studies on Paracheyletia bakeri (Acarina: Cheyletidae). Fla. Entomol. 50:117-123.

Knapp, J.L. 1981. Integrated pest management in Florida citrus. Misc. Publ. Entomol. Soc. Amer. 12(2):91-98.

Litvinova, M.N., O.H. Aleshina and T.A. Deysadez. 1979. Report on testing of an entomopathogenic preparation based on Hirsutella thompsonii obtained from the USA. Scientific Res. Rep., All Union Scientific Res. Inst., Moscow, USSR. pp. 13-18.

Luck, R.F., J.C. Allen and D. Baasch. 1980. A systems approach to research and decision-making in the citrus ecosystem, pp. 366-396. In: C.B. Huffaker (Ed.), New technology of pest control. John Wiley & Sons, N.Y.

McCoy, C.W. 1976. Leaf injury and defoliation caused by the citrus rust mite Phyllocoptruta oleivora. Fla. Entomol. 59:403-410.

McCoy, C.W. 1977. Resurgence of citrus rust mite populations following application of methidathion. J. Econ. Entomol. 70:748-752.

McCoy, C.W. 1979. Migration and development of citrus rust mite on the spring flush of Valencia orange. Proc. Fla. State Hort. Soc. 92:48-51.

McCoy, C.W. 1981. Pest control by the fungus Hirsutella thompsonii, pp. 499-512. In: H.D. Burgess (Ed.), Microbial control of insects and mites. Vol. 2. Academic Press, London.

McCoy, C.W., and L.G. Albrigo. 1975. Feeding injury to the orange caused by the citrus rust mite, Phyllocoptruta oleivora. Ann. Entomol. Soc. Amer. 68:289-297.

McCoy, C.W., R.F. Brooks, J.C. Allen and A.G. Selhime. 1976a. Management of arthropod pests and plant diseases in citrus agroecosystems. Proc. Tall Timbers Conf. Ecol. Animal Control Habitat Mgt. 8:1-17.

McCoy, C.W., R.F. Brooks, J.C. Allen, A.G. Selhime and W.F. Wardowski. 1976b. Effect of reduced pest control programs on yield and quality of Valencia orange. Proc. Fla. State Hort. Soc. 89:74-77.

McCoy, C.W., R.C. Bullock and R.A. Dybas. 1982. Avermectin B_1: A novel miticide active against citrus mites in Florida. Proc. Fla. State Hort. Soc. 95:51-56.

McCoy, C.W., and T.L. Couch. 1982. Microbial control of the citrus rust mite with the mycoacaricide, Mycar. Fla. Entomol. 65:115-126.

McCoy, C.W., P.L. Davis and K.A. Munroe. 1976c. Effect of late season fruit injury by the citrus rust mite, *Phyllocoptruta oleivora* on the internal quality of Valencia orange. Fla. Entomol. 59:335-341.

McCoy, C.W., and A.M. Heimpel. 1980. Safety of potential mycoacaricide, *Hirsutella thompsonii*, to vertebrates. Environ. Entomol. 9:47-49.

McCoy, C.W., A.J. Hill and R.F. Kanavel. 1975. Large-scale production of the fungal pathogen *Hirsutella thompsonii* in submerged culture and its formulation for application in the field. Entomophaga 20:229-240.

McCoy, C.W., and R.F. Kanavel. 1969. Isolation of *Hirsutella thompsonii* from the citrus rust mite, *Phyllocoptruta oleivora* and its cultivation on various synthetic media. J. Invertebr. Pathol. 14:386-390.

McCoy, C.W., R.I. Sailer and J.L. Knapp. 1980. History and establishment of the parasite *Prospatella lahorensis*, for the biological control of citrus whitefly in central Florida. Proc. Fla. State Hort. Soc. 93:47-49.

McCoy, C.W., and A.G. Selhime. 1971. Influence of some natural enemies on black scale, *Saissetia* spp., in Florida. J. Econ. Entomol. 64:213-217.

McCoy, C.W., and A.G. Selhime. 1977. The fungus pathogen, *Hirsutella thompsonii* and its potential use for control of the citrus rust mite in Florida. Proc. Intern. Citrus Congr. 2:521-527.

McCoy, C.W., A.G. Selhime, R.F. Kanavel and A.J. Hill 1971. Suppression of citrus rust mite populations with application of fragmented mycelia of *Hirsutella thompsonii*. J. Invertbr. Pathol. 17:270-276.

Muma, M.H. 1955a. Factors contributing to the natural control of citrus insects and mites in Florida. J. Econ. Entomol. 48:432-438.

Muma, M.H. 1955b. Lady beetles found on citrus in Florida. Fla. Entomol. 38: 117-124.

Muma, M.H. 1955c. Phytoseiidae (Acarina) associated with citrus in Florida. Ann. Entomol. Soc. Amer. 48:262-272.

Muma, M.H. 1958. Predators and parasites of citrus mites in Florida. Proc. 10th Intern. Congr. Entomol. 4:633-648.

Muma, M.H. 1965a. Biological control of mites on subtropical fruit trees. Proc. 2nd Intern. Congr. Acarol. 196:373-382.

Muma, M.H. 1965b. Populations of common mites in Florida citrus groves. Fla. Entomol. 48:35-46.

Muma, M.H. 1967. Biological notes on *Coniopteryx vicina*. Fla. Entomol. 50:285-293.

Muma, M.H. 1969. Biological control of various insects and mites on Florida citrus. Proc. 1st Intern. Citrus Symp. 2:863-869.

Muma, M.H., and D.W. Clancy. 1961. Parasitism of purple scale in Florida citrus groves. Fla. Entomol. 44:159-165.

Muma, M.H. and A.G. Selhime. 1971. *Agistemus floridanus*, a predatory mite, on Florida citrus. Fla. Entomol. 54:249-258.

Muma, M.H., A.G. Selhime and H.A. Denmark. 1975. An annotated list of predators and parasites associated with insects and mites on Florida citrus. Fla. Agr. Expt. Sta. Bull. 634B, 46 pp.

Pratt, R.M., and W.L. Thompson. 1953. Spray programs, varieties, and weather conditions in relation to six-spotted mite and purple mite infestations. Proc. Fla. State Hort. Soc. 66:65-69.

Reinert, J.A. 1976. Citrus blackfly control by foliar treatments of dooryard citrus. Proc. Fla. State Hort. Soc. 89:365-366.

Riehl, L.A., R.F. Brooks, C.W. McCoy, T.W. Fisher and H.A. Dean. 1980. Accomplishments toward improving integrated pest management for citrus, pp. 320-363. In: C.B. Huffaker (Ed.), New technology of pest control. John Wiley & Sons, N.Y.

Sailer, R.I. 1983. History of insect introductions, pp. 15-38. In: C.L. Wilson and C.L. Graham (Eds.), Exotic plant pests and North American agriculture. Academic Press, N.Y.

Santos, L.M., S. Gravena and L.C. Donadio. 1981. Efeitos de productos quimicos e do fungo *Hirsutella thompsonii* no acaro da ferrugern do citros, *Phyllocoptruta oleivora*. An. Estac. Exper. Boguim 25:106-120.

Schroder, W.J., and J.B. Beavers. 1977. Citrus root weevils in Florida. Identification, biology and control. Proc. Intern. Soc. Citricult. 2:498-500.

Selhime, A.G. 1979. Biological control of citrus blackfly in south Florida. Proc. Fla. State Hort. Soc. 92:32-33.

Selhime, A.G., and R.F. Brooks. 1977. Biological control of some armoured scale insects on citrus. Proc. Intern. Soc. Citricult. 2:475-478.

Selhime, A.G., and M.H. Muma. 1966. Biology of *Entomophthora floridana* attacking *Eutetranychus banksi*. Fla. Entomol. 49:161-168.

Selhime, A.G., M.H. Muma and D.W. Clancy. 1963. Biological, chemical, and ecological studies on *Aleurodothrips fasciapennis* in Florida citrus groves. Ann. Entomol. Soc. Amer. 56:709-712.

Selhime, A.G., M.H. Muma, W.A. Simanton and C.W. McCoy. 1969. Control of Florida red scale in Florida with the parasite *Aphytis holoxanthus*. J. Econ. Entomol. 62:954-955.

Spears, A.T., and W.W. Yothers. 1924. Is there an entomogenous fungus attacking the citrus rust mite in Florida? Science 60:41-42.

Stamper, J.H., and J.C. Allen. 1979. A model of the daily photosynthetic rate in a tree. Agric. Meteorol. 20:459-481.

Stirling, G.R., and R. Mankan. 1977. Biological control of parasites of citrus by natural enemies. Proc. Intern. Soc. Citricult. 3:843-847.

Tarjan, A.C. 1961. Attempts at controlling citrus burrowing nematode using nematode trapping fungi. Proc. Soil Crop Sci. Soc. Fla. 21:17-36.

Urueta, E.J. 1980. Control de acaro *Retracrus elaeis* mediante el hongo *Hirsutella thompsonii* e inhibition de este por dos fungicidas. Rev. Columb. Entomol. 6: 25-31.

van Brussels, E.W. 1975. Interrelations between citrus rust mite, *Hirsutella thompsonii* and greasy spot on citrus in Surinam. Landbouwproefstation Surinam Agr. Sta. Bull. 98, 43 pp.

Watson, J.R. 1926. Citrus insects and their control. Univ. of Fla. Agr. Exp. Sta. Bull. 183:293-423.

Weiser, J., and M.H. Muma. 1966. *Entomophthora floridana* n.sp., a parasite of the Texas citrus mite, *Eutetranychus banksi*. Fla. Entomol. 49:155-159.

Whitcomb, W.H., T.D. Gowan and W.F. Buren. 1982. Predators of Diaprepes abbreviatus (Coleoptera: Curculionidae) larvae. Fla. Entomol. 65:150-158.

Wolcott, G.N. 1952. Control of soil inhabiting crops of Puerto Rico. J. Agr. Univ. Puerto Rico 34:333-337.

Woodruff, R.E. 1979. Florida citrus weevils (Coleoptera: Curculionidae). Fla. Dept. Agr. Cons. Serv., Div. Plant Ind., Entomol. Circ. No. 202.

Yen, H. 1974. Isolation of the filamentous fungus Hirsutella thompsonii from Phyllocoptruta oleivora. Acta Entomol. Sinica 17:225-226.

Yothers, W.W., and A.C. Mason. 1930. The citrus rust mite and its control. USDA Tech. Bull. 176:1-5.

BIOLOGICAL CONTROL IN SAN JOAQUIN VALLEY VINEYARDS

**Donald L. Flaherty, Lloyd T. Wilson,
Vernon M. Stern and Hiroshi Kido**

Interest in the biological control of vineyard pests increased in the late 1950's. By that time grape leafhopper, Erythroneura elegantula, had developed considerable resistance to the new synthetic organic insecticides, and there were accompanying serious biological upsets of spider mites and mealybugs. Biological control became an important part of intensive studies to develop a practical grape IPM program for San Joaquin Valley vineyards during the 1960's and 70's. In this paper we discuss the status and limits of biological control of major grape pests in the San Joaquin Valley and suggest research needed to enhance its further utilization.

GRAPE LEAFHOPPER

Grape leafhopper is the most common insect pest of grapes in the San Joaquin Valley. Economic damage is caused by leaf chlorophyll reduction, vine defoliation, marring of the surface of table grapes with excrement, and leafhoppers annoying pickers at harvest (Jensen & Flaherty 1981a).

Doutt & Nakata (1965, 1973) reported that large acreages of grapes planted near streams and rivers, where wild blackberries (Rubus ursinus and R. procerus) flourish, seldom require control for grape leafhopper. This is because of the activity of a minute wasp, Anagrus epos, which parasitizes the eggs of grape leafhopper and blackberry leafhopper, Dikrella californica, a noneconomic species whose eggs are present throughout the year on wild blackberries. Their studies showed that Anagrus overwinters in Dikrella eggs. Overwintering wasp populations tend to be largest in blackberries along rivers that have an overstory of sheltering trees. When the blackberries produce new foliage in February, heavy oviposition by Dikrella leafhoppers is stimulated. The parasite population increases enormously on the eggs so that by late March and early April there is widespread dispersal of newly-produced Anagrus adults. Fortunately, their dispersal occurs at the same time that grape leafhopper begins to

ISBN 0-12-357030-1

lay eggs in vineyards. Doutt et al. (1966) reported that vineyards located within 5.6 km of an established blackberry refuge will benefit immediately from the immigrant parasites. Vineyards distant from refuges may not show Anagrus activity until midsummer or later.

Commercial Blackberry Refuges

Efforts to establish effective blackberry refuges near commercial vineyards have not been successful. Doutt & Nakata (personal communication) determined that the output of Anagrus from a R. procerus refuge they established declined from an average of 2.5 parasitized Dikrella eggs per leaf in 1966 to 0.8 in 1967 to 0 in 1968. This decline was attributed to a lack of sheltering trees, as occurs in natural blackberry systems along rivers and streams. In full sun, R. procerus vineyards fare even worse as producers of Anagrus. Usually, their refuges are irrigated on vineyard schedules, which are inappropriate for blackberries, and with the resulting dry conditions Dikrella populations apparently die out. Observations by Flaherty & Kido (unpublished data) in 1980-81 suggested that dry conditions are, indeed, unfavorable for Dikrella populations. Comparing R. procerus brambles growing under various moisture regimes, they found Dikrella absent (leaf and D-Vac samples) in brambles growing in pastures where winter rains were the only source of moisture. Inoculation with Dikrella adults in April 1980 of one of these bramble patches resulted in a very low and short-lived summer population while nearby brambles growing in irrigated pastures supported much higher populations throughout the year.

Williams (1984) speculated, based on data from natural blackberry systems, that R. ursinus is clearly the preferable blackberry species for artificial refuges. To evaluate the effects of shading and berry species on Dikrella and Anagrus production, Flaherty & Kido established a study refuge in 1981. The study site is located near Kettlemen City on the west side of the San Joaquin Valley, about 120 km from the natural systems along the east side of the valley. Their aim was to have a refuge in an area where Anagrus did not, initially at least, influence Dikrella populations to the extent that the effects of blackberry species and shading would not be obvious. R. ursinus, California blackberry, is a native species which grows luxuriantly in shady, moist conditions. R. procerus, Himalaya blackberry, is a native of Europe that grows well in both full sun and shade. Adjacent, 3x8 m, lath-shaded and unshaded plantings of both berry species were established and inoculated with Dikrella adults in 1981. The refuges are irrigated independently of a large, adjacent vineyard, using a microjet, low volume system. Monthly leaf counts of Dikrella nymphs and eggs (parasitized and unparasitized) were initiated in January 1982.

Approximately three times as many Dikrella were produced in shaded refuges of both species than in nonshaded refuges. It is unclear from the data whether R. ursinus is more productive than R. procerus. Anagrus activity in the refuge has been erratic and its production low. Data suggest that, despite very high Dikrella levels, Anagrus activity may be short-circuited by periods of very low leafhopper egg production, particularly during the critical winter period. Studies are needed to determine if the periodic low production of Dikrella eggs is inherent in the environment of artificial refuges, and whether our thinking on effective overwintering of Anagrus has been wrongly locked into parasitism of Dikrella eggs in wild blackberries. Species of leafhoppers which overwinter as diapausing eggs on other host plants where blackberries flourish need consideration as important hosts of Anagrus. Such relationships would suggest that Anagrus emerges in late winter from other hosts to parasitize and build up on Dikrella eggs and subsequently spill over in early spring onto eggs of grape leafhopper in nearby wild grapes and vineyards. The following section in this paper lends credence to the importance of other hosts.

Other Plants as Refuges

Anagrus epos has been reported to attack several other leafhopper species. In Canada, McKenzie & Beirne (1972) found A. epos effective in controlling a grape-feeding leafhopper, Erythroneura ziczac, and they suggested augmenting natural control by planting wild rose and apple to harbor rose and apple leafhoppers, which are overwintering hosts of Anagrus.

In 1980, high parasitization of grape leafhopper eggs by Anagrus was observed in vineyards near Easton, Fresno County, where natural or developed blackberry refuges are not present. In nearby French prune orchards, heavy parasitism by Anagrus epos of another leafhopper species, Edwardsiana prunicola, was also observed. Kido et al. (1983) presented data showing a correlation between adult Anagrus dispersion in early spring from prune orchards and parasitism of grape leafhopper in nearby vineyards. These studies suggested that French prune refuges may be more productive and easier to manage than blackberry refuges. Prune refuge management trials are now being established in various areas of the San Joaquin Valley.

Limitations of Anagrus in Vineyards

Commercial dependence on Anagrus activity is influenced by grape variety and its production usage. For example, high populations of grape leafhopper are tolerated on Thompson Seedless vines destined for raisin or wine production and control can depend upon Anagrus. However, only low populations of leafhoppers are tolerated on Thompson Seedless vines cultured for fresh market production, making it more difficult to depend upon Anagrus instead of applying chemicals for control. Excessive spotting (leafhopper droppings) determines the economic damage level on table grapes and this occurs long before any reduction in sugar or yield can be measured (Jensen & Flaherty 1981a). Such spotting usually starts with the second brood. Potential damage depends on the length of exposure which varies with harvest date. Early maturing table grape varieties, such as Perlette and Cardinal, are exposed only a short time so can tolerate fairly high populations and advantage is usually taken of Anagrus activity. Late varieties, such as Emperor and Calmeria, accumulate spotting for more than three months and only very low populations are tolerated. Grapes with deep red color obscure spotting and have higher population tolerances than white grapes. Spots cannot be seen at all on fully ripened Ribier variety, a black table grape, although the berries may be so coated with leafhopper droppings as to feel rough. Jensen & Flaherty (1981) discuss economic levels of grape leafhopper in the presence and absence of Anagrus activity.

Anagrus is highly effective on Thompson seedless vines. If parasitized eggs are found in the first brood in a Thompson Seedless vineyard, the possibility of the grape leafhopper population remaining within tolerance limits is good, even in table grape vineyards. Anagrus is considerably less effective on Emperor and Ribier vines, particularly the latter. Grape leafhopper populations may increase to enormous numbers on them and some wine varieties despite the presence of Anagrus activity early in the first brood. Reasons for its effectiveness on Thompson Seedless, a variety harvested mid-season, are not clear, but observations indicate that the smooth leaf surface favors egg stinging, and early vine maturity is less favorable for grape leafhopper egg laying. Late varieties such as Emperor and Ribier have hairy leaf surfaces and produce newly matured leaves late into the season and this favors leafhopper egg laying. Late season irrigation in late variety vineyards has also been observed to favor grape leafhopper.

Impact of the Variegated Grape Leafhopper on IPM

The variegated grape leafhopper, Erythroneura variabilis, is the principle vineyard pest in desert areas of southern California. Infestations are especially severe in the Coachella Valley where high temperatures allow rapid development and up to six generations yearly (Barnes 1981). Until recently, San Joaquin Valley grape growers had to contend only with grape leafhopper. Now, however, they will also have to deal with the variegated grape leafhopper whose populations have exploded and spread rapidly since 1981. It poses a more serious pest of grapes than the related grape leafhopper species. Heavy defoliating populations are now widespread in Fresno and Tulare Counties. Defoliation has occurred even in many Thompson Seedless vineyards where grape leafhopper historically has been under good control by Anagrus. Studies by Kido et al. (1984) suggested that the variegated grape leafhopper in such vineyards is less successfully controlled by

Anagrus, making chemical control necessary. Chemical control measures are going to cause serious outbreaks of secondary pests, such as spider mites and mealybugs. Grape IPM in the San Joaquin Valley has been dealt a serious setback and adjustment will have to be made.

Andris et al. (1983) showed that variegated grape leafhopper is presently easier to control with pesticides than grape leafhopper, which has developed considerable resistance to a wide range of insecticides (Andris et al., unpublished data). Data and observations indicate that grape leafhoppers may increase after treatment with chemicals that effectively control the variegated species. However, these studies gave leads as to how timing of application may take advantage of Anagrus action when dealing with both species. Briefly, treatments for the variegated species should be delayed until the second brood to allow the action of Anagrus on grape leafhopper. When needed, variegated grape leafhopper should be controlled by a material which allows continued Anagrus action. Candidate materials for selective treatment include endosulfan, phosalone, methomyl, and propargite (Andris et al. 1983, unpublished data). Surprisingly, the latter material, which is an acaricide, gave good control of spider mites, reduced variegated leafhopper populations, and allowed Anagrus to effectively control grape leafhopper.

Parasites of Grape Leafhopper Adults

Studies are planned to expand our knowledge about the management of parasites of leafhopper adults. In 1983, up to 40% of the grape leafhopper adults were parasitized by Aphelopus comesi, a dryinid wasp, in one of our study vineyards. Cate (1975) reported that in one instance up to 72 percent were parasitized. He speculated that vineyard cultural practices can be altered to favor its survival and parasitism. A. comesi apparently does not attack variegated grape leafhopper. Exploration for a dryinid species that does needs consideration.

SPIDER MITES

Spider mites have been serious pests of grapes in the San Joaquin Valley for several decades. Reasons for this development have not been fully documented, but it is believed that the use of synthetic organic insecticides has upset natural controls. Serious spider mite problems in vineyards can be reduced, however, by judicious use of pesticides.

Two spider mite species are commonly abundant on grapes in the San Joaquin Valley: Pacific spider mite, Tetranychus pacificus, which deserves serious consideration, and Willamette spider mite, Eotetranychus willamettei, whose populations only occasionally become large enough to cause concern. The twospotted mite, T. urticae, is rarely found on grapes in the Valley.

The most important natural enemy of spider mites in vineyards is the predatory mite, *Metaseiulus occidentalis*, which under certain conditions (abundant prey and favorable temperatures) has enormous capacity for population increase. Flaherty & Hoy (1971) and Flaherty & Huffaker (1970) discussed predator-prey distribution patterns in vineyards, availability of alternate prey, and viticultural practices that affect spider mite population levels as other important factors which determine the effectiveness of *M. occidentalis*.

Distribution Patterns

Estimating mite abundance using standard monitoring procedures is particularly tedious, since mites are extremely small, often numerous, and therefore difficult and time-consuming to count. Flaherty et al. (1981a) proposed a presence-absence monitoring technique based on predator-prey distribution patterns that allows estimation of spider mite abundance and predictions of predator effectiveness without having to resort to spider mite and predator counts. More recently a quantitative presence-absence monitoring technique has been developed for spider mites on cotton (Wilson et al. 1983) and on almonds (Wilson et al. 1984, Zalom et al. 1984).

The canopy size and structure of grapevines poses considerable problems for monitoring predators and their prey. They have defined habitat preferences and as such, depending on the area of the canopy sampled, considerably different and sometimes conflicting control estimates will likely result. Guidelines for use in deciding how best to monitor predators and prey are a necessary part of the grape IPM system. Ongoing research efforts are aimed at studying predator-prey (including alternate prey) distributional patterns on grapevines for the development of a quantitative presence-absence monitoring technique (Flaherty et al., unpublished data).

Alternate Prey

Besides spider mites, *M. occidentalis* preys upon other mites, such as eriophyids, tydeids, and possibly tarsonemids. These prey help to support and maintain *M. occidentalis* when spider mites are scarce. Tydeid mites are particularly important in this respect (Flaherty & Hoy 1971). Without sufficient numbers of tydeid mites in late summer and fall, poorly distributed populations of predatory mites were observed to overwinter. A well distributed predator population in the spring was found to be the key ingredient for spider mite control. Knop & Hoy (1983a, b) discuss the biology, food requirements, and the effects of pesticides on *Homeopronematus anconai*, a tydeid mite commonly found in San Joaquin Valley vineyards. Flaherty & Hoy (1971) increased tydeid and *M. occidentalis* numbers in vineyards with pollen applications. Their erratic results may have been caused by detrimental effects of sulfur applications for powdery mildew, *Uncinula necator*, control on tydeid mites as shown by Calvert & Huffaker (1974), who also increased tydeid and predator numbers with pollen applications. Both Calvert & Huffaker (1974) and Flaherty & Hoy (1971) speculated on possible benefits of commercial pollen applications or the planting of cover crops that produce wind borne pollen.

Willamette spider mites are much less serious pests of grapes than Pacific spider mites in San Joaquin Valley vineyards. Only occasionally are treatments justified. Flaherty et al. (1981a) recommend that Willamette spider mites be considered as important alternate prey for *M. occidentalis*.

Viticultural Practices and Spider Mite Problems

Flaherty et al. (1981a) point out that viticultural practices influence spider mite outbreaks. They place considerable emphasis on correcting problems concerned with vineyard vigor, dusty conditions, and vineyard water stress, all of which greatly influence Pacific spider mite outbreaks. They also discuss how treatments for other vineyard pests may disturb natural controls of spider mites. Recently, Knop & Hoy (1983b) report that their laboratory studies were consistent with the studies of Calvert & Huffaker (1974), who demonstrated that sulfur was toxic to tydeids in a noncommercial vineyard. They also speculated from their laboratory studies that triadimefon, which is replacing sulfur for powdery mildew control, would have considerably less effect on tydeids. This has not been demonstrated as yet in commercial vineyards, but our observations in 1983 suggested that in triadimefon-treated vineyards we may expect other changes in mite fauna. The grape erineum mite, *Colomerus vitis*, normally controlled by sulfur applications for powdery mildew control, increased to damaging numbers in a few triadimefon-treated vineyards. This eriophyid mite may cause economic damage, or, if its control is not necessary, it may provide additional alternate prey for *M. occidentalis*.

Observations in 1983 also suggested that a predaceous mite species, *Amblyseius* nr. *hibisci*, previously rare in commercial vineyards, was abundant in some vineyards where triadimefon replaced sulfur. *Amblyseius hibisci* is commonly found in wild grapes and citrus groves where sulfur is not applied. Kennett & Flaherty (1974) and Tanigoshi (personal communication) reported that *A. hibisci* is an important component in the biological control of citrus red mite, *Panonychus citri*, and citrus thrips, *Scirtothrips citri*. We are currently studying whether the shift from sulfur to triadimefon in vineyards will affect mite fauna to the extent that changes in monitoring and management procedures as recommended by Flaherty et al. (1981a) will be required.

Other Predators of Spider Mites

Predaceous insects and spiders are generally considered ineffectual natural enemies of spider mites; it is assumed they appear too late in the season or increase too slowly. However, their contribution to natural control in vineyards should not be discounted. For example, observers are often impressed with how quickly six-spotted thrips, *Scolothrips sexmaculatus*, destroys Pacific spider mite populations. Their inconsistency in vineyards apparently is related to periodic absences of spider mite prey (Flaherty et al. 1981a).

The amount of aid given by other predators (lacewings, spiders, nabid bugs, ladybird beetles) is little known in vineyards. Individually they appear to exert little control, but collectively they could be important. For example, observations in vineyards and studies in the laboratory by Flaherty & Hoy (unpublished data) revealed that western flower thrips, *Frankliniella occidentalis*, feeds on Pacific spider mite eggs and may actually affect the pest's population growth in vineyards.

Mass Releases of Predaceous Mites

A few consulting entomologists in the San Joaquin Valley release *M. occidentalis* in vineyards as part of their services. Usually, the releases are made in early summer to control increasing Pacific spider mite populations. Some growers report benefits and reduced spider mite treatments. With this in mind, Flaherty & Peacock (Tulare County farm advisors) set up a replicated study with a grape grower who had been contracting release services with a sulfur-resistant strain (Hoy & Standow 1982). Sixteen ten-acre Thompson Seedless vineyards were used in the study of *M. occidentalis*. Releases were made by the consultant in half of the vineyards, chosen at random, on June 22, 1982 when weekly leaf samples indicated increasing Pacific spider mite populations. Weekly sampling was continued until September 21. Predator-prey distribution counts were recorded to compare spider mite and predator populations in the release and nonrelease vineyards.

The data (Flaherty & Peacock, unpublished data) showed no significant difference in predator-prey distributions between the release and nonrelease vineyards prior to the release or at any time thereafter. Moreover, five of the release and five of the nonrelease vineyards were treated with an acaricide, further indicating no benefit from the releases.

It was concluded that there are too many problems to overcome in the development of a timely release of *M. occidentalis* for the control of increasing Pacific spider mites in early summer. For example, if the spider mite population is too low at the time of release, the predators are likely to starve and the effort is wasted. If the spider mite population is high and increasing rapidly, predator survival is likely, but effective distribution of the predator population would not develop fast enough to prevent serious vineyard damage and need for treatment. Starvation, however, may become less of a problem if tydeids increase in vineyards where triadimefon replaces sulfur for powdery mildew control. Earlier releases then could take advantage of the alternate food source to enhance predator distribution with respect to later increasing Pacific spider mites. As to growers seeing benefits from summer releases, it is our opinion that it is more psychological than factual. Growers are likely to tolerate higher spider mite populations and treat less after they have already paid for predator release services.

We suggest that the development of a fall release program to help correct Pacific spider mite imbalance may be more fruitful than early summer releases for control. Flaherty & Huffaker (1970) showed that late-season predator activity in vineyards is an essential ingredient in spider mite balance. In their studies, an October release of M. occidentalis reared on two-spotted mites on soybeans grown outdoors gave an excellent biological control the following spring and summer. Hoy & Flaherty (1970, 1975) established the importance of late-season diapause induction for successful overwintering of M. occidentalis populations. A fall release, we reason, would minimize the timing and survivorship problems because immediate control of Pacific spider mite is not a factor and overwintering predators require little food. It would be a matter of harvesting the predators from soybeans after mid-September to insure the release of a diapausing predator population in vineyards. It is not construed here that the logistics of rearing and release to achieve an effective distribution pattern of predators in the vineyard is without problems. However, the methods of releasing and rearing developed by Hoy et al. (1982a, b) hold promise that commercial releases could be economical.

WESTERN GRAPE SKELETONIZER

The western grapeleaf skeletonizer (GLS), Harrisina brillians (Lepidoptera: Zygaenidae), was originally distributed through the southwestern United States (excluding California) and in Mexico's Sonora, Chihuahua, Coahuila, and Aguascalientes States (Stern et al. 1983). It was first found in California near San Diego in 1941 severely defoliating native wild grapes, Vitis girdiana, in canyon areas (Clausen 1961). In a short time GLS became a serious pest in commercial vineyards. By 1943, crop loss in some San Diego County vineyards reached 40 to 60 percent (Lange 1944). The late instar larvae are voracious feeders causing sunburn of fruit. When foliage is scarce these larvae feed on the grape clusters, causing bunch rot which destroys the entire cluster.

A state-imposed eradication program, using cryolite dust to control GLS and 2,4-D to kill native wild grapes was tried, but it was not successful. In 1950, the Department of Biological Control at the University of California, Riverside, initiated a program to search for natural enemies of GLS. A number of parasites were introduced but only two, Apanteles harrisinae and Sturmia harrisinae, became established. A very virulent granulosis virus, which became an important component in the successful biological control of GLS, was also accidentally introduced (Clausen 1961).

Surveys in San Diego County in 1982 and 1983 (V.M. Stern & D.L. Flaherty, unpublished data) revealed that it is now necessary to treat GLS in commercial vineyards. Abandoned untreated vineyards and backyard vines are severely defoliated in this area. Symptoms of GLS virus infection were not observed in the survey. Curiously, the survey revealed a total absence of GLS colonies in wild grapes, except where they were in close proximity to heavily infested commercial vines.

The first GLS infestation in the San Joaquin Valley was found at Kerman (Fresno County) in June 1961 (Clausen 1961). New infestations appeared thereafter throughout the San Joaquin Valley. Despite eradication attempts GLS became well established in Fresno and Tulare Counties by 1977. It now occurs in most areas of the southern San Joaquin Valley and in certain locations in the Sacramento Valley. We have observed 100 percent crop loss in some poorly managed vineyards in Tulare and Fresno Counties. A large number of home properties in the San Joaquin Valley have grapevines in backyards and these, as well as wild grapes, Vitis californica, along rivers and streams, have been devastated.

Chemical Control Studies

Although GLS has increased the pesticide load in the San Joaquin Valley, particularly in wine and raisin grapes, it is easily controlled by a number of insecticides, several of which are compatible with established integrated pest management programs (Stern et al. 1983). Cryolite sprays can be used to control both GLS and omnivorous leafroller (OLR), Platynota stultana, where OLR is a pest. The inorganic insecticide has a long residual action and will generally provide season-long control of GLS unless new moth flights enter the vineyard from poorly-managed vineyards or heavily infested wild grapes nearby. Thus far, there is no evidence that the late May to early June cryolite treatment for OLR and/or GLS has caused outbreaks of spider mites, leafhoppers, or mealybugs. Bacillus thuringiensis preparations give good control of GLS but often require more than one treatment. These preparations are not harmful to natural enemies and do not cause outbreaks of nontarget pests (Jensen 1969). Naled and methomyl are short-lived materials. They are used late in the season to prevent excessive leafhopper excrement from spotting table grapes and also kill adults that fly into the eyes and nostrils of grape pickers. Both give excellent control of GLS, which has urticating hairs that sting pickers at harvest. Methomyl has been noted to increase spider mites, but since it has a very short residual toxicity, it can be somewhat selective (Hoy et al. 1979). Currently, there are no effective biological control agents of GLS in commercial vineyards in the San Joaquin Valley. Growers have no recourse other than to use chemicals.

Parasite Studies

In 1980 through 1983 about 85,000 Apanteles and 18,000 Sturmia parasites were brought from San Diego County and released by V.M. Stern in the San Joaquin Valley. The releases were mainly in wild grapes along rivers and streams. Only a few parasite recoveries have been made at the release sites. Moreover, V. Stern & D. Flaherty have reservations about these particular parasites having the ability to hold GLS populations at sufficiently low levels to be compatible with commercial grape production. Between 1981 and 1983, samples of fourth and fifth instar larvae were taken from each of the two annual generations of GLS in three abandoned vineyards in San Diego County. These samples showed a combined parasitization by Apanteles and Sturmia of about 13 percent from six generations, far less than the percent parasitization (42% in 1953 and 62% in 1954) reported by Clausen (1961). These abandoned vineyards have not been harvested or treated with chemicals for more than 20 years. During the three year sampling period they were about 90 percent defoliated from the first generation larvae and nearly 100 percent defoliated from the second generation larvae. In the last generation of each sampling year most of the larvae died in the early instar stages from starvation. No evidence of the GLS virus was present in these vineyards. Clausen (1961) commented that the virus must be credited with the major role in reducing GLS populations to low levels and exterminating many small infestations. Perhaps at that time the virus was more widespread and had reduced GLS populations to levels that made GLS more manageable by the parasites. This may account for the higher parasitization level reported by Clausen (1961) than that found by the current authors. The current absence of virus in the abandoned vineyards in San Diego County needs explanation as does the absence of observable GLS populations in nearby wild grapes.

Virus Studies

The initial survey of the GLS problem in San Diego County in the late 1940's failed to reveal the presence of any disease affecting larval populations (Clausen 1961). However, a key problem that developed in the mass rearing of GLS larvae for parasite propagation and release in San Diego County was the occurrence of GLS granulosis virus epizootics in the insectary. The virus disease can almost certainly be attributed to unintentional introduction of contaminated host material from Arizona and Sonora, Mexico, since it was later found in field-collected larvae from both areas (Clausen 1961). High density GLS infestations in San Diego County became increasingly rare after 1952 due to virus epizootics (Clausen 1961). Smith et al. (1956) showed that Apanteles and Sturmia were involved with the transmission of the virus and suggested these parasites, and other insects, could become contaminated by walking on leaves smeared with the diarrhetic discharges of the diseased larvae. Transovarial transmission of the virus was also strongly suspected.

The GLS virus is extremely virulent. A sample was still highly effective after 26 years under refrigeration (V.M. Stern, personal communication). Tests were also conducted to determine if the virus is transmitted from one generation to the next. Healthy fourth and fifth instars were fed grape leaves dipped in an aqueous suspension containing pulverized virus-infected dead larvae. Surviving males and females were mated and taken to an untreated vineyard where egg laying was observed. Fifth instar larvae were also collected from a virus-free vineyard in San Diego County and mated females from these larvae were used as controls. Healthy females produced an average of 275 eggs while diseased females produced an average of 75 eggs. The virus-infected females scattered eggs on the leaves while normal females deposited eggs in clusters. Less than one percent of the larvae of diseased parents survived and pupated while 97% of the larvae from healthy parents pupated. Mortality of larvae arising from virus-infected adults was about 98 percent in the first to third instars.

Tests were conducted to determine whether both sexes were carriers of the virus from one generation to the next. The results showed that both male and female carry the virus from one generation to the next. Mortality of eggs and larvae was highest when both sexes had been exposed to the virus in the larval stage of the previous generation.

In 1980-81 small areas of devastated wild grapes along rivers and canals were sprayed with the virus east of Visalia in the San Joaquin Valley. The treated sites were determined by the availability of roads adjacent to the river banks. The sprayed sites are estimated to be a small fraction of the wild grapes along the rivers, creeks and irrigation canals in the areas. By 1982 and all through 1983 the skeletonizer population in the entire wild grape area east of Visalia was reduced to about 99% of the pretreatment population. Unsprayed wild grapes along the Tule River east of Porterville continued to be devastated by GLS. This and other data show that the GLS virus is well established in wild grapes east of Visalia and is providing excellent biological control. Moth flights out of the wild grapes into surrounding commercial vineyards were very heavy in 1978-81. By 1982-83 the flights essentially ceased.

Our data confirm that this virus is extremely virulent and has the potential to be molded into an area-wide biological control program, including wild grapes, backyard vines, and commercial vineyards. A continuing program is planned for establishing the virus in commercial vineyards.

OMNIVOROUS LEAFROLLER

Omnivorous leafroller (OLR), Platynota stultana (Lepidoptera: Tortricidae), has been a major pest of grapes in the San Joaquin Valley since the 1960's. OLR larvae reduce grape yields and quality by feeding on the flowers and developing berries, and especially by allowing entry of bunch rot organisms that damage the clusters (Kido et al. 1981). Leaf feeding by OLR is generally minor.

A number of insect parasites have been recorded attacking OLR larvae in vineyards. However, Kido et al. (1981) reported that parasites seldom accounted for more than ten percent mortality even in the presence of very high worm populations. Fortunately, OLR is effectively controlled with cryolite, a stomach poison that has not been observed to cause disruptions of other vineyard pests. Preparations of Bacillus thuringiensis are also used selectively for the control of OLR, but their short residual action makes them less effective than cryolite. Control with these stomach poisons is best achieved by bloom sprays that thoroughly cover the clusters.

Contact insecticide treatments (carbaryl) are recommended after heavy vine foliage and berry size restrict good coverage, but these treatments can be disruptive. There is a need to encourage more effective biological control of OLR to supplement selective chemicals.

Kido et al. (1981) suggested that sulfur applied to control powdery mildew is detrimental to Trichogramma sp. and this accounts for its very low activity on OLR eggs in vineyards. The replacement of sulfur with triadimefon may encourage better parasitism by Trichogramma sp. and other OLR parasites. However, Stern & Bowen (1963) reported that Trichogramma sp. that attacks OLR and other lepidopterous pests on other crops does not diapause. This may account for its low numbers from year to year in the San Joaquin Valley. Its activity is short-circuited by the absence of host eggs during the winter in the San Joaquin Valley in contrast to hot southern desert valleys where host eggs are present throughout the year. Mass release of Trichogramma sp. to augment biological control in vineyards has not been studied.

GRAPE MEALYBUG

Grape mealybug, Pseudococcus maritimus, is primarily a pest of table grapes, particularly Emperor and Ribier grapes whose bunches make contact with vine bark and become infested. Before the 1940's occasional mealybug losses occurred in vineyards; even though bothersome, infestations were mostly spotty and frequently disappeared the following year. Increasing and more persistent populations were noted in the late 1940's. Extensive use of DDT and other synthetic pesticides to control grape leafhoppers apparently had disrupted mealybug natural enemies (Flaherty et al. 1981b).

Control for grape mealybug generally consists of dormant season treatment with parathion and oil. Flaherty et al. (1982) showed that this treatment did not disrupt spider mites. Unfortunately, grape mealybug is developing resistance to parathion. This poses a serious problem for IPM in table grape vineyards since these studies showed that alternate materials are highly disruptive.

Little work has been done to determine the effectiveness of natural enemies of grape mealybug. They must be responsible for keeping populations at low levels because a few mealybugs can be found in all vineyards. Flaherty et al. (1982) reported that parasitic wasps were not reared from mealybug samples taken from their study vineyard that was heavily treated with pesticides, whereas in another Emperor vineyard that had a minimal insecticide program, parasitism was 46 percent, and mealybug activity was below the economic level of two percent infested bunches. Clausen (1924) reported rearing five primary parasites from the grape mealybug in the Fresno area. He also reported that in the late summer and fall of 1919 parasitism exceeded 90%. However, it is not clear whether he considered the mealybug populations to be effectively controlled.

Even less is known about the effectiveness of predators of grape mealybug in vineyards. Flaherty et al. (1982) observed a cecidomyiid fly larva attacking mealybug egg masses late in the season in the above mentioned lightly treated Emperor vineyard. *Cryptolaemus montrouzieri* (the mealybug destroyer) has been observed attacking mealybugs in Napa Valley vineyards but not in the San Joaquin Valley. Green lacewing adults, frequently abundant on grapevines harboring mealybugs, are attracted to mealybug honeydew, but to what degree their egg laying and subsequent control of mealybugs is influenced by the presence of honeydew is not known.

In the absence of research information, and the fact that the grape mealybug problem is most likely caused by the need to treat table grapes, such as Emperor and Ribier, for grape leafhopper, there is little to suggest for augmenting biological control at the present time. One approach that needs study is the development of treatment programs which leave untreated or minimally treated areas in table grape vineyards to encourage the activity and spread of natural enemies.

GRAPE LEAFFOLDER

Grape leaffolder (GLF), *Desmia funeralis* (Lepidoptera: Pyralidae), is a grape pest in the central and southern San Joaquin Valley, the degree of its infestation varying greatly from year to year. In the worst years less than half of the vineyards have economic infestations. GLF causes injury in the larval stage by rolling and feeding on the leaves, reducing photosynthetic function, and exposing clusters to sunburn. GLF larvae may feed on fruit, but economic damage usually occurs only with massive, late-season infestations. Selective chemicals for control include cryolite and preparations of *Bacillus thuringiensis*. The latter are particularly effective if applications are properly timed (Jensen & Flaherty 1981b).

The most commonly observed natural enemy of GLF is the larval parasite, *Bracon cushmani* (Jensen & Flaherty 1981b). Parasitism ordinarily is in the range of 30 to 40 percent but often is higher. A number of other parasites, including two tachinid species, attack GLF but are not effective (Jensen & Flaherty 1981b).

GRAPE DISEASES

Applicable information is not available for the biological control of grape diseases. However, treatment for disease control may influence the biological control of arthropods. The effects of sulfur for powdery mildew were discussed in the sections on spider mites and omnivorous leafroller. Flaherty et al. (1982) showed that dormant treatment with dinitro compounds (dinoseb) for control of the fungus *Phomopsis viticola* significantly reduced *M. occidentalis* populations with concurrent increases in spider mite populations. These same studies also showed that dormant treatment with sodium arsenite for Spanish measles and/or *Phomopsis* control was not disruptive. Other commonly used fungicides (benomyl) should also be studied for possible disruptions.

NEMATODES

Like grape diseases applicable information on the biological control of nematodes is not available. The effect of nematicide treatments on arthropods has received little attention. However, years of observations suggest that the recently banned nematicide DBCP caused little if any disruptions of arthropods in vineyards. Studies are needed to determine whether this will hold for replacement materials.

SUMMARY

Biological control is an important component of grape IPM in the San Joaquin Valley. Many vineyards require little or no treatment for grape leafhopper, spider mites, and mealybugs because of highly effective natural enemies. An egg parasite, *Anagrus epos*, is of particular importance because its activity reduces the need to control grape leafhopper with contact insecticides that often cause outbreaks of other pests. Attempts to augment *A. epos* by planting blackberry refuges near vineyards in areas of the Valley where its overwintering leafhopper host (*Dikrella* sp.) is absent have been unsuccessful. Recent studies suggest that French prune trees make more productive refuges than blackberries.

Natural enemies are not effective against two lepidopterous pests (omnivorous leafroller and western grapeleaf skeletonizer) and only partially effective against grape leaffolder. Nevertheless, necessary chemical controls are accomplished without disrupting natural enemies in vineyards by using selective stomach poisons, including cryolite and *Bacillus thuringiensis* preparations. Studies suggest that a granulosis virus holds promise for selective control of western grapeleaf skeletonizer.

The variegated grape leafhopper, formerly restricted to southern California vineyards, has become a pest in the San Joaquin Valley and has dealt a serious setback to grape IPM. Studies have shown that *A. epos* does not control the variegated species as well as it does the grape leafhopper species. Treatments for variegated grape leafhopper are necessary in many vineyards where grape leafhopper historically has not been a pest. However, studies also suggested that the variegated species may be selectively controlled with chemicals that do not disturb the activity of *A. epos* on grape leafhopper populations.

RECOMMENDATIONS

Grape integrated pest management in the San Joaquin Valley would be improved by augmentation of biological control, including importation of natural enemies of new pests. Moreover, the adaptation of recent developments in monitoring and prediction technologies, which take into account natural controls, would increase the utilization of biological control in commercial vineyards. The following outline is a list of ongoing (*) and recommended research projects for grape IPM in the San Joaquin Valley. The success of this research lies in the continual updating and implementation of applicable information in commercial vineyards.

Grape Leafhopper and Variegated Leafhopper

1.* Determine within and between vine distributions of the two species, for use in development of a presence-absence sampling plan as affected by parasites, grape variety, and market destination (raisin, table, and wine grapes).
2.* Develop and improve *Anagrus* output from blackberry and French prune refuges.
3.* Develop predictive models for both leafhopper spp. which incorporate natural enemies and the impact of cultural practices and chemical control.
4. Import natural enemies of variegated leafhopper.

Spider Mites

1. * Determine within and between vine distribution patterns, as affected by predation and chemical control, for use in developing a presence-absence sampling plan.
2. * Develop a predictive model for Pacific mite which incorporates predation, alternate prey, and the impact of cultural practices and chemical control.
3. Determine the efficacy of augmentative releases of predaceous mites as affected by timing.
4. Augment tydeid mites as alternate prey for *M. occidentalis* with pollen applications in vineyards.

Western Grapeleaf Skeletonizer

1. * Determine GLS granulosis virus efficacy in commercial vineyards.
2. Import natural enemies.

Omnivorous Leafroller

1. Import and augment natural enemies of omnivorous leafroller.

Grape Mealybug

1. Develop treatment programs which leave untreated or minimally treated areas in table grape vineyards to encourage the activity and spread of grape mealybug natural enemies.
2. Develop a quantitative monitoring program which incorporates economic and natural enemy predictive components.

Grape Leaffolder

1. Develop a quantitative monitoring program which incorporates economic and natural enemy predictive components.

Diseases and Nematodes

1. Expand studies of the effects of chemical control on nematodes and diseases of arthropods.
2. Expand investigations for the potential utilization of biological control of nematodes and diseases.

REFERENCES

Andris, H., R. Coviello, W. Barnett, W. Peacock and D. Flaherty. 1983. Variegated grape leafhopper: A new pest in Valley orchards. Sun-Diamond 1983 (April-May): 33-34.

Barnes, M.M. 1981. Variegated grape leafhopper of southern California, pp. 137-139. In: D.L. Flaherty, F.L. Jensen, A.N. Kasimatis, H. Kido and W.J. Moller (Eds.), Grape pest management. Univ. Calif., Berkeley., Agric. Sci. Publ. 4105.

Calvert, D.J., and C.B. Huffaker. 1974. Predator (*Metaseiulus occidentalis*)-prey (*Pronematus* spp.) interactions under sulfur and cattail pollen applications in a non-commercial vineyard. Entomophaga 19:361-369.

Cate, J.R. 1975. Ecology of *Erythroneura elegantula* Osborn (Homoptera: Cicadellidae) in grape agroecosystems in California. Ph.D. thesis, Univ. of California, Berkeley.

Clausen, C.P. 1924. The parasites of *Pseudococcus maritimus* (Ehrhorn) in California (Hymenoptera, Chalcidoidea). Part II. Biological studies and life histories. Univ. Calif. Publ. Entomol. 3:253-288.

Clausen, C.P. 1961. Biological control of western grapeleaf skeletonizer (*Harrisina brillians* B. and McD.) in California. Hilgardia 31:613-638.

Doutt, R.L., and J. Nakata. 1965. Overwintering refuge of *Anagrus epos* (Hymenoptera: Mymaridae). J. Econ. Entomol. 58:586.

Doutt, R.L., and J. Nakata. 1973. The *Rubus* leafhopper and its egg parasitoid: An endemic biotic system useful in grape pest management. Environ. Entomol. 2: 381-386.

Doutt, R.L., J. Nakata and F.E. Skinner. 1966. Dispersal of grape leafhopper parasites from a blackberry refuge. Calif. Agric. 20(10):14-15.

Flaherty, D.L., and M.A. Hoy. 1971. Biological control of Pacific mites and Willamette mites in San Joaquin Valley vineyards: Part III. Role of tydeid mites. Res. Popul. Ecol. 8:80-96.

Flaherty, D.L., and C.B. Huffaker. 1970. Biological control of Pacific mites and Willamette mites in San Joaquin Valley vineyards. I. Role of *Metaseiulus occidentalis*. II. Influence of dispersion patterns of *Metaseiulus occidentalis*. Hilgardia 40:267-330.

Flaherty, D.L., M.A. Hoy and C.D. Lynn. 1981a. Spider mites, pp. 111-125. In: D.L. Flaherty, F.L. Jensen, A.N. Kasimatis, H. Kido and W.J. Moller (Eds.), Grape pest management. Univ. Calif., Berkeley, Agric. Sci. Publ. 4105.

Flaherty, D.L., F.L. Jensen, W.L. Peacock and L. Bettiga. 1981b. Grape mealybug, pp. 160-4. In: D.L. Flaherty, F.L. Jensen, A.N. Kasimatis, H. Kido and W.J. Moller (Eds.), Grape pest management. Univ. Calif., Berkeley, Agric. Sci. Publ. 4105.

Flaherty, D.L., W.L. Peacock, L. Bettiga and G.M. Leavitt. 1982. Chemicals losing effect against grape mealybug. Calif. Agric. 36(5/6):15-16.

Hoy, M.A., and D.L. Flaherty. 1970. Photoperiodic induction of diapause in a predaceous mite, *Metaseiulus occidentalis*. Ann. Entomol. Soc. Amer. 63:960-963.

Hoy, M.A., and D.L. Flaherty. 1975. Diapause induction and duration in vineyard-collected *Metaseiulus occidentalis*. Environ. Entomol. 4:262-264.

Hoy, M.A., W.W. Barnett, W.O. Reil, D. Castro, D. Cahn, L.C. Hendricks, R. Coviello and W.J. Bentley. 1982a. Large-scale releases of pesticide-resistant spider mite predators. Calif. Agric. 36(1):8-10.

Hoy, M.A., D. Castro and D. Cahn. 1982b. Two methods for large scale production of pesticide-resistant strains of the spider mite predator *Metaseiulus occidentalis* (Nesbitt) (Acarina: Phytoseiidae). Z. angew. Entomol. 94:1-9.

Hoy, M.A., D. Flaherty, W. Peacock and D. Culver. 1979. Vineyard and laboratory evaluations of methomyl, dimethoate, and permethrin for a grape pest management program in the San Joaquin Valley of California. J. Econ. Entomol. 72: 250-255.

Hoy, M.A., and K.A. Standow. 1982. Inheritance of resistance to sulfur in the spider mite predator Metaseiulus occidentalis. Entomol. Exp. Appl. 31:316-323.

Jensen, F.L. 1969. Microbial insecticides for control of grape leaffolder. Calif. Agric. 23(4):5-6.

Jensen, F.L., and D.L. Flaherty. 1981a. Grape leafhopper, pp. 98-110. In: D.L. Flaherty, F.L. Jensen, A.N. Kasimatis, H. Kido and W.J. Moller (Eds.), Grape pest management. Univ. Calif., Berkeley, Agric. Sci. Pub. 4105.

Jensen, F.L., and D.L. Flaherty. 1981b. Grape leaffolder, pp. 147-154. In: D.L. Flaherty, F.L. Jensen, A.N. Kasimatis, H. Kido, and W.J. Moller (Eds.), Grape pest management. Univ. Calif., Berkeley, Agric. Sci. Publ. 4105.

Kennett, C.E., and D.L. Flaherty. 1974. Pest management of citrus and red mite in citrus in the southern San Joaquin Valley. San Joaquin Valley Agr. Res. Ext. Cent. Publ. 27 pp.

Kido, H., D.L. Flaherty, W.W. Barnett and H.L. Andris. 1981. Omnivorous leafroller, pp. 126-136. In: D.L. Flaherty, F.L. Jensen, A.N. Kasimatis, H. Kido and W.J. Moller (Eds.), Grape pest management. Univ. Calif., Berkeley, Agric. Sci. Pub. 4105.

Kido, H., D.L. Flaherty, D.F. Bosch and K.A. Valero. 1983. Biological control of grape leafhopper. Calif. Agric. 37(5/6):4-6.

Kido, H., D.L. Flaherty, D.F. Bosch and K.A. Valero. 1984. The variegated grape leafhopper in the San Joaquin Valley. Calif. Agric. 38(1/2):31-2.

Knop, N.F., and M.A. Hoy. 1983a. Biology of a tydeid mite, Homeopronematus anconai (n. comb.) (Acari: Tydeidae), important in San Joaquin Valley vineyards. Hilgardia 51:1-30.

Knop, N.F., and M.A. Hoy. 1983b. Factors limiting the utility of Homeopronematus anconai (Acari: Tydeidae) in integrated pest management in San Joaquin Valley vineyards. J. Econ. Entomol. 76:1181-1186.

Lange, W. H. 1944. The western grapeleaf skeletonizer, Harrisina brillians, in California. Calif. Dept. Agric. Bull. 32:98-104.

McKenzie, L.M., and B.P. Beirne. 1972. A grape leafhopper, Erythroneura ziczac (Homoptera: Cicadellidae), and its mymarid (Hymenoptera) egg parasite in the Okanagan Valley, British Columbia. Can. Entomol. 104:1229-1233.

Smith, O.J., K.M. Hughes, P.H. Dunn and I.M. Hall. 1956. A granulosis virus disease of the western grapeleaf skeletonizer and its transmission. Can. Entomol. 88:507-515.

Stern, V.M., and W. Bowen. 1963. Ecological studies of Trichogramma semifumatum, with notes on Apanteles medicaginis, their suppression of Colias eurytheme in southern California. Ann. Entomol. Soc. Amer. 56:358-372.

Stern, V.M., D.L. Flaherty and W.L. Peacock. 1983. Control of western grapeleaf skeletonizer (Lepidoptera: Zygaenidae), a new grape pest in the San Joaquin Valley, California. J. Econ. Entomol. 76:192-195.

Williams, D.W. 1984. Ecology of a blackberry-leafhopper-parasite system and its relevance to California grape agroecosystems. Hilgardia 52:1-32.

Wilson, L.T., D. Gonzalez, T.F. Leigh, V. Maggi, C. Foristiere and P. Goodell. 1983. Within-plant distributions of spider mites (Acari: Tetranychidae) on cotton: A developing implementable monitoring program. Environ. Entomol. 12: 128-134.

Wilson, L.T., M.A. Hoy, F.G. Zalom and J.M. Smilanick. 1984. Sampling mites in almonds: I. Within-tree distribution and clumping pattern of mites with comments on predator-prey interactions. Hilgardia 52:1-13.

Zalom, F.G., M.A. Hoy, L.T. Wilson and W. Barnett. 1984. Sampling mites in almonds: II. Presence-absence sequential sampling for Tetranychus mite species. Hilgardia 52:14-24.

ALFALFA: STATUS AND CURRENT LIMITS TO BIOLOGICAL CONTROL IN THE EASTERN U.S.

Kenneth V. Yeargan

As a basis for understanding biological control of alfalfa pests, it is worthwhile to briefly review production practices and major pest groups of this crop. This article is limited in scope to forage alfalfa grown east of the Rocky Mountains in the U.S. Alfalfa seed production is a specialized enterprise which differs in agronomic practices and pest problems from forage alfalfa. Furthermore, most alfalfa seed production is concentrated in the western U.S., even for cultivars commonly grown for forage in the eastern part of the country.

ALFALFA PRODUCTION

Alfalfa has been grown as a forage crop in the U.S. since about 1850, but it was not until the introduction and development of winter-hardy cultivars in the early 1900's that significant production was begun east of the Mississippi River (Hansen & Davis 1972). During the 1950's, a major expansion occurred in alfalfa production in the north central U.S., while production in the remainder of the country changed little. Wisconsin now has the nation's largest production and, except for California, the top five alfalfa producing states are located in the northcentral part of the U.S. (Anonymous 1982). All regions of the country except the southeastern states have substantial alfalfa production.

Forage alfalfa is a perennial crop, with individual stands usually persisting for three to seven years, or occasionally longer. It may be grown as a pure stand or mixed with desirable forage grasses. Within each growing season it is periodically harvested; the number of harvests usually ranges from two or three per season in the northern U.S. to four or more further south.

Several characteristics of the alfalfa ecosystem are relevant to a consideration of pest management practices, and to biological control in particular. Although it is a relatively stable habitat as compared with cultivated fields used for annual crops,

ISBN 0-12-357030-1

the system is periodically disturbed during the growing season by harvests. During cuttings, especially in mid-summer, the habitat may be abruptly changed (albeit temporarily) from a relatively mesic environment to a more xeric one. Previously moderate temperatures may suddenly become extremely high (Pinter et al. 1975). Pests and biological control agents must cope with these changes.

In contrast with many crops, such as fruits and vegetables, alfalfa has relatively low cash value per hectare, and cosmetic damage is of little concern. Thus, residual pest populations sufficient to perpetuate populations of biological control agents may be permissible within the system. Also, in many row crops, beneficial organisms must recolonize fields each season from surrounding undisturbed habitats, because the fields are plowed annually. Despite these considerations, pesticides are sometimes needed in alfalfa production. Although alfalfa producers may tolerate more cosmetic damage than would a tomato or apple grower, the forage producer is not simply concerned with gross yield. Alfalfa is a forage of choice primarily because of its high protein content, and certain pests may affect protein content at lower population levels than would be required to reduce overall quantitative yield. The total amount of pesticides used per year on alfalfa, however, is considerably lower than that for most other crops (Anonymous 1983).

Stehr (1982) formulated several questions that should be considered when assessing the feasibility of biological control for a particular crop-pest system, such as tolerable injury, cash value, and persistence of the ecosystem. Overall, the alfalfa ecosystem possesses several characteristics which historically have been associated with successful biological control (Southwood 1977).

MAJOR PEST GROUPS

The major pest groups affecting alfalfa production in the eastern U.S. will be discussed briefly, and references for further reading on weeds, plant pathogens, and nematodes will be provided. Biological control efforts in alfalfa to date, however, have been directed primarily toward insect pests. Therefore, after the following synopses of other pest groups, the remainder of this chapter will focus on biological control of insect pests.

Forage alfalfa is subjected to stress from a variety of pests during the several years that a given stand exists. Pest problems, and their management, during the first year (stand establishment) often differ from those occurring later in the life of a stand. During the establishment year, individual alfalfa plants are relatively small, and the producer's primary goal is to obtain a uniform, dense stand of plants. Thereafter, emphasis shifts to producing profitable yields of quality forage, which entails maintenance for several years of a reasonable stand of vigorous plants.

Weeds

Effects of weeds on alfalfa production and methods of weed control were reviewed by Peters & Peters (1972). Weed control in alfalfa consists largely of cultural control techniques and use of herbicides. Cultural control of weeds includes use of clean seed, selection of planting date, use of companion crops (small grains) at planting, and selected timing of mowing. Herbicides have been effective in some cases, but not in others. Problems encountered with herbicides include phytotoxicity to alfalfa plants under certain conditions and the inability to control weedy grasses in mixed alfalfa-grass stands without killing or injuring the desirable forage grass. Biological control of weeds in alfalfa has not been seriously pursued.

Plant Pathogens

Alfalfa diseases were comprehensively reviewed by Graham et al. (1979). Their discussion of diseases caused by biotic agents includes description of symptoms, causal organisms, epidemiology, and control measures. Alfalfa diseases in the eastern U.S. may be caused by viruses (alfalfa mosaic), bacteria (bacterial wilt), or fungi (anthracnose, common leafspot, *Phytophthora* root rot), but the majority of the alfalfa disease problems of the region are caused by fungi. Plant pathogen control in alfalfa consists primarily of use of resistant cultivars (when available) and cultural control techniques, such as altering planting date or site, timely harvesting, and rotating crops. Probably the greatest progress in developing alfalfa resistance to diseases has been made with bacterial wilt, anthracnose, and *Phytophthora* root rot. Use of pesticides for disease control in alfalfa has been limited, primarily due to economic considerations. There is renewed interest in biological control of plant pathogens (Barnes et al. 1981), but this approach has rarely been implemented (Papavizas & Lumsen 1980).

Nematodes

Major nematode pests of alfalfa also were reviewed by Graham et al. (1979). They indicated that alfalfa stem nematodes, root-knot nematodes, and root-lesion nematodes are the most economically important. In addition to direct injury to the alfalfa plant, some nematode species can cause increased incidence and severity of bacterial or fungal disease. For example, rootknot nematodes (*Meloidogyne* spp.) can increase the incidence of bacterial wilt and *Fusarium* wilt. Effectiveness of available control tactics varies according to the type of nematode. For example, crop rotation is ineffective for those nematode species with wide host ranges, such as root-lesion nematodes. Resistant cultivars are available for some species, (alfalfa stem nematode) but not for others (rootlesion nematodes). Nematicides are effective against certain nematodes, but cost of treatment has been an important deterrent to their widespread usage. Naturally-occurring antagonists of nematode pests have been investigated for several decades, but biological control of nematodes seldom has reached the stage of practical application (Mankau 1980, Sayre 1980). This is due to several reasons, including the limited number of trained nematologists working on the subject and the virtual lack of organized efforts to seek antagonists in the original habitats of pest nematodes in order to import the antagonists for trials in problem areas (Mankau 1980; see also Van Gundy, this volume).

Insect Pests

Major insect pests of alfalfa were reviewed by App & Manglitz (1972) and Gyrisco (1958). Since the latter review, the blue alfalfa aphid (*Acyrthosiphon kondoi*) was discovered in California (Gonzalez et al. 1978) and has moved steadily eastward. The insect fauna of alfalfa is strikingly diverse. Pimentel & Wheeler (1973) recorded 591 arthropod species from New York alfalfa, many of which were either incidental inhabitants or beneficial species. Several insect species can cause serious damage to alfalfa; these include foliage feeders such as the alfalfa weevil (*Hypera postica*), sap feeders such as the potato leafhopper (*Empoasca fabae*), and root feeders such as the clover root curculio (*Sitona hispidulus*). Control tactics vary according to pest species, but include the use of insecticides, cultural control, resistant alfalfa cultivars, and biological control.

STATUS OF BIOLOGICAL CONTROL OF INSECT PESTS

While some 35% of the economically important insect pests in the U.S. are of foreign origin (Sailer 1978), these exotic species are not equally distributed among crops. For example, virtually all insect pests of soybean in this country are indigenous species which have adapted to that crop (Kogan 1980). On the other hand, several of the major insect pests of alfalfa in the eastern U.S. are of foreign origin. Examples include the alfalfa weevil and several species of aphids. Such pests have traditionally been the targets of classical biological control efforts, although DeBach (1974) and others have pointed out that this technique also may be effective against indigenous pests. Alfalfa in the eastern U.S. also is attacked by several indigenous pests, most notably the potato leafhopper, but classical biological control programs have not been attempted for most of these species.

Classical Biological Control

Hagen et al. (1976) briefly reviewed classical biological control of alfalfa pests. Extensive biological control programs have been developed for the alfalfa weevil, certain aphid species, and the alfalfa blotch leafminer (*Agromyza frontella*) in the eastern U.S. More recently, natural enemies have been imported for potential use against the clover root curculio and selected mirid plant bugs. The status of these programs will be reviewed individually, followed by a brief discussion of the status of knowledge on indigenous natural enemies in alfalfa.

Alfalfa weevil During the 1950's the alfalfa weevil became the most important insect pest of alfalfa in the eastern U.S., and it remains so in many areas. It occurs throughout the alfalfa producing regions, where it is essentially a univoltine insect. Damage is caused primarily by larval feeding on foliage and growth tips; most damage occurs prior to the first harvest, but may continue after that harvest under certain conditions (if the insect's phenology is delayed relative to that of the plant). The adult undergoes aestival diapause and becomes active again during autumn. Oviposition occurs in dead or living plant tissue during fall, winter, and/or spring, but most occurs during the spring. Various control tactics have been directed at the egg, larval, and adult life stages.

The most recent complete bibliography of the alfalfa weevil was compiled by Wood et al. (1978). Classical biological control efforts against the alfalfa weevil in the eastern U.S. were reviewed by Brunson & Coles (1968) and Dysart & Day (1976). More recently, Day (1981) discussed the status of biological control of this pest in the northeastern U.S., an area where much effort was concentrated early in the program for biological control of the alfalfa weevil in eastern North America. He concluded that in the states north of Virginia and West Virginia, and east of Ohio, successful biological control had greatly reduced alfalfa weevil populations and generally maintained them below economic injury levels. That conclusion was based largely on increasing incidence of parasitism during the mid-1960's and a concurrent decline in damage caused by the alfalfa weevil in the region.

Of the several species which have been released for biological control of the alfalfa weevil, Day (1981) attributes most of the success in the northeastern U.S. to two species: *Bathyplectes anurus*, which attacks larval stages, and *Microctonus aethiopoides*, which attacks adult weevils. Much of the literature prior to the mid-1970's on parasites of the alfalfa weevil dealt with another species, namely *Bathyplectes curculionis*. More recent data, however, indicate that *B. anurus* has replaced *B. curculionis* as the predominant parasite of alfalfa weevil larvae at sites where both species have coexisted for several years. Day (1983) demonstrated this with data collected from study sites in Pennsylvania and New Jersey between 1969 and 1981. A similar shift has occurred at a study site near Lexington, Kentucky since 1970. Latheef et al. (1979) reported that *B. curculionis* caused about five times more mortality than *B. anurus* at that site between 1972 and 1976. Since then, *B. anurus* has steadily increased in abundance relative to *B. curculionis*, and during 1983, weekly samples of alfalfa weevil larvae revealed that more than 90% of the larval parasitism was by *B. anurus* (J.C. Parr & K.V. Yeargan, unpublished data). Day (1983), Yeargan (1979) and others have discussed the biology of *Bathyplectes* species: *B. anurus* possesses several characteristics which might explain its apparent competitive advantage over *B. curculionis*. Those characteristics include much higher fecundity, faster development, and less frequent encapsulation by the host. Furthermore, a substantial percentage of *B. curculionis* females have nonfunctional ovaries, while this peculiar phenomenon is relatively rare in *B. anurus* (Day 1983).

Currently, the primary emphasis in classical biological control of the alfalfa weevil is on transporting effective parasites within the U.S. for inoculative releases in those areas where they are rare or absent. This recolonization effort, which is being conducted by USDA-APHIS and cooperating agencies, had by 1983 resulted in the release of more than 11 million parasites in 33 states.

In recent years an entomopathological phenomenon has occurred which may have long-term implications on classical biological control of the alfalfa weevil. Although the geographical origin of the causative pathogen(s) is presently unknown, epizootics of fungal disease have occurred over much of the eastern U.S. in recent years. Part of the problem in determining the original source of the pathogen(s) is due to the difficult taxonomy of the group of fungi involved. When Harcourt et al. (1974) first reported the occurrence of the disease in North America from sites in Ontario, Canada, it was thought to be caused by *Entomophthora phytonomi*, a species which had long been known to decimate clover leaf weevil (*Hypera punctata*) populations. Many authors now refer to the pathogen of *H. postica* as *Erynia* sp. The disease affects only larval stages of the alfalfa weevil, and was found to be the key factor influencing intragenerational population trends in eastern Ontario (Harcourt et al. 1977). Epizootics have since been reported as far south as Georgia (Gardner 1982), and as far west as the Great Plains states of the U.S. (Puttler et al. 1978). Nordin et al. (1983) discussed factors affecting the phenology of epizootics of this disease, and briefly reviewed the taxonomic status of the pathogen. In general, epizootics appear to occur relatively early in the phenology of the host larval population in the northern U.S., but later (with respect to host phenology) at more southerly locations. The relative synchrony of pathogen and host may influence the degree of host population suppression, and thus explain why the pathogen is presently more effective in southeastern Canada and the extreme northern U.S. than elsewhere. Current research includes efforts to develop methods for eliciting earlier epizootics in the central U.S. (G.C. Brown, personal communication). The possible interactions between established insect parasites, such as M. aethiopoides and B. anurus, and the pathogen remain to be determined.

Aphids The pea aphid (Acyrthosiphon pisum) and the spotted alfalfa aphid (Therioaphis maculata) have been recognized for many years as aphid pests of alfalfa in the eastern U.S., although damage by the latter has been most serious in the southwestern U.S. As noted earlier, the blue alfalfa aphid recently has become a serious pest of alfalfa in California (Gonzalez et al. 1978), and was first reported east of the Mississippi River in 1983 in Georgia, Kentucky and Louisiana (M.B. Stoetzel, personal communication). All three species are thought to be of foreign origin. Resistant cultivars of alfalfa played a minor role in the control of the spotted alfalfa aphid, although introduced natural enemies also may have contributed to its decline in pest status. The pea aphid is heavily attacked in many areas of the U.S. by parasites (especially Aphidius ervi and Aphidius smithi), but it is difficult to assess their impact on this pest, because its population dynamics also may be influenced by predation, disease, weather, and other factors (Angalet & Fuester 1977). In fact, several researchers in western North America have concluded that predators are primarily responsible for much of the biological control of aphid populations in alfalfa (Frazer et al. 1981, Neuenschwander et al. 1975). The possible role and utility of entomopathogens in the biological control of aphids in alfalfa also is being investigated, especially in Australia (Milner et al. 1982).

Research on classical biological control of aphids in alfalfa is continuing in many parts of the world, with much of the renewed effort apparently spurred by recent accidental introductions of the blue alfalfa aphid into Australia and the U.S. Aeschlimann (1981) surveyed aphid parasites in the Mediterranean region with the goal of finding species that would successfully attack the blue alfalfa aphid in Australia. Researchers in California (Gonzalez et al. 1978) also are attempting biological control of this species. It is likely that similar efforts will follow in the eastern U.S. if the blue alfalfa aphid becomes an economically damaging pest there.

Alfalfa blotch leafminer The alfalfa blotch leafminer, an agromyzid fly native to Europe, was first detected on alfalfa in the U.S. in Massachusetts in 1968 (Miller & Jensen 1970). It has since spread as far as North Carolina, Ohio, and southeastern Canada (Hendrickson & Plummer 1983). This insect mines alfalfa leaves prior to the first and second harvests, and it may reduce digestible dry matter and crude protein content of the forage (Byers & Valley 1981). Other researchers, however, have found it to be of minor economic importance (MacCollom et al. 1982, Thompson 1981).

The alfalfa blotch leafminer is attacked in the eastern U.S. by European species of parasites (primarily *Dacnusa dryas* and *Chrysocharis punctifacies*) which were introduced during the 1970's. Certain native parasites of *Liriomyza* spp. leafminers also attack the alfalfa blotch leafminer. Hendrickson & Plummer (1983) showed that in Delaware the indigenous parasites consistently killed about one third of the alfalfa blotch leafminer population over a seven-year period. They also showed that the introduced parasites have recently doubled the amount of parasitism caused by native parasites alone, and that average densities of the alfalfa blotch leafminer have concurrently declined to levels well below currently suggested economic injury levels.

Clover root curculio The clover root curculio is one of several *Sitona* spp. which attack leguminous plants in the eastern U.S. and elsewhere, but it is the species most consistently associated with alfalfa in this country (Morrison et al. 1974). The annual life cycle of the clover root curculio is similar to that of the alfalfa weevil which was discussed earlier, and feeding damage to alfalfa occurs primarily in the spring. Unlike alfalfa weevil larvae, however, clover root curculio larvae feed on alfalfa roots and nodules. Because the damage is subterranean, it is less noticeable than that caused by the alfalfa weevil. Nonetheless, several studies have shown that root damage caused by this pest can be severe, especially in young alfalfa stands, and may also involve increased incidence of root diseases (Dickason et al. 1968, Godfrey 1984, Underhill et al. 1955). Presently, the only insecticide registered for this pest is limited in use to an application at planting. No chemical control measures are available for its control in established alfalfa, and the subterranean habits of the larvae in an uncultivated, perennial habitat make this tactic difficult to develop. Biological control of the clover root curculio, however, appears feasible.

Early attempts at biological control of *Sitona* spp. in North America were primarily in Canada against the sweet clover weevil (*S. cylindricollis*). The results of those efforts are not well documented (Hagen et al. 1976). It should be noted that a biological control program aimed at a similar species, *S. discoideus*, is underway in Australia, where that accidentally introduced species damages alfalfa and other *Medicago* spp. Aeschlimann (1980) reported that 20 species of *Sitonia* were found on *Medicago* spp. in the Mediterranean region, and that these were attacked by several species of predators, parasites, and entomopathogens. Recent classical biological control efforts have been aimed at *S. discoideus* in Australia and *S. hispidulus* in the eastern U.S.

An egg parasite, *Anaphes diana* (formerly *Patasson lameerei*), was imported into the U.S. from Europe by USDA personnel in the mid-1970's and was released in Delaware, Idaho, and Kentucky. Field establishment to date, however, appears uncertain with only rare, tentative recoveries having been made. The lack of sufficient (Sitonia spp.) host egg populations during the summer in or near alfalfa fields in the eastern U.S. may be a limiting factor.

More recently, a multivoltine strain of Microctonus aethiopoides from Morocco has been imported into the U.S. by USDA personnel. The phenology of this strain differs from that of the bivoltine strain of M. aethiopoides introduced from Europe against the alfalfa weevil. The European M. aethiopoides rarely attacks the clover root curculio in the field. Recoveries of M. aethiopoides have been made from clover root curculio adults at field release sites in Kentucky, and this biological control effort is continuing. A strain of M. aethiopoides from Morocco has been successfully established on S. discoideus populations in Australia (Cullen & Hopkins 1982). Because of this, and the apparent success of European M. aethiopoides in the biological control of the alfalfa weevil in the eastern U.S., there is reason for optimism about the potential success of Moroccan M. aethiopoides against the clover root curculio. A number of questions remain to be answered, however, including that of reproductive isolation between Moroccan M. aethiopoides and the European strain already established in the eastern U.S. on the alfalfa weevil.

Mirid plant bugs Plant bugs are generally considered to be serious pests of alfalfa which is grown for seed production, because of the damage which they inflict on the flowers and developing seeds (App & Manglitz 1972). Several indigenous species of Lygus and Adelphocoris are associated with both forage and seed alfalfa in the U.S., as is the alfalfa plant bug (Adelphocoris lineolatus), which was probably introduced from Europe.

The reason for interest in biological control of these plant bugs on seed alfalfa is obvious. On forage alfalfa, however, these insects are generally thought to be minor pests (but see Walstrom 1983). Nonetheless, there is a less obvious reason for attempting their biological control in forage alfalfa. Mirid plant bugs are sometimes serious pests of other crops, such as cotton, and they are known to migrate from alfalfa to other habitats, especially following harvests (Poston & Pedigo 1975). Thus, biological control of these plant bugs in forage alfalfa could have benefits beyond those potentially accruing to this single crop. Four species of braconid parasites have been imported into the eastern U.S. from Europe to be released against the alfalfa plant bug and closely related mirids, but these have not yet been established (W.H. Day, personal communication).

Indigenous Natural Enemies

Numerous studies have demonstrated the richness of the indigenous beneficial arthropod fauna of forage alfalfa in the eastern U.S. (Culin & Yeargan 1983a, Howell & Pienkowski 1971, Los & Allen 1983, Pimentel & Wheeler 1973). Although this fauna is diverse, certain groups are particularly noteworthy because of their consistent occurrence in high numbers. Among the most abundant predaceous taxa are certain beetles (coccinellids and carabids), hemipteran bugs (nabids and anthocorids), and spiders (*Tetragnatha laboriosa*).

Several indigenous pests of alfalfa are attacked by an array of presumably co-evolved parasites and pathogens. For example, Roberts et al. (1977) reported that more than 15 species of primary parasites were associated with nine lepidopteran species on alfalfa in Illinois, and noted that more entomopathogens were associated with the caterpillars on alfalfa than on soybean. Lepidoptera larvae rarely cause significant damage to alfalfa in the eastern U.S. On the other hand, few parasites have been reported to attack the potato leafhopper on alfalfa. This native insect, which is sometimes a serious pest of alfalfa, migrates northward in the eastern U.S. each year from the Gulf Coast states. The possible impact of generalist predators or entomopathogens on potato leafhopper populations is largely unknown. Many other indigenous species of insects are occasional pests of alfalfa (App & Manglitz 1972), but they cause relatively little damage to forage alfalfa in the eastern U.S. While it is tempting to attribute their minor pest status to the action of indigenous natural enemies, this has not been experimentally demonstrated.

As with other crops, it has proven difficult to ascertain the role of indigenous natural enemies in the population regulation of specific pests. Several studies have reported predator/prey or host/parasite associations in alfalfa, but these usually have involved laboratory feeding trials (Martinez & Pienkowski 1982) or serendipitous field observations (Wheeler 1973) for predators, and simple host-rearing studies for parasites. These represent useful steps toward defining possible pest/natural-enemy relationships in forage alfalfa, but the demonstration of actual impact of natural enemies on pest population dynamics in alfalfa remains largely a subject of future research. In a recent study, involving the use of serological techniques and field population density estimates, Andrus (1984) attempted to quantify the impact of a predaceous hemipteran species on two selected herbivore species.

IMPACT OF CROP AND PEST MANAGEMENT TACTICS ON BIOLOGICAL CONTROL

Introduced or indigenous natural enemies of alfalfa pests may be affected by crop management practices, such as harvesting, or by certain pest management tactics, such as pesticide application. The severity of such effects is likely to be related to the timing of management activities with respect to the phenology of the natural enemies. For example, Bartell et al. (1976) showed that diapausing *B. curculionis* cocoons were relatively resistant to penetration by insecticides as compared with cocoons on nondiapausing individuals. Thus, insecticide applications made when diapausing individuals constitute the majority of that parasite's population should have little impact on its survival. Recent studies have determined the impact of selected insecticide applications on species composition and diversity of certain predaceous arthropods in alfalfa (Culin & Yeargan 1983b, Los & Allen 1983). These and similar studies are but the first step toward understanding the potential repercussions of management tactics in the alfalfa ecosystem.

There are few, if any, reports of attempts to experimentally demonstrate secondary outbreaks or pest resurgence in forage alfalfa following pesticide application. Such studies in other crops (Morrison et al. 1979) have provided strong circumstantial evidence that natural enemies were suppressing pest populations, and also indicated possible consequences of particular pesticide applications.

While it is obvious that broad-spectrum insecticides will likely affect insect predators and parasites, more subtle interactions may occur among management practices and biotic components of the alfalfa ecosystem. Wolfson & Yeargan (1983) found that spring applications of the herbicide metribuzin in Kentucky resulted in higher larval populations of the alfalfa weevil, but the cause is unknown. Norris et al. (1984) recently reported a similar phenomenon with the Egyptian alfalfa weevil (*Hypera brunneipennis*) in California; increases in larval weevil numbers were not dependent on the type of herbicide used. It is reasonable to suspect that fungicides applied to alfalfa for plant disease control might adversely affect fungal entomopathogens. Hostetter et al. (1983) investigated this possibility by using one of the few fungicides currently registered for common leaf spot control on alfalfa. They found that this particular fungicide (cupric hydroxide) had no effect on the incidence of *Erynia*-caused mortality in larval alfalfa weevil populations. Alfalfa plant resistance to certain insects (notably aphids) has been developed, but possible interactions between plant resistance and natural enemies have not been frequently investigated. Potential effects on natural enemies range from direct physical and chemical impact of glandular hair secretions (Shade et al. 1975) to three-trophic level interactions involving secondary plant compounds, as have been reported for resistant cultivars of certain other crops (Campbell & Duffey 1979). Clearly, numerous possible interactions between various forage alfalfa management tactics and biological control agents remain to be investigated.

SUMMARY AND RECOMMENDATIONS

Classical biological control programs for most major groups of alfalfa pests other than insects are either nonexistent or the current research and development efforts are in an early stage. Current research on biological control of plant pathogens and nematodes may prove practical, but there is little indication at this time of serious efforts to accomplish biological control of weeds in alfalfa. Concerns about host specificity of potential biological control agents of weeds in this crop, such as herbivorous insects or plant pathogens, dictate a conservative approach to this problem. Furthermore, with the (apparently imminent) advent of effective herbicides for control of weedy grasses in pure alfalfa stands, and the existence of effective herbicides for control of broadleaf weeds, there may be little impetus for biological control of weeds in alfalfa in the near future. There is evidence that classical biological control has been successful against a few insect pests of alfalfa in parts of the eastern U.S., notably the alfalfa weevil and the alfalfa blotch leafminer. Current programs for the redistribution of the most effective biological control agents of the alfalfa weevil to other parts of the U.S. appear to be worthwhile, but caution should be used to avoid inadvertent spread of hyperparasites. Two other likely candidates for increased classical biological control efforts in the near future are the clover root curculio and the blue alfalfa aphid. Both are introduced pests which are known to have effective natural enemies on other continents.

One of the biggest challenges to effective biological control of alfalfa pests is that of truly integrating biological control with other management tactics. The effectiveness of introduced and indigenous natural enemies may be affected by other management activities. Currently available economic injury levels are at the core of the decision process in most integrated pest management programs. These thresholds seldom take into account the presence or absence, much less magnitude, of natural enemy populations in the system. To minimize unnecessary and potentially disruptive pesticide applications, economic thresholds are needed which take into account pest density reductions that will occur from predictable biotic factors if no other immediate action is taken. Conversely, the consequences to biological control agents of various alternative pest management tactics need to be defined. To accomplish this, it must be determined whether biotic factors are regulating particular pest populations, and if so, which are the key factors. Procedures can then be determined to reduce adverse effects on important natural enemies by the selection and timing of appropriate management activities, such as harvesting or economically-justified pesticide applications.

Much is already known about the composition of the fauna and flora of the alfalfa ecosystem in the eastern U.S. Much remains to be learned about the interactions among these components, and the influence of various crop and pest management tactics on them.

REFERENCES

Aeschlimann, J.P. 1980. The Sitona species occurring on Medicago and their natural enemies in the Mediterranean region. Entomophaga 25:139-153.

Aeschlimann, J.P. 1981. Occurrence and natural enemies of Therioaphis trifolii Monell and Acyrthosiphon pisum Harris on lucerne in the Mediterranean region. Acta Ecol. Ecol. Appl. 2:3-11.

Andrus, S.M. 1984. A study of Reduviolus americoferus (Carayon) predation on two pests of alfalfa, Empoasca fabae (Harris) and Acyrthosiphon pisum (Harris). Ph.D. thesis, Virginia Polytech. Inst. State Univ.

Angalet, G.W., and R. Fuester. 1977. The Aphidius parasites of the pea aphid Acyrthosiphon pisum in the eastern half of the United States. Ann. Entomol. Soc. Amer. 70:87-96.

Anonymous. 1982. Agricultural Statistics, 1982. USDA. 566 pp.

Anonymous. 1983. Pesticides, pp. 4-13. In: Inputs, outlook and situation. IOS-2, October, 1983. Econ. Res. Serv., USDA.

App, B.A., and G.R. Manglitz. 1972. Insects and related pests, pp. 527-554. In: C.H. Hansen (Ed.), Alfalfa science and technology. Amer. Soc. Agron., Madison, WI.

Barnes, G.L., C.C. Russell, W.D. Foster and R.W. McNew. 1981. Aphelenchus avenae, a potential biological control agent for root rot fungi. Plant Dis. 65: 423-424.

Bartell, D.P., J.R. Sanborn and K.A. Wood. 1976. Insecticide penetration of cocoons containing diapausing and non-diapausing Bathyplectes curculionis, an endoparasite of the alfalfa weevil. Environ. Entomol. 5:659-661.

Brunson, M.H., and L.W. Coles. 1968. The introduction, release, and recovery of parasites of the alfalfa weevil in eastern United States. USDA Prod. Res. Rpt. 101, 12 pp.

Byers, R.A., and K. Valley. 1981. Losses in digestible dry matter and crude protein in alfalfa caused by the alfalfa blotch leafminer. Melsheimer Entomol. Ser. 31:8-13.

Campbell, B.C., and S.S. Duffey. 1979. Tomatine and parasitic wasps: Potential incompatibility of plant antibiosis with biological control. Science 205:700-702.

Culin, J.D., and K.V. Yeargan. 1983a. Comparative study of spider communities in alfalfa and soybean ecosystems: Foliage-dwelling spiders. Ann. Entomol. Soc. Amer. 76:825-831.

Culin, J.D., and K.V. Yeargan. 1983b. The effects of selected insecticides on spiders in alfalfa. J. Kan. Entomol. Soc. 56:151-158.

Cullen, J.M., and D.C. Hopkins. 1982. Rearing, release, and recovery of Microctonus aethiopoides imported for the control of Sitona discoideus in southeastern Australia. J. Austr. Entomol. Soc. 21:279-284.

Day, W.H. 1981. Biological control of the alfalfa weevil in the northeastern United States, pp. 361-374. In: G.C. Papavizas (Ed.), Biological control in crop production. Allanheld, Osmun, Montclair, N.J.

Day, W.H. 1983. Incidence of nonfunctional ovaries in Bathyplectes anurus and Bathyplectes curculionis, parasites of the alfalfa weevil in the northeastern United States. Environ. Entomol. 12:1125-1128.

DeBach, P. 1974. Biological control by natural enemies. Cambridge Univ. Press, London. 323 pp.

Dickason, E.A., C.M. Leach and A.E. Gross. 1968. Clover root curculio injury and vascular decay of alfalfa roots. J. Econ. Entomol. 61:1163-1168.

Dysart, R.J., and W.H. Day. 1976. Release and recovery of introduced parasites of the alfalfa weevil in eastern North America. USDA Prod. Res. Rpt. 167, 61 pp.

Frazer, B.D., N. Gilbert, V. Nealis and D.A. Raworth. 1981. Control of aphid density by a complex of predators. Can. Entomol. 113:1035-1041.

Gardner, W.A. 1982. Occurrence of Erynia sp. in Hypera postica in central Georgia. J. Invertebr. Pathol. 40:146-147.

Godfrey, L.D. 1984. Effects and interactions of the clover root curculio, Sitona hispidulus, alfalfa weevil, Hypera postica and root rot fungi in the alfalfa agroecosystem. Ph.D. Dissertation, Univ. of Kentucky, Lexington.

Gonzalez, D., W. White, J. Hall and R.C. Dickson. 1978. Geographical distribution of Aphidiidae imported to California for biological control of Acyrthosiphon kondoi and Acyrthosiphon pisum. Entomophaga 23:239-248.

Graham, J.H., D.L. Stuteville, F.I. Frosheiser and D.C. Erwin. 1979. A compendium of alfalfa diseases. Amer. Phytopathol. Soc., St. Paul, MN. 65 pp.

Gyrisco, G.G. 1958. Forage insects and their control. Annu. Rev. Entomol. 3: 421-448.

Hagen, K.S., G.A. Viktorov, K. Yasumatsu and M.F. Schuster. 1976. Biological control of pests of range, forage, and grain crops, pp. 397-442. In: C.B. Huffaker and P.S. Messenger (Eds.), Theory and practice of biological control. Academic Press, N.Y.

Hansen, C.H., and R.L. Davis. 1972. Highlights in the United States, pp. 35-51. In: C.H. Hansen (Ed.), Alfalfa science and technology. Amer. Soc. Agron., Madison, WI.

Harcourt, D.G., J.C. Guppy and M.R. Binns. 1977. The analysis of intrageneration change in eastern Ontario populations of the alfalfa weevil, Hypera postica. Can. Entomol. 109:1521-1534.

Harcourt, D.G., J.C. Guppy, D.M. MacLeod and D. Tyrrell. 1974. The fungus Entomophthora phytonomi pathogenic to the alfalfa weevil, Hypera postica. Can. Entomol. 106:1295-1300.

Hendrickson, R.M., and J.A. Plummer. 1983. Biological control of the alfalfa blotch leafminer in Delaware. J. Econ. Entomol. 76:757-761.

Hostetter, D.L., B. Puttler, J.L. Huggans, R.E. Pinnell and S.H. Long. 1983. Effect of the fungicide Kocide on the entomopathogenic fungus Erynia phytonomi of the alfalfa weevil, Hypera postica in Missouri. J. Econ. Entomol. 76:619-621.

Howell, J.O., and R.L. Pienkowski. 1971. Spider populations in alfalfa, with notes on spider prey and effect of harvest. J. Econ. Entomol. 64:163-168.

Kogan, M. 1980. Insect problems of soybeans in the United States, pp. 303-325. In: F.T. Corbin (Ed.), World soybean research conference II: Proceedings. Westview Press, Boulder, CO.

Latheef, M.A., J.C. Parr and B.C. Pass. 1979. Factors affecting the survival of Kentucky populations of the alfalfa weevil, Hypera postica. Environ. Entomol. 8:1032-1036.

Los, L.M., and W.A. Allen. 1983. Abundance and diversity of adult Carabidae in insecticide-treated and untreated alfalfa fields. Environ. Entomol. 12:1068-1072.

MacCollom, G.B., G.L. Baumann, N.L. Gilroy and J.G. Welch. 1982. Alfalfa blotch leafminer, Agromyza frontella, effects on alfalfa in Vermont. Can. Entomol. 114:673-680.

Mankau, R. 1980. Biological control of nematode pests by natural enemies. Annu. Rev. Phytopathol. 18:415-440.

Martinez, D.G., and R.L. Pienkowski. 1982. Laboratory studies on insect predators of potato leafhopper eggs, nymphs, and adults. Environ. Entomol. 11:361-362.

Miller, D.E., and G.L. Jensen. 1970. Agromyzid alfalfa leafminers and their parasites in Massachusetts. J. Econ. Entomol. 63:1337-1338.

Milner, R.J., R.S. Soper and G.G. Lutton. 1982. Field release of an Israeli strain of the fungus Zoophthora radicans for biological control of Therioaphis trifolii. J. Austr. Entomol. Soc. 21:113-118.

Morrison, D.E., J.R. Bradley and J.W. Van Duyn. 1979. Populations of corn earworm and associated predators after applications of certain soil applied pesticides to soybeans. J. Econ. Entomol. 72:97-100.

Morrison, W.P., B.C. Pass, M.P. Nichols and E.J. Armbrust. 1974. A bibliography of the Sitona species. Ill. Nat. Hist. Survey, Biol. Notes 88, 24 pp.

Neuenschwander, P., K.S. Hagen and R.F. Smith. 1975. Predation of aphids in California's alfalfa fields. Hilgardia 43:53-78.

Nordin, G.L., G.C. Brown and J.A. Millstein. 1983. Epizootic phenology of Erynia disease of the alfalfa weevil, Hypera postica, in central Kentucky. Environ. Entomol. 12:1350-1355.

Norris, R.F., W.R. Cothran and V.E. Burton. 1984. Interactions between winter annual weeds and Egyptian alfalfa weevil in alfalfa. J. Econ. Entomol. 77:43-52.

Papavizas, G.C., and R.D. Lumsden. 1980. Biological control of soilborne fungal propagules. Annu. Rev. Phytopathol. 18:389-413.

Peters, E.J., and R.A. Peters. 1972. Weeds and weed control, pp. 555-573. In: C.H. Hansen (Ed.), Alfalfa science and technology. Amer. Soc. Agron., Madison, WI. 812 pp.

Pimentel, D., and A.G. Wheeler. 1973. Species diversity of arthropods in the alfalfa community. Environ. Entomol. 2:659-668.

Pinter, P.J., N.F. Hadley and J.H. Lindsay. 1975. Alfalfa crop micrometeorology and its relation to insect pest biology and control. Environ. Entomol. 4:153-162.

Poston, F.L., and L.P. Pedigo. 1975. Migration of plant bugs and the potato leafhopper in a soybean-alfalfa complex. Environ. Entomol. 4:8-10.

Puttler, B., D.L. Hostetter, S.H. Long and R.E. Pinell. 1978. Entomophthora phytonomi, a fungal pathogen of the alfalfa weevil in the Mid-Great Plains. Environ. Entomol. 7:670-671.

Roberts, S.J., W.K. Mellors and E.J. Armbrust. 1977. Parasites of lepidopterous larvae in alfalfa and soybeans in central Illinois. Great Lakes Entomol. 10: 87-93.

Sailer, R.I. 1978. Our immigrant insect fauna. Bull. Entomol. Soc. Amer. 24:3-11.

Sayre, R.M. 1980. Promising organisms for biocontrol of nematodes. Plant Dis. 64:526-532.

Shade, R.E., T.E. Thompson and W.R. Campbell. 1975. An alfalfa weevil larval resistance mechanism detected in Medicago. J. Econ. Entomol. 68:399-404.

Southwood, T.R.E. 1977. The relevance of population dynamics theory to pest status, pp. 35-54. In: J.M. Cherrett and G.R. Sagar (Eds.), Origin of pest, parasite, disease, and weed problems. Blackwell Sci. Publ., London.

Stehr, F.W. 1982. Parasitoids and predators in pest management, pp. 135-173. In: R.L. Metcalf and W.H. Luckmann (Eds.), Introduction to insect pest management. John Wiley & Sons, N.Y.

Thompson. L.S. 1981. Field evaluation of insecticides for control of the alfalfa blotch leafminer and its effects on alfalfa yield in Prince Edward Island. J. Econ. Entomol. 74:363-365.

Underhill, G.W., E.C. Turner and R.G. Henderson. 1955. Control of the clover root curculio on alfalfa with notes on life history and habits. J. Econ. Entomol. 48:184-187.

Walstrom, R.J. 1983. Plant bug damage to first-crop alfalfa in South Dakota. J. Econ. Entomol. 76:1309-1311.

Wheeler, A.G. 1973. Studies of the arthropod fauna of alfalfa. V. Spiders (Araneida). Can. Entomol. 105:425-432.

Wolfson, J.L., and K.V. Yeargan. 1983. The effects of metribuzin on larval populations of alfalfa weevil, Hypera postica. J. Kan. Entomol. Soc. 56:40-46.

Wood, K.A., E.J. Armbrust, D.P. Bartell and B.J. Irwin. 1978. A bibliography of the alfalfa weevil, Hypera postica (Gyllenhal), and the Egyptian alfalfa weevil Hypera brunneipennis (Boheman). Ill. Agr. Exp. Sta. Spec. Publ. 54, 92 pp.

Yeargan, K.V. 1979. Ovipositional rate, fecundity, and longevity of Bathyplectes stenostigma. Environ. Entomol. 8:150-153.

COTTON: STATUS AND CURRENT LIMITATIONS TO BIOLOGICAL CONTROL IN TEXAS AND ARKANSAS

James R. Cate

Biological control in cotton may seem to be a contradiction. Cotton in the mid-1960's received an estimated 40-50% of all insecticides applied to agricultural crops in the U.S. Eichers et al. (1970) projected this estimate to be 60 to 75% when states that do not produce cotton were removed from the computations. There are numerous ways insecticides are misused and used to excess, exemplifying the "pesticide syndrome" (Doutt & Smith 1971). Unilateral dependence upon insecticidal control has failed to various degrees throughout the south and southwest, most notably in southern Texas and northeastern Mexico where a cotton production area of 700,000 acres was brought to the verge of economic collapse (Adkisson 1969).

The crisis in the cotton economy, as is often the case during periods of adversity, brought about significant positive change. Cotton production in Texas began a shift away from the unilateral dependence upon insecticides (Frisbie & Adkisson, this volume; Walker et al. 1978). While Arkansas cotton production has not gone to the brink of collapse as has production in South Texas, the chemical cart has most certainly, on occasion, run away with the ecological horse.

Current crop production systems in Texas and Arkansas are notable because they are accepted and used by producers. Secondarily, they address a complete crop production system. It is most likely this last characteristic which has allowed success. The systems provide two different solutions to the integrated pest management of boll weevil, *Anthonomus grandis*, and *Heliothis* spp. (*H. virescens*, the tobacco budworm, and *H. zea*, the cotton bollworm). Biological control is an integral part of the management of *Heliothis* populations, but would be negated except for the use of very carefully timed controls for boll weevil. Insect pest management in cotton in Arkansas and Texas uses cultural practices, weather advantages, biological control and effectively applied insecticides. Since biological control of insects in cotton is such an integral part of the crop production system, it will be necessary to describe these systems.

ISBN 0-12-357030-1

THE IPM SYSTEM

The present IPM system in cotton did not become what it is by some well coordinated plan. It has evolved over the last 35 years, and incorporates strategies recommended at the turn of this century. The intermeshing of insect control tactics has evolved to suit the production practices and the sequential appearance of a number of pests in the United States. Cotton bollworm was present very early as both a pest of cotton and corn, two crops grown together in the south in the 18th and 19th centuries. Since there was no chemical remedy for it, it was endured along with other perturbations of nature. In 1894, the boll weevil made its appearance in south Texas and began its spread eastward and northward. Again, a ready chemical remedy was not available so it too was endured. However, damage by boll weevil was much more acute than that inflicted by Heliothis causing producers to seek an effective remedy, which eventually was found in the form of an inorganic chemical, calcium arsenate. Before a chemical remedy was discovered, however, Mally (1901) suggested production practices which would lessen the weevil's damage. These cultural controls were: 1) uniform planting dates to assure rapid growth of a crop of uniform age, 2) use of varieties that matured early, 3) management of the crop so as to obtain early harvest and field sanitation, 4) growing the crop on the most productive lands, leaving out of production marginal areas unlikely to produce vigorous growth and early maturity. Isely & Baerg (1924) recommended limited use of insecticides by "spot-treatment" and "field edge treatment" only when boll weevil populations were large enough to warrant control.

In 1917, pink bollworm, Pectinophora gossypiella, was found in the United States. During the 1950's, it caused extensive losses. Investigations at the time suggested practicing: 1) field sanitation, including shredding and burial of stalks immediately after harvest and destruction of other crop residue such as weed, waste fruit and ginning refuse, 2) maintenance of high soil moisture to facilitate decomposition of residues, 3) uniform planting over a short interval to obtain thick and vigorous stands, 4) avoidance of excessive fertilization and 5) uniform and early termination of crop growth. Thus, the early recommendations for crop maintenance for management of boll weevil and pink bollworm were similar and enforced the production of a determinate cotton crop. These production guidelines became a part of the philosophy and recommendations for IPM in cotton in both Texas and Arkansas. The result of these recommendations has been the evolution toward a shorter, more abruptly disturbed agroecosystem. This strategy provided virtually complete control of pink bollworm so that it is rarely a pest in either state today, and this cultural control of pink bollworm is mandated by law in Texas.

This discussion of the "evolved" IPM system in cotton in Arkansas and Texas sets the stage for consideration of biological control. For it is this routinely and annually perturbed production system in which predators and parasitoids must colonize and survive. The phenology of occurrence of the pests is shown in Fig. 1 (from Frisbie & Bohmfalk 1982).

BIOLOGICAL CONTROL IN COTTON

The important points to be made about biological control of boll weevil are the potential for classical biological control and the control strategies for boll weevil which permit biological control of Heliothis. Biological control of boll weevil is not practiced in Texas or Arkansas except to the extent that native natural enemies inflict unmeasured mortality. This mortality is not predictable in its occurrence, nor is it generally acknowledged except in experimental trials. However, considerable effort has been made to locate and identify potential, exotic natural enemies of the boll weevil. Additionally, studies of the ecology of the indigenous parasite Bracon mellitor are being pursued, as is the ecology of an omnivorous, imported predator, the imported fire ant (Solenopsis invicta).

Control of boll weevil (a key pest as defined by Smith & van den Bosch 1967) in Texas and Arkansas remains dependent upon insecticides. Control may be achieved by application of insecticides during the growing season based upon an economic threshold or by carefully timed insecticide applications made early or late in the growing season. In Texas, the economic threshold is 15-25% punctured flower buds (squares). In Arkansas, this threshold ranges from 1.0 to 1.5 punctured squares per foot of row, depending upon crop maturity. Insecticides applied during the season for boll weevil or other pests destroy entomophagous arthropods which help hold in check the populations of the secondary pests, Heliothis (Dinkins et al 1970, Ewing & Ivy 1943, Gaines 1955, Lincoln & Leigh 1957, Shepard & Sterling 1972a). For this reason, the generally recognized strategy in cotton is to apply these insecticides to control boll weevil at times during the season which will minimally affect the entomophagous arthropods.

One boll weevil control strategy is Fall Reproductive/Diapause Control (Brazzel et al. 1961, Rummel et al. 1975). In this instance insecticides are applied for control of pre-diapause boll weevils at the time the crop is mature and thereafter at 10-14 day intervals until frost. Generally, no more control is required before the next season's crop is mature. This type of control must be practiced on an areawide basis with virtually 100% compliance to be effective. This control strategy has not proven effective in the boll weevil's southern range.

Another alternative is the short-season cotton strategy which utilizes two to three insecticide applications made at the time flower buds first appear on the plant (Parker et al. 1980, Walker & Niles 1971). These applications, if timed properly, kill the colonizing overwintering adult weevils. If relatively rapid-

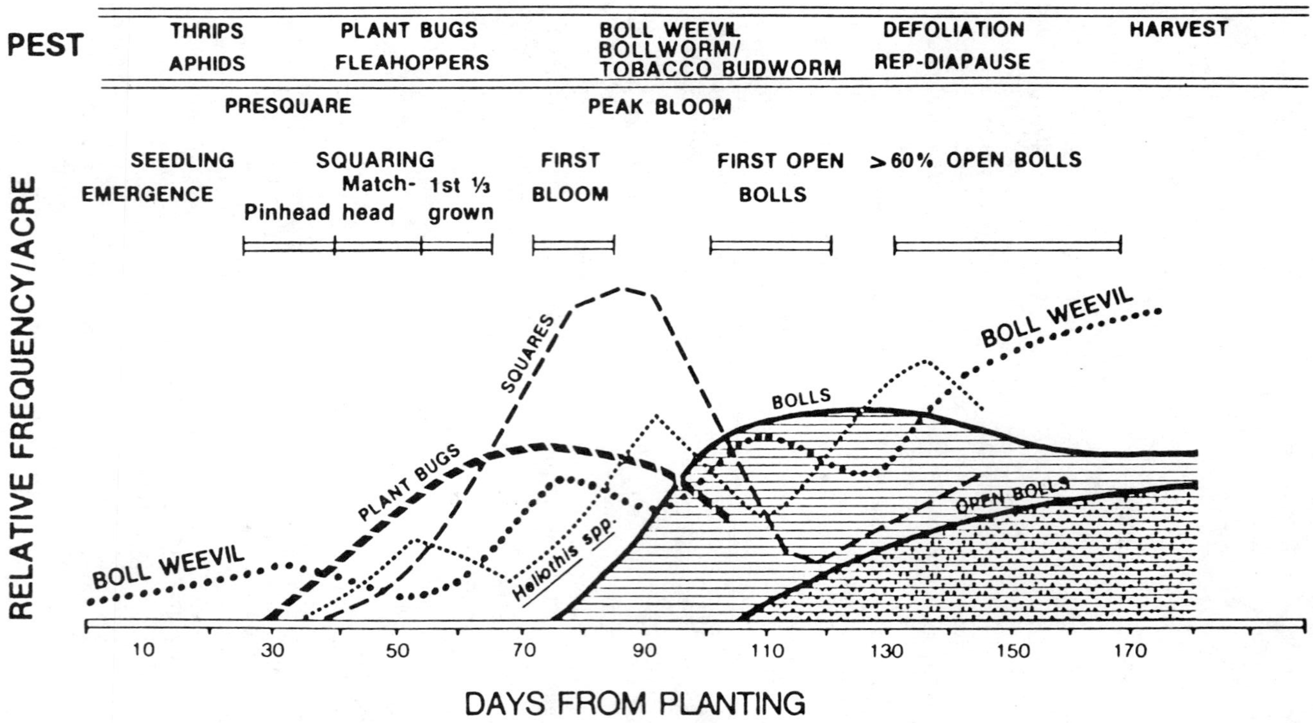

Fig. 1. Phenology of cotton growth and of the occurrence of insect pests on Texas cotton (after Frisbie & Bohmfolk 1982).

fruiting cottons are grown, a crop can be matured before weevil populations increase to damaging levels.

Most recently, in the Rolling Plains of Texas, producers have agreed to practice "uniform, delayed planting" of the crop so that most boll weevils emerge from diapause and die before food (flower buds) is available (Boring 1974, Slosser 1978). This tactic, practiced on an areawide basis, has eliminated insecticide use for boll weevil in some years and drastically reduced the number of treatments in other years (Boring & Fuchs 1980). Elimination of insecticide treatments for boll weevil has consequently eliminated applications for *Heliothis*.

The Fall Reproductive/Diapause Control and in-season applications of pesticides are used in Arkansas. The emphasis in Arkansas is to refrain from insecticide use until absolutely necessary, as indicated by predetermined economic threshold levels.

BIOLOGICAL CONTROL OF THE BOLL WEEVIL

Entomologists and botanists of the 1890's and 1900's made considerable efforts to locate and identify other hosts and natural enemies of boll weevil in the United States, Mexico, Cuba, and Central and South America (Cross & Chesnut 1971, Hunter & Pierce 1912, Pierce et al. 1912). By 1912, 55 entomophagous arthropods attacking boll weevil had been recorded (Pierce et al. 1912). A review by Cross & Chesnut (1971) revealed that during the 59 years elapsing since the review of Pierce et al. (1912) three new parasitoids were found, two of which are native to the U.S. and only facultatively associated with the boll weevil (Cross & McGovern 1969). The other parasitoid *Catolaccus* (*Heterolaccus*) *grandis* (Pteromalidae) is widely distributed in Mexico (J.R. Cate, unpublished data; Cross & Mitchell 1969). Although several efforts have been made, no effective exotic natural enemy of the boll weevil has been imported and permanently established in the U.S.

W.D. Hunter transported a colony of parasitic mites, *Pediculoides ventricosus*, from Professor A.L. Herrera's laboratory in Coaxuila, Mexico to Texas in the fall of 1902 without success (Hunter & Hinds 1905). A botanist, O.C. Cook, discovered an ant in Guatemala that attacked adult boll weevils, stung them, and carried them to provision their nests. He imported 89 colonies of this "kelep ant", *Ectatoma tuberculatum* (Ponerinae), and released them near Victoria, Texas (Cook 1904, 1905). Laboratory studies showed that the ant could tolerate the cold seasons normally encountered in the Victoria area, but they never established. Cook suggested that the heavy, clay soils of the Victoria area were very different from the well drained, limestone soils in which the ant existed in its indigenous habitat and gave this as the cause of its failure to establish. Wheeler (1904) proposed that the ant failed to survive because of inherent behavioral inflexibility.

Berry (1947) imported two parasites of *Anthonomus vestitus* (a small weevil about one-half as large as boll weevil which develops in flower buds of cultivated cotton in Peru and Columbia) between 1941 and 1945. These parasites, *Triaspis vestiticida* and *Bracon vestiticida*, parasitize boll weevil larvae, but failed to become permanently established (Annand 1946). *Bracon kirkpatricki* and *B. greeni* have been imported from Africa and India, respectively. These parasites of pink bollworm have been found to parasitize the boll weevil in laboratory culture. Field releases of *B. kirkpatricki* have been made, but it too failed to become permanently established (Cross et al. 1969).

W.H. Cross (personal communication) has made several explorations into Mexico, Central America and South America to locate host plants and natural enemies of the boll weevil. He and co-workers imported *Catolaccus grandis* and *Urosigalphus monotonus* (referred to as *U. schwarzi* by Cross) (Braconidae). *U. monotonus* failed to mate in the quarantine laboratory and consequently was never released to the field. *C. grandis* was released in Mississippi and produced a five-fold population increase during the season, but failed to survive the winter. Calvert (1974) collected *U. monotonus* from Guatemala and Mexico, but none of the parasitoids survived in quarantine. Since 1976, we have conducted foreign explorations for natural enemies of boll weevil and its nearest relative, *A. hunteri* (Burke & Cate 1979, 1983). The following species are being cultured now in our laboratory: *C. grandis*, *C. hunteri*, *Heterospilus megalopus*, *H. annulatus*, *B. compressitarsis* and *U. monotonus*. Releases are being made near College Station and Brownsville, Texas.

Considerable attention has been given indigenous natural enemies of the boll weevil since Townsend's initial studies (Townsend 1895). Qualitative studies conducted in the United States provide a long list of entomophagous arthropods attacking some stage of the boll weevil (Bondy & Rainwater 1935, Chesnut & Cross 1971, Cross & Chesnut 1971, Fenton & Dunnam 1929, Grossmann 1931, Howard 1897, Pierce 1908, Pierce et al. 1912). Surveys for parasites of the boll weevil have not been made systematically during the season; however, relative abundance has been noted (Chesnut & Cross 1971, Hunter & Pierce 1912). In general, changes occur due to the year and locality, but *B. mellitor* constitutes 75-80% of all parasites reared from boll weevils. *Aliolus curculionis*, *Eurytoma gossypii*, *Catolaccus hunteri*, *Zatropis incertus* and *Eupelmus cyaniceps* also occur in significant numbers.

Parasitism by native parasites is the product of a complex association of natural enemies, their insect hosts and host plant species (Pierce et al. 1912). The host associations of *B. mellitor* in Texas, Louisiana, Mississippi, and Arkansas (Pierce et al. 1912) and in southeast Texas (Tillman 1985) have been determined. Host insects include representatives of the Lepidoptera and Coleoptera which occur concurrently with the appearance of fruiting structures on their respective host plants. The parasitoid overwinters on insect hosts in *Grindelia squarrosa*, *Heterotheca* latifolia and Datura stramonium. Host plant fruiting cycles suggest a sequence of host plants which could maintain B. mellitor throughout the year, and host insects occur at almost the same time as the fruiting periods for the plants. The entire phenological sequence is, as expected, altered appreciably between years due to vagaries of the weather.

While arthropod enemies of the boll weevil are known and frequently reported, no quantitative studies have been completed, with the exception of a recent life table analysis made in West Texas (Bottrell 1976b). In this particular study, where B. mellitor was the dominant parasite, 44% real mortality attributable to parasites occurred with weevil populations of 46% infested squares. With 15% infested squares, 4% real mortality was attributable to parasites.

In addition to arthropod enemies of the boll weevil, a few pathogens have been reported (McLaughlin 1965, 1969). They are the sporozoans Glugea gasti and Mettesia grandis. The bacterium, Serratia marcescens (McLaughlin 1969) and the fungi, Aspergillis flavus and A. parasiticus (Moody et al. 1975) also have been reported to infect the boll weevil. Protozoans are limited for use in biological control because they must be ingested to cause infection. However, McLaughlin (1966) demonstrated that a feeding stimulant (Daum et al. 1967) would successfully induce infection in boll weevils by M. grandis. Subsequently, he kept boll weevil populations below economic levels in field cages with M. grandis and G. gasti incorporated in food baits (McLaughlin 1967, McLaughlin et al. 1968). McLaughlin et al. (1968) demonstrated that field populations could be infected by this method, but no evaluations of the degree of population regulation were made. The disease organisms overwintered in infected diapausing adults and mortality of hibernating weevils increased.

The newly-invaded habitat of the boll weevil offered other natural deterrents. Forty-three species of birds (Howell 1907, 1909) and the Fowler's toad (Roach 1973) utilize adult boll weevils as food.

Efforts have been made to increase the effectiveness of certain indigenous arthropod enemies of the boll weevil. Pierce (1908) found that percentage parasitism could be increased by collecting parasites from one location and releasing them in another; and that increases in percentage parasitism could be obtained by collecting fallen squares and allowing the parasites to emerge from them in screen wire cages that permitted the parasite's escape without releasing the weevils.

Pierce et al. (1912) suggested that parasite populations could be developed on alternate host insects in and about cotton fields by encouraging the growth of plant species that attract and support hosts of these parasites. As an added measure he suggested that these wild or cultivated "nursery" plants be destroyed or mowed at critical times so as to encourage the parasites to seek new hosts in cotton. While Pierce and his co-workers reported increases in percentage parasitism by all of these methods, their evidence was inconclusive.

Recent studies show that the pheromone, grandlure, can concentrate 70% or more of the overwintering generation of boll weevil in 5 to 15% of a cotton field when population densities are not high (Bottrell 1972). This method might be used to attract *B. mellitor* into cotton fields early in the season when densities of the boll weevil are too low to attract the parasite from its native hosts.

McGovern & Cross (1976) increased the effectiveness of *B. mellitor* by the use of the Frego bract character in commercial cottons. Parasitism was increased from 7% in normal bract cottons to 56% in Frego bract cottons.

In some cotton varieties boll weevil-infested buds do not abscise completely, leaving infested buds hanging on the plant with the developing larva within. Field studies show that parasitism by *B. mellitor* in these "hanging squares" is 2- to 3-fold greater than in buds collected from the ground (Adams et al. 1969; J.R. Cate, unpublished data). *B. mellitor* attacks third instar larvae or pupae of the boll weevil, which in normally abscised buds are on the ground. However, with the "hanging square" varieties, the susceptible stage of the host is retained on the plant where the parasitoid prefers to search. This distinct searching behavior was seen in cage studies where hosts were placed either on the top or the bottom of the cage, no more than 0.5 meter apart (O'Neil & Cate 1985).

The imported fire ant, *Solenopsis invicta*, provides control of boll weevil in some years where it occurs in eastern Texas (Sterling 1978, Sterling et al. 1984). Jones & Sterling (1979) were able to utilize these omnivorous predators to reduce damage in the crop from 38.8 to 16.6%. In circumstances where it can be utilized, it may be an effective predator. It is indiscriminant in preying on parasitized and non-parasitized larvae of *Anthonomus fulvus* in buds of *Callirhoe involucrata*; thus, it must be considered a potential threat to the native parasitoid populations and any successfully established parasitoids.

All of the important parasites and predators of *A. grandis* in the United States probably have been identified. They are predominantly facultative parasites of the boll weevil, which depend on native weevil species for their survival. With the exception of one study conducted in West Texas (Bottrell 1976a), no studies designed to determine the effectiveness of indigenous natural enemies in the absence of pesticides have been completed.

BIOLOGICAL CONTROL OF *HELIOTHIS* SPP.

An excellent treatment of biological control of *Heliothis* spp., both by conservation and importation, has been given by Johnson et al. (1985). These authors have taken a broad perspective examining the current status of biological control in soybean and cotton. This discussion will not address the prospects or history of classical biological control although it appears to be a very viable alternative. There are a number of arthropod species which constitute relatively consistant members of a guild of predators which inhabits cotton fields in Arkansas and Texas (Ridgway & Lingren 1972a). Populations of each species and the relative abundance among the species may vary from year to year and from location to location. However, the trend is for an increase in numbers early in the season, peaking about the time of peak squaring (Dinkins et al. 1970, Schuster & Boling 1974, Smith & Stadelbacher 1978). Populations of entomophagous arthropods colonize cotton during the spring, coming from natural vegetation and winter grain crops. This colonization appears to be related to changes in availability of extrafloral nectar which peaks at peak bloom. This reservoir of natural enemies is not unlimited in numbers since, while one early season application of insecticide has only two weeks' effect, repeated applications at this time can eliminate the colonizing population (Dinkins et al. 1970, Ridgway et al. 1967, Shepard & Sterling 1972b). These reservoirs are generally depleted by mid-July (Dinkins et al. 1970, Schuster & Boling 1974).

The parasitoid fauna of Heliothis larvae is diverse (Ridgway & Lingren 1972). In concert with the predator species, these parasitoids add an element which, in aggregate, may act in a density dependent manner. Ables et al. (1983) showed an inverse relationship between the number of predators and number of Heliothis larvae per acre. The problem is to predict the impact, or potential impact, of these natural enemies so that decisions can be made regarding the need for insecticides or the wisdom of relying upon mortality from predators and parasites. Sterling established "inaction levels" for natural enemy populations (McDaniel & Sterling 1979). Because of complexities that go into the decision process, several researchers have used computer simulations. The question is whether to model, with intricate detail, the behavior of each predator and parasitoid, or seek some holistic, but simple description of this combined predation/parasitism. Hartstack et al. (1975) utilized a simulation of the aggregate response in the MOTHZV-2 simulation. The relationship used is:

$$M = 1-\exp \frac{-0.693\ P_n}{NS}$$

where, M = Probability of mortality of Heliothis spp. eggs or larvae,

P_n = Number of effective predators,

N = Number of effective predators required to cause a 50% mortality of prey, and

S = Relative search area (1=first square to 6=immature plant).

This relationship takes into account the rapid increase in area to be searched as the plant grows and the decreased efficiency of the predator as Heliothis larvae mature. After this simplicity, Hartstack et al. (1975) improved the model by establishing age distribution classes and assigning different indices of relative predation efficiency to each predator or predator group. The model, using data from 19 cotton fields predicted survival of large Heliothis larvae with an R-value of 0.674 using a linear relationship.

Hartstack et al. (1975) also established a trichotomus decision chart to aid in the decision to use insecticides or not depending on natural enemy presence and potential impact on the pest population (Fig. 2). The number of Heliothis eggs is plotted against the number of "effective natural predators", with decision line at about one "effective natural predator" per two Heliothis eggs. The key to this decision aid is the determination of the "normalized" relative efficiency indices for predation of Heliothis eggs and larvae (Ables et al. 1983, Table 1), using prey consumed per predator per day. While considerably better data are required to produce more realistic indices, this approach seeks to provide a decision tool for IPM in cotton utilizing natural enemies. This "natural enemy element" will have to be factored into the economic threshold before biological con-

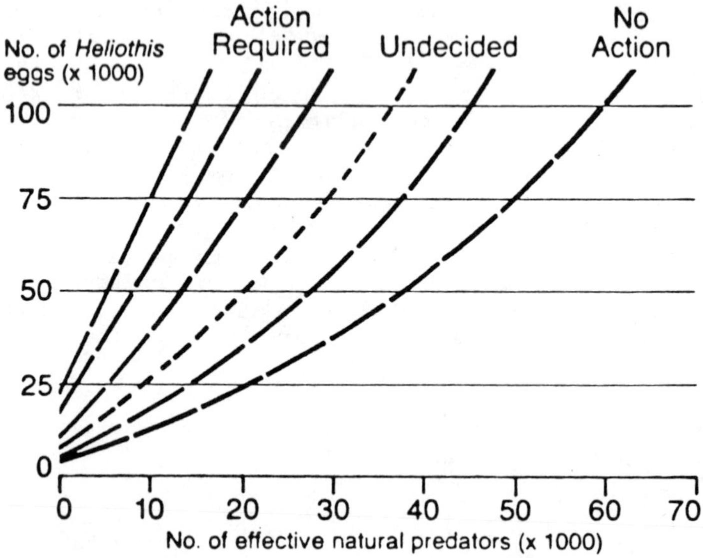

Fig. 2. Decision chart showing the number of naturally-occurring predators and Heliothis spp. eggs per acre of cotton. A required corrective action is expected when the number of large Heliothis spp. larvae reaches or exceeds 2500/ acre (reproduced from Ridgway et al. 1983).

trol truly will be used in IPM in cotton. Computerized simulation models may allow instantaneous evaluation of these complex components. Parasites are hardly considered in this decision process. Lists of parasitic species have been accumulated (Fletcher & Thomas 1943, Harding 1976, Lewis & Brazzel 1968, Quaintance & Brues 1905, Shepard & Sterling 1972b, Whitcomb & Bell 1964), yet neither the abundance of a particular parasitoid nor of the aggregate of all species has been utilized in the IPM decision process.

Conservation of the parasitoid and predator complex of Heliothis has been the focus of Harris & Phillips (1984), Mueller (1983) and Mueller & Phillips (1983). These scientists seek to conserve natural populations of Microplitis croceipes and Cardiochiles nigriceps, two effective and host specific parasitoids of Heliothis. The concern for these parasitoids and the predator guild caused Mueller et al. (1984) to question the wisdom of destroying weed hosts (Stadelbacher 1981) which support the first spring generation of Heliothis and their natural enemies.

The natural enemies discussed above are utilized in Texas as part of the short-season cotton (low-input) system. This ap-

Table 1. Relative efficiency indices for predation on Heliothis spp. eggs and larvae by eight predator groups.[a]

	Index for Egg Predation Factor	Index for Larval Predation Factor
Scymnus	0.21	0.22
Other Coccinellidae[b]	2.70	0.69
Other Coleoptera[c]	0.23	0.086
Orius	0.31	0.40
Geocoris	0.92	2.30
Other Hemiptera[d]	0.69	1.00
Chrysopa	2.50	2.90
Spiders	0.49	0.35

[a] Original data (comsumption of eggs or larvae/predator/day) were adjusted to reflect factors for each predator group relative to each other (i.e., a higher value represents a higher efficiency); due to normalization, the total for all factors is eight, which is the total number of predator groups (modified from Ables et al. 1983).

[b] Primarily Hippodamia, Coleomegilla, and Cycloneda.

[c] Primarily Collops spp., Notoxus spp., and small Carabidae.

[d] Primarily Nabidae, Reduviidae, Pentatomidae (Asopinae), and Miridae.

proach seeks to manage the crop as to enhance the abiotic mortality of immature stages of *Heliothis* by minimizing the fertilizer and irrigation applied to the cotton. The approach also seeks to maximize the numbers and action of the natural enemy component. Most significantly, this has come about because of a change in perspective among producers regarding *Heliothis* (Walker et al. 1978). At one time it was perceived that an application of an insecticide resulted in a positive and, at the very worst, a neutral impact. However, today many Texas and Arkansas producers perceive both a positive and a negative impact. Experience has shown that an insecticide application can increase the insect damage to the crop.

In Arkansas, the Areawide Bollworm Management Community (BMC) Program has been the conceptual vehicle to alter producer practices. After three years of intensive study of bollworm in Arkansas, Phillips and coworkers (Frisbie et al. 1983; Nicholson et al. 1983, 1984; Phillips et al. 1981) determined that: 1) long-range bollworm migration had not contributed to bollworm outbreaks; 2) June populations of bollworms in cotton were sufficient to produce the population levels that were subsequently measured; 3) treatment thresholds could be adjusted upward; 4) considerable progress had been made in developing a bollworm forecasting network; and 5) it was possible to utilize as much as 12,000 acres of cotton as a single field for a bollworm suppression system. They concluded that mismanagement of insect control procedures was responsible for bollworm problems.

The program leaders sought first to establish a cultural system for cotton which would minimize pest problems. Practices to avoid were: 1) planting into unfavorable situations increases problems with thrips, delaying the crop and increasing the possibility of bollworm and boll weevil problems; 2) using excess nitrogen fertilizer aggravates plant bug and bollworm problems; 3) planting thick stands aggravates plant bug and bollworm problems; 4) indiscriminate use of herbicides can delay the crop, exposing it to the more difficult-to-control bollworm and boll weevil. In both states, the goal is to produce a suitable yield within a single fruiting cycle.

After raising the treatment threshold for bollworm it followed that the normally-occurring June population of *Heliothis* should not be treated. This action preserved natural enemies that might help control the more significant August populations. Careful field scouting informed the producer and BMC Program personnel of the pest insect populations. Over 90% of the producers chose to participate in the program.

It was decided after the first year that *Heliothis* population buildup could be delayed from July to August by control of the June populations. Commercial formulations of *Bacillus thuringiensis* and *Baculovirus heliothis* were suited for this since they selectively kill *Heliothis* rather than associated parasitoids and predators. Although *B. thuringiensis* and *B. heliothis* gave only 25-50% control, this selective mortality plus the indigenous natural enemies provided economic suppression. With various seasonal variations this program worked well and reduced insecticide usage by 80%. During 1978 the program was less effective, but still reduced pesticide applications by about 50%. In 1983 an economic evaluation showed an overall reduction from 8.9-4.1 treatments (Scott et al. 1983). Savings to producers averaged $11.00 per acre. Additionally, these types of community structured educational programs aided the introduction of new technology to the producer.

BIOLOGICAL CONTROL OF PLANT BUGS

The tarnished plant bug, Lygus lineolaris, and the cotton fleahopper, Pseusatomoscelis seriatus, commonly cause crop damage early in the season, and as a consequence, cause early, disruptive insecticide applications. Biological control is not utilized for either pest. However, foreign exploration is now under way and parasitoids are being sought from related host mirids in eastern Africa. A native nymphal parasite, Leiophron pallipes, seems to have virtually no effect in cotton or other crops (Scales 1973). Surveys for indigenous parasitoids of cotton fleahopper show that only two parasitize eggs; Anaphes anomocerus (Gahan 1937) and Erythmelus psallidis (Ewing & Crawford 1939). These seem to have little effect in commercial cotton. Foreign exploration may hold potential, but taxonomic problems with the species of Phyliniae (Miridae) first must be resolved.

LIMITATIONS TO BIOLOGICAL CONTROL

Biological control by conservation of indigenous natural enemies seems to be utilized to a reasonable degree in Texas and Arkansas. That is not to say that practices might not be improved. However, this can occur only after we know more about the potential of the various predators and parasites. In the cases of Heliothis and boll weevil it is imperative that the ecology of the indigenous natural enemies be better understood.

Non-crop refugia for these natural enemies are very important. It is necessary to know what and where these refugia are, and to gain enough knowledge of the population dynamics to know the impact of seasonal weather variations. We must be able to predict with a high degree of certainty the potential for control from the indigenous fauna. We must understand the recruitment response which attracts these natural enemies to the pest. Once understood, recruitment rates may be predicted. With enough understanding, recruitment might be manipulated. The entomophagous adults have requisites for survival and reproduction which should be identified and provided, if possible. These requisites may be a particular physical environment, a particular host combination, nutrients for reproduction, or any number of other factors. These requisites may be provided by altering the nature of the crop planted. For example, the "hanging square" character could be bred into commercial cottons to enhance parasitism by B. mellitor.

If these indigenous species are to be utilized, their population performance and regulatory effect must be factored into the decision process which is used to determine treatment need. The value of a predator or parasite, or perhaps a complex of predators or parasites, must be determined. This is undoubtedly a dynamic interface which probably can best be handled by computer simulation and projection. This information can be gained only by detailed quantitative population studies which measure the regulatory capacity of a beneficial species. In the course of this research, it will be necessary to know the role of the predator/parasite in respect to other biotic and abiotic mortality factors. This role must be determined in terms of contribution to irreplaceable mortality, not percentage parasitism.

Nearly every successful classical biological control program has been the result of the activity of a very limited number of host specific natural enemies. I feel that this likely will be the key to predictable utilization of the indigenous species. Most likely a very limited number of species will be most important. It may be that the research can be limited to these species. Because of their host specificity, it may be that host-specific parasites such as Cardiochiles or Apanteles may hold the best hope for predictable utilization in IPM of cotton.

The aquisition of exotic entomophagous species holds considerable potential for our introduced pests. In the case of boll weevil, B. mellitor or the other native parasitoid species simply have not provided an acceptable degree of control. Yet, there exists a number of species which are better searchers, have shorter life cycles and have behaviors which appear to be otherwise advantageous to parasitism of boll weevil. An example is the parasitism of boll weevil larvae on the ground by the species reported from Mexico.

It is possible that natural enemies of exotic insects closely related to our pests could be utilized. This was done in the case of Telenomus alsophilae, an egg parasite of a geometrid pest in forest, Oxydia trychiata. The host of origin was Aslophila pometaria (Drooz et al. 1977). We are attempting this in the biological control of boll weevil with the near relative Anthonomus hunteri. This approach may be the solution to biological control of Lygus and cotton fleahopper. It may be a valid approach for Heliothis.

Classical biological control in cotton is a special case, since the crop is grown as an annual. It may be even more difficult because the crop pest protection system has evolved to the more abruptly disturbed short-season cotton system. Consideration must be given to enhancing the ecological continuity of the cotton agroecosystem. One approach would be to utilize the exotic natural enemies in the alternate hosts of the pests. This is a distinct possibility with the plant bugs which migrate seasonally to cotton from the herbaceous flora surrounding the fields. Introduction of effective natural enemies into these flora could reduce the pest to levels that would be insignificant to the crop. At the same time these natural floras would provide the perennial stability not inherent to the annual row crop. It is possible also that these natural flora could be the refugia from which these exotic natural enemies colonize the crop, should the pests move to cotton.

An approach that we have taken with boll weevil parasitoids is to grow the crop as a perennial in South Texas to enhance the chance for colonization and survival of the introduced species. With successful colonization of the exotic species in perennial cotton, it may be possible to practice conservation measures to make it effective in commercial production. This conservation methodology will have to be developed much as it would for a native species.

Biological control in annual crops is a challenge for the classical biological control approach. Instead of declaring it impossible, new concepts should be developed to bring about success. Augmentation by inundative release of predators, parasites or pathogens may hold great commercial potential in cotton. However, in the case of parasites and predators, effective and economical rearing methodology must be developed. Artificial diets are a primary need for economic rearing of entomophagous species. Lastly, a plea must again be made for the development of accurate and dynamic economic thresholds for pest control decisions. This must be done to: 1) establish the pest status; 2) justify research aimed at control of the species; 3) provide the decision-maker the criteria upon which to recommend the application of an insecticide or to withold an application in deference to the entomophagous arthropod complex.

RECOMMENDATIONS

Recommendations for implementing biological control as a component of integrated control in cotton could be summarized by simply seeking to know as much about the ecology and behavior of the natural enemies as we do of the pests. However, more specific elements of this theme can be identified:

1. Identify the key parasites and/or predators which do, or potentially, affect population regulation of pests of cotton.
2. The population dynamics of the natural enemies must be understood in regard to natural host and plant refugia so that colonization rates and phenology can be manipulated and predicted.
3. Requisites for reproduction and survival of immature and adult parasitoids and predators must be identified.

4. The recruitment processes which cause entomophagous species to enter cotton and be efficaceous must be identified.
5. The population performance of the entomophagous species must be factored into the decision process which determines when or whether pesticides are applied.
6. Develop effective artificial diets for parasitoids and predators.
7. Actively seek exotic natural enemies of the cotton pests, possibly utilizing exotic natural enemies of hosts closely related to the pest.
8. Assure proper release and colonization efforts so that exotic species are given every opportunity to establish. This includes making releases in many diverse ecological sites and on all host plant species.
9. Develop new concepts and approaches to enhance the role of classical biological control in annual crop systems.
10. Accurately determine economic thresholds for all pests and, in some cases, combinations of pests.
11. Develop a stable funding system which would be available for research in biological control.

REFERENCES

Ables, J.R., J.L. Goodenough and A.W. Hartstack. 1983. Entomophagous arthropods, pp. 103-127. In: R.L. Ridgway, E.P. Lloyd and W.H. Cross (Eds.), Cotton insect management with special reference to the boll weevil. USDA Agric. Handb. 589.

Adams, C.H., W.H. Cross and H.C. Mitchell. 1969. Biology of Bracon mellitor, a parasite of the boll weevil. J. Econ. Entomol. 62:889-895.

Adkisson, P.L. 1969. How insects damage crops. Conn. Agr. Exp. Sta. Bull. 708: 155-164.

Annand, P.N. 1946. Report of the chief of the bureau of entomology and plant quarantine. USDA, Agr. Res. Admin. 1944-45. Washington D.C. 12 pp.

Berry, P.A. 1947. Anthonomus vestitus and its natural enemies in Peru, and their importation into the United States. J. Econ. Entomol. 40:801-804.

Bondy, F.F., and C.F. Rainwater. 1935. Boll weevil and miscellaneous cotton insect investigations. S.C. Agric. Exp. Sta. Bull. 48, 100 pp.

Boring, E.P. 1974. Planting date and early maturity importance in boll weevil control. Tex. Agr. Ext. Serv. News Rel., Feb. 12. 2 pp.

Boring, E.P., and T.W. Fuchs. 1980. Delayed uniform planting--a new weapon in the boll weevil war. Texas Agric. Ext. Serv. Ann. Rept. 1980, 4 pp.

Bottrell, D.G. 1972. New strategies for management of the boll weevil. Tex. Agric. Exp. Sta., Dep. Entomol. Tech. Mimeo. Rep. 19 pp.

Bottrell, D.G. 1976a. Biological control agents of the boll weevil, pp. 22-25. In: Proceedings, conference of boll weevil suppression, management and elimination technology. Memphis, TN., USDA ARS-S-71.

Bottrell, D.G. 1976b. The boll weevil as a key pest, pp. 5-8. In: Proceedings, Conference on boll weevil suppression, management, and elimination technology. Memphis, TN., USDA ARS-S-71.

Brazzel, J.R., T.B. Davich and L.D. Harris. 1961. A new approach to boll weevil control. J. Econ. Entomol. 54:723-730.

Burke, H.R., and J.R. Cate. 1979. A new species of Mexican *Anthonomus* related to the boll weevil (Coleoptera: Curculionidae). Ann. Entomol. Soc. Amer. 72:189-192.

Burke, H.R., and J.R. Cate. 1983. Descriptions of the larva and pupa of *Anthonomus hunteri* and comparison with *Anthonomus grandis* (Coleoptera: Curculionidae). Proc. Entomol. Soc. Wash. 85:456-462.

Calvert, D.J. 1974. Report on the search for natural enemies of *Anthonomus grandis* Boh. associated with plant hosts of the tribe Gossypieae (*Cienfuegosia rosei* Fryxell, *Gossypium* spp., and *Hampea* spp.) in Guatemala and Mexico. Univ. Calif., Dept. Entomol. Sci., Div. Biol. Cont. Tech. Rep. Mimeo. 21 pp.

Chesnut, T.L., and W.H. Cross. 1971. Arthropod parasites of the boll weevil, *Anthonomus grandis*: 2. Comparisons of their importance in the United States over a period of thirty-eight years. Ann. Entomol. Soc. Amer. 64:549-557.

Cook, O.F. 1904. Report on the habits of the kelep, or Guatemalan cotton boll weevil ant. USDA Bur. Entomol. Bull. 49, 15 pp.

Cook, O.F. 1905. The social ogranization and breeding habits of the cotton protecting kelep of Guatemala. USDA Bur. Entomol. Tech. Ser. Bull. 10, 55 pp.

Cross, W.H., and T.H. Chesnut. 1971. Arthropod parasites of the boll weevil, *Anthonomus grandis*: 1. An annotated list. Ann. Entomol. Soc. Amer. 64:516-527.

Cross, W.H., and W.L. McGovern. 1969. New parasites, *Zatropis perdubius* and *Megaselia aletiae*, of the boll weevil, *Anthonomus grandis*. Ann. Entomol. Soc. Amer. 62:674.

Cross, W.H., and H.C. Mitchell. 1969. Distribution and importance of *Heterolaccus grandis* as a parasite of the boll weevil. Ann. Entomol. Soc. Amer. 62: 235-236.

Cross, W.H., W.L. McGovern and H.C. Mitchell. 1969. Biology of *Bracon kirkpatricki* and field releases of the parasite for control of the boll weevil. J. Econ. Entomol. 62:448-454.

Daum, R.J., R.E. McLaughlin and D.D. Hardee. 1967. Development of the bait principle for boll weevil control: Cottonseed oil, a source for the boll weevil. J. Econ. Entomol. 60:321-325.

Dinkins, R.L., J.R. Brazzel and C.A. Wilson. 1970. Seasonal incidence of major predaceous arthropods in Mississippi cotton fields. J. Econ. Entomol. 63:814-817.

Doutt, R.L., and R.F. Smith. 1971. The pesticide syndrome-diagnosis and suggested prophylaxis, pp. 3-15. *In*: C.B. Huffaker (Ed.), Biological control. Plenum Press, London.

Drooz, A.T., A.E. Bustillo, F.G. Fedde and V.H. Fedde. 1977. North American egg parasite successfully controls a different host genus in South America. Science 197:390-391.

Eichers, T., P. Andrilenas, H. Blake, R. Jenkins and A. Fox. 1970. Quantities of pesticides used by farmers in 1966. USDA/ERS Agr. Econ. Rep. 179, 89 pp.

Ewing, K.P., and E.E. Ivy. 1943. Some factors influencing bollworm populations and damage. J. Econ. Entomol. 36:602-606.

Ewing, K.P., and H.J. Crawford. 1939. Egg parasites of the cotton fleahopper. J. Econ. Entomol. 32:303-305.

Fenton, F.A., and E.W. Dunham. 1929. Biology of the cotton boll weevil at Florence, S.C. USDA Tech. Bull. 112, 76 pp.

Fletcher, R.K., and F.L. Thomas. 1943. Natural control of eggs and first instar larvae of Heliothis armigera. J. Econ. Entomol. 36:557-560.

Frisbie, R.E., and G.T. Bohmfalk. 1982. Management decisions, pp. 37-39. In: Bolmfalk, G.T., R.E. Frisbie, W.L. Sterling, R.B. Metzer and A.E. Knutson (Eds.), Identification, biology and sampling of cotton insects. Tex. Agric. Ext. Ser. Bull. B-933, 43 pp.

Frisbie, R.E., J.R. Phillips, W.R.A. Lambert and H.B. Jackson. 1983. Opportunities for improving cotton insect management programs and some constraints on beltwide implementation, pp. 521-557. In: R.L. Ridgway, E.P. Lloyd and W.H. Cross (Eds.), Cotton insect management with special reference to the boll weevil. USDA Agric. Handb. 589.

Gahan, A.B. 1937. Two new chalcidoid egg parasites (Eulophidae and Mymaridae). Proc. Entomol. Soc. Wash. 39:266-269.

Gaines, R.C. 1955. Effect on beneficial insects of three insecticide mixtures applied for cotton insect control in 1954. J. Econ. Entomol. 48:477-478.

Grossman, E.F. 1931. Control of the cotton boll weevil by insect enemies. Science 69:362.

Harding, J.A. 1976. Heliothis spp.: Parasitism and parasites plus host plants and parasites of the beet armyworm, diamondback moth and two tortricids in the lower Rio Grande Valley of Texas. Environ. Entomol. 5:669-671.

Harris, V.E., and J.R. Phillips. 1984. The effect on predatory arthropods of mowing spring weed hosts of Heliothis spp. Environ. Entomol. 13:(In Press).

Hartstack, A.W., J.A. Witz and R.L. Ridgway. 1975. Suggested applications of a dynamic Heliothis model (MOTHZV-1) in pest management decision making, pp. 118-122. In: Proc. of the 1973 Beltwide Cotton Prod. Res. Conf. Phoenix, AZ.

Howard, L.O. 1897. The Mexican cotton boll weevil (Anthonomus grandis Boheman). USDA Div. Entomol. Circ. 18, 8 pp.

Howell, A.H. 1907. The relation of birds to the cotton boll weevil. USDA Bur. Biol. Survey Bull. 29, 30 pp.

Howell, A.H. 1909. Destruction of the cotton boll weevil by birds in winter. USDA Bur. Biol. Survey Circ. 64:5.

Hunter, W.D., and W.E. Hinds. 1905. The Mexican boll weevil. USDA Bur. Entomol. Bull. 51:143-150.

Hunter, W.D., and W.D. Pierce. 1912. The Mexican cotton-boll weevil: A summary of the results of the investigations of this insect up to December 31, 1911. Senate Doc. 305, 188 pp.

Isely, D., and W.J. Baerg. 1924. The boll weevil problem in Arkansas. Ark. Exp. Sta. Bull. 462, 29 pp.

Johnson, S.J., H.N. Pitre, J.E. Powell and W.L. Sterling. 1985. Control of Heliothis spp. by conservation and importation of natural enemies. In: S. Johnson, E.G. King and J.R. Bradley (Eds.), Theory and tactics of Heliothis population management: I. Cultural and biological control. South. Coop. Ser. Bull. (In Press).

Jones, D., and W.L. Sterling. 1979. Manipulation of red imported fire ants in a trap crop for boll weevil suppression. Environ. Entomol. 8:1073-1077.

Lewis, W.J., and J.R. Brazzel. 1968. A three-year study of parasites of the bollworm and tobacco budworm in Mississippi. J. Econ. Entomol. 61:673-676.

Lincoln, C., and T.F. Leigh. 1957. Timing insecticide applications for cotton insect control. Ark. Agr. Exp. Sta. Bull. 588, 47 pp.

Mally, F.W. 1901. The Mexican boll weevil. USDA Farm. Bull. 130, 30 pp.

McDaniel, S.G., and W.L. Sterling. 1979. Predator determination and efficiency on Heliothis virescens eggs in cotton using P^{32}. Environ. Entomol. 8:1083-1087.

McGovern, W.L., and W.H. Cross. 1976. Affects of two cotton varieties on levels of boll weevil parasitism (Col.: Curculionidae). Entomophaga 21:123-125.

McLaughlin, R.E. 1965. Some relationships between boll weevil, Anthonomus grandis Boheman and Mattesia grandis McLaughlin (Protozoa: Neogregarinida). J. Invertebr. Pathol. 7:464-473.

McLaughlin, R.E. 1966. Infection of the boll weevil with Mattesia grandis induced by a feeding stimulant. J. Econ. Entomol. 59:909-911.

McLaughlin, R.E. 1967. Development of the bait principle for boll weevil control. II. Field-cage tests with a feeding stimulant and the protozoan Mattesia grandis. J. Invertebr. Pathol. 9:70-7.

McLaughlin, R.E. 1969. Glugea gasti sp.n., a microsporidian pathogen of the boll weevil, Anthonomus grandis. J. Protozool. 16:84-92.

McLaughlin, R.E., R.J. Daum and M.R. Bell. 1968. Development of the bait principle with a feeding stimulant and the protozoan, Mattesia grandis (Neogregarinida) and a microsporidian. J. Protozool. 128:168-174.

Moody, D.S., S.C. Munoz, R.E. Pettit and D.G. Bottrell. 1975. Relative toxicity of the fungus Aspergillus flavus to the male the female boll weevil. Tex. Agric. Exp. Sta. Prog. Rep. 3307, 2 pp.

Mueller, T.F. 1983. The effect of plants on the host relations of a specialist parasitoid of Heliothis larvae. Entomol. Exp. Appl. 34:78-84.

Mueller, T,F, and J.R. Phillips. 1983. Population dynamics of Heliothis spp. in spring weed hosts in southeastern Arkansas: Survivorship and stagespecific parasitism. Environ. Entomol. 12:1846-1850.

Mueller, T.F., V.E. Harris and J.R. Phillips. 1984. Theory of Heliothis (Lepidoptera: Noctuidae) management through reduction of the first spring generation: A critique. Environ. Entomol. 13:(In Press).

Nicholson, W.F., J.R. Phillips, J.L. Bernhardt and M. Cochran. 1983. An assessment of community-wide bollworm management programs. Ark. Farm Res. 1983(Jan/Feb):3.

Nicholson, W.F., J.R. Phillips and D.R. Johnson. 1984. Bollworm management communities in Arkansas. Ark. Coop. Ext. Serv. Bull. EL663, 6 pp.

O'Neil, R.J., and J.R. Cate. 1984. Competition between Bracon mellitor Say and Catolaccus grandis (Burkes) for their host Anthonomus grandis Boheman, a laboratory study. Entomophaga 29:(In Press).

Parker, R.D., J.K. Walker, G.A. Niles and J.R. Mulkey. 1980. The short-season effect in cotton and escape from the boll weevil. Tex. Agric. Exp. Sta. Bull. 1315, 45 pp.

Phillips, J.R., W.F. Nicholson, T. Teague, J. Bernhardt and T.F. Mueller. 1981. Community insect management programs, pp. 44-49. In: Proc. Cotton Biological Control Conf. Dallas, TX.

Pierce, W.D. 1908. Studies of parasites of the cotton boll weevil. USDA Bur. Entomol. Bull. 73:631-663.

Pierce, W.D., R.A. Cushman and C.E. Hood. 1912. The insect enemies of the cotton boll weevil. USDA Bur. Entomol. Bull. 100, 99 pp.

Quaintance, A.L., and C.T. Brues. 1905. The cotton bollworm. USDA Bur. Entomol. Bull. 50, 155 pp.

Ridgway, R.L., and P.D. Lingren. 1972. Predaceous and parasitic arthropods as regulators of Heliothis populations. USDA South. Coop. Ser. Bull. 169, 92 pp.

Ridgway, R.L., P.D. Lingren, C.B. Cowan and J.W. Davis. 1967. Populations of arthropod predators and Heliothis spp. after applications of systemic insecticides to cotton. J. Econ. Entomol. 60:1012-1016.

Roach, E.R. 1973. The effects of the boll weevil eradication experiment on certain wildlife species. M.S. Thesis, Mississippi State Univ., Mississippi State.

Rummel, D.R., D.G. Bottrell, P.L. Adkisson and R.C. McIntyre. 1975. An appraisal of a 10-year effort to prevent the westward spread of the boll weevil. Bull. Entomol. Soc. Amer. 21:6-11.

Scales, A.L. 1973. Parasites of the tarnished plant bug in the Mississippi Delta. Environ. Entomol. 2:304-306.

Schuster, M.F., and J.C. Boling. 1974. Phenology of early and mid-season predatory and phytophagous insects in cotton in the Lower Rio Grande Valley of Texas. Tex. Agric. Exp. Sta. Misc. Publ. 1133, 31 pp.

Scott, D., M. Cochran and W.F. Nicholson. 1983. Evaluation of farm level benefits of bollworm management communities. Ark. Farm Res. 1983(Nov/Dec):3.

Shepard, M., and W. Sterling. 1972a. Effects of early season applications of insecticide on beneficial insects and spiders in cotton. Tex. Agric. Exp. Sta. Misc. Publ. 1045, 14 pp.

Shepard, M., and W. Sterling. 1972b. Incidence of parasitism of Heliothis spp. (Lepidoptera: Noctuidae) in some cotton fields of Texas. Ann. Entomol. Soc. Amer. 65:759-760.

Slosser, J.E. 1978. The influence of planting date on boll weevil management. Southwest. Entomol. 3:241-246.

Smith, J.W., and E.A. Stadelbacher. 1978. Predatory arthropods: Seasonal rise and decline of populations in cotton fields in the Mississippi Delta. Environ. Entomol. 7:367-371.

Smith, R.F., and R. van den Bosch. 1967. Integrated control, pp. 295-340. In: W.W. Kilgore and R.L. Doutt (Eds.), Pest control--biological, physical, and selected chemical methods. Academic Press, N.Y.

Stadelbacher, E.A. 1981. Role of early season wild and naturalized host plants in the build-up of the F_1 generation of Heliothis zea and H. virescens in the Delta of Mississippi. Environ. Entomol. 10:766-770.

Sterling, W.L. 1978. Fortuitous biological suppression of the boll weevil by the red imported fire ant. Environ. Entomol. 7:564-568.

Sterling, W.L., D.A. Dean, D.A. Fillman and D. Jones. 1984. Naturally-occurring biological control of the boll weevil (Col.: Curculionidae). Entomophaga 29: 1-10.

Tillman, P.G. 1985. Host associations of Bracon mellitor Say (Hymenoptera: Braconidae) in native vegetation in central Texas. Ph.D. Dissertation, Texas A&M Univ., College Station.

Townsend, C.H.T. 1895. Report of the Mexican cotton boll weevil in Texas (Anthonomus grandis Boh.). Insect Life 7:295-309.

Walker, J.K., and G.A. Niles. 1971. Population dynamics of the boll weevil and modified cotton types. Tex. Agric. Exp. Sta. Bull. 1109, 14 pp.

Walker, J.K., R.E. Frisbie and G.A. Niles. 1978. A changing perspective: Heliothis on short-season cottons in Texas. Bull. Entomol. Soc. Amer. 24:385-391.

Wheeler, W.M. 1904. Some further comments on the Guatemalan boll weevil ant. Science 20:766-768.

Whitcomb, W.H., and K.O. Bell. 1964. Predaceous insects, spiders, and mites of Arkansas cotton fields. Ark. Agric. Exp. Sta. Bull. 690, 84 pp.

SOYBEAN:
STATUS AND CURRENT LIMITS TO BIOLOGICAL CONTROL IN THE SOUTHEASTERN U.S.

Merle Shepard and D.C. Herzog

The hectarage of soybean in the southeastern U.S. is modest in comparison to the midwestern production area. Milder winters and longer growing seasons in the Southeast provide a favorable environment for increases in the density and diversity of pests associated with soybean. Concurrently, communities of beneficial arthropods and entomopathogens flourish in this environment and often maintain pest populations below economically damaging levels. By contrast to other field crops (cotton, tobacco, vegetables), insecticide applications to the soybean crop are not as frequent and the crop has not had a long history of insecticide use and misuse. These factors, combined with the ability of soybean to withstand considerable defoliation before yield reduction is sustained (Turnipseed 1972a), allows for the development of abundant natural enemy populations. Depending on the particular geographical location within the southeastern U.S., insecticidal control may not be necessary. Thus, a suitable setting is provided for the development of effective management programs which include and maximize the use of indigenous and exotic biological control agents.

Populations of a relatively innocuous defoliating caterpillar, the green cloverworm, Plathypena scabra, colonize the crop early in the growing season and serve as food for developing communities of arthropod predators, parasites and pathogens. These natural enemies are present when economically important noctuid pests such as the corn earworm, Heliothis zea, soybean looper, Pseudoplusia includens, and velvetbean caterpillar, Anticarsia gemmatalis, invade the crop. Other important insect pests of soybean in the southeastern U.S. include the southern green stinkbug, Nezara viridula; threecornered alfalfa hopper, Spissistilus festinus; and bean leaf beetle, Cerotoma trifurcata. The latter also transmits bean pod mottle virus. The Mexican bean beetle, Epilachna varivestis, has diminished in importance in the Southeast in the past few years, but the lesser cornstalk borer, Elasmopalpus lignosellus, has become more important, especially in areas with dry sandy soils. A review of soybean insects which

ISBN 0-12-357030-1

attack soybean foliage, stems, roots, nodules, and pods is presented by Turnipseed & Kogan (1976).

The successful development of soybean insect management programs in the southeastern U.S. resulted largely due to support provided through the "Huffaker Project" by the National Science Foundation (Huffaker 1980). With funding provided by State Agricultural Experiment Stations and extramural grants, a substantial body of information describing the soybean agroecosystem has been generated. Major expansion of the soybean crop in the southeastern U.S. coincided with increased interest in the philosophies of integrated pest management during the 1970's. A logical extension of the research momentum gained through the Huffaker Project was further enhanced through support from the U.S. Environmental Protection Agency and the U.S. Department of Agriculture to the "Consortium for Integrated Pest Management" (CIPM) Project under the leadership of P.L. Adkisson of Texas A&M University. This program focused more sharply on the development of comprehensive, unified, economically and environmentally sound systems of pest management which could be implemented in a practical manner. One of the major objectives of this project was maximum utilization of indigenous biological control agents and the search for, and importation and evaluation of introduced species.

There has been considerable interest in the development of an understanding of the role of indigenous natural enemies in soybean, and limited foreign exploration has been undertaken. From these modest beginnings, some successes have been registered. Details of biological control agent introductions will be presented later, but a major constraint to expansion of classical biological control in soybean has been lack of support, coordination and interest. This is somewhat surprising considering that current control tactics for major insect pests of soybean are limited to the use of chemical insecticides. Although the Peer Review Panel for the CIPM Project recommended suspension of foreign exploratory activities, the need for classical biological control has been subsequently recognized at the national level and a Southern Regional Research Project (S-192) was subsequently developed which focuses specifically on the introduction and use of exotic natural enemies.

The scope of this review will be confined to the present status and limits to the use of indigenous and introduced natural enemies as control agents for major insect pests of soybean in the southeastern U.S. Efforts at biological control of nematodes and plant pathogens in soybean are extremely limited. Biological control of weeds in soybean is adequately described by Charudattan (this volume) and will not be recounted here.

INDIGENOUS NATURAL ENEMIES

Predators

Studies dealing with predators of insect pests of soybean have been largely qualitative in nature, with lists of natural enemies (Buschman et al. 1977, Neal & Whitcomb 1972, Whitcomb et al. 1972), descriptions of their relative and seasonal abundance (Shepard et al. 1974), and response to insecticides (Brown & Goyer 1982; Price & Shepard 1977, 1978; Shepard et al. 1977; Turnipseed 1972b; and others). It is generally accepted that suppression of populations of major insect pests of soybean results from the action of communities of indigenous arthropod predators. Herzog et al. (1985a) suggested that predators which are generalists in feeding behavior, with relatively long generation times and low reproductive capacity, act as buffers against pests which first colonize the crop. Mortality, therefore, is inflicted principally on eggs and early life stages of pests. However, densities of most predatory species increase as the cropping season progresses. Although direct numerical response by predators to changes in the densities of pest populations normally is not observed, there is substantial mortality at all prey densities due to predators (see also Stimac & O'Neil, this volume).

Destruction of predator populations by applications of insecticides normally results in resurgence of the target pest population (Brown & Goyer 1982, Shepard et al. 1974, Turnipseed 1972b); however, it is usually not possible to separate the effects of pesticides on predators from effects on other natural enemies (entomopathogens and parasites).

Predatory species have been estimated to number over 1,000 (Whitcomb 1973), but it is probable that only a few of the most efficient, presumably most abundant, species exert a significant impact on pest populations. In South Carolina the most abundant groups of predators are the nabids, geocorids and spiders (Turnipseed 1972b). Autoradiographic methods have demonstrated that these predators frequently fed on noctuid eggs and larvae in the field (McCarty et al. 1980). The seasonal abundances of these groups have been documented (Shepard et al. 1974) and sequential sampling models have been developed for classifying populations of nabids and geocorids into broad relative abundance categories (Waddill et al. 1974). To date, information on predator population densities has not yet been formally incorporated into soybean pest management decision-making; however, efforts to do so are underway (Stimac & O'Neil; Wilson, this volume).

The polyphagous nature of predators makes quantification of the role of a single species difficult. It is probable that the impact of predatory species will have to be determined as a community or complex, rather than trying to determine the role of individual predators isolated in cages or by direct observation. Although recent research by Elvin et al. (1983), Naranjo (1983),

O'Neil (1984) and Reed et al. (1984) increased our understanding of the role of predatory arthropod communities in the soybean agroecosystem, there still is a great need to use the available information concerning predator abundance, behavior and efficiency to refine our estimates of damage thresholds. For example, while surveying a field for insect pest abundance, simultaneous counts of beneficial species could be made. By adjusting damage thresholds to incorporate potential pest mortality due to action of predators, the decision to apply chemical controls might be postponed or avoided altogether.

Precise determination of the interaction of each predator species with pests and with other species of predators seems unlikely to be worked out in the near future. Meanwhile, methods are needed with which will allow rapid determination of the impact of the total predator community within individual fields. Already in other ecosystems counts of predators are being considered as inputs to the IPM decision-making process (see Cate; Wilson, this volume). This may be done by developing predator-prey ratios or modifying sequential sampling models with flexible thresholds (Shepard & Grothusen 1984).

Given the rich community of indigenous predators which can reside in soybean, the most logical first step in developing an effective biological control program is conservation. This has been emphasized in most soybean insect management programs in place in the field. Benefits from this approach would be more likely to occur in areas where annual immigrants, such as the velvetbean caterpillar in Florida, do not dominate insect control activities. Even with the case of the _A. gemmatalis_, use of selective chemicals such as the insect growth regulator diflubenzuron, applied once early in the season at low rates, have been shown to be sufficiently effective to provide season-long control. This promises to be a viable approach to the conservation of predators and parasites which could be severely affected by the use of less selective, but faster acting, chemicals which are needed when dense velvetbean caterpillar populations develop (J.D. Chapin, personal communication). Use of minimal rates of other pesticides, or properly timing their application, should help to avoid the development of resistance in the pest and favor continued natural enemy activity.

Parasites

Like with the predators, there is a rich community of indigenous parasites associated with insect pests of soybean (Dietz et al. 1976, Harper et al. 1983, McCutcheon & Turnipseed 1981, and others). Assessment of the incidence of parasitism is easier to determine than evaluation of predator-induced mortality. To determine levels of parasitism, host larvae are field collected, brought to the laboratory and provided food until parasites emerge. However, a quantitative method of evaluating the impact of parasites on pest populations has not been developed and estimates of the incidence of parasitism may be biased (Marston 1980, Shepard et al. 1983a). The seasonal and geographical occurrence of parasites of soybean insect pests have been described by Harper et al. (1983).

With few exceptions, entomologists generally agree that indigenous parasites alone are not capable of regulating pest populations in soybean because pest outbreaks often occur. With few exceptions, parasitism of most pest species is low (<20%). For example, parasitism of A. gemmatalis was 2% or less in South Carolina (McCutcheon & Turnipseed 1981) and 5% or less in Louisiana and Florida (D.C. Herzog, unpublished data). McCutcheon & Turnipseed (1981) found only 3.7% parasitism of P. includens larvae in South Carolina while parasitism often reached 30% or more in Louisiana and Florida (Burleigh 1972; D.C. Herzog, unpublished data). Other researchers have obtained similar results although Jones (1979) and Yeargan (1979) found up to 50% of the eggs of N. viridula were parasitized, principally by Trissolcus basalis.

Although indigenous parasites may not provide control of most of the target pests, they are important members of the beneficial arthropod community and efforts to conserve them should be continued. It is likely that importation of parasites for control of some soybean insect pests could be at least partially successful. This is evident from results of recent parasite introductions which will be discussed later.

Meanwhile, much work is needed to decipher the individual roles of the major parasite groups and determine their response to pesticides currently used on soybean. Timing of pesticide applications, using more selective materials or changing formulations may aid in parasite conservation.

Mass rearing of parasites for augmentative programs has not received much attention in soybean. Martin et al. (1976) demonstrated that Trichogramma pretiosum was effective against Heliothis spp. in small scale field experiments. Although these results were encouraging, we know of no other attempts at mass rearing of Trichogramma in augmentation programs for control of noctuid pests of soybean in the field. There may be opportunities for increasing the incidence of parasitism using kairomones (Lewis et al. 1975; Lewis & Nordlund, this volume), or supplemental foods for parasitic insects (Hagen et al. 1971).

Although indigenous parasites attack pest populations, their host-regulating characteristics have not yet been demonstrated in soybean. However, it is generally recognized that they are reliable sources of mortality against several pest species and by acting in concert with predators and entomopathogens, insect pest populations may be held in check. Additionally, augmentation of their numbers by mass release of laboratory-reared parasites offers possibilities. The need for increased emphasis on the classical biological control approach (importation) has been suggested as a viable approach for field crops such as soybean (Jones et al. 1983).

Entomopathogens

The natural enemies most closely tied to decreases in insect pest densities are the entomopathogens. Many entomologists consider this group of organisms to be the most important with respect to regulation of high density populations of lepidopterous pests. Harper et al. (1983) presented a review of the geographical and seasonal distribution of entomopathogens in soybean.

The soybean crop provides a particularly favorable environment for epizootics of fungal pathogens because of the closed crop canopy which develops late in the growing season when populations of important pests are more likely to occur at high densities (Carner et al. 1975). All major lepidopterous pests are infected by the fungus Nomuraea rileyi. The fungus Entomophthora gammae is often responsible for decimation of large populations of P. includens in the field (Burleigh 1972). Another important Entomophthora sp. often prevents outbreaks of H. zea, as does a nuclear polyhedrosis virus (G.R. Carner, personal communication). The use of the commercial virus formulation, Elcar®, may be responsible for providing inoculum for virus epizootics reported from North Carolina, South Carolina and Alabama (Harper et al. 1983).

There are several other pathogens reported on important soybean insect pests but their impact on pest populations is considered to be minimal. However, there are excellent opportunities for manipulating entomopathogens as biocontrol agents against insect pests of soybean. They often are easier to culture and manipulate than are predators or parasites. Many pathogens can be applied with conventional pesticide application equipment and initiation of epizootics in noctuid defoliators has shown promise (Ignoffo et al. 1976, Sprenkel & Brooks 1975).

Although natural epizootics of entomopathogens occur commonly, there often is crop damage before disease arrests the pest population. Because the factors which cause initiation and spread of these diseases are not understood, outbreaks of entomopathogens are not predictable at this time.

INTRODUCED NATURAL ENEMIES

Predators, Parasites and Entomopathogens

The importation and use of exotic natural enemies in soybean has been limited. However, several promising parasites and pathogens have been identified and with the relatively few tactics available for control of soybean insects, this approach offers exciting possibilities.

Greene & Shepard (1974) reported that successful biocontrol by a reduviid predator, Sycanus indigator, from India was limited by poor survival in the field and low rates of consumption of P. includens larvae. A reduviid, Pristhesancus papuensis, from

Australia (Shepard et al. 1982) produces only limited mortality of lepidopterous pests because it does not possess enough desirable characteristics to make a significant impact on pest populations (R.E. McWhorter & M. Shepard, unpublished data). Eocanthocona furcellata, an asopine imported from Thailand, has been shown to be inferior to the indigenous Podisus maculiventris (Tseng 1982). Importation of predators for use in soybean has been limited to these examples. However, there are possibilities for importation of other species of exotic predators such as small carabids, especially those in the genera Calleida and Lebia, which are abundant in Brazil and other areas. Introduction and establishment of these groups could yield good results.

The most outstanding example of the use of an introduced parasitoid for control of an insect pest of soybean is the eulophid parasite, Pediobius foveolatus, from India against the Mexican bean beetle (Sailer 1976, Shepard & Robinson 1976, Stevens et al. 1975). During the past several years, E. varivestis has diminished in importance in the Southeast. Releases of this parasite in Alachua County, Florida, have resulted in recoveries as far north as Georgia. It is no coincidence that programs of P. foveolatus releases in Florida and South Carolina have been followed by reduced population density of the Mexican bean beetle in these states. Releases of P. foveolatus in South Carolina have resulted in complete control of E. varivestis in snap bean plantings with a history of heavy infestation by this pest (M. Shepard & R.E. McWhorter, unpublished data). E. varivestis often colonizes and develops dense populations on snap bean before moving to soybean. Release of P. foveolatus in a snap bean "trap crop" is a workable approach to control of this pest. Mexican bean beetle control by P. foveolatus, however, can only be obtained through annual releases of the parasite in the southeastern U.S. since it does not overwinter in most soybean growing areas.

Surveys of parasites of A. gemmatalis in the southeastern U.S. have revealed that parasitism of this pest is very low (McCutcheon & Turnipseed 1981). Introduction and establishment of an exotic parasite, Euplectrus puttleri, has been reported for A. gemmatalis. The parasite was collected and introduced from Colombia (Puttler et al. 1980), was released and has spread several miles away from initial release sites in south Florida (Waddill & Puttler 1980). Because A. gemmatalis has not been found to overwinter in soybean-producing areas of the Southeast, this parasite may potentially play an important role in reducing numbers of A. gemmatalis in overwintering reservoirs from which moths eventually migrate northward each year.

Foreign exploration for parasites of A. gemmatalis has produced Microcharops bimaculata, Corsoncus magus, and Zelomorpha sp. from Costa Rica, and Glyptapanteles sp. from Ecuador. These species have been released in south Florida and recovered from A. gemmatalis larvae (V.H. Waddill, personal communication). A concerted effort to establish a parasite such as M. bimaculata, the most abundant species in Brazil and Costa Rica, could result in long-term suppression of this serious pest, and will probably be the species of choice for release into southeastern soybean production areas where this pest occurs (Herzog et al. 1985a, b).

There are other imported parasites which are currently being assessed as to their biological control potential (Jones et al. 1983). Microplitis demolitor from Australia parasitizes H. zea and P. includens (Shepard et al. 1983b) and has been found to overwinter in the Southeast (J.F. Grant, personal communication). Apanteles ruficrus, a soybean looper parasite also from Australia, has been shown to recycle in the field in South Carolina, (J.P. Parkman & S.G. Turnipseed, personal communication), and Palexorista laxa, a tachnid parasite from Kenya, parasitizes up to 80% of H. zea larvae in field cages. Excellent host-finding capabilities and high reproductive capacity make P. laxa a promising component of the natural enemy complex (J.F. Grant & M. Shepard, unpublished data).

A nuclear polyhedrosis virus which infects P. includens has imported from Guatemala, introduced into fields in Louisiana and Arkansas and has persisted in the soil until the following year (Livingston & Yearian 1972). A virus (AgNPV) of the velvetbean caterpillar has been collected in Brazil (Allen & Knell 1977, Carner & Turnipseed 1977). Moscardi et al. (1981) found this virus highly effective against A. gemmatalis in the field at low dosage rates; however, this virus has not shown promise for establishment in the field for long-term suppression (Beach et al. 1984). F. Moscardi (personal communication) has organized programs for production of the nuclear polyhedrosis virus in Brazil using small screen house production units. This may be investigated as a viable approach in the southeastern U.S. if approval of its use can be obtained and the procedure proves to be cost effective. A major drawback to the commercial production of most viruses is limited market due to their host specificity.

EVALUATION OF NATURAL ENEMY EFFECTIVENESS

Whether biological control agents are indigenous or introduced, the determination of their impact on the target pest is one of the most important, yet difficult, problems in biological control. Several reviews of the subject have been published (DeBach et al. 1976, Huffaker & Kennett 1969, Kiritani & Dempster 1973). Although most of the reviews deal with more stable ecosystems such as orchards, the review of Grant & Shepard (1984) focused on field crops. The sparse literature related to assessment of the role of natural enemies in regulating soybean insect pest populations was reported by Pedigo et al. (1983).

In general, evaluation techniques for natural enemies involve either 1) comparison methods involving addition, exclusion, or interference with natural enemy action, or 2) censusing populations of pests and natural enemies in the field to establish possible correlations. Direct observation, precipitin tests, radioisotope labelling each provide some insights into the action of certain predators. Irrespective of the approach used, studies of how predators, parasites and diseases act in concert, or by themselves, to influence populations of pests has produced many problems and has produced only circumstantial information in many cases.

Parasites

Although many species of parasites have been collected and described from insect pests of soybean (see review by Harper et al. 1983), except for one example (P. foveolatus against E. varivestis), there are no reports of suppression of a soybean insect pest in the field due solely to the action of parasites. However, most entomologists believe that parasites have some impact on populations of insect pests of soybean. The major reason for lack of information on the quantitative interaction between parasites and their hosts is absence of methods for testing this interaction. The approach used to determine the incidence of parasitism normally includes field collections of hosts which are brought to the laboratory for observation of parasite emergence.

A clear example of the potential for using imported parasites was reported by Sailer (1976), Shepard & Robinson (1976) and Stevens et al. (1975), who released the imported eulophid, P. foveolatus, to reduce field populations of the Mexican bean beetle. This was clearly a cause and effect relationship. Other parasites have been introduced and at least one, E. putterli, is established on populations of the velvetbean caterpillar in South Florida (Waddill & Puttler 1980). The impact of this parasite on A. gemmatalis populations has not yet been determined.

These examples illustrate the potential for the classical biological control approach against insect pests of soybean. Parasite introductions have received relatively little attention as have methods for assessing their impact on target pests. There are good opportunities for the importation of parasites in light of recommendations provided by Hoy (this volume) and by using improved methods for determining the incidence of parasitism (Van Driesche 1983).

Predators

Many of the techniques used to evaluate the action of predators in soybean were developed in cotton (Lingren et al. 1968, Roach et al. 1979, Whitcomb & Bell 1964). With non-mobile prey such as eggs, rates of predation in the field can be rather easily determined by attaching eggs to leaves using egg albumen or allowing caged moths to oviposit on leaves. After exposure of eggs to natural communities of predators remaining eggs may counted at some later time (Bushman et al. 1977, Reed et al. 1984). The major constraint to the use of this method is that predator species responsible for the absence of exposed eggs cannot be discerned. Some insights into this were provided by Elvin et al. (1983), Whitcomb & Bell (1964) and others, who made direct observations of predation in the field. It is obvious that this method suffers from severe time constraints and limited observations imposed by continuous surveillance of predator-prey interactions. Attempts at avoiding direct observation have led researchers to the use of radionucleotides, such as ^{32}P (Bushman et al. 1977, McCarty et al. 1980, McDaniel et al. 1978), and serological methods (Gardner et al. 1981, Ragsdale et al. 1981).

By far, the most widely used method of assessing the role of arthropod predators in soybean has included the use of field cages (Barry et al. 1974; Elvin et al. 1983; Reed et al. 1984; Richman et al. 1980; Waddill & Shepard 1974, 1975; and others). Major drawbacks to this technique include possible changes in microclimate created by the cages, changes in the behavior of arthropods (predator and/or prey), and restricted movement of prey and predators.

There are numerous reports of results from predator-prey studies conducted in the laboratory. But results from these studies <u>cannot</u> be related to predator-prey interactions in the field. A review of techniques used for evaluating predators in field crops has been presented by Grant & Shepard (1984).

Entomopathogens

The effects of epizootics of entomopathogens in soybean are often very dramatic. This group of organisms often plays an important role in regulating populations of lepidopterous pests, although it is sometimes difficult to determine mortality caused by pathogens which attack early life stages of the insects. This is largely a sampling problem. The rate of development of an epizootic can usually be easily followed by sampling pest populations at close intervals (2-3 days) in the field. Diseases impact most often on populations of lepidopterous pests such as the soybean looper, corn earworm, green cloverworm, and velvetbean caterpillar. Collections of larvae, which are taken to the laboratory and provided food until mortality occurs at which time disease assessment is made, can provide further evidence of the incidence of disease in the field. Pedigo et al. (1983) reviewed the current work on assessing the role of pathogens in regulating insect pest populations in soybean.

Determination of the role of diseases in regulating soybean insect pest populations is particularly important in a classical biological control program using entomopathogens. In those instances where a pathogen is introduced from a foreign country, it has been relatively easy to assess its role after application to the field. However, factors which are necessary for an epizootic to be initiated in the field have yet to be determined.

LIMITS TO AND RECOMMENDATIONS FOR THE USE OF NATURAL ENEMIES IN SOYBEAN

Compatibility of biological control in soybean insect management has been reviewed in some detail in Herzog et al. (1985a), and particularly with respect to host plant resistance in Herzog & Funderburk (this volume). One of the major constraints to the use of biological control agents in soybean is lack of knowledge of how indigenous natural enemies can be considered in the making of pesticide treatment decisions. There is still much to be done in terms of quantitative definition of the role of major beneficial species. Also, methods are needed for determining how total communities of predators, parasites and pathogens impact on pests. There is a need to develop standardized sampling methods which could include predator "guilds" or predator "units" as described by Wilson (this volume). Although it is difficult to assess the population-regulating role of major beneficial groups, information which does exist should be incorporated into sampling programs in a general manner such that action thresholds could be slightly modified upward when an abundance of natural enemies are encountered. For example, the presence of six small (<2.5 cm) *H. zea*/row-m can trigger an insecticide application. The same pest density with abundant populations of nabids or geocorids (two major predatory genera), as determined by sequential sampling (Waddill et al. 1974), could perhaps relax the threshold to 8-10 pest larvae/row-m.

1. Although studies have been carried out which deal with the feeding rates of different predators, the ultimate test of their benefit lies in the ability of those species to reliably and consistently remove a given number of pests of a given size. Methodologies or techniques for assessing efficiency of natural enemies are needed.
2. Historically, importation of natural enemies for control of insect pests of soybean has received only minimal support. Further, there are those who believe that because we do not understand the role of indigenous natural enemies, no others should be imported. However, there is clear need for an increase in the support and effort required to import, rear, release and evaluate natural enemies of several major insect pests of soybean, irrespective of the origin of the pest. A clear example of the success of this approach is the control of the Mexican bean beetle (from Mexico) with *P. foveolatus*, (from India).
3. Foreign exploration efforts for natural enemies should be closely coordinated among scientists involved so that the likelihood of finding natural enemies for several pest species will be increased. This could be accomplished by establishing a regional or national biological control committee which could set priorities and evaluate programs. After natural enemies are imported and mass rearing techniques

developed, they should be released in several geographical locations to increase the probability of establishment.

4. In the area of microbial control, an organized program for importation, culture, application and evaluation should be developed. There are good opportunities for finding more virulent strains of pathogens. Methods for testing these pathogens should be standardized so that results can be compared from different areas of the country and testing can take place under different environmental conditions.
5. There is a need for developing disease simulation models in order to help identify those factors which result in initiation of epizootics and spread of inoculum and to make predictions about biological control capabilities.
6. Research on selection and use of chemicals which are favorable for, or at least not as detrimental to, natural enemy population development should be continued. This approach could include the application of minimum effective dosage rates, alteration of formulations or methods of application, and determining optimal timing of application.
7. Successful augmentation of natural enemies necessarily includes improvement of mass rearing facilities and techniques. Research in this area is currently receiving only minimal support.
8. Priority should be given to integrating biological control with other tactics such as host plant resistance and chemical control. Often there is too little attention given to the conservation of natural enemies when chemical pesticides are applied. There are excellent opportunities for using entomopathogens against insect pests and thereby insuring conservation of indigenous or imported predators and parasites.

REFERENCES

Allen, G.E., and J.D. Knell. 1977. A nuclear polyhedrosis virus of Anticarsia gemmatalis: I. Ultrastructure, replication and pathogenicity. Fla. Entomol. 60:233-240.

Barry, R.M., J.H. Hatchett and R.D. Jackson. 1974. Cage studies with predators of the cabbage looper, Trichoplusia ni, and corn earworm, Heliothis zea in soybeans. J. Ga. Entomol. Soc. 9:71-78.

Beach, R.M., G.R. Carner and S.G. Turnipseed. 1984. Field efficacy and persistence of a nuclear polyhedrosis virus of the velvetbean caterpillar in soybeans. J. Agric. Entomol. 1:296-304.

Brown, D.W., and R.A. Goyer. 1982. Effects of a predator complex on lepidopterous defoliators of soybean. Environ. Entomol. 11:385-389.

Burleigh, J.G. 1972. Population dynamics and biotic controls of the soybean looper in Louisiana. Environ. Entomol. 1:290-294.

Bushman, L.L., W.H. Whitcomb, R.C. Hemenway, D.L. Mays, N. Ru, N.C. Leppla and B.J. Smittle. 1977. Predators of velvetbean caterpillar eggs in Florida soybeans. Environ. Entomol. 6:403-407.

Carner, G.R., M. Shepard and S.G. Turnipseed. 1975. Disease incidence in lepidopterous pests of soybeans. J. Ga. Entomol. Soc. 10:99-105.

Carner, G.R., and S.G. Turnipseed. 1977. Potential of a nuclear polyhedrosis virus for control of the velvetbean caterpillar in soybean. J. Econ. Entomol. 70:608-610.

Deitz, L.L., J.W. Van Duyn, J.R. Bradley, R.L. Rabb, W.M. Brooks and R.E. Stinner. 1976. A guide to the identification and biology of soybean arthropods in North Carolina. N.C. Agr. Exp. Stn. Tech. Bull. 283, 265 pp.

DeBach, P., C.B. Huffaker and A.W. MacPhee. 1976. Evaluation of the impact of natural enemies, pp. 255-285. In: C.B. Huffaker and P.S. Messenger (Eds.), Theory and practice of biological control. Academic Press, New York.

Elvin, M.K., J.L. Stimac and W.H. Whitcomb. 1983. Estimating rates of arthropod predation on velvetbean caterpillar larvae in soybeans. Fla. Entomol. 66:319-330.

Gardner, W.A., M. Shepard and R. Noblet. 1981. Precipitin test for examining predator-prey interactions in soybean fields. Can. Entomol. 113:365-369.

Grant, J.F., and M. Shepard. 1984. Techniques for evaluating predators for control of insect pests in field crops. J. Agr. Entomol. 1:(In Press).

Greene, G.L., and M. Shepard. 1974. Biological studies of a predator, Sycanus indigator. II. Field survival and predation potential. Fla. Entomol. 57:33-38.

Hagen, K.S., E.F. Sawall and R.L. Tassen. 1971. The use of food sprays to increase the effectiveness of entomophagous insects. Proc. Tall Timbers Conf. on Ecol. Animal Contr. by Habitat Mgt. 2:59-81.

Harper, J.D., R.M. McPherson and M. Shepard. 1983. Geographical and seasonal occurrence of parasites, predators and entomopathogens, pp. 7-19. In: H.N. Pitre (Ed.), Natural enemies of arthropod pests in soybean. South. Coop. Ser. Bull. 285.

Herzog, D.C., J.L. Stimac, D.G. Boucias and V.H. Waddill. 1985a. Compatibility of biological control in soybean insect management. In: P.L. Adkisson and Shijun Ma (Eds.), Proc. Joint U.S. Nat. Acad. Sci./Chinese Acad. Sci. Symp. on Biologicial Control of Insects. Sci. Sinica (In Press).

Herzog, D.C., J.W. Todd and W.H. Whitcomb. 1985b. The velvetbean caterpillar. In: L.D. Newsom (Ed.), Integrated management of insect pests of soybean. Wiley Interscience. (In Press).

Huffaker, C.B. 1980. New technology of pest control. John Wiley & Sons, N.Y. 580 pp.

Huffaker, C.B., and C.E. Kennett. 1969. Some aspects of assessing efficiency of natural enemies. Can. Entomol. 101:425-447.

Ignoffo, C.M., M.L. Marston, D.L. Hostetter, B. Puttler and J.V. Bell. 1976. Natural and induced epizootics of Nomuraea rileyi in soybean caterpillars. J. Invertebr. Pathol. 27:191-198.

Jones, W.A. 1979. The distribution and ecology of pentatomid pests of soybean in South Carolina. Ph.D. dissertation. Clemson Univ., S.C.

Jones, W.A., S.Y. Young, M. Shepard and W.H. Whitcomb. 1983. Use of imported natural enemies against insect pests of soybean, pp. 63-77. In: H.N. Pitre (Ed.), Natural enemies of arthropod pests in soybean. South. Coop. Ser. Bull. 285.

Kiritani, K., and J.P. Dempster. 1973. Different approaches to the quantitative evaluation of natural enemies. J. Appl. Ecol. 10:323-330.

Lewis, W.J., R.L. Jones, D.A. Nordlund and A.N. Sparks. 1975. Kairomones and their use for management of entomophagous insects. I. Evaluation for increasing rates of parasitization by Trichogramma spp. in the field. J. Chem. Ecol. 1:343-347.

Lingren, P.D., R.L. Ridgway and S.L. Jones. 1968. Consumption by several common arthropod predators of eggs and larvae of two Heliothis species that attack cotton. Ann. Entomol. Soc. Amer. 61:613-618.

Livingston, J.M., and W.C. Yearian. 1972. A nuclear polyhedrosis virus of Pseudoplusia includens (Lepidoptera:Noctuidae). J. Invertebr. Pathol. 19:107-112.

Marston, N.L. 1980. Sampling parasitoids of soybean insect pests, pp. 482-504. In: M. Kogan and D.C. Herzog (Eds.), Sampling methods in soybean entomology. Springer-Verlag, N.Y.

Martin, P.B., P.D. Lingren, G.L. Greene and R.L. Ridgeway. 1976. Parasitization of two species of Plusiinae and Heliothis spp. after releases of Trichogramma pretiosum in seven crops. Environ. Entomol. 5:991-995.

McCarty, M.T., M. Shepard and S.G. Turnipseed. 1980. Identification of predaceous arthropods in soybeans by using autoradiography. Environ. Entomol. 9: 199-203.

McCutcheon, G.S.. and S.G. Turnipseed. 1981. Parasites of lepidopterous larvae in insect resistant and susceptible soybeans in South Carolina. Environ. Entomol. 10:69-74.

McDaniel, S.G., L.L. Keeley and W.L. Sterling. 1978. Radiolabeling Heliothis virescens eggs by ^{32}P injection of adult females. Ann. Entomol. Soc. Amer. 7-1: 432-434.

Moscardi, F., G.E. Allen and G.L. Greene. 1981. Control of the velvetbean caterpillar by nuclear polydedrosis virus and insecticides and impact of treatments on the natural incidence of the entomopathogenic fungus Nomuraea rileyi. J. Econ. Entomol. 74:480-485.

Naranjo, S.E. 1983. Influence of a weed complex in soybean on the predatory, ovipositional and plant feeding responses of Geocoris punctipes Say. M.S. Thesis, Univ. of Florida, Gainesville.

Neal, T.M., and W.H. Whitcomb. 1972. Odonata in the Florida soybean agroecosystem. Fla. Entomol. 55:107-114.

O'Neil, R.J. 1984. Measurement and analysis of arthropod predation on velvetbean caterpillar, Anticarsia gemmatalis Hubner. Ph.D. Dissertation, Univ. of Florida, Gainesville.

Pedigo, L.P., H.N. Pitre, W.H. Whitcomb and S.Y. Young. 1983. Assessment of the role of natural enemies in regulation of soybean pest populations, pp. 39-48. In: H.N. Pitre (Ed.), Natural enemies of arthropod pests in soybean. South. Coop. Ser. Bull. 285.

Price, J.F., and M. Shepard. 1977. Striped earwig, Labidura riparia, colonization of soybean fields in response to insecticides. Environ. Entomol. 6:679-683.

Price, J.F., and M. Shepard. 1978. Calosoma sayi: Seasonal history and response to insecticides in soybeans. Environ. Entomol. 7:703-707.

Puttler, B., G. Gordh and S.H. Long. 1980. Bionomics of Euplectrus puttleri, new species, an introduced parasite of velvetbean caterpillar, Anticarsia gemmatalis, from South America. Ann. Entomol. Soc. Amer. 78:28-35.

Ragsdale, D.W., A.D. Larson and L.D. Newsom. 1981. Quantitative assessment of the predators of Nezara viridula eggs and nymphs within a soybean agroecosystem using an ELISA. Environ. Entomol. 10:402-405.

Reed, T., M. Shepard and S.G. Turnipseed. 1984. Assessment of the impact of arthropod predators on noctuid larvae in field cages in soybean. Environ. Entomol. 13:954-961.

Richman, D.B., R.C. Hemenway and W.H. Whitcomb. 1980. Field cage evaluation of predators of the soybean looper, Psuedoplusia includens (Lepidoptera: Noctuidae). Environ. Entomol. 9:315-317.

Roach, S.H., J.W. Smith, S.B. Vinson, H.M. Graham and J.A. Harding. 1979. Sampling predators and parasites of Heliothis species on crops and native host plants, pp. 132-145. In: W.L. Sterling (Ed.), Economic thresholds and sampling of Heliothis species on cotton, corn, soybeans and other host plants. South. Coop. Series Bull. 231.

Sailer, R.I. 1976. Organizational responsibility for biological control research and development inputs, pp. 10-19. In: Organized programs to utilize natural enemies of pests in Canada, Mexico, United States. Papers presented at the Joint Symposium. Nat. Mtg. Entomol. Soc. Amer., Nov. 30-Dec. 4, 1975. New Orleans. USDA/APHIS 81-28, 23 pp.

Shepard, M., G.R. Carner and S.G. Turnipseed. 1974. Seasonal abundance of predaceous arthropods in soybeans. Environ. Entomol. 3:985-988.

Shepard, M., G.R. Carner and S.G. Turnipseed. 1977. Colonization and resurgence of insect pests of soybean in response to insecticides and field isolation. Environ. Entomol. 6:501-506.

Shepard, M., and J. Grothusen. 1984. A simple calculator program for flexible sequential sampling of insects. Bull. Entomol. Soc. Amer. 30:**(In Press).**

Shepard, M., N. Marston and G. Carner. 1983. Sampling predators, parasites and entomopathogens, pp. 20-31. In: H.N. Pitre (Ed.), Natural enemies of arthropod pests in soybean. South. Coop. Ser. Bull. 285.

Shepard, M., R.E. McWhorter and E. King. 1982. Life history of Pristhesancus papuensis Stal (Hemiptera: Reduviidae). Can. Entomol. 114:1089-1094.

Shepard, M., J.E. Powell and W.A. Jones. 1983. Biology of Microplitis demolitor, an imported parasitoid of Heliothis spp. and the soybean looper, Pseudoplusia includens. Environ. Entomol. 12:641-645.

Shepard, M., and J. Robinson. 1976. Suppression of Mexican bean beetles in soybeans by the imported parasitoid, Pediobius foveolatus. Sumter Area Agri. Dec. Proj. Rept. 49-53.

Sprenkel, R.K., and W.M. Brooks. 1975. Artificial dissemination and epizootic initiation of Nomuraea rileyi, an entomogenous fungus of lepidopterous pests of soybeans. J. Econ. Entomol. 68:847-851.

Stevens, L.M., A.L. Steinhauer and J.R. Coulson. 1975. Suppression of Mexican bean beetle on soybeans with annual innoculative releases of Pediobius foveolatus. Environ. Entomol. 4:947-52.

Tseng, J.M. 1982. Comparative response of Eocanthocona furcellata (Wolff) and Podisus maculiventris (Say) to temperature and photoperiod and their relative potential as beneficial predators. Ph.D. Dissertation, Univ. of Florida, Gainesville.

Turnipseed, S.G. 1972a. Response of soybeans to foliage losses in South Carolina. J. Econ. Entomol. 65:224-229.

Turnipseed, S.G. 1972b. Management of insect pests of soybean. Proc. Tall Timbers Conf. Ecol. Animal Cont. Habitat Manag. 4:189-203.

Turnipseed, S.G., and M. Kogan. 1976. Soybean entomology. Ann. Rev. Entomol. 21:247-82.

Van Driesche, R.G. 1983. Meaning of "percent parasitism" in studies of insect parasitoids. Environ. Entomol. 12:1-12.

Waddill, V.H., and B. Puttler. 1980. Euplectrus puttleri established on the velvetbean caterpillar, Anticarsia gemmatalis, in South Florida. Environ. Entomol. 9:781-782.

Waddill, V., and M. Shepard. 1974. Potential of Geocoris punctipes (Hemiptera: Lygaeidae) and Nabis spp. (Hemiptera: Nabidae) as predators of Epilachna varivestis (Coleoptera: Coccinellidae). Entomophaga 19:421-426.

Waddill, V., and M. Shepard. 1975. A comparison of predation by the pentatomids, Pedisus maculiventris (Say) and Stiretrus anchorago (F.) on the Mexican bean beetle, Epilachna varivestis (Mulsant). Ann. Entomol. Soc. Amer. 68:1023-1027.

Waddill, V.H., B.M. Shepard, S.G. Turnipseed and G.R. Carner. 1974. Sequential sampling plans for Nabis spp. and Geocoris spp. on soybeans. Environ. Entomol. 3:415-419.

Whitcomb, W.H. 1973. Natural populations on entomophagous arthropods and their effect on the agroecosystem, pp. 150-169. In: F.G. Maxwell and F.A. Harris (Eds.), Proceedings of the summer institute on biological control of plant insects and diseases. Miss. Univ. Press, Jackson.

Whitcomb, W.H., and K. Bell. 1964. Predaceous insects, spiders and mites of Arkansas cotton fields. Ark. Agric. Exp. Stn. Bull. 690, 83 pp.

Whitcomb, W.H., H.A. Denmark, A.P. Bhathar and G.L. Greene. 1972. Preliminary studies on the ants of Florida soybean fields. Fla. Entomol. 55: 129-142.

Yeargan, K.V. 1979. Parasitism and predators of stink bug eggs in soybeans and alfalfa fields. Environ. Entomol. 8:715-719.

DISCUSSION: Status and current limits to biological control in citrus, grapes, alfalfa, cotton and soybean systems.

DISCUSSION LEADER: W.A. Allen

The crops selected for discussion at this Conference, while not all among the most important economically, do give a good range of ecological diversity. Citrus represents an evergreen perennial with a high degree of ecological stability, thereby providing good opportunities for biological control. Grapes are a long term perennial crop with moderate disruption when the leaves are lost each year. Alfalfa is a short term perennial with frequent disruptions from periodic cuttings. Cotton and soybean are annual crops, planted on bare ground each year, where the development of pest populations is strongly influenced by the vagility of the pests and their natural enemies, and the availability of refuges for the pests during non-crop periods.

In the discussion of these crops, it was clearly evident that natural enemies are important suppressive agents for arthropod pests in all of the agroecosystems. This was true not only for the more stable perennial crops, but also for the annual row crops. Although with the annual crops there must be migration of the pests and their natural enemies into the crops each year, it was agreed that natural enemies play a significant role in arthropod suppression in both the annual crops themselves and in the non-crop habitats of the pests.

Although it was unanimously agreed that natural enemies are important suppressive agents, it is very difficult to precisely assess the actual impact of natural enemies, and we have relatively poor quantification of their importance. Mortality from the more specific parasites and diseases can be measured by periodic collection of field samples, but it is much more difficult to quantify mortality caused by the general predators. For some pests on some crops, specific parasites are very important in pest suppression, but on other crops, such as soybean, it appears that general predators cause much more significant mortality on some pests than the specific parasites.

Another common theme was the difficulty of predicting suppression of pest populations by parasites, predators, and epizootics of pathogens. Prediction is easier on perennial crops, where many of the key pests and their natural enemies remain in the crop habitat year after year, but much needs to be learned about the environmental factors that influence pest outbreaks and the effectiveness of natural enemies. Prediction is even more difficult with the annual crops where chance invasion by pests and natural enemies is influenced by arthropod populations in surrounding crops and vegetation, and pest suppression must take place within a period of about six months. There was agreement that much progress had been made in prediction capabilities, but much more needs to be done in the future.

A common limitation cited throughout the discussion of biological control on the various crops centered around the many types of disruptions. Insecticides are recognized as the most common causes of disruption because it is difficult to achieve selectivity between arthropod pests and their arthropod natural enemies. However, in the discussions of citrus and alfalfa, fungicides were major causes of disruption. On grapes, sulfur caused disruption of some pests, but a more serious breakdown in biological control resulted from the introduction of a new species of leafhopper which, although closely related to the grape leafhopper, is less susceptible to attack by the parasite Anagrus. With cotton, the introduction of corn into the cotton growing areas of Texas was predicted to be the next serious disruption for cotton pest control.

It appears that one of the past difficulties with biological control has been the unrealistic dream of a totally stable and harmonius system of agriculture with few or no disruptions. In reality, agriculture must be regarded as an ecological system with disruption being the rule rather than the exception. In modern agriculture, cultivars, irrigation and cultural methods are constantly changing, with favorable and unfavorable affects on biological control. Although pesticides are disruptive, it must be accepted that modern agriculture is a chemically oriented system. If biological control is going to be developed only for use in non-chemically oriented agricultural systems, its implementation and utility will be limited. However, biological control is still highly important and much needed in modern agriculture, but special efforts must be made to integrate biological control into these artificial and disrupted systems. Those working in biological control must recognize that a totally harmonius balance will seldom be attained with agricultural crops, but that biological control can make important contributions in these ecologically "messy" systems. Much greater progress will be made when it is accepted that biological control is an important, but not the only, component of pest control.

In the discussions of the status of biological control in the different crops, there was much talk about the lack of funds, and how little is really known about all the factors influencing the effectiveness of natural enemies. However, it must be recognized that biological control has been an accepted discipline for 100 years, and it has received considerable funding in past pest management projects. It must must also be recognized that the interactions within even the simplest cropping systems are very complex and even with mammoth funding, we could not answer all pertinent questions. Since funds for research are always going to be limited and the needs for research unlimited, it will be essential to prioritize research needs if meaningful accomplishments are going to result from biological control funding. The most critical research may not be the most interesting or easily completed, but that research will likely give the greatest results, and thus insure future funding.

One unfortunate funding difficulty results from the termination of funds soon after biological control programs start to show success. When this happens, the full success of the program cannot be fully documented and placed on record. The utility and value of biological control will only be appreciated when resources are made available to document the full success of the programs. If funds are not made available for statistical and economic analyses of the successes, future funding for biological control will be seriously handicapped.

Index

C

M

Q

R

T